Eating and Being

Eating and Being

A History of Ideas about Our Food and Ourselves

STEVEN SHAPIN

The University of Chicago Press
Chicago and London

The University of Chicago Press, Chicago 60637
The University of Chicago Press, Ltd., London

Published 2024
Printed in the United States of America

33 32 31 30 29 28 27 26 25 24   1 2 3 4 5

ISBN-13: 978-0-226-83221-0 (cloth)
ISBN-13: 978-0-226-83222-7 (e-book)
DOI: https://doi.org/10.7208/chicago/9780226832227.001.0001

Published with support of the Susan E. Abrams Fund.

Library of Congress Cataloging-in-Publication Data

Names: Shapin, Steven, author.
Title: Eating and being : a history of ideas about our food and ourselves / Steven Shapin.
Description: Chicago : The University of Chicago Press, 2024. | Includes bibliographical references and index
Identifiers: LCCN 2024011756 | ISBN 9780226832210 (cloth) | ISBN 9780226832227 (ebook)
Subjects: LCSH: Food habits—History. | Diet—History.
Classification: LCC GT2850 .S524 2024 | DDC 613.2—dc23/eng/20240402
LC record available at https://lccn.loc.gov/2024011756

♾ This paper meets the requirements of ANSI/NISO Z39.48-1992 (Permanence of Paper).

With Appreciation:
Lisa Haushofer and Paolo Savoia

"Digestion, n. The conversion of victuals into virtues."
—Ambrose Bierce, *The Devil's Dictionary*

"Erst kommt das Fressen, dann kommt die Moral."
—Berthold Brecht, *Die Dreigroschenoper*

CONTENTS

INTRODUCTION

What's for Dinner?

Thinking and Eating, Thinking about Eating

Tonight, I am making Tuscan bean soup for dinner. For my wife and me, this is a light dinner. Maybe there will be bruschetta before and cheese after. The soup isn't a dish I was familiar with as I was growing up. I came across the recipe about twenty years ago, perhaps in a newspaper, perhaps shown to me by an Italian friend. I make the soup from time to time, and it is now quite an ordinary thing for me to do. I like the bean soup, and my wife likes it too. Usually, that's as much as I would say if someone asked me why I'm going to have an Italian bean soup for dinner. But this is the beginning of a book about what we *think* as we decide what to eat and then eat. It is a work of history: cultural history, the history of ideas, including the ideas belonging to science and medicine. *Eating and Being* reaches back into the past to describe fundamental changes in how we think about our food and its fate in our bodies. But it starts and ends with the present-day ordinary, the taken-for-granted about our food, what it is and what it does. And this soup can serve as ordinary enough.

The bean soup dinner is, then, an occasion to take stock of what came to my mind as relevant, or potentially relevant, as I chose to make it and to then to eat it. I find that the list of such considerations might, in principle, become unmanageably long. And that's because it's hard to think of many aspects of our knowledge, our culture, and our social and political life, which are not implicated in either everyday or special-purpose thinking about food and eating. There are now very many books that pronounce on what we ought and ought not eat in order to secure health and long life; many other books judge what's good-tasting, economical, authentic; and still others trace connections between what we eat, planetary well-being, and the social and political orders. There's so much to be said about food, because thinking about food links to thinking about practically everything else.

Thinking about food can therefore count as a big deal, to do with big things: celebration or criticism of modernity, celebration or criticism of the globalized order, part of a plan to avoid disease and live long. But in truth, this big-deal thinking happens at the margins of the everyday and the taken-for-granted. One quite ordinary consideration, for instance, is the time of year. The edible seasonal and the local are now much commended, but my wife and I just tend to like cooling things in the summer and warming things in the winter. We live in New England, and, as I write it's November, so the bean soup and bruschetta seem about right. I'm aware of beans as the butt of flatulence jokes. My mother used to say that certain foods either "agreed" or "disagreed" with her, but that's not a form of words that I now ordinarily use, and at the risk of Too Much Information, neither my wife nor I find bean soup physiologically troubling. What is ordinarily on our minds is taste. We like the taste of the bean soup, although with only the two of us sitting down to dinner, we don't go on about taste: maybe "that's good" or maybe "needs more salt." Over many years of living and eating together, our tastes have converged, though only to some extent. But I have learned to respect my wife's tastes: she's English and likes Marmite; I don't. And she's kind enough to let me make liver and onions—which I like and she doesn't—as long as she doesn't have to eat it. We both know and accept the wisdom of the expression "There's no arguing about taste" or, feeling fancy, of the Latin form *De gustibus non est disputandum*. On rare occasions, we might try to tell each other what it was about the taste of the soup that struck us as good or not so good, maybe that it *didn't* have enough salt. But we would usually presume that our judgments were subjective. We would be wary about thinking that we were referring to anything but the crudest facts about the nature of the soup or about its chemical constituents, and neither of us would offer our taste judgments as universal standards. We like it, others may not, and we tend to leave it at that.

I am fortunate in having *choice* about what I am going to eat, and so do many other people in modern Western societies. Yet a wide-ranging ability to choose was, until fairly recent times, confined to a small part of the population, while choice remains restricted or nonexistent for shamefully large numbers of people in the modern world. Scarcity and hunger—sometimes punctuated by harvest-time glut—were historically the normal state of affairs and for millions and remain so today, and so were localism and seasonality. (There were many ways of preserving food known since antiquity—drying, salting, smoking, and the fermentation of fresh foods into longer-lasting wine, beer, cider, cheese, and pickled vegetables—but the nineteenth-century development of canning, refrigeration, and global food transport systems changed things in fundamental ways.) For people such as me and my wife, the decision to eat what is seasonal and what is locally produced may be a studied gesture toward connoisseurship and may be meant to produce a warm inner glow of environmental responsibility, but for much of history there was no option, and the same is true for many present-day people.

Tonight's bean soup is, or is intended to be, Italian. When I was a boy, the only Italian foods with which I was familiar were pizza—in my neighborhood then known as "tomato pie"—and tinned spaghetti and meatballs. For many people, food choice is bound up with religion as well as ethnicity and habit. To my knowledge, my father and mother never ate anything like a Tuscan bean soup. Being secular Jews, they would not have been bothered by the pieces of pork in it, while my mother's mother would have found it an abomination. Growing up as I did in mid-twentieth-century

America, the standard domestic fare was either of the meatloaf, mash, and Jell-O sort or the ethnic stuff my mother learned from her mother: stuffed cabbage, chopped liver, borscht, and, of course, full-fat chicken soup. (My father cooked once or twice a year, just to demonstrate "how it ought to be done," while I do virtually all the cooking in my little family. The increasing ordinariness of men doing some or all of the domestic food preparation is yet another story about the changing cultural meanings of food and eating.) I still very much like the Eastern European sorts of things my mother prepared, though I don't make them very much these days. My tastes began to run toward the cosmopolitan and the eclectic very early on. While still living at home, I secretly sloped off to have sweet-and-sour pork and egg foo yong at Cantonese restaurants. My father feared that such foods were dangerous, but I found that they did me no harm. And as soon as I left home for college, edible exoticism became my new normal. It has continued that way ever since, partly as a result of living an international life and partly because of whatever social and cultural processes have turned out millions of Westerners who, like me, are comfortably off and have globally adventurous tastes.

What I Eat and Who I Am

The present-day reach and resonances of food thinking and food choices are evident and pervasive. There is, for example, the phenomenon of "foodie" culture, and there is also the gathering puritanical backlash against the foodies, the irritated recommendation that we just get on with things and *eat*, which is itself a longstanding and richly significant cultural gesture: "eat to live, don't live to eat."[1] (Dietary asceticism and the condemnation of "luxury" and self-indulgence have a long history.) Claims to personal and collective identity through food can be both subtle and precise: in my present neighborhood, it says a lot whether you grind your own coffee beans, buy a tin of preground, pour boiling water over a spoonful of instant, or patronize Starbuck's or the local nonfranchise coffeehouse. It may signify something whether you find the Spanish tripe stew *callos madridleños* tasty or disgusting; whether you are a vegetarian, a vegan, or an unashamed carnivore; and whether you take dinner at 5:30 or at 8:00. Then, there is a number of what might be called *shibboleth foods*, those that go some way toward defining social, racial, or national identity, and, among these, those that are much liked by group members and widely found to be inedible or pointless by outsiders: poutine, lutefisk, bottarga, stinky tofu, menudo, the Swedish fermented herring delicacy surströmming, the equally challenging Icelandic fermented shark dish called hákarl, the haggis of Scotland, the Pennsylvania Dutch scrapple, and even the Welsh leek that was the object of raucous byplay among the representatives of the four British nationalities in Act 5 of Shakespeare's *Henry V*.[2]

I have no such relationship to tonight's bean soup or indeed to any Italian foods. Growing up, I knew only a few people of Italian heritage, and what they ate, as I've since learned, was that distinctive cuisine now called Italian American. And as I've also learned, there is no such thing as Italian cuisine in Italy, only those regional foods and preparations that have managed to travel over much of the country. The closest my mother came to Italian cooking was anointing a wedge of iceberg lettuce with some bottled "Italian" salad dressing. I've now been to Italy very many times. I like Italian food enormously, and I have even taught about the history and sociology of taste in the marvelously named University of Gastronomic Sciences in Piedmont, but the appeal of Tuscan bean soup to me is just an everyday expression of personal preference. It long ago stopped having any particular association with Italianness. (If I want to make a retro gesture toward who I really am—assuming I could stably know such a thing—I suppose that my foods of choice might be a New York–style hotdog, with mustard and sauerkraut, or that most American of foods, spaghetti with meatballs—which I now only pretend to dislike because of its supposed inauthenticity—something that "real Italians" do not recognize.) But my choice of the Tuscan bean soup *does*, after all, testify to something about who I am, as do the yu hsiang fish I made the day before and the braised oxtail I will probably cook tomorrow, just because I am the kind of person who picks and chooses among the world's cuisines and tastes and who fancies making those things part of his normal domestic world. This is authentic and is normal *for me* and for people *like me*.

Another feature of the bean soup that would have been unexceptional in the past but now marks the soup as something out of the ordinary for many moderns is that it is made from scratch or mostly so. I made the chicken stock from leftover bones and vegetables, there are garlic and some pancetta bought from a shop, the thyme or rosemary comes from my tiny urban garden, and the cannellini beans are either the dried things soaked overnight or dumped out of a can. This is the kind of cooking widely celebrated these days as "traditional," "authentic," or "good" but is increasingly rare in modern American society. Commercially concocted processed food is the norm. Many people no longer know how to cook something like a bean soup, and there may be a bit of self-conscious virtue signaling in recovering lost or endangered modes of food preparation, though I deny that this describes my state of mind. Branded foods have not been around forever, the techniques of canning go back to the nineteenth century, and the rise of prepared, ready-made, and fast food happened during my own lifetime. (As a boy, I was an enthusiastic consumer of the newly introduced frozen TV dinners and chicken potpies.) The bean soup stands outside the stream of ordinary life so profoundly influenced by the role of big commercial concerns in the food supply, including the importance of their voices—through advertising, sponsorship, and political action—in shaping what counts as nutritional knowledge. A from-scratch, largely vegetable soup that used to be—as *potage*—one of the most common European sources of sustenance now can appear as a gesture of nostalgia or as some sort of studied ideological disengagement from modern foodways. I claim that nothing of this sort occurred to me as I decided on tonight's dinner, but as it's said, I would say that, wouldn't I?

Edible Expertise

Still, I insist that tonight's bean soup is nothing very special: it's just a way of getting the historical story going and signaling its rootedness in the just-thisness of everyday thinking about everyday eating. I start, as I should, with the concrete and the ordinary, and then I follow edible things as commonplace as the soup and what we think about them, wherever the story goes in our intellectual life and especially over historical time spans. Familiarity, propriety, digestibility, nutritional value, taste, cost, availability, and political and environmental beliefs are just a start. I'm now well on the wrong side of middle age, and people in my age group are supposed to inform our eating practices by medical concerns, which tends to mean something like taking account of the latest scientific evidence. Again, I think I know—though without great confidence—what medical and nutritional expertise has to say about such things as the bean soup or, more accurately, about its *constituents* and their individual physiological effects. I've heard that beans contain fiber and that fiber might be good for your colon and good for lowering your cholesterol, though as I write I seem to recall that those claims have recently been disputed and that I might have read that there are different sorts of fiber (soluble and nonsoluble, perhaps), one apparently better for you than another. Now that I think about cholesterol, there is some pancetta in the soup, and there must be a decent amount of cholesterol in that. I could possibly look this up on the internet, but I'm not often bothered. Anyway, I've been taking statins for years, and while the doctors say that the drug reduces "bad" cholesterol, I'm unsure whether that is LDL or HDL. I've repeatedly heard that lowering cholesterol happens "when combined with the proper diet and exercise." I have a friend whose doctor told him that the statins actually work pretty well alone, without the discipline of eating properly and without jogging, so I'm not going to worry too much about the pancetta. In this contest of expertise, I am for the most part a noncombatant: I don't know *what* to believe, and I'm usually more interested in watching the passing parade of nutritional experts.

Some of the people who come to our house complain that I don't put enough salt in the food. I think I have become used to undersalting things over many years, and I also think I got into that habit through trying to prevent my hypertensive mother—who adored salty foods—from overdoing it, setting her a good example, as it were. So, right now I don't know whether my salt thing has to do with taste, habit, deference to medical expertise about high blood pressure, or maybe a bit of all of these. But even in this matter, I am inconsistent. My wife and I like the taste and texture of Maldon sea salt, and during New England tomato season exceptions are made. The bean soup has a chicken stock base; I don't imagine that the real nutritional experts have a lot to say about chicken soup these days. That sort of thing belongs firmly in the world of folk wisdom, and you can pretty much tell that someone *isn't* a genuine expert if they spout off about the virtues of chicken soup. Nevertheless, my grandmother was one of those wise women who regarded it as a panacea, and no amount of academic training entirely expels your grandmother's wisdom. When friends are ill, my instinct is to cook them up some chicken soup, a gesture that is not notably informed by nutritional theory or facts. In all this, and even though I have an outdated biology and chemistry background, I am recognizably a typical member of early twenty-first-century Western culture. I have my tastes and my habits; I have a certain amount of knowledge about what I am about to eat; and I know, or believe, that there is expertise out there I could access if I wanted. Thinking about and acting on nutritional expertise for me and many others is as occasional as it is important.

Knowing about Constituents

So far, I've talked about the bean soup from a personal point of view: what I know (or think I know), what I might reflect on if I chose to reflect, what I like, and what I'm accustomed to. But there's a massive institutional fact about things such as the bean soup that intrudes itself these days into practically everyone's thinking. I'm probably going to use a package of dried great northern beans for tonight's soup. The verbiage on the pack tells me the net weight, who made the product (though not where the beans were grown), and who I can phone if I'm aggrieved enough to have "questions or comments." If I had chosen canned beans—which I often do—I would have been told that they contain some calcium chloride and some "disodium EDTA (preservative)." I almost always ignore that information: I know next to nothing about what these substances are or why they are in the can, though skeptical as I am about many things, I vaguely assume that they would not be there if there were serious expert doubts about their innocuousness. (The government wouldn't allow that, would it?)

If I turn over the package of beans, I encounter eloquent testimony to one quite recent way we can now collectively know about our food: the Nutrition Facts label (figure I.1). (In truth, I don't usually look at these labels, and I don't know how many people do and adjust their consumption accordingly.) In a serving reckoned as 35 grams of the dried beans, there is no "total fat" (therefore, no saturated or trans fat), and there is nothing in the way of cholesterol (which I understand to be a good thing). There are 22 grams of total carbohydrate and 7 grams of dietary fiber, 8 grams of protein, less than 1 gram of sugar, 60 mg of calcium (4 percent of daily needs), 1.9 mg of iron (10 percent of daily needs), and 480 mg of potassium (also 10 percent of daily needs). Sadly, the beans offer nothing in the way of vitamin D, and no other vitamins are mentioned. There is also—listed first on the label—the fact that each serving contains 120 calories. Calories, however, are a bit different from protein, carbohydrates, fats, vitamins, and minerals, since they are a measure of the *power* or *energetic capacity* of the food and not a constituent in the way the other things are. Calories tell you how much energy is produced by metabolizing the beans, while the other information says what's *in* the beans, the nutrient substances that make them up. I now know the calorie content of the beans, but I haven't bothered to figure out the calories in the bean *soup*. I *assume* it's comparatively low, but then there's the issue of an "average serving," and I haven't a clue how my typical soup intake relates to that.

The expert knowledge on the label is vouched for by the national government and has been in the United States since the Nutrition Labeling and Education Act of 1990.[3] The American government, and other governments around the world that now mandate

GREAT NORTHERN BEANS

Nutrition Facts

13 servings per container
Serving size 1/4 Cup Dry (35g)

Amount Per Serving
Calories 120

	% Daily Value*
Total Fat 0g	**0%**
Saturated Fat 0g	**0%**
Trans Fat 0g	
Cholesterol 0mg	**0%**
Sodium 0mg	**0%**
Total Carbohydrate 22g	**8%**
Dietary Fiber 7g	**25%**
Total Sugars <1g	
Includes 0g Added Sugars	**0%**
Protein 8g	**16%**
Vitamin D 0mcg	0%
Calcium 60mg	4%
Iron 1.9mg	10%
Potassium 480mg	10%

* The % Daily Value (DV) tells you how much a nutrient in a serving of food contributes to a daily diet. 2,000 calories a day is used for general nutrition advice.

INGREDIENTS: BEANS, GREAT NORTHERN, MATURE SEEDS, RAW

Figure I.1. Nutrition Facts label on a can of great northern beans

broadly similar labels, channel and add their authority to that of nutritional expertise. The state gives public voice to that expertise and selects the people and the institutions claiming expertise that will have their credibility buttressed by state power. That heightened credibility is all the more important because laypeople have no way of knowing about the chemical constituents of their food and the specific physiological consequences of those constituents except by way of external expertise. If a person should say, for example, that some food is warm or cold or that it is wet or dry, that's the sort of thing that can be known directly: it is accessible to our senses; there is no need for experts. If it is said that our food is fatty (in the ordinary usage of the term), that too can be directly known (we can often see the fat), and the same is true for such categories as sweetness, sourness, and bitterness, which we can taste. It is true that there is expert vocabulary to name and measure fatty and sweet chemical constituents and even an expert numerical measure of the spiciness of chilis. But when it is said that our food does what it does to our bodies because it contains protein, complex carbohydrates, or monounsaturated fats or that it delivers a certain number of calories, our knowledge is *mediated*. We know these things not directly, through our own observation of their characteristics, but instead indirectly, by finding trustworthy technical experts and then trusting them. In that sort of way, the Nutrition Facts label and the more widely distributed language inscribed on it is an index of the relation between different sorts and sources of knowledge.

"You Are What You Eat"

The information on the Nutrition Facts label and the forms of knowledge associated with it do not, however, come close to defining what we know and feel about food. Our beliefs about the nature of food, the everyday practices mobilized around food, and when, where, how, and with whom we eat are all so much bound up with the fabric of our lives and with our individual and collective identities that it's hard to give an account of who we are without also telling a story about what and how we eat. As I began writing this book, I reluctantly considered that its title might be "You Are What You Eat," but I resisted using that title because many people would consider it a tired cliché. And even when I hit upon something better, I still thought that the cliché was worth initial discussion. My topic here is the historically changing relationships between eating and identity, between what we think about our food and what we think about ourselves, about eating and being. And that topic requires engagement with the supposed cliché.

"You are what you eat" is a modern commonplace, while versions of the sensibility—if not this exact linguistic form—can be found as far back in time as you like. Sentiments generally linking eating and modes of selfhood belong to many forms of folk wisdom. Urging careful attention to diet as a major contributor to good health and a cause of disease, a seventeenth-century English book of practical medical advice rehearsed what was said to be a then-common popular saying: "we our selves have had ourselves upon our trenchers."[4] In Shakespeare's *Othello* (Act 3, scene 4), Emilia cynically announced that men "are all but stomachs, and we all but food," though here the gendered specificity was intended. A Spanish proverb had it that "the belly rules the mind"; "the way you cut your meat reflects the way you live" is attributed to Confucius; and a proverb from Lombardy observed that "la minestra de l'vezin l'e pusee buna" (your local soup tastes better than foreign stuff). Traditional sayings causally relating some specific features of character to specific modes and manners of eating are legion: "The full belly does not believe in hunger"; "Forbidden fruit is the sweetest"; "The poor seek food for the stomach, and the rich, the stomach for food"; and "A full belly makes for an empty head."[5] Some commentators have plausibly suggested that the sentiment linking eating and being has religious roots, as in the Christian sacrament transubstantiating bread and wine into the body and blood of Christ. The transformations of food into self, of inanimate stuff into the vessels of immortal spirit and rational thought, were religious mysteries long before they were scientific problems to be investigated and, later, hocus-pocus to be dismissed.

The more recent genealogy of the present-day English saying is worth considering. In 1825, the French gourmand Jean Anthelme Brillat-Savarin wrote that if you told him what you eat, he'd tell you what you are—"Dis-moi ce que tu manges, je te dirai ce que tu es"—and in 1850 the German philosopher Ludwig Feuerbach wrote that "Der Mensch ist, was er ißt" (the German form, much catchier than "man is what he eats").[6] The Frenchman was saying something about social identity and cultural meaning—about *distinction*—and the German was offering a slogan for the politically charged ideology of scientific materialism. In the 1870s, physiologists citing Brillat-Savarin glossed the tag using the emerging vocabulary of nutrition science: "In physiology, one demonstrates that that which nourishes are the materials furnished by foods, materials which, absorbed in the digestive tract, must reconstitute the blood: one demonstrates also that it is the blood that provides for the maintenance of all the organs."[7] The particular sources of these chemical materials—whether from cow, pig, fish, grains or vegetables—matter little; what counts is the identity and quantity of the nutrients, the chemically specific body-making and body-fueling substances.

Through the first part of the twentieth century, and as it became part of the contemporary vernacular, the understanding of the tag "you are what you eat" tended to follow this scientific sensibility. In 1940 Victor Lindlahr, an American osteopath, popular nutrition writer, and aggressive hawker of such patent medicines as the fiber-based laxative Serutan ("nature's" spelled backward), published a best-selling account of the macronutrient, vitamin, mineral, and calorie contents of different foods: its title was *You Are What You Eat*. Lindlahr had been using the phrase from 1923 as part of a meat-marketing campaign: "Ninety per cent of the diseases known to man are caused by cheap foodstuffs," and it is said that it was Lindlahr who propelled the tag into English popular usage.[8]

That vividly colored soft-covered book was on my shelves when I was a boy, and I recall being fascinated by its graphic representations of the chemical makeup of various foodstuffs (figures I.2 and I.3). The book boosted my very early inclination toward scientific materialism: this is what human beings *really* were: so much protein; so much fat; so much calcium, iron, and magnesium; and quite a lot of water. Though the facts were presented not by the federal government but instead by a dubiously reputable nutritional huckster, the figures in Lindlahr's book became part of my vernacular for thinking about foods and eating. An ounce of dried navy beans contains 300 "Sherman units" of vitamin B_1 (I was supposed to take in about 12,000–16,000 of these units per day),[9] 145 mg

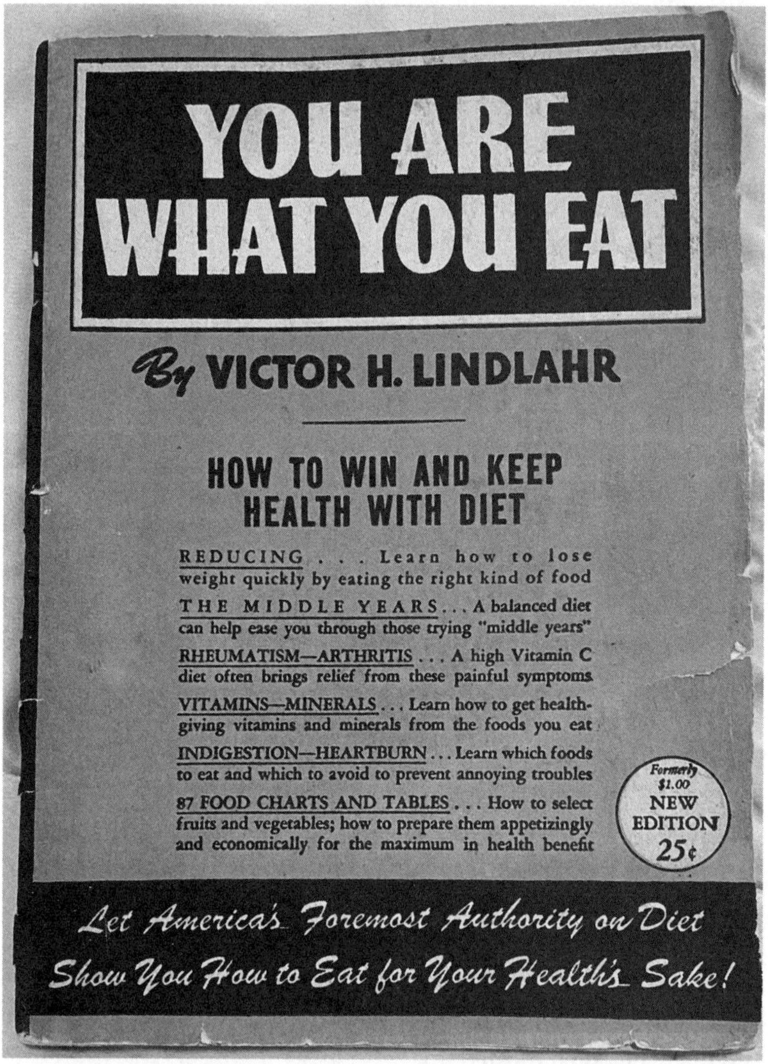

Figure I.2. Cover of Victor Lindlahr, *You Are What You Eat* (1940). The book is said to have sold half a million copies. Lindlahr had a radio show from 1936 to 1953 in which the phrase also figured.

of phosphorus (I was told that I needed 1,980 mg), 22.5 percent of my daily protein requirement, and 1.8 percent of the necessary fat intake and, if eaten in the form of canned baked beans, it delivers 28 calories. Expertly constituted categories of relevant chemical constituents have changed since the 1940s—for example, Lind-

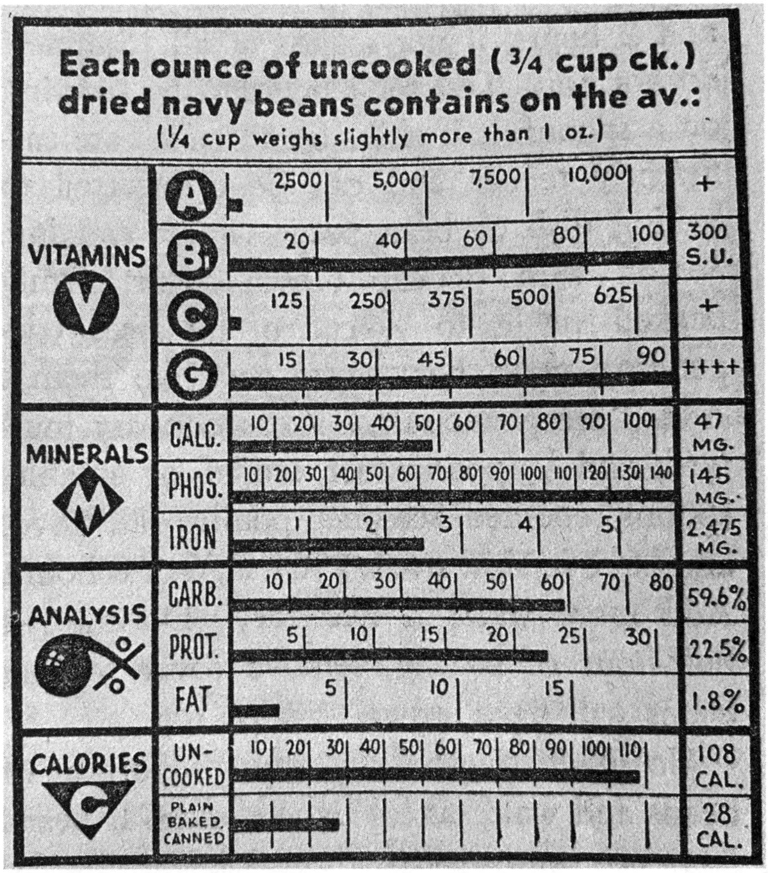

Figure I.3. Victor's Lindlahr's graphic representation of food constituents in navy beans

lahr's book did not subdivide the notion of fat, and no distinction was made between refined and complex carbohydrates—but it was the idiom of Lindlahr and of many other popular nutritional writers from the late nineteenth century that was the widely shared vocabulary for appreciating what might be meant by "you are what you eat."

We now believe such things as this: if you consume more than a certain quantity of calories, you gain weight; if you take too much sugar and refined carbohydrates, you develop diabetes; too much sodium can lead to hypertension; and too little in the way of fresh fruit and vegetables increases your cancer risk. And if you consume the right quantity and proportion of these constituents, you have your best chance of a long and healthy life. The notion of constituents—identified by chemical science and given functional significance by physiological science—is now not only a widely distributed part of our nutritional knowledge, filtered down from technical expertise; it has also been vernacularized as categories used identify what kind of people we are and what we are doing. We are "watching our calories," we are "on low-sodium diets," and we are "minding our cholesterol." We appreciate that the management of constituents bears on the risk profiles for a range of diseases and that it may be an important determinant of how long we live. Supplemented by more recent knowledge about the health and environmental effects of, for instance, high fructose corn syrup and the ratio of omega-3 to omega-6 fatty acids in beef, we may also use the language of constituents to position ourselves as moral actors in an environmentally challenged world and to oppose the consolidated commercial and political institutions that deliver to our tables and our car cup-holders foods with constituents widely said to be both bad for us and bad for an increasingly endangered planet. Much modern morality is expressed through these sorts of scientific idioms.

That way of understanding what our food is and how it becomes us is as historically recent as it is consequential. Whatever we believe about our foods and their effects on our nature belongs to the practices of self-making. Our beliefs about our foods give us a vocabulary to say who we are, why we are the way we are, whether we are acting prudently and morally, and what accounts for the ways of life that do or do not suit us. This occurs on both an individual and a collective level and allows us to say quite a lot both about our personal character and the character of the groups to which we may belong: Spaniards or Germans, soldiers or scholars, the posh or the proletariat. And yet, this language and associated concepts have changed radically over past centuries. We moderns have in hand very different idioms for understanding the relationship between our food and ourselves. What we think about what we eat is bound up with what we think we are.

Ordinary Eating and Ordinary Being

Eating is and always has been woven into the fabric of our lives. This does not mean that eating is "more important" than other things we think about and other social and cultural practices in which we engage, but eating is uniquely pervasive and resonant. We fear death intermittently, and we die once; our thoughts may turn to God—less often perhaps in my part of the modern world than in other sectors of society or in the past—but there are many days when even people accounting themselves religious do not reflect on the Creator and the afterlife. We think of sex often and may have it occasionally. Childbirth may be experienced several times in a woman's life, and children are cared for until they leave home. We are ill from time to time and experience the care of medical experts on only some of those occasions. People intermittently go to war and participate in politics and the public sphere. But if we are fortunate we eat several times a day, often with others. Eating choreographs our life. If we are hungry, pangs quickly remind us of the body's demand for sustenance; feasts and banquets are used to celebrate the harvest, the stations of the liturgical year, the anniversaries of civic institutions, the changes of political regimes, and the smaller personal life markers of births, deaths, weddings, and such personal achievements as are deemed worthy to be so commemorated. In modern as well as premodern arrangements, social intimacy may proceed by way of shared feedings, from meeting for drinks to dinner-and-a-movie to coming-around-for-a-cup-of-coffee. Job interviews, when they become serious, commonly culminate with a dinner for the candidate and her future colleagues; venture capital decisions are forged over lattes, brunch, or Asian soup noodles in Silicon Valley eateries; and American presidential campaigns require candidates to make a photographically recorded circuit through tables of ethnic edibles: pizza, barbecued ribs, tacos, knishes, and kielbasa. We find out about each other as we feed together, we adjust the manner and matter of our feedings to the occasion, and we can conceive of the social order in terms of who eats what, with whom, when, and how. We feed ourselves at the table and, at the same time, we build, maintain, and modify our social relations.[10]

Eating and Being describes long-term historical changes in thinking about our food and offers a way to understand aspects of what "modernity" is. In the late nineteenth and early twentieth centuries, one appealing narrative about modernization identified it with the rise of science and with the disruption of historical links between the worlds of "is" and "ought," between the descriptive and the prescriptive, between the proper domains of science and of morality. This was a story about modernization as secularization, about the "disenchantment of the world" (as the sociologist Max Weber had it). Facing the world as it really was—that is, as naturalistic nineteenth-century science had revealed it to be—modern people knew better than to ask the technical expert about the *meaning* of the world or about how one *ought* to behave in it.[11] Scientific expertise was not moral expertise; it was a mistake to think otherwise. And knowing how things worked in the natural world gave scientists no authority to say what virtue was or how people should conduct themselves in society.[12]

Now, consider this story from the point of view of the modern medical expert. We appreciate that the modern physician possesses genuine expertise about the human body: how it works and how it can be maintained in health or, when diseased, restored to health. And indeed, we turn to physicians for expert counsel on how we *ought* to behave if we want to stay healthy, live long, and recover from illness. Some of that expertise has to do with food and drink and may come from physicians—go easy on the salt, cut down on your weekly units of alcohol—but diet no longer has a major place in medical education, and much of our expert-derived knowledge of nutrition does not now come from research done in medical schools. Yet, physicians' residual "ought" authority presumes no special moral rights because it is thought that the desire for health and longevity is, for all practical purposes, universally distributed. Who would *not* prefer health to disease, a long life to a short one, and more rather than less bodily efficiency?

We expect our physicians to dispense expert advice on what

is *good for us*, not on what is *good*; we think of doctors and related scientists as experts on bodily health, not on virtue. They speak about risk factors, not about sin. True, there are physicians called psychiatrists whose services may be called on to make us less unhappy, more content with our lot, and maybe also—as a byproduct of making us happier and more adjusted—better spouses, parents, or friends, but few of these practitioners now present themselves as moral counselors, and most now intervene into the mental and emotional spheres by way of drugs. Happiness may follow from swallowing a prescription pill, but even now, few think that virtue can be pharmacologically compounded.

How We Eat Now

The general disjunction of *is* expertise from *ought* expertise as well as the special place of medicine in that disjunction is a recent historical development. The focus of this book is on medicine, not science, though from the nineteenth century natural scientific concepts and practices profoundly shaped both expert and lay understandings of what food was and how it affected the human body in health and disease. The first several chapters introduce a pervasive and long-lasting medical culture whose proper name was "dietetics" and in which descriptions of what was *good for you* amounted at the same time to prescriptions about what was *good*. Dietetics was central to medicine, but the relations between medical and moral thought were once quite different from what they now are. Medicine and morality had different recognized boundaries and different bearings on each other. Medicine was once understood to encompass conceptions of virtue, and moral thought was once considered to encompass the practical management of the body, its routines, ingoings, and outgoings. The *power* of modern medicine is celebrated—its ability to diagnose and to heal—while the *scope* of modern medicine is much restricted from what it was in past centuries.

Dietetic thinking flourished for a remarkable span of time, from antiquity through the medieval period and into the seventeenth and eighteenth centuries. Then, dietetic thinking declined, its place in medicine partly assumed in the mid to late nineteenth century by emerging *nutrition science*, with its different concepts and different bases of authority flowing from chemical and physiological laboratories. This book documents both the culture in which dietetics was integral to medicine and practical morality and the circumstances in which its vocabulary dropped out of expert culture. Much of the book aims to recover dietetic language and sensibilities and to display the bases of their authority and cultural penetration. At the same time, *Eating and Being* encourages open-mindedness about what might be intelligibly meant by the decline or even death of a body of beliefs and practices that had been so well entrenched in the culture.

Eating and Being is not prescriptive. It is not a call to the dietary barricades, though I am aware of ways in which those who condemn modern foodways might find in the book material to support their criticisms. My aim instead is to ask what sort of people we now are and how we make ourselves up as people of that sort. I want to illuminate, if not answer, such questions through a genealogy of thinking about food, eating, and our conceptions of what it is to be a person. I encourage curiosity about taken-for-granted aspects of modern selfhood through a historical story about how modern sensibilities came to be. This is an exercise in description, not in nostalgia. And if some readers find premodern arrangements to some extent attractive, I ask them whether, after all is said and done, they would rather live, eat, and manage our bodies as we now do or as we used to centuries ago. And then I would remind them that nostalgia, which is now thought of just as a wistful hankering after a better past, was once the *disease* of homesickness, related to *melancholia* and in pressing need of medical treatment.[13]

My brief reference to the Nutrition Facts label and the vocabulary of food *constituents* pointed to historical changes in the conditions of *knowledge* we may have about food and the consequences of different regimes of consumption. The dietetic culture described in the first several chapters of *Eating and Being* presumed very different capacities of the human *senses* in knowing what our food was and what its effects were when taken into our bodies. I've noted that we now possess our knowledge of food constituents *by courtesy*, through locating and trusting external expertise, but the historical changes discussed here also involved significant shifts in the role of the senses in securing knowledge of the alimentary environment. Given the massive stress on the *visual sense* not merely as a source of knowledge but also as a model for knowledge ("I see"; "You are blinded to the facts"), it may be unsettling to engage with past cultures in which the much-derided sense of *taste* and even the sensory experiences of digestion might be accounted as reliable probes into the natures and powers of edible things. And that is one consideration giving force to traditional sentiments that you are (or should be) your own physician: the prudent person should possess a degree of self-knowledge that, if good enough, would avoid dependence on physicians and that physicians, if and when they are consulted, would need to know. These differences in the distribution of knowledge between lay patient and expert physician, as well as the attendant differences in sensory access to the edible environment, were among the many ways in which long-term historical changes affected the nature of medical understanding and of the medical encounter. In the saying "you are what you eat," the premodern "you" and the modern "you" experience the edible world through different sensory modes, and their different knowledges of that world reflect those changes in mode.

Finally, some brief notes on sources and how they are used here. An aim of *Eating and Being* is to uncover aspects of the history of *ideas* about food, the body, and knowledge of food and the body and to tell a story about historical transformations in these ideas. The sources I rely on are inevitably *texts* about such things, inevitably because it is almost impossible to retrieve past thought from other mediums or sets of artifacts. That said, I use texts in a somewhat different way than is common in much of the history of ideas genre. My purpose is not to talk about the unique thoughts of an author or group of authors, not to detail the heterogeneity in thought at any one time, and not to establish the single thinker, or group of thinkers, who instigated changes in thought. Rather, I am concerned with large-scale intellectual transformations, notably those marked by the submergence or disappearance of dietetic categories from about the end of the seventeenth century to the early nineteenth century and the rise of nutrition science from the end of the nineteenth century to the mid-twentieth century. However, *Eating and Being* engages with and stresses substantial *stability* and *homogeneity* of dietetic thought during the many centuries from antiquity through the early modern period. Indeed, I offer that relative stability and homogeneity as a focused topic. What accounted for this stability? What does this stability say about the power and reach of dietetic thought?

Describing a stable and homogeneous culture is not the usual task taken up by historians of ideas. More common is a focus on innovation, idiosyncrasy, ruptures, discontinuities, and variations. I make no apologies for abandoning that focus in favor of describing bodies of ideas and sentiments that may frankly reek of the commonplace, the commonsensical, and the self-evident, ideas and sentiments that belonged not to unique thinkers but instead were the everyday possessions of the culture in which they circulated. Renaissance and early modern dietary texts embraced and commented on recognized ancient sources; they drew on each other and often frankly *copied* each other. To that extent, the selection of which specific texts to recruit to display aspects of dietetic sensibilities is occasionally arbitrary: one text will often do as well as another, and I introduce readers to a range of such books mainly to establish the ubiquity and the cultural reach of the dietetic commonplace. There have now been scholarly attempts to identify distinct periods in the tone and substance of dietetic writing from the Renaissance through early modernity, the periods associated with shifts in the audience for such texts, changes in attitudes toward ancient authority, and developments in medical thought into the seventeenth century. I have no reason to deny temporal variation over such a long period of time and all sorts of variation within the large body of dietetic writing, but significant stability remains, and that stability can be recognized and interpreted.[14] *Eating and Being* has a bias toward texts known in the English-speaking world, even though much of the literature the book treats was the legacy of Greek and Roman antiquity. Dietetic sentiments known in English circulated widely in other major European languages, and many of the texts known in Britain were translated from ancient languages and other European vernaculars. Again, it is possible to make separate studies of dietetics in non-Anglophone settings, but it is also possible and legitimate to discuss ideas that were broadly shared throughout much of European culture. It is that samey-ness and apparently commonsensical character that merits consideration, that ultimately makes the decline or disappearance of dietetics interpretatively *interesting*, and that makes dietetics and its historical changes consequential for present-day self-understanding.

Eating and Being aims to describe and to interpret, but it also aims to *evoke* by offering readers a sense of what people in the past might have felt like as they thought about their food and themselves. The richness and the occasional layering of historical detail are intended to build a picture of what it was like to inhabit the culture of traditional dietetics, to accept many of its deliverances as self-evident.[15] Apart from the last sections of the book, this is a story about the unfamiliar past, indeed a past that many consider as irretrievably (and happily) lost. But I have written this book partly as an invitation to conceive of ourselves—our thinking about what our food is and what we are—as the uppermost sediment of layered pasts, a surface through which supposedly past sentiments intermittently intrude, one in which some elements of the past were never completely submerged. This is a work of cultural history, but it begins and ends with *us*, with our own largely unquestioned sentiments about what food is and about who we are, about our eating and our being.

1

The Words and Ways of Traditional Dietetics

Those with physiques that are fleshy, soft, and red, find it beneficial to adopt a rather dry regimen for the greater part of the year. For the nature of these physiques is moist. Those that are lean and sinewy, whether ruddy or dark, should adopt a moister regimen for the greater part of the time, for the bodies of such are constitutionally dry. Young people also do well to adopt a softer and moister regimen, for this age is dry, and young bodies are firm.
—Hippocrates, "Regimen in Health"

Barley in its own nature is cold, moist, and drying, but it has something purgative from the juice of the husks. . . . As to animals which are eatable, you must know that beef is strong and binding, and hard of digestion, because this animal abounds with a gross thick blood. . . . Lambs' flesh is lighter than sheep's, and kids' than goats,' because they do not abound with so much blood, and are more moist. . . . Cheese is strong, heating, nourishing and binding; it is strong because it is nearest to a creature's origin; it is nourishing because the fleshy part of the milk remains in it; it is heating because it is fat; binding, because it is coagulated by fig juice or rennet.
—Hippocrates, "Regimen, II"

Beginnings

These passages talk about foods, about human bodies, and about what happens in human bodies when foods are taken into it. They treat the specific nature of different foods in connection with the specific nature of different bodies and with their place in the human life course. The characteristics of foods are described using the same vocabulary as that used to describe the characteristics of bodies. The passages come from texts that embody and communicate medical and nutritional expertise, though the warrants of that expertise are not made explicit. What kind of texts are these? Who wrote them? Who read them, quoted them, and offered them as evidence of authority on the matters at issue? Occasionally, the passages support the rightness of inference from foods to bodily effects by a sort of analogical reasoning: cheese is nourishing *because* its milky origin positions it close to the origin of the consuming human body; it heats the body *because* it is fat, it being common knowledge that fat feeds fire. Texts such as these erect authoritative expert knowledge on top of a body of ordinary knowledge and ordinary modes of inference, even though the voice articulating this special knowledge speaks from wide experience of such matters. It is the voice of someone whose wisdom flows not just from great insight but also from great familiarity with many cases of things eaten and persons eating.

Compare passages such as these to the facts and theories on display in the Nutrition Facts label described in the introduction. There are evident similarities. Both offer ways of understanding the nature of foods and the nature of bodies; both belong to stories about the fate of foods in the body and their contributions to the maintenance of health and the avoidance of disease; both represent modes of expertise, telling people what they may be presumed not to know about such things. That said, the differences between these passages and present-day sensibilities are massive, marking a gulf between the alien exotic of the distant past and the self-evident familiar of the contemporary. The extracts talking about the nature of barley and cheese, about dry diets and dry constitutions, come from Greek antiquity. They were plucked from the Hippocratic Corpus, a body of about sixty medical treatises, most written between about 430 and 330 BCE and possibly associated with medical schools in Cnidus (in what is now southwestern Turkey) and on the island of Cos, just off the Turkish coast. They were composed by a number of authors; perhaps none were written by a historical individual called Hippocrates of Cos (ca. 460–370 BCE). The styles of the tracts vary, and the views expressed in some treatises seemingly conflict with those in others.

Nevertheless, the figure of Hippocrates eventually became associated with what it meant in Europe to be a knowledgeable, observant, and caring physician, seeking natural rather than supernatural causes of disease. The body of Hippocratic writings, later extended and codified by Roman and Arabic writers, remained central to Western medical thought and practice through the Middle Ages and far beyond. This was the ancient origin of ways of thinking about the body, its food, its environment, its states of health and disease, and patterns of living that bore upon health and disease. These modes of thinking persisted for centuries, not unchanging (for nothing about culture is) but remarkably secure, folded into a wide range of thought about nature and about human nature. It has been said that the Hippocratic writings represent both the most enduring and the most resonant body of medical thought ever produced. And it could also be plausibly said that the career of broadly Hippocratic ways of thinking about food and the human body—both their remarkable persistence and their ultimate dissolution—is a perspicuous site for telling the story of how aspects of the modern cultural order came to be. Hippocratic thinking is a starting point for stories about both change and persistence. How do we now think about what we eat, how do we now think about what sorts of beings we are, and how we have come to know the things we now know about the relations between eating and being?

The history of thought about the nature of aliment and the nature of human beings related here pertains to the West. A broadly Hippocratic stream of thought persisted for centuries, widely distributed in European settings and eventually in the New

World and in globally distributed cultures with European roots. As that medical culture developed after Greek and Roman antiquity, significant contributions were made by Islamic physicians and philosophers, and it was extended and modified through contacts between the West and Asia, Africa, and, after the late fifteenth century, the New World of the Americas. Non-Western cultures had elaborate bodies of thought about food, health, disease, and human nature whose origins were indigenous. These non-Western modes were distinct from European traditions even while tempting commentators in one cultural tradition to translate the terms of the one into the recognized vocabulary of the other. Those non-Western schemes were systematic, complex, and culturally resonant. They informed practical medical advice and dietary choice, explained health and disease, were resources in making sense of what people were like—in their similarities and differences—and how they came to be that way, and, beginning around the sixteenth and seventeenth centuries, non-Western traditions made some inroads into Western medical thought and popular sensibilities. The history told here is about *Western* thinking over a span of history; it does not offer a global, all-encompassing account of ideas about eating and being, although it is possible to bring the questions and sensibilities informing this account to bear on such non-Western thought as Indian Aryuvedic medical thought and traditional Chinese medicine.[1]

Regimen and Right Order

Contained in the titles of several treatises in the Hippocratic Corpus was the notion of *regimen*: "Regimen," "Regimen in Acute Diseases," and "A Regimen in Health." The word "regimen" is of Latin origin and became familiar in French and English around the fourteenth or fifteenth centuries. In the original Greek, the relevant word was "diet" (δίατα, *diaita*), which, in English, became cognate with both regimen and hygiene, the latter gesturing at the goddess Hygeia, the daughter of Aesculapius, the god of medicine, and herself the personification of health.[2]

What did diet then mean? First, it referred here, as it continued to do for many centuries, not just to the management of food and drink and especially not just to this management as a way of losing or maintaining weight. Diet, like the notions of regimen and hygiene, was understood as an everyday form of *government*, *discipline*, or *management*. A *regime* (sometimes given in Renaissance and early modern English as a *regiment*) was a set of arrangements for governing a state or an institution, but the notions of diet, regime, and regimen were also used to talk about the personal management of one's *body* and one's *self*. It was a type of discipline that served the goal of securing well-being. Diet for the Greeks was a framework for understanding and regulating a wide range of human behavior, as it provided rules for managing human life: *how to live*, how to live *well*, and how to live *so as* to be well.[3]

Diet (or regimen) included the practices of eating and drinking, but the notion also took in *all* ordered manners of living that bore upon the thriving of the human body and avoiding or managing the range of distempers to which the body was subject. Diet spoke about what was healthful, what was prudent, what was virtuous, and what was *normal* and did so with sensitivity to individuality and to circumstance. A diet was a way of living day by day, through the seasons of the year, as one moved through the world and through the stages of life. Diet took note of the physical environment in which you habitually lived and of changes in that environment as the seasons passed. The folding together of rules for nourishing the body with those counseling prudent and right living is, in general and for us, an index of the antique. The restrictions of meaning that now give us to understand that diet is *just* about feeding were put in place much later, for the most part during the nineteenth and twentieth centuries.

Second, the formal practice known as *dietetics* belonged to the art and science of medicine. Dietetics was generally considered the most important part of medical thought and advice, counting as such for the ancient Greeks and Romans and continuing through the early modern period, from about the middle of the sixteenth century to the end of the eighteenth century.[4] Dietetics stood alongside and sometimes above the medical modes that told you what was wrong with you (*diagnosis*) and what was likely to happen to you in the course of a disease (*prognosis*). Dietetics was often present in the medical practice that aimed to restore you to health when you were ill (*therapeutics*), which included the management of food and drink as well as the prescription and administration of drugs (purges, emetics, etc.). And dietetics might provide justification for bleeding (phlebotomy), periodically removing a portion of your blood. There was much to medical thought and care that did not significantly involve dietetics: setting fractures, healing spear wounds, operating for bladder stones. But dietetics was central to what physicians knew, what they did, and what their patients expected of them. Dietetics was intended to counsel you how you should live so as to maintain yourself in health and to avoid illness. It focused attention on causes that were reckoned to be internal and that affected the whole of your body.[5] Dietetic advice was offered to those who were ill, but the major medical role of dietetics was prevention. A sentiment attributed to Hippocrates was "Let food be thy medicine and medicine be thy food."[6] In the early modern period, some medical writers distinguished between food and medicine, but it was a distinction that also served to establish food's fundamental importance: food is "chang'd into our Nature," while medicine "changes our Nature."[7]

Dietetic sensibilities held that if you lived well, you would remain healthy. There were, however, exceptions to the claimed effectiveness of dietetic advice. Physicians in antiquity and for many centuries thereafter did *not* claim that they could prevent all forms of ill health or that they could restore a disordered body to health in all circumstances: there were conditions about which it was understood nothing could be done.[8] If, for example, you had ignored sound dietetic advice for too long or violated its precepts too seriously, you could not expect even the most skilled physician to save you. You had put yourself beyond medical help. You could fall off your horse and break your leg, and that was no fault of the dietetic physician. Or epidemic diseases—illnesses that literally came upon (*epi-*) the people (*-demos*) in a specific place and time—could afflict you, and their incidence was beyond physicians' control, though even in such cases those who had lived in accordance with good dietetic principles might expect to be less gravely afflicted. Epidemic diseases were widely understood to be caused by a pathological constitution of the air, the water, or some other aspect of the physical environment,[9] and Hippocratic treatises were interested in diseases that tended to occur at specific times of the year, for example, clinically similar fevers occurring during the autumnal equinox or at the rising of the Pleiades constellation of stars.[10] But even if you did succumb to accident or to epidemic disease, proper regimen (here used therapeutically) often offered you the best chance of recovery. With good fortune, you were ill or injured only now and then. For the most part, you were considered to be healthy—that is, in a normal condition for someone of your sort—and this is one way to appreciate why dietetics reached so far into the texture of everyday life and why it occupied such a central place in both medical practice and lay culture. A modern editor of the Hippocratic Corpus wrote that however much the various treatises differed in detail, "all these works are written under the conviction that medicine is merely a branch of dietetics,"[11] and historians of medicine have observed that there were some physicians in antiquity who wrote as if medicine was just dietetics.[12]

The vocabulary of dietetics extended far into many areas of cultural and social life. Modified and subtly adapted to a range of changing circumstances—notably including the introduction into Europe of exotic foodstuffs from Asia and the New World—its basic frameworks for giving accounts of the body and its aliment continued to circulate in expert medical culture into the early modern period before they were challenged in the eighteenth and nineteenth centuries by newer orientations emerging from reformed natural philosophy, chemistry, and physiology. Ancient authority continued to be confidently invoked by physicians into the seventeenth and eighteenth centuries, while the basic forms of dietetic sentiment remained current in much vernacular culture long after that. That is why—with due care and sensitivity to *relevant* variation—expressions of dietetic thought and examples of dietetic practice can be plucked from a wide range of temporal settings and why sources of dietetic thought widely separated in time and cultural setting share concepts, vocabularies, anecdotes, evidence, and acknowledged sources of authority.[13] The counsels of dietetics often appear commonsensically proverbial, and indeed many of them *do* belong to the stock of proverbial wisdom, sticking in the minds of adherents to whom the categories of dietetics were *familiar*, and in many cases linguistic fragments of dietetics continued to circulate in lay culture long after that conceptual system lost its authority among medical experts.[14] And that is a basis on which a historian of ancient medicine noted the remarkable persistence of dietetic thought. The fact that dietetic culture "belonged to everyone," not just to medical experts, does much to account for its continuance in the culture, "essentially unchanged through the centuries."[15] This relative stability has made dietetics largely uninteresting to historians of medicine and science, oriented as they tend to be to conceptual *change* and especially to the sorts of radical changes that can be placed in a progressive genealogy leading to modernity. For that reason, the durability of dietetics has signaled that, so to speak, "nothing interesting is going on," that medicine had yet to be illuminated by the spark of science. Yet, this same persistence bears other interpretations, including those that look for the reasons why dietetic culture lasted so long and why it left its mark on so many seemingly disconnected areas of culture and social practice.

Everyone cared about remaining healthy and avoiding disease, and many features of everyday life bore upon securing those goals. To some observers in antiquity and later, the notion of the physician as a life manager established a parallel between the role of the physician and that of the trainer or mentor, both "in charge of" how their subjects conducted their daily affairs.[16] For the physician, this included an aspect of what might be called moral management. To live *healthily* was to live *rightly*. Historians observe that for the ancient Greeks, "a healthy life is a moral obligation. . . . Health becomes a responsibility and disease a matter for possible moral reflection. . . . Correct and incorrect diet could determine health and disease, and because it was under human control, the choice of diet gave a moral dimension to health and sickness." If you suffered ill health, the cause might well be the imprudent way you conducted your life.[17] So, traditional dietetics—from antiquity through the early modern period—occupied much the same terrain as moral culture and took some of its authority from the way in which advice on health and advice on a virtuous life coincided. At the same time, the authority of dietetics was partly borrowed from its place in an overall philosophy of nature. What were things made of? What were their qualities, powers, and tendencies? And what was the place of the human body and its vicissitudes in general understandings of how nature worked?

There are several reasons why this long-lasting body of thought can be called *traditional*. First, there was an aura surrounding dietetics that represented it as deriving not just from medical expertise but also from *experience*, the sort of experience that professional experts might have in large measure but in its basic forms was available to all. There was often a family resemblance between the deliverances of experts and the sentiments belonging to lay culture, hence speaking with the authority of *tradition*. Second, for a very long period of time, the authority of dietetics was referred back to respected ancient sources so that from the late Middle Ages through much of the early modern period, Greek and Roman expertise itself counted as tradition, this in a culture where ancient knowledge was widely presumed to be superior. From the seventeenth century, the widespread devaluation of ancient scientific and medical authority meant that attaching the notion of tradition to dietetics might serve to mark its inferiority. Third, there was always something about dietetics—from antiquity through at least the eighteenth century—that joined up the medical and the moral, the *is* and the *ought*, expert knowledge and self-knowledge, and this joining up resembles bodies of thought that modernity tends to identify as commonsensical, premodern, and *traditional*. And finally, from the point of view of modern criticisms, dietetics is definitely not *nutrition science*, the laboratory-based expert culture developing from the nineteenth century that is said to have replaced it, superseded it, and revealed it to be mere folkish nonsense. In that sense, the notion of tradition is opposed to what is counted as scientific, expert, and effective.

Both the authority and the cultural reach of dietetic culture were considerable, but that authority was never total, and the notion of expertise in dietetics was often, and consequentially, contested.[18] There is a sense in which you could say that the categories and sensibilities of traditional dietetics were common possessions of European societies—with qualifications, it is fair to suppose that essentially everyone knew its fundamental categories and shared many of its sensibilities—and there is another sense in which distinctions should be made. The overwhelming majority of dietary texts were male-oriented and written by men. Dietetic writers presumed a readership whose members could *choose* their housing, their manner of living, and the nature and quantity of their food and drink. Those considerations do not, however, specify the identity of those who actually provided medical care, possessed recipes for medical therapeutics, or prepared the foods that were said to make for health or disease. Educated males dominated dietetic writing. Males who belonged to the lower orders of society were frequently *talked about* in dietetic writing, though they themselves rarely *talked* in ways that historians can unproblematically access. Women commonly delivered a wide range of medical care—in both elite and ordinary social classes—and a recognition of women's presence is necessary for any account aiming to describe medical care, how and by whom it was provided. Historians in recent years have written about the medical recipes and therapeutic practices of some elite women. An engagement with the categories of traditional dietetics is evident in that female culture, and women are intermittently treated in dietary texts as *objects* of practice, with their constitutional distinctions sometimes discussed.[19] The body of traditional dietetic writing was overwhelmingly produced by men (of a certain social standing), and was oriented to the bodies and circumstances of men (again, of a certain social standing). That said, there is evidence that both women and men of lower social standing were familiar with much of the basic vocabulary of traditional dietetics. In other words, dietetic texts were written and read by elite men, but there is a sense in which dietetics did not exclusively *belong* to them.

Aliments and Elements: The World, the Body, and the Body's Food

Twenty-first-century medicine is enfolded within the basic concepts of present-day natural science. We now know that the natural world is made up of chemical compounds, themselves made up of elemental atoms, and we acknowledge that our food and our bodies are also so compounded. Insofar as we subscribe to this science in its materialist modes, we go on to attribute mental and emotional phenomena to the movements and configurations of chemical compounds. That much is largely taken for granted by the modern educated classes to the extent that many people assume that the science-medicine connection originated only in fairly recent times, identifying relevant "science" with the nineteenth-century emergence of the germ theory of disease, antisepsis, the understanding of intermediary metabolism, and so on. But what counts as medical knowledge has *always* been a part of whatever was recognized as reliable natural knowledge, even as conceptions differed about what indeed was *natural* and about what sorts of things one could securely *know*. Indeed, the ability to maintain health and to cure disease has often been taken as a gauge of the quality of natural knowledge. Some doctors (and patients) in the past were generally satisfied with the knowledge informing medical practice and with the overall efficacy of that practice; others were not, looking forward to better scientific knowledge that would eventually enable more effective medical practice.[20] "Where the philosopher ends, there the physician begins," it was said in antiquity and through the early modern period. The saying was attributed to Aristotle, and its meaning was twofold: first, that natural philosophy had to be the foundation of medicine, and second, that medicine, *unlike* speculative philosophy, was always and rightly put to the test of practice.[21] A common early modern English term for medicine was "physick," deriving from the Greek *physis* (φύσις), or nature.

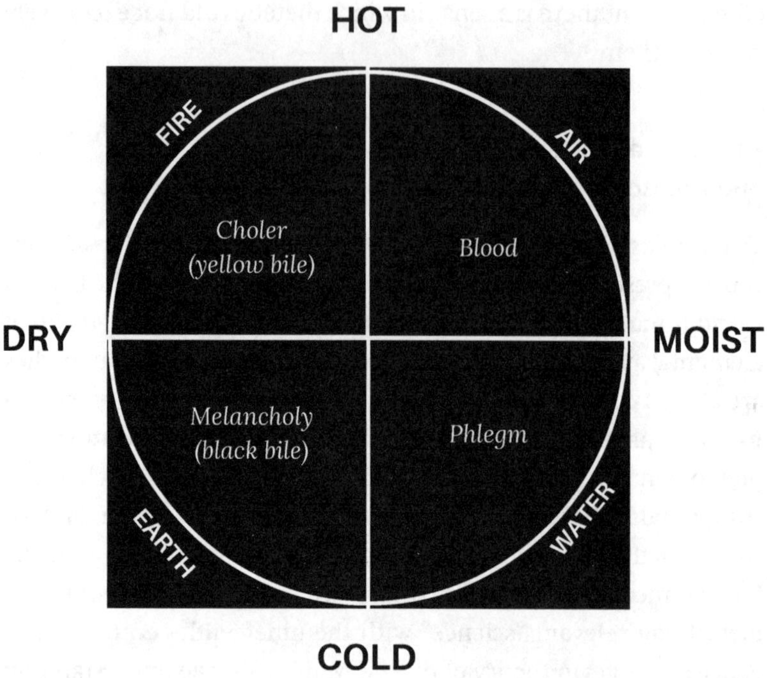

Figure 1.1. Elements, qualities, humors

The natural philosophical concepts informing traditional dietetics were radically different from those of modern science and modern medical practice, different in content, the conditions for coming to knowing them, and their capacity for self-making (figure 1.1). For the Greeks, there were four elements making up everything in the terrestrial world: earth, air, water, and fire. (The existence of a fifth element, the aether—or "quintessence"—was posited for the composition of heavenly bodies.) These elements were present in pre-Socratic philosophy, for example, in the thought of the Greek Sicilian philosopher Empedocles in the fifth century BCE. They were elaborated in Plato's *Timaeus* (ca. 360 BCE) and in Aristotle's tract *On Generation and Corruption* (ca. 350 BCE), and they remained as fundamental cosmological concepts through the Middle Ages and the Renaissance. Associated with each of the elements was a pair of *qualities*: fire was accounted warm and dry, earth was cold and dry, air was warm and moist, and water was cold and moist. (These were the fundamental, or *primary*, qualities, and they were invoked to give accounts of such *secondary* qualities as color, taste, and odor.)[22] This basic vocabulary made sense of a wide range of natural phenomena, including how things tended to move to their *natural places*, and its plausibility rested in part on the observation of ordinary terrestrial phenomena, for example, how a piece of wood when burned dissociated into its elemental constituents: smoky air and fire ascending upward, droplets of water momentarily forming on the burning wood, and earthy ashes remaining behind. Actual terrestrial things were understood ordinarily to be *mixtures* of the different elements, while, of course, ordinary water was mostly elemental water, ordinary soil was mostly elemental earth, and so on.

The elements and the qualities were mobilized in accounts of the human body.[23] And here, inscribed within the cosmological scheme, were body's four *humors*: in Greek, *chymoi* (χυμοί); in Latin, *humorem*. The humors were the basic bodily fluids, each elaborated from digested food and each respectively associated with a specific organ (brain, gallbladder and liver, spleen, veins and arteries). The humors coursed through the body's vessels, performing a range of vital functions. The Hippocratic Corpus identified these humors as blood, phlegm, yellow bile, and black bile (or melancholy): "These make up the nature of [the] body," and it was the presence and relative proportion of the humors that defined health and disease and that figured in describing the causes of well-being and possible remedies of ill health.[24] The humors were central to dietetic medicine and also situated the body—its properties and processes—within philosophical frameworks for understanding *what everything in the world was like*. The humors testified to both the fundamental *similarities* of all human bodies and consequential *differences* between individuals and between types of people. And they were a medically pertinent way of appreciating relations between you and the natural world, where the *world* included those things you took into yourself as food and that ultimately *became you*.

The most enduring and influential accounts of the humors and their role in health and disease were produced in the second century CE by the Greco-Roman physician Galen (129–ca. 216 CE). Galen codified and developed Hippocratic legacies, and it was his writings that remained the authoritative reference for later medical commentators, ultimately becoming part of vernacular ways of understanding the body and its management.[25] So, according to Galen, "the elements from which the world is made are air, fire, water and earth. . . . The humors from which animals and humans are composed are yellow bile, blood, phlegm and black bile."[26] Each of the humors was present in all individual human bodies, but their relative proportion differed from one individual to another, and those differences accounted for much about varying body types, their disposition to different sorts of disease, the propriety for them of different sorts of food and drink, and, when they were ill, the pertinence of different medical regimens. No ordinary terrestrial object was thought to be purely elemental, so none of the actual observable fluids of the human body was thought to be a pure form of any of the humors; all were considered as differing humoral admixtures. (It was, for instance, understood that the actual blood found in your veins and arteries was a mixture in which the humor called "blood" dominated but contained some of each of the other humors.)[27] Evidence for the existence of the set of humors might be grounded in observation of how ordinary blood clotted when a physician removed a portion from the body, visibly dissociating into its constituent humors, with the darkest part corresponding to black bile, the serum to yellow bile (choler), the light matter at the top to phlegm, and the red fluid bits to blood itself.[28] Abstract as the humoral scheme seemed, it—like the scheme of elements—was grounded in accessible sensory experience.

The humors, like the elements, bore their associated pairs of qualities, and each humor was matched with one of the elements and its pair of qualities: blood was warm and moist (like air), yellow bile was warm and dry (like fire), phlegm was cold and moist (like water), and black bile was cold and dry (like earth). Each of the humors was a normal component of the human constitution; none was in itself good or bad. So, the vocabulary of humors and their associated qualities enabled physicians and laypeople to appreciate human *variability*. According to which of the humors tended to "dominate" in individuals (their *krasis*), all persons innately possessed a certain "temperament" (alternatively known as "complexion" or "constitution"). Individual temperament was an aspect of innate nature, but it was appreciated that transactions with external nature—including those bits of nature taken into the body—could maintain and modify innate human nature.[29]

The Nature of Constitutions and the Constitution of Nature

The presence, balance, imbalance, and qualitative state of the humors were fundamental resources in explaining health and disease. But the humors also figured in accounts and explanations of human *character*. The temperaments in which one or other humor dominated were named accordingly, so *people* were described as sanguinary (the humor of blood dominating), bilious or choleric (yellow bile dominating), phlegmatic, and melancholic, each constitution possessing its characteristic bodily and psychological states.[30] You might also—imperfectly but importantly—recognize temperament from a person's superficial appearance, especially those of the skin and face.[31] The concern here is with human beings, but some later commentators extended the vocabulary of humors to describe the characteristics of different sorts of beasts and within a type of beast the characteristics of its varieties. Of horses, for instance, a seventeenth-century aristocratic English equestrian wrote that "*White Horses* are Flegmatick, and so participate of the Element of *Water*, and therefore are Dull and Heavy jades," while the sorrel horse was reported to be "of the Element of *Fire*, and therefore is full of Mettle, Hot, and Fiery."[32] Humors figured in accounts of what beasts were like, but some of those beasts were commonly taken in as human food, and then the humoral qualities of animals might bear upon the humoral qualities of the people who ate them.[33]

In this way, the vocabulary used to describe what nature was like and what bodies were like was available to give an account of the sorts of *psychological* and *emotional* dispositions belonging to different sorts of people. The temperaments—phlegmatic, melancholic, etc.—were not just ways of describing people in whom one or another humor was dominant, and they were not, as the Hippocratic Corpus had it, merely ways of accounting for health and disease. They were also a vocabulary that identified and accounted for character or personality. "The mind's inclinations follow the body's temperament," Galen said, and that formulation continued to be authoritative into the early modern period.[34] Historians note that "the coupling of right diet and virtue was an essential part of Galenic philosophy. Proper regimen balanced the temperament of the body and its parts, and with them the psychic functions."[35] The qualities associated with the humors were the causal basis for people being the *kinds* of people they were: their disposition to be brave or cowardly, to prefer solitude or company, to look on the bright side of life or to view things with foreboding, to be quick or slow to anger. You could give a substantial account of what kind of person you were dealing with through knowing that individual's humoral temperament. This was a self-making resource whose basic vocabulary was forged in Greek and Roman antiquity but became more pervasive, more integrated into everyday life, and more culturally resonant in the centuries that followed.

It was a vocabulary formally owned by medical experts, but much of it was held in common by both educated and uneducated laypeople. Late sixteenth- and early seventeenth-century London theater audiences, in both the cheap stands and the expensive seats, were presumed to be familiar with the vocabulary of humors and temperaments and their capacity to describe human character. A popular genre of Elizabethan and Stuart theater was the *comedy of humors* in which characters represented their dominant personality type or humor, with the best known early instances including George Chapman's *An Humorous Day's Mirth* (1597) and Ben Jonson's pair *Every Man in His Humour* (1598) and *Every Man Out of His Humor* (1599). As the introduction to the latter explained,

In every humane bodie
The choller, melancholy, flegme, and bloud,

By reason that they flow continually
In some one part, and are not continent,
Receive the name of Humors. Now thus farre
It may by Metaphore applie it selfe
Unto the generall disposition,
As when some one peculiar qualitie
Doth so possesse a man, that it doth draw
All his affects, his spirits, and his powers
In their conflunctions all to runne one way,
This may be truly said to be a Humor.[36]

Shakespeare pervasively used the temperaments and humors to allow players to understand the sorts of characters they were playing and to allow playgoers to understand the nature, dispositions, and likely conduct of these characters. Later *comedies of manners* (for example, by Molière, Wycherley, Congreve, and Sheridan) incorporated many of the sensibilities of the earlier humors genre to construct its stock characters (misanthrope, miser, rake, shrew, etc.).[37] Character setting by way of humors was understood in a physical-literal mode, but there were also many early modern figurative usages. The humors and their states were metaphorically extended from the body to the body politic and to cultural bodies. Religious heresies, for example, were sometimes likened to a "peccant humor"—humors that were corrupt or deranged—and Bacon diagnosed the "distempers of learning" also as following from "peccant humors."[38]

The humors and their qualities were resources for self-making. Each quality corresponded to ordinarily available sensory experience, all four being accessed by the uninstructed sense of touch. (You can *feel* things that are hot, cold, moist, and dry.)[39] Sensory experience and metaphysical frameworks buttressed each other: the philosophical claim that blood, for example, had the quality of warmth was supported by ordinary sensation of blood's heat. The qualities were *accessible* and were also applicable in parsing personality. Each of the cosmological qualities—carried by the humors and expressed in the temperaments—was a way of recognizing the psychological characteristics of *people*. That is, the qualities describing the physical makeup of things in the world also described specifically human attributes. Things in the world (rocks, trees, barley, and pork) had qualities, and people had the same range of qualities. The humors performed a character-making and character-recognizing role, and so too did the qualities marking humoral constitution. So, it was sensible to say that a *person* was cold in manner, hot-tempered, possessing a dry wit, and, although moistness is a little harder, the Tory "wets" condemned by Margaret Thatcher's Conservative government (those lacking in neoliberal fervor) were preceded in the seventeenth century by more lax Quakers called "wets" (alternatively "gays"), deemed deficient in religious enthusiasm.[40] Attaching cosmological qualities—to things and to people—was for these reasons supported by the deliverances of ordinary sensation, and that too lent credibility to this way of describing what things and people were like. You might say, with some qualifications, that from late antiquity through the early modern period, essentially everybody—literate or not—was familiar with the fundamental categories of dietetic medicine and that essentially everybody could use these categories to make sense of bodies, minds, and dispositions; to explain, excuse, praise, and blame; and to form robust appreciations of what different sorts of people were likely to do and why they were likely to do it.

From a medical point of view, in antiquity and as long as this vocabulary circulated, there was no notion of a single state of normalcy, of body or of mind. The normal condition of a melancholic person veered toward coldness and dryness (psychologically prone to depression and anxiety), of a phlegmatic person toward coldness and moistness (sluggish and slow), of a choleric person toward dryness and warmth (argumentative and envious), and of a sanguinary person toward warmth and moistness (optimistic and valiant). The sanguinary was as close to the notion of an ideal temperament as this scheme came. A late sixteenth-century treatise on the temperaments approved blood as "the best of all humors," praised the sanguinary complexion as "nearest and likest to the best," and judged that "no state of body . . . is better or more commendable then this."[41] Blood, at the same time, was also understood to be specially labile, with its quantity in the body having critical consequences for the overall state of health, hence the importance of *bleeding*, or bloodletting, the medically pervasive periodical removal of a quantity of blood to reduce bodily heat and the seasonal timing of this bleeding.[42]

The medical excellencies of the sanguinary temperament resonated with the sensibilities of chivalric masculinity, but the sanguinary temperament also had its extremes and attendant pathologies, and it did not simply define normalcy. An individual in whom all of the humors were in due proportion was referred to as "well-tempered." In the sixteenth century, Rabelais so accounted Panurge—"of a well-balanced temperament, constitutionally sound"[43]—and Shakespeare's Brutus was just this sort of person: "His life was gentle, and the elements / So mixed up in him that Nature might stand up / And say to all the world 'This was a man'" (*Julius Caesar*, Act 5, scene 5). In the thinking of many physicians even into the nineteenth century, the temperamental categories permitted even finer gradations, such as the bilious-sanguinary and the melancholic-phlegmatic. And still other hybrid temperamental types could be identified if it was thought that two humors (or even more) tended to share dominance in a specific individual.[44]

The balance of humors that marked out one individual from another also distinguished groups of people living in different environments. In "Airs Water Places," Hippocrates was concerned with explaining why the race of Scythians (presumed to live toward the north of the Black Sea) came to be as they were. They manifested little physical variation one from the other; their bodies were "gross, fleshy, . . . moist and flabby"; both men and women were fat and hairless; and they were virile and prolific. The causes of this condition were present in the environment in which they lived: the seasons of the year showed little variability, and they breathed "a moist, thick atmosphere, [and drank] water from ice and snow." The food they consumed also bore upon humor-based character, the qualities of their aliment being similarly marked by the physical environment. And the people of a region were directly influenced by the constitution of local waters, airs, and soils; the direction and strength of winds; the ambient temperature; changes or stability in these over the course of a year; and the people's habitual manner of living. Did the people, for example, work hard or little? Were they heavy or light eaters and drinkers? Did their water come from collected rain or from melted snow and ice? A variable physical environment tended to make for a violent and hot-headed temperament, and this is why Europeans were said to be constitutionally more violent and courageous than Asians, who inhabited a more temperate, less changeable climate, "for in general you will find assimilated to the nature of the land both the physique and the characteristics of the inhabitants."[45]

The constitutional qualities of women were understood to differ from those of men, though medical and philosophical writings were not universally in agreement about what those differences were. Aristotle's opinion was that women were moister and colder than men, though he acknowledged that earlier philosophers thought they were hotter.[46] The Hippocratic treatise *Diseases of Women* judged that "a woman has hotter blood, and for this reason she herself is hotter than a man."[47] But into the early modern period, female tendencies toward the moist and cold seem to have been widely accepted. The French physician Lazare Rivière, for instance, considered that condition to be "infallible," citing both ancient warrant and abundant everyday knowledge: "'Tis obvious and discernible to every eye, that by their animal actions males are hotter, they being much stronger, and of greater ability to labour, but females dull and slow." Men's pulse and breathing were stronger, their voices were "fuller and more intense," and their physical strength was manifest, all testifying to their "greater heat."[48] The generalization was, however, acknowledged to be subject to exceptions. *Particular* women were said to be dry and hot, and circumstances and stages of life cross-cut generalization: a young woman might be hotter than an older man. And there were physicians in the sixteenth and seventeenth centuries who rejected the notion that women were indeed moist and cold.[49]

The humors also linked nature to human natures in temporal

dimensions. The qualities varied with the seasons of the year and with the stages of human life. Spring was the season of the warm and moist, summer of the warm and dry, autumn of the cold and dry, and winter of the cold and moist. As the human body aged it moved in the same sequence, from the warmth and moistness of childhood to the cold and dry of old age.[50] There were some historical disparities in how the boxes of temperamental schemes were filled in. By the seventeenth century at least, the links between the temperaments, the qualities, and the seasons became, to a degree, variable. Some writers, for example, assigned melancholy to winter, while others located it in autumn.[51] Nevertheless, the overall stability of these ways of thinking about nature and human nature was remarkable: cultural historians have judged that the scheme of the temperaments—founded on the humors, qualities, and elements—was "one of the longest-lived and, in some respects, one of the most conservative" blocs of European culture.[52]

Both the state of the physical environment and the state of your maturing body bore upon the decisions to be taken about transactions with the environment, including what was good to eat and drink. Your stomach was part of the fabric of your body, but it also participated in your body's life course and was affected by the rhythms of the seasons. The environment governed you, and if you were prudent you learned to adapt yourself to annual changes in the environment and to the natural changes in your aging body. The humors, the variability in their preponderance from one person to another, their changes during the life course, and their relationship to varying physical environments and to annual rhythms constituted a rich framework for understanding the natures of different people: what they did and what they ought to do at different times and places and in different circumstances.[53]

The Spiritual Body

The elements and qualities were fundamental to traditional dietetic medicine, but there were other vocabularies used to talk about the nature and processes of the human body and about its food. Prominent among these was the notion of the *spirits*. The spirits appeared in Galen's writings, and their natures and roles were elaborated by Latin and Arabic authors later in antiquity and in the medieval period, persisting in both medical and lay culture well into the eighteenth and nineteenth centuries. While the generic existence and vital significance of the spirits were widely accepted, there were disagreements about many aspects of their identity and how they functioned.[54] The doctrine emerging from Galen and later commentators considered spirits as mobile vapors, exceedingly fine yet ultimately corporeal—"airey, thin and clear substances," according to a late seventeenth-century dietary text.[55] Spirits acted as intermediaries between the material body and the immaterial soul, powering vital functions.[56] Spirits were of three distinct sorts: *natural*, *animal*, and *vital*. Schemes differed with respect to what each type of spirit did, where and how it was produced, and how it moved in the body, but a typical Renaissance version understood the ultimate source of spirits as food and air, compounded in the making of blood. The natural spirits were produced in the liver and then dispersed through the veins, serving nutrition and reproduction; the vital spirits—generated in the left ventricle of the heart—diffused vital heat and the basic principles of life through the arteries, keeping body and soul together; and the animal spirits—elaborated in the brain and distributed through the hollow nerves—powered movement and underpinned appetite and imagination. The spirits went where their functions were called upon by momentary vital activities, so the body's ability to perform its various operations—mental as well as physical—was dependent on the supply of spirits and whether their powers were then available to do what was required.

The spirits were a potent resource for talking about physical connections between the vital and mental phenomena of the human body and about the ultimate source of human life in the divine *spiritual* domain. Some writers through the early modern period invoked the distinct varieties of spirit, while others used "the spirits" in a lumping fashion as protean principles of vitality.[57] Spirits and humors were distinct categories for describing bodily affairs, but some accounts spoke of transactional connections between them. Medical conceptions claimed, for example, that the humor of bile was left over when aliment had been digested and assimilated to pure animal spirits, and this offered a substantial link between the nature of the body's food and psychological dispositions.[58] The body's endowment of spirits was part of its natural constitution, though as with the humors, the spirits' balance and integrity were affected by how people lived, including what they ate. Some foods were thought to be more conducive than others to the elaboration and rectification of spirits. Food was the material substrate "out of which the humours and spirits (which be the incensors and stirrers forward of the minde) obtaine and receive their nature," wrote a sixteenth-century Dutch physician. You should "use the most exquisite diet . . . to the end that the meates and nourishments . . . may make the spirits pure, sincere, and perfect."[59] In this way, the substance of your food was a substrate for the quality and quantity of your spirits, and the balance of spirit-producing qualities in your food bore upon the balance of spirits in your body. The spirits were treated pervasively in late medieval and Renaissance medical and philosophical writings that offered accounts of body, soul, and human vitality, and although there were challenges to the notion of spirits in the seventeenth and eighteenth centuries, many dietetic texts continued to comment on how various foods and environmental circumstances affected the spirits and how the working of the spirits bore upon a range of human activities.

As with the humors and the temperaments, the vocabulary of spirits belonged to the vernacular domain as well as to medical and philosophical expertise. Many sorts of people had some knowledge of the spirits as part of their general cultural equipment, well-educated people certainly and probably too the ill-educated and many of those not educated at all. People knew—in more or less detail—that there were such things as the spirits, what sorts they came in, what functions each of them served, how they affected passions and personalities, and how they were in turn affected by patterns of human behavior, including dietary behavior. In the 1530s, Rabelais's *Gargantua and Pantagruel* indicated that an understanding of the spirits was part of what might be called the educated vernacular. The physician Rondibilis advised the libertine Panurge on the management of sensuality, explaining how the spirits were elaborated and how they moved through the body.[60] Shakespeare's audiences were likewise assumed to know about the spirits as well as about the humors and temperaments, as when in *Love's Labour's Lost* (Act 4, scene 3) Shakespeare referred to "the nimble spirits in the arteries," presuming some sort of understanding of these vessels as the conduits of the vital spirits,[61] or when Juliet (*Romeo and Juliet*, Act 4, scene 1) was urged to take her sleeping potion: "When, presently, through all thy veins shall run / A cold and drowsy humour, which shall seize / Each vital spirit." John Donne's early seventeenth-century poem "The Extasie" considered readers to be familiar with the doctrine of the spirits: "As our blood labours to beget / Spirits, as like soules as it can, / Because such fingers need to knit / That subtile knot, which makes us man."[62] Invocations of the spirits were pervasive in all sorts of cultural scenes, and those who encountered the notion of spirits were presumed to know something about how dietary regimen bore upon their role, as they were presumed to appreciate the role of the spirits in self-making and self-presentation.

The species of spirits were also treated in ethical, moral, and religious writings, as in tracts by the Italian humanist philosopher Marsilio Ficino in the late fifteenth century and the French physician Jean Fernel in the late sixteenth century.[63] The English divine Thomas Burnet, delivering the natural theology lectures endowed by Robert Boyle, explained the physiological effect of Adam and Eve's original sin as the disturbing effect on the animal spirits of eating the forbidden fruit.[64] In 1691 the cleric Gilbert Burnet referred Boyle's occasional religious scruples to "depressions or weaknesses of the Animal Spirits," itself caused by imperfect dietetic management: "the want of Nourishment or Free Air or Exercise, or pleaseing Circumstances."[65] The spirits were frequently invoked in biographical exercises as a resource for describing *kinds of people*, why certain people were the way they were. The early eighteenth-century biographer of the Cambridge Platonist Henry

More described the philosopher's aetherial body as marked by "a mighty Purity and Plenty of the Animal Spirits," their purity maintained by a rigorously temperate regimen.[66] The spirits, like the humors and temperaments, belonged to a medical vocabulary that was used causally to link diet and bodily and mental states, a vocabulary that was part of vernacular as well as expert culture. It was used to make sense of what people were like, how they came to be that way, and how dietary regimen might bear upon health and disease, upon character and its changes. Physicians and philosophers were familiar with the language of spirits, their nature and their role. But so too were many other sorts of people as they accounted for what human beings were like, how they came to be that way, and how human nature might relate to the food that people ate.

Tuning the Body: Health and Harmony

It was medically and physiologically good for your qualities and humors to be in a state of balance. The Hippocratic Corpus noted of the humors that "health is primarily that state in which these constituent substances are in the correct proportion . . . and are well mixed."[67] Galen wrote that "health is characterized by the equality and symmetry of these humours. Diseases occur when the humours decrease or increase contrary to what is usual in terms of quantity [and] quality."[68] Health was harmony—a balance of qualities and of humors—and disease was imbalance. And when you were diseased, your aim was to be restored to health by the correction of imbalance. The scheme was simple to state in principle, but in practice and when brought to bear on individual bodies, it allowed for much modification, qualification, and adaptation to individuals' makeup and circumstances.[69] People differed in temperament, and each temperamental type had its characteristic state of normalcy. In warrior and chivalric cultures, there was a disposition to highly value the sanguinary temperament, but it was understood that the other temperaments also flowed from their own characteristic states of humoral balance (*eucrasia*, or "good mixture") and that each was also prone to its characteristic pathological imbalance (*dyscrasia*, or "bad mixture"). So, for a person with a choleric temperament, normalcy was a tendency toward the hot and dry, and the besetting pathological tendencies were toward extremes of the hot and the dry. Traditional dietetics acknowledged the individuality of health and ill health. Disease was an altered state of that balanced normalcy that was *you*.[70]

Your patterns of feeding bore upon humoral balance. Galen said that "when we make proper use of foods in recipes, the attendant humours follow," and so, while humoral balance or imbalance might be given by nature—as a parental endowment—it was importantly managed through way of life and aliment, the proper qualities of the food making for the proper balance of humors and their attendant qualities.[71] In these matters, dietetic advice could be as complex and sinuous as it was cosmologically grounded and self-evidently prudent. Melancholic persons had their own form of temperamental normalcy and for that reason might be recommended to take food and drink matching (or, as it was said, "agreeing with") that normalcy. For example, a cold and dry person when in health should use aliment tending toward the cold and dry, since that was their normal condition. However, since the pathologies to which melancholics were prone were characterized by the *extremes* of cold and dry, such people when not in health might be advised to adopt a diet of *opposites*, that is, food and drink whose qualities tended to warmth and moistness. The Hippocratic "Regimen in Health" was clear about this sort of thing: those whose constitutions were naturally dry, for example, should keep to a moister dietary practice; in advising on regimen, attention should be paid to "age, season, habit, land, and physique"; and care must be taken to "counteract the prevailing heat or cold."[72] Here the long-honored principle was the cure of disease by opposites (*contraria contrariis curantur*). The Hippocratic tract "Breaths" aphoristically summarized much medical counsel "in a single sentence": "Opposites are cures for opposites. Medicine in fact is subtraction and addition, subtraction of what is in excess, addition of what is wanting."[73] This intricate dietetic choreography had to proceed from judgments about whether the persons concerned were indeed in a normal state of health or whether it was deemed that they were ill or about to become ill. Balance was key, but that balance was understood dynamically and circumstantially. The point of balance shifted with judgments about a person's momentary state of health but also with annual rhythms, with the stages of life, and with the specific circumstances bearing on the person. And it might also change based on to whom judgment belonged and to whom choice pertained.

Traditional dietetic writings were concerned with *choice* about what to eat and drink but also where to situate your house, what quantity and type of exercise you took, and so on. But the presumption of choice picked out—usually without special comment—certain sorts of people while excluding many others from consideration. Dietary choice was pertinent to people who were able to choose. Taken as a whole, from antiquity to modernity and in much of the modern world, people lived (and continue to live) in conditions of scarcity and constraint. Simply put, most people had and have little choice about what to eat and drink. Gender roles were, of course, involved—women typically had far less choice than men—but gender was heavily cross-cut by social standing and economic circumstance: wealthy women might exercise more dietary choice than poor men. When the Hippocratic "Regimen in Health" advised that in winter people should eat as much as possible and that their meat and fish should be prepared by roasting, the presumption was that readers were in a position to follow this sort of advice.[74] When Galen counseled that apricots were slightly better food than peaches, that harsh-tasting apples were recommended when the temperament was excessively hot, and that both bone marrow and brain should not be eaten in excess, he was also taking as a matter of course the ability to choose among such things.[75] Dietetic counsel recommended the best type of massage, advised whether it was better in summer to walk slowly or vigorously, distinguished the consequences of eating barley rather than wheat, drinking red and sweet wine versus white and dry, and commented on the virtues of domesticated rather than wild rabbits, whether the loin or the rear leg of the rabbit was healthiest, and whether the rabbit's rear leg ought to be roasted or boiled.

A concise summary of ancient dietetics by a historian of medicine noted that "the man who wishes to live in accordance with the physician's requirements" and to adjust diet to constitution, age, time of life, and season "must have time at his disposal and be rich, in order to do everything he should. For only one who is not completely absorbed in any activity and can afford to do as he wishes can avoid every excessive exertion, always eat what is good for him at the proper time, stay quietly at home when it is hot, or be more active in sport when it is cold. Only the rich and independent, therefore, can live in a completely healthy manner. The average person can do only as much for health as his manner of life allows."[76] On the positive side, rich persons were able to choose and, if they liked, could choose to devote their lives to the perfection of health, while on the less happy side such dedication made for a manner of living that was rigid and anxious. In practice, the dietetics of antiquity and its later forms were caught up in distinctions between different sorts of people, between tensions bearing on the ideal and the real, and between the requirements of health and the demands and expectations of ordinary civic life.

Similar tensions were in play in decisions about *who knew best* about a healthy course of life. Did such knowledge belong solely or mainly to medical experts? Did laypeople also possess relevant knowledge? What were the proper relations between your knowledge of your own body and the knowledge that dietary experts might have about your body? Expertise could be commended on several grounds. First, it was claimed that good medical practice was informed by and followed from good natural philosophy, that is, knowledge of what the world was like, including knowledge of the fundamental makeup of the human body and of all other natural bodies. That is what was meant when a Hippocratic treatise said that any one ignorant of the "primary constituents" of the body would be incapable of "administering to a patient suitable treatment."[77] Yet, at the same time the physicians of antiquity and their historical successors also claimed authority on the basis of their large empirical *experience*. They had attended many cases, including many like yours. They knew what illnesses people like you tended to suffer from, how people like you tended to fare as these illnesses took their course, and how people like you might best avoid affliction and, when ill, what regimes tended to be most effective in curing or mitigating disease. Both philosophy and experience supported the idea that a physician could offer effective counsel to the patient in health and in disease.

Knowing Yourself and How to Do It

Physicians' claims to superior knowledge flowed from both philosophy and experience. Yet, at the same time, traditional dietetics repeatedly recommended that *you should be your own physician*, that you knew best about how to secure your health and cure your disease. A Hippocratic tract asserted that "a wise man should consider that health is the greatest of human blessings, and learn how by his own thought to derive benefit in his illnesses."[78] The saying "After the age of thirty every man is either a fool or a physician," attributed to Roman emperor Tiberius, was endlessly repeated well into the nineteenth century, though other versions put it at twenty, forty, fifty, or sixty years of age.[79] That is, if you were attentive and wise enough, time would grant you vast experience of the vicissitudes of your own body, and if you did not learn that self-management, then you made it clear to others that you were either ignorant or imprudent. You could and should learn enough to be your own physician.

Medicine importantly included self-knowledge, subjects' understandings of their own bodies and minds. And that insistence on self-knowledge was a constant and continuing accompaniment to the deliverances of medical expertise. In the first century CE, the Roman encyclopedist and medical writer Aulus Cornelius Celsus recommended that "above all things everyone should be acquainted with the nature of his own body": you ought to learn what sort of body you had and what regimen best suited it.[80] Toward the beginning of the second century CE, the Greco-Roman historian and Stoic moralist Plutarch, writing about how to preserve one's health, reckoned that everyone could and should know their own body and its responses: “For he has no sense, and is both a blind and lame inhabitant of his body, that must learn these things from another, and must ask his physicians whether it is better with him in winter or summer; or whether moist or dry things agree best with him. . . . For it is necessary and easy to know such things by custom and experience. . . . We ought also to know what things are cognate and convenient to our bodies, and be able to direct a proper diet to any one upon any change of weather or other circumstance.”[81] One of the sayings supposedly inscribed on the Temple at Delphi, dedicated to Apollo, the divine patron of medicine, was the injunction “Know thyself” (*gnothi seauton*, γνῶθι σεαυτόν). This might mean that you should know your human limitations, but it was also cited in the dietetic sense that you were meant to know—better than anyone else—what worked for you, what agreed and disagreed with you, and how to listen to your body’s urgings and warnings in health and disease. Authority on the human body—considered in general and in the abstract—might belong to the physician, but authority on a particular patient’s body was distributed between the physician and the inhabitant of that body.

A question always at issue in traditional dietetics and on which both the layperson and the physician might have a view was whether a particular human body *was* in a state of health or of disease or even whether an apparently healthy body was *about* to become ill. When did you need to consult a physician or, once consulted, rigorously to follow a physician’s advice? One potential answer in traditional dietetics implied that a layperson’s dependence on the physician’s advice was almost constant. After all, since dietetics belonged overwhelmingly to preventive medicine, and since dietetic advice encompassed such a wide range of transactions between you and the environment, you might plausibly think of your relationship with your physician as intimate and durable. The dietetic physician would then be in a position to monitor aliment, exercise, the conditions and modes of travel, and so on.

But another answer was also contained in traditional dietetics and proceeded from the same sensibility that pronounced that you should “know yourself”: that you should attend carefully and continuously to the effects of quotidian transactions between your body and its environment, edible and physical. Hippocratic and later medical texts offered fine-textured accounts of the rules for healthy living while at the same time cautioning against a too-strict observance of these rules for people in a normal state of health. If you adhered too rigidly to dietary rules, the occasional and unavoidable deviations might have serious consequences: "In a restricted regime the patient makes mistakes, and thereby suffers more. For this reason in health too an established regimen that is rigidly restricted is treacherous, because mistakes are more hardly borne." Given a choice and when in health, you should opt for the more liberal regime and should not be too particular in following dietary rules.[82] The first pages of Celsus's treatise on medicine offered a widely cited version of this sensibility. The Roman physician advised when you should put yourself under a physician's care and when you need not do so:

> A man in health, who is both vigorous and his own master, should be under no obligatory rules, and have no need, either for a medical attendant, or for a rubber and anointer. His kind of life should afford him variety; he should be now in the country, now in town, and more often about the farm; he should sail, hunt, rest sometimes, but more often take exercise; for whilst inaction weakens the body, work strengthens it; the former brings on premature old age, the latter prolongs youth. . . . It is well also at times to go to the bath, at times to make use of cold waters; . . . to avoid no kind of food in common use; to attend at times a banquet, at times to hold aloof; to eat more than sufficient at one time, at another no more; to take food twice rather than once a day, and always as much as one wants provided one digests it.[83]

Continually invoked through the early modern period, this became known as the Rule of Celsus and was drawn on as an authoritative argument against people who were too fastidious about their health, too ready to place themselves under expert care, and too inclined to use their health as an excuse for why they could not undertake the normal obligations of civic life.[84] Perfect health and longevity, while clearly to be valued, were not the only things that mattered in life, and it was both morally wrong and practically unsociable to allow regimen to trump other civic responsibilities. Plato's *Republic* told a story about a man called Herodicus who was offered up as a cautionary example: he monitored his health so exactly that he lived as if he were perpetually ill, "unfit for the business of life, suffering the tortures of the damned if he departed a whit from his fixed regimen." Herodicus *did* live a very long time, and his illness was not imaginary, but his precise and punctilious regimen wasn't worth it: his old age was unhappy and unproductive. Too much attention to the care of the body makes for trouble; it's not good for you even if it happens to extend life: "For living in perpetual observance of his malady, which was incurable, [Herodicus] was not able to effect a cure, but lived through his days unfit for the business of life, suffering the tortures of the damned if he departed a whit from his fixed regimen." No one should want to live like that, and if you did live like that, you paid a heavy civic price.[85]

The Rule of Celsus implied that persons who were in fact capable of the range of activities appropriate to their station and role in life counted as healthy, in no need of the counsel and direction of external medical expertise. The ability to carry on normal activity just *was* the state of health. While it was, of course, proper to closely monitor your condition and therefore to "be your own physician," there was no need and no justification for anyone else to tell you that you might be at risk of illness. Such a person should get on with life: it was good to do so and it posed no danger. When you were actually ill you should, of course, make getting well your chief concern, but when you were actually well you should think of *action*—of doing things in the world—and not of the precarious state of your body.[86] The grounds of knowledge of how your body worked and of the effects on your body of different forms of aliment and activity were sufficiently available to you. Expert dietetic knowledge contained and prescribed a large dose of self-knowledge.

Dietetic sensibilities were pervasive and historically tenacious.

They had great temporal stability and great reach into cultural and social life. Dietetics was stabilized in large part because its concepts, sensibilities, evaluations of right conduct, and counsels of practical action were at once medical (what you should do to maintain health, cure disease, and live long) and moral (what you should do if you aimed to live a proper life and count as a prudent person). In dietetics, the *good for you* occupied much the same cultural terrain as the *good*. Dietetics was part of the repertoire of thought, action, and judgment that allowed people in the past to recognize different human types, to form expectations of what those types were like and what they were likely to do. Dietetics allowed you to present yourself as a certain sort of person and to signal to others what they might expect of you. Dietetics characterized the body's aliments as resources for self-making, those bits of the edible environment that made you what you were and that might change what you were. For all that, traditional dietetics failed. Its vocabulary and the practices in which that vocabulary was embedded were increasingly challenged from the seventeenth century, ultimately disappearing from almost all expert medical writing by the nineteenth century. And when traditional dietetic culture ultimately lost its authority and was supplanted by other ways of understanding, relations between eating and being also changed. The career of dietetics tracks the making of the modern.

2

Medicine, Morality, and the Fabric of Everyday Life

The Natural and the Moral

The categories of traditional dietetics straddled the domains of the descriptive and the prescriptive, the natural and the moral. They offered ways of traveling from *is* to *ought* and back again. That capacity allowed traditional dietetics to sink roots deep into everyday cultural and social life. The notion of "the natural" is itself poised between description and prescription. To say that something is *natural* is to say that it belongs to the normal state of affairs; it is warranted, justified, moral, or good. Similarly, the *unnatural* is the abnormal, outside the order of things, immoral, or bad. Contrast, for example, the sense in ordinary speech of *natural appetites* and *unnatural acts*. That is why nature was and remains a potent resource for legitimation and justification.[1] The special cultural place of traditional dietetics is indexed by the inscription of the natural and the unnatural at its center. This was the scheme of *The Six Things Non-Natural*. The systematic formulation of the non-naturals emerged from about the thirteenth century. Well into the eighteenth century, dietetic texts strove to attribute the idea to antiquity and it was probably assembled and systematized from material scattered in the Hippocratic Corpus and especially in Galen's tracts on the arts of medicine in his *Hygiene* (*De sanitate tuenda*) and in the work of the eleventh-century Uzbek polymath Avicenna, whose *Canon of Medicine* was itself a compendium of Galenic medicine and Aristotelian philosophy. That is to say, the idea of the nonnaturals was old enough to count as traditional wisdom, with later commentators burnishing its authority through attribution to specific Greek and Roman medical writers.[2]

The *nonnaturals* were so called in contrast to what were considered the *natural* endowments of the human body: the temperament passed on to you at birth by your parents, the anatomical form of your organs, and your original supply of spirits. The nonnaturals were also contrasted with the so-called *contra-naturals*, or things against nature: birth defects, injuries, and pathological events other than those brought about by your own actions. The things nonnatural were those transactions with the world that in principle were under your volitional control, which had substantial consequences for the state of your body and mind and about which you might receive and possibly act on rational advice.[3] It followed that your bad management of these things could only be *your fault*: "Our selves therefore we are most to blame for our Maladies, whose unhappy disorders they inseparably follow, as the Shadow doth the Body."[4]

By the sixteenth century, the list of nonnaturals was broadly standardized, even in its sequence, and that list and sequence circulated well into the eighteenth century as physicians urged knowledge of the nonnaturals and attention to their importance in maintaining health and curing disease.[5] The nonnaturals were familiar in literate culture and, possibly like the vocabulary of humors and temperaments, figured in articles of prudence among the uneducated:

1. Airs, waters, and places: Where it is best to live and to situate your house, how to move about in the physical environment, and what features of that environment bear upon health.
2. Food and drink: What to eat; when, how much, at what stages of life and at what times of year; how foods should be prepared; and in what order certain foods are to be taken and in what combinations with other foods.

3. Motion and rest: What sorts of exercise ought to be taken at various times of day, year, and stages of life.
4. Sleeping and watching: When, how, how long, and in what positions and orientations to sleep and what to do before going to bed and on waking.
5. Evacuations: How to manage urination, defecation, and the body's fluid economy. Counsel on sexual activity presumed a male readership and in this context was geared to ejaculation and to the economy of spirits, while there was a background of understandings of female sexuality.
6. The passions of the soul: How to regulate what would now be called the emotions, such as anger, jealousy, and sadness; what aspects of regimen bear on the passions; how passions and ways of living affect each other; and how to make the right dietetic choices to achieve a desired state of the passions.

Few dietetic texts in the sixteenth and seventeenth centuries were *not* organized in some significant way by lists of the nonnaturals, and many tracts of practical ethics—how to behave, how to raise children—were similarly structured.[6] The nonnaturals worked as organizing principles for medicine *and* for practical morality. They allowed people to give advice and to appreciate the soundness of received advice about what was good for their bodies and, at the same time and using the same categories, what counted as prudent and virtuous conduct.

People might be instructed to site their houses facing east and far from marshy areas, never to eat ripe melons, avoid meals and dishes containing too many different ingredients, sleep with the head facing north and not more than nine hours a night, walk vigorously after dinner and rest after supper, move the bowels regularly and ideally in the morning, and not give way to rage or envy. It was presumed that the rational soul was and ought to be master of the emotions. To be the servant of one's emotions rather than their governor was demeaning. The texture of dietetic advice on the nonnaturals reeked of prudential common sense. Its counsel was invariably to seek temperance and moderation, the mean between extremes in each of the categories. The second pronouncement said to have been inscribed on the Temple of Apollo at Delphi was "Nothing in excess" (*meden agan*, μηδὲν ἄγαν). A temperate climate was the best; eating and sleeping in moderation was the road to health. Patterns of bodily movement should be neither too indolent nor too violent. Regular and timely evacuation and moderate sexual release were good, as was a due command of the emotions, being wrathful when occasions called for it but always maintaining rational mastery of the passions. Hippocratic aphorisms commended moderation in all aspects of regimen even while considering efforts to follow moderation too punctiliously itself a form of excess, a sensibility often supported by invoking the Rule of Celsus that indexed moderation to circumstance.[7] This was the sort of advice that you could find endlessly repeated not just in the dietetic texts of the Renaissance and early modern period—"Too much of anything is an enemy to Nature"—but also in the mass of practical ethical tracts on how to be a "governor," a "prince," a "gentleman," a civic actor and how to bring up your son to be one of these.[8] Moderation made for health and moderation was not only a way to virtue but also itself a virtue. Moderation was a cultural prize belonging to both the medical and the moral orders. An English Puritan moralist writing in the early seventeenth century about dietary restraint praised "holy moderation."[9]

Things Nonnatural: Taking the Air

The air was everywhere; it varied in its makeup and was a constituent of the physical environment in which you lived and that you unavoidably took in. Yet, exposure to air, specifically to different sorts of air, was something that traditional dietetics reckoned you could and should do something about. You could move around, resituate your house, and go out or not go out in certain atmospheric conditions and at certain times of day. Airs—like waters and other aspects of the physical environment—were to an extent considered to be under your volitional control, and dietetic advice counseled you about exposures, presuming again that you had choice about such things.[10] It was vitally important to choose proper exposure to airs. “Such as is the air,” wrote the English divine and scholar Robert Burton in his 1621 *Anatomy of Melancholy*, “such be our spirits; and as our spirits, such are our humours.”[11] The air was not just about you; it was something that made you. Your bodily matter was a compound of airs and aliments. You are what you breathe as well as what you eat.

What sorts of airs were healthful, and what sorts might do you harm? The most common conception of healthful air was that it ought to be *pure* and *temperate*. “Consyderynge it doth compasse us round aboute, and we do receive it in to us,” a sixteenth-century Welsh physician wrote: it “can not be to[o] clene and pure . . . for we lyve by it as the fysshe lyveth by the water.”[12] Wholesome air, like wholesome water, was free of contaminating matter. The purity or impurity of air was something you were presumed to know through your own senses, and you might also be familiar with some medical consequences of bad air, such as difficulties in breathing. A seventeenth-century English court physician noted that the best air is free of pollutions: “Serene, moved and stirred with the Winds, breathing sweetly with pleasant Gales and sometimes moistened with wholesome showers.”[13] Impure air made for an impure and disordered body. You should keep the air pure in your own home, a counsel more difficult to observe in houses with open fireplaces. You should fling open bedroom windows when it was fair outside so as to let out accumulated exhalations; you should get up and about, though you shouldn’t go out too early or stay out too late, as you would then be exposed to air not yet purified by the heating action of the sun.[14] You should ideally locate your house on a good soil—perhaps gravel mixed with clay—far from swamps and marshy effluvia, your windows should face north or east (and winds from those quarters were often accounted the most wholesome), you should plant a pleasant and sweet-smelling garden near the house, and you should consume only the purest water, nothing stale or stinking. You could do little if anything about the state of outside airs, but dietetic writers reflected on where the airs were best and, therefore where it was best to live. The airs most conducive to health were those of temperate climates, not subject to extremes in their qualities or rates of change. Air should be "neither too hot, nor too cold, neither too dry, nor too moist," and the English congratulated themselves on inhabiting what they regarded as just such a climate, with just such temperate airs.[15]

Putrefaction was something you could smell in the air (and also in the water) and whose exhalations and vapors indexed dangerous alterations of the air's composition, just as the absence of stench marked wholesomeness. A "sweet thin Aire" actively works its health-making way on a disordered body: it "perfumes and purifies an unwholesome body; cherisheth the heart, makes a lively pulse, and much encreaseth the vital spirits"; it "sharpens the appetite, and helps digestion." Pathological consequences flowed from air infected with "putrid vapours and noxious exhalations."[16] A putrid smell indicated putrid matter in the air; it was a clear warning of medical danger, signaling the risk of putrefaction in the human body. Shakespeare's audiences were presumed to know about the causal relation between bad air and ill health, as in *The Tempest* (Act 2, scene 2) when Caliban urges that, "All the infections that the sun sucks up / From bogs, fens, flats [swamps], on Prosper fall, and make him / By inch-meal a disease!" The great London plague of 1665 was widely ascribed to bad air, and its possible remedy was "the use of all such rational helps as might cleanse and rectifie the Air."[17] (Varieties of *agues*—notably the fevers later recognized as malaria—were associated with impure air, especially those emerging from swamps and marshy places, and indeed, in the mid-eighteenth century the British later imported the Italian term *mala aria*, literally "bad air." The miasmas arising from bad water and rotting organic matter were identified as sources of atmospheric infection and disease risk into the nineteenth century.) That is one reason why there was a *medical* market for perfumes, scented flowers and woods, rose water and vinegar water, musk, pomanders and silver vinaigrettes, and other substances that masked putrid smells. It wasn't just a matter of countering disgust; aromatics were understood to act as medical prophylactics, a resource for maintaining health.[18]

Of course, if you were ill—suffering from a surfeit of one or another humor—you should move yourself to a corrective climate. If you were subject to an excess of choler you should find a moist and dry air, if phlegmy you should remove to a dry and warm air, and so on.[19] As urbanization proceeded through the eighteenth and nineteenth centuries, one ground on which cities were deemed unhealthy was their bad air, receiving so many unnatural exhalations from slaughterhouses, breweries and distilleries, badly arranged cemeteries, and the concentrated breaths of densely packed and debauched populations.[20] Even modern improvements in the heating of houses—as they developed through the eighteenth century—might be identified as another source of urban bad air, injuring rather than improving the health of people who could afford domestic warmth and, through the smoke produced, endangering their neighbors.[21] People who worked in the city should, if they could, endeavor to sleep in the country. In general, however, traditional dietetics commended the wholesomeness of the air of your home country: that is what you were accustomed to, and that is what helped form your constitution.[22] Even in the mid-eighteenth century, a popular tract on health gave the home-air sentiment poetic form:

> [C]ustom moulds
> To every clime the soft Promethean clay;
> And he who first the fogs of breath'd
> (So kind is native air) may in the fens
> Of Essex from inveterate ills revive
> At pure Montpelier or Bermuda caught.[23]

Travel might, for these reasons, prove dietetically dangerous: unaccustomed foreign airs could be just as unwholesome as native airs were good.[24]

Good airs made for good health in mind as well as in body, and the effects of airs on mental and moral characteristics were evident at collective as well as individual levels. "Close, thick and moist" airs rendered the spirits "more dull, heavy and indisposed," wrote a seventeenth-century English court physician: "So considerable is the *Climate* and *Air* in relation to our Being that it not only changeth and altereth our Bodies, but also our *Minds* are wrought upon by it: in as much as the *wit*, *inclinations* and *manners* of a people, are different upon this score."[25] The English boasted that English air bred English virtue. Having an "excellent Temperature," it was partly responsible for the valor of its inhabitants.[26] The constitution of English air was, however, fragile; it could go bad due to either seasonal and momentary environmental changes or the effects on its makeup of human activity. Bad air could even make for a diseased social order. John Evelyn, writing to the king just after the Restoration about the problems brought on by London fogs, passed on medical opinion that a badly constituted air made for a badly disordered polity: the "*Aer* it selfe is many times a potent and great disposer to Rebellion." The character of different airs made people differently disposed to political stability. You might be able to do something about politically disordering air by, for example, planting hedges around London to sweeten its air, so rendering the city's people more docile.[27] Dietetic medicine advised individuals to take great care of the airs in which they lived. Air was self-making in body and in mind, and the same sort of dietetic counsel was offered to rulers of the state.

Things Nonnatural: Sleeping and Waking

Air management was dietetically important: air surrounded you everywhere and at all times. You took in air every time you breathed. You could control some aspects of exposure to airs when you were awake, but you continued to take in air—both by breathing and (it was widely believed) through the pores of your skin—when you were asleep. Sleeping too was an activity under only partial control. You might decide when, how, and where to sleep, but many things happened to you when you were not awake. And so traditional dietetic medicine addressed itself to a series of volitional acts you might take about those things that were not wholly under volitional control.

The thrust of dietetic advice was, again, to commend temperance and moderation. Sleep was necessary, of course, but you should sleep neither too much nor too little, and you should not lie in bed too late or rise up too early. Extremes of sleeping—or of "watching"—damaged body and mind. In typical sixteenth- and seventeenth-century formulations, "moderate sleep refresheth the spirits, increaseth natural heat, helps concoction, gives strength to the body, pacifies anger and calmes the spirits, gives a relaxation to a troubled mind." But excessive sleep "dulls the spirits," "fills the head with superfluous moisture and clouds the brain," while staying up too late and taking insufficient sleep "debilitates nature," "heats the body, dryes, and exasperates Choler; in time extinguisheth natural heat, breeds Rhumes [and] Crudities," which is most serious for "thin leane bodies," but an afternoon nap every now and then was refreshing if you were momentarily tired, if you were old, or if the weather was unusually hot.[28] A commonly prescribed bed regimen was eight hours in the summer and nine hours in the winter, though some writers reckoned that six or seven hours was sufficient for adults, and others said that individual temperament trumped any secure generalization.[29] You should sleep when nature intended that you sleep—that is, at night—and both fashionable waking past proper bedtime and sleeping when the sun was up were disapproved on medical as well as social grounds. On the other hand, "watchings that are moderate"—that is, staying up just a little late—could be healthful in certain circumstances, "a help to distribute Aliment and promote the emission of excrements."[30]

Dietetic advice specified much more than the *duration* of sleep and when you should take it. The position of the sleeping body was also addressed. Some writers said that there were medical reasons why you should sleep on your belly and not on your back: a belly-down position would strengthen the "naturall heate of your stomacke," "make a better digestion," and avoid undue heating of the kidneys.[31] Others disagreed, cautioning, again on medical grounds, that sleeping on your belly was not in general a good idea, that you should do it only when suffering indigestion, and that you ought to sleep with your head elevated (which helped digestion and prevented regurgitation). Advice was given in great detail. In repose, the head should be a foot and a half or two feet higher than the feet. You should change posture during sleep (aiding "concoction"), and you should start sleeping on your left side before flipping over and sleeping on the right, while other dietetic experts reversed the preferred postures and orders.[32] You should go to bed with your stomach neither too full nor too empty and in a relaxed and easy state of mind. If you do not, you risk unsettling dreams and a broken sleep: "when you lay by your *cloaths*; lay aside also your *business*, *care* and *thoughts*; and let not a wandring phancy prevent your rest; or awake you before due time."[33] You should get up in the morning "with myrth, and remember God." You should then have a good stretch, a cough, a spit, a bowel movement, and, perhaps later, take a walk in your garden or play a game of tennis "to augment natural hete" before the first meal of the day. You should go to bed with your stomach neither too full nor too empty and retire in the same contented mood as you arose.[34] Dietetic advice included the nature of the bed and bedclothes: you *should* sleep in a proper bed, since there were medical risks attending open-air sleeping as well as sleeping on the ground, in any damp place, or in any newly washed room. The bed should be soft but not too soft, ideally a feather bed covered by a quilt. The bedding should be clean and well aired, since stinky bedclothes signaled medically dangerous fumes, "for Bedding receives the vapours and sweaty moisture, that comes forth from bodies lying in them; which if they be not purified by Air or Fire: they will contract an ill scent, and are then unwholesom to lie in."[35]

Traditional dietetic writers were confident in their counsel about how you ought to sleep, even as they were well aware of the ways in which circumstance could trump rule. For example, if you were a prince or if you were intended to become a governor, then moderation in all aspects of sleep was still advised, but you had to accustom your body to be ready for change and action at a moment's notice, ready to serve the state in any emergency. King James I cautioned his son Henry, Prince of Wales, to learn in this and in the management of all the nonnaturals that "your diet may be accommodate to your affaires, & not your affaires to your diet."[36] Sometimes you just had to stay up all night and sleep with your head below your feet in a damp place and without the comfort and medical virtue of a feather bed. And for civic actors who had to live like that, there was no virtue and no legitimate conception of health in adhering rigidly to a rule of moderation.

Things Nonnatural: Moving About and Letting Go

From antiquity through the early modern period, dietary writers advised how the body should be moved and exercised, and here too moderation significantly defined both what was good for you and what was good. Moderate movement set the fluids into healthful motion, guarded against humoral imbalance, increased natural heat, helped move food through the digestive process and encouraged the expulsion of excrement, promoted the movement of bodily vapors and gaseous wastes through the pores of the skin, and added vivacity to the whole body. Even quite brisk forms of exercise, judiciously taken, were said to render a man "strong, active, [and] mettlesome," equipped to fulfill all the normal obligations of life and rise to the occasion when great physical strength and agility were called for.[37] Extremes of both inactivity and vigorous exercise were warned against. Idleness made the body flabby and vulnerable to injury; encouraged flatulence, constipation, and the retention of waste matter; made for stagnant humors, glandular obstructions, and weak nerves; and bred lust. "The great spring and origin of *lust* is *Idleness*," said a practical ethical writer.[38] Prolonged inactivity was especially dangerous to women, who, "mostly leading sedentary lives," lose "color and vivacity of countenance."[39] Violent forms of exercise were, in most circumstances, to be avoided. Dietary texts condemned "vehement and immoderate" activities: "they provoke sweat," "over-much heat the body," and, "opening the Pores of respiration," risk the infiltration of "infected aire" into the body.[40] Some early modern dietary texts specifically commended "the play with the little ball, which we call Tenise," saying that it was greatly approved by Galen, while Francis Bacon's studies of longevity advised against all exercises that are "too Nimble, and Swift"—specifying running, fencing, and tennis—because of their deleterious effects on the spirits, making them "more Eager, and Predatorie," ultimately shortening life. But Bacon approved running, jumping, and riding, as they stirred up "a good strong Motion" of the spirits.[41] The caution against violent exercises was amplified in the case of pregnant women: they were to avoid dancing, running, and riding as well as "leaping," though here the concern was specifically to prevent miscarriage.[42]

As with the other nonnaturals, dietary advice was molded to circumstance and setting. Different sorts of activity were recommended for different sorts of people at different stages of life and at different times of year. The aged should take it easy so that the "small heate which they have" not be lost. Light massage with a linen cloth was appropriate for the old, especially when taken in the morning, and "they must be combed, and cherished up with fine delights." Phlegmatic persons were in particular need of exercise, but cholerics were carefully warned against violent forms, especially in the summer and after sunrise.[43] Moderate walking was good exercise for many sorts of people "and may be used in hot months, specially of cholericke persons." And even the most wholesome and temperate forms of exercise ought to be taken in the right atmospheric conditions, "in a good and wholesome aire" free of noxious "vapours and mists."[44] Exercising and eating had to be temporally coordinated, though—here as elsewhere—authors differed about just how this should be done. Was exercise best before or after eating? Did proper exercise depend on which of the day's meals was involved? "After dinner sit a while, and after supper walke a mile" was a widely quoted English proverb, but there were dietary writers who reckoned that walking after dinner was innocuous, while others recruited Hippocratic authority to say that exercise should always be taken before eating and that doing these things in the reverse order corrupted the humors.[45] If you were constitutionally weak and "short winded"—and here those who lived a scholarly life were in mind—there was helpful dietary advice adapted to your constitution: such persons "must use loud reading, and disputations, that thereby their winde pipes may be extended and their pores opened."[46] And all such violent exercises as "running, leaping, and coyting" were judged by a humanist writer to be "vile for schollers," though some dietary writers were much in favor of scholarly tennis.[47]

The sorts of exercise deemed medically good were commonly those *appropriate* to your standing and role in life. And this was notably the case in dietetic as well as practical ethical advice for those destined to be governors and soldiers or, in general, to lead life on a public stage. Dancing was considered a "moderate exercise," good for health and good for forming a "*graceful* motion of the body." No nation "civil or barbarous," except—an English Restoration writer noted—for dour Puritans, neglected to express legitimate joy through dancing, "which makes it seem a *sprout of the Law of Nature*." Dancing was widely judged to be good for the body, the mind, and the civil state. But dance forms that were health making were moderate in execution: there should be no jumping about and no overly elaborate dance steps, with some moralists taking a dim view of the intricate steps then in fashion.[48] There was, however, qualified and circumstanced approval for more strenuous activities, such "manly exercises" as "running," "leaping," and "vaulting" and also "wrestling," "fencing," and "playing at the caitche or tennise, . . . palle maille [a form of croquet], & such like other faire & pleasant field games." For the prince, hunting, hawking, and above all horse riding were commended, "for it becommeth a Prince best of anie man, to be a faire and good horse-man."[49] The "man of action" must be *capable* of physical action, able to swing into vigorous motion when it was required.

Even so, it was recognized that there were risks to body and to reputation attending gentlemanly participation in strenuous physical activity. The participants in Castiglione's conversations about the formation of the Renaissance courtier generally commended the propriety of riding, leaping, running, and wrestling while worrying that he should not be seen to do such things in public unless he was confident that he could do them well and without embarrassment.[50] "Rough & violent exercise" could be unseemly as well as dangerous and was likely to make the gentleman look like a fool or a peasant, activities such as "the foot-ball" he should play only if he wanted to injure or be injured, and "such tumbling trickes [that] onely serve for Comedians & Balladines [i.e., theatrical dancers and tricksters], to win their bread with."[51] The healthy gentleman should be bold, but he must not be a buffoon.

The nonnatural category of *evacuations* encompassed gaseous, liquid, and solid releases from the body: perspiration, urination, defecation, flatulence, and also—often treated obliquely and delicately—the fluid economy of the sexual act. Releases were commonly sorted into two types: those deemed "*Benigne, or Profitable*"—notably, the male semen—and those that were "*unprofitable*, of which the Body hath no use," urine and feces, but also phlegm, nasal mucus, earwax, and sweat.[52] Excrements represented various stages of the transformation of food into blood and the subsequent extraction from the blood of materials required for vital actions. What was left over after these extractions had to be removed from the body, and health depended upon its elimination in due quantities at due times and in due order. The excrement of the stomach was ordure, that of the liver was urine, and that of the veins was sweat. In males after each "concoction" of the blood was performed, yielding its proper excrement, the remaining portion of blood was carried to the genitals, where it was transformed into semen. In women, blood was similarly transformed by the breasts into milk, "which is nothing else but bloud twise concocted."[53]

In each case, health and a prudent or virtuous life depended on *moderation* in arranging evacuations, where the volitional management of these matters was especially charged with moral and reputational meaning. So, it was advised that defecation and urination should be ruled by natural urges: you should go when nature called. Even though you could, on occasion and within limits, willfully delay release, you should strive to align your will with natural inclinations. Similarly, you could *strain* in trying to eliminate waste when the time was not ripe and when the natural urge was weak, but doing so was likely to lead to ill health or injury. "Few things conduce more to health than keeping the body regular," an eighteenth-century popular medical text observed, and while some commentators acknowledged that civil obligations might in certain cases require irregularity, few disagreed with the medical and moral virtue of fixed times for defecation. You should ideally produce a stool once a day, and advice was offered how to do so, for example, before bed or after rising early in the morning and taking a walk. Remaining too long in bed discouraged bowel movement because of both the recumbent posture and the warmth of the bedclothes.[54] You should appreciate the price to be paid for irregularity. Retaining fecal matter in the body beyond its due time breeds "noxious Fumes and putrid Vapours." Kept in too long, feces harden, heating the body, upsetting the stomach, ruining the appetite, and producing headaches and dull spirits.[55] Similar risks attended irregularity in urinating: if performed too often or not often enough, the bladder may be damaged and, in extreme cases, results in "*Feaver*, a great pain and speedy death."[56] Foods not evacuated in a timely manner "hinder concoction, whilst putrid vapours exhale from thence to the Stomach and neighboring parts, and so offend the head."[57] Constipation might have even more serious consequences: inflammation of the eyes, nervous conditions, melancholy, and apoplexy.[58] A major remedy for closeness was dietary, while the taking of mineral waters and purgative drugs to alleviate constipation was a major part of medical therapeutics. That was the sense of the notion of "taking physic": announcing that you were "taking physic" in early modern society was a well-understood excuse for being socially unavailable, and some physicians could be intelligibly criticized for administering heroic doses of industrial-strength purgatives.

The opposite of volitional closeness was frequent and excessive release of body waste. "Immoderate Evacuations" exhaust the body and lead to a variety of diseases. John Locke's tract on education fell in with standard advice to observe regularity in going to stool and cautioned against being "very loose," as people who evacuated excessively "have seldom strong Thoughts, or strong Bodies." Locke was a physician and knew how often purgatives were taken to counter constipation, but he believed that rational thought and the vigorous exercise of will might bring more effective relief. You must first try hard to make regularity a habit, especially going to stool just after breakfast, or you might see whether taking an innocuous pipe of tobacco, which was widely believed to bring on a movement, worked for you.[59] There was, however, a degree of dissent about the dangers of "looseness," and the most popular eighteenth-century medical text declared that diarrhea was "not to be considered as a disease, but rather as a salutary evacuation." Diarrhea ought never to be stopped unless it appeared to drain the body's strength or was brought about by circumstances such as a cold or an obstructed perspiration.[60]

Flatulence was a sign of incomplete concoction and was a condition making for ill health. The production of a certain amount of gas was normal, but medical attention was directed toward immoderate windiness. All sorts of illnesses might follow: headache, earache, melancholy, nightmares, vertigo, pleurisy, dropsy, heart palpitations, and pains in the testicles or uterus.[61] Excessive stomach fumes and putrid vapors had several causes: some writers considered them the result of nervous conditions, noting that "strong and healthy people are seldom troubled with wind." The production of these fumes might have something to do with a physical weakness of the gut, while most dietary writers thought that wind was caused by eating the wrong sorts of foods—cabbages, beans, turnips, milk, cider, eel—in the wrong amounts, in the wrong order, or at the wrong time.[62] Dietetic texts offered long lists of foods to be avoided because of their capacity to produce wind, and prescriptions were offered of carminative drugs that could alleviate it. But once those vapors were elaborated, little good could come from keeping them in. In a Renaissance and early modern society that increasingly policed the public release of body wastes—spittle, nasal mucous, urine, feces—the fart was recognized as a social solecism, while at the same time its free release was medically encouraged. A widely circulated verse storehouse of practical dietary wisdom proclaimed that

> Great harmes have growne, and maladies exceeding,
> By keeping in a little blast of wind:

So *C[r]amps*, *Dropsies*, *Collickes* have their breeding
And *Mazed Braines* for want of vent behind:
Besides we find in stories worth the reading,
A certain *Romane Emperour* was so kind,
Claudius by name, he made a Proclamation,
A *Scape* to be no losse of reputation.[63]

More famously, Benjamin Franklin, while recognizing "the odiously offensive Smell accompanying such Escapes," cautioned that all kinds of diseases followed from retaining wind "contrary to Nature."[64] Farting was good because expelling the stomach's putrid vapors was good. Unduly confined, those fumes might themselves produce many other kinds of bodily and mental ills. But it was also good, as far as possible, to *prevent* the production of gas or at least more gas than inevitably arose from a normal diet and in normal circumstances.[65] The task was to balance the dictates of health and civic duty: moderation in farting was the way to do that.[66]

Dietary texts treated other forms of bodily evacuation in less detail. Some modes of sweating were, of course, understood as involuntary, but you could *choose* whether to exercise violently enough to induce profuse sweat. Writers diverged on the virtue and wholesomeness of sweating. The opening of the skin's pores was recognized to be risky, but this danger might be set against the purifying effects of free sweating, and both advantages and disadvantages should be adapted to the time of year and temperature.[67] There was also expert advice on how to manage spit, nasal mucus, and earwax.[68] Menstrual fluids were a matter of medical concern, sometimes considered as a corrupt humor, sometimes as pure, and widely viewed as a natural regulatory response of the female body to *plethora* (or a fluid excess), but it was not considered that menstrual flow could come under volitional control, so there was no question of advising on managing its course.[69] And, of course, nondietetic medical measures—drugging and especially bleeding—were importantly directed toward maintaining or restoring the body's fluid balance.[70] "Universall evacuation," a seventeenth-century physician explained, "is the cleansing of the whole body from superfluous humours by purging, vomiting, sweating, opening a veine, scarification, friction, bathing, &c."[71]

Spitting and sniveling counted as health problems and, increasingly, as problems for reputation and moral order, but nowhere were the categories of *the good* and *the good for you* so powerfully coupled as in dietetic treatment of sexual release, where the focus of dietary writing was overwhelmingly on the evacuation of *male* sexual fluids. Semen was treated as an excrement, though one of a very special and precious sort. It was the basis of male generative potency, the source of life, the "most *spiritual part*" of the blood's concoction "whose excellency is exceedingly improved by its elaboration in the *Testicles*," the evaluative opposite of such excrements as sweat and snot.[72] Men should take very great care in its release with respect to time of life, time of year and day, frequency, and the manner and circumstances of discharge. The sexual act—*venery* in early modern usage—was divinely ordained, and neither medical authority nor commonsense judgment would say that it was bad except for those few celibates who dedicated their purified bodies to God, and even here the consequences of denial for bodily health and longevity were debated.[73] "Moderate Venerie"—dietary writers repeatedly declared—preserved (male) health: it opened the pores, exhilarated "the heart and wit," mitigated "anger and fury," and aided in the overall balance of bodily fluids and qualities.[74] Many evidently believed that sexual release had curative powers.

Early modern physicians generally preferred to be coy about such things: "There are some diseases Cured by it; But I am bid silence."[75] Others were more frank, listing any number of diseases and disorders—from catarrh to cataracts—for which moderate venery was specifically directed.[76] Persistent prevention of the release of male seed "causeth heaviness or dulness of the body, and if it be corrupted stirs up grievous accidents."[77] There was a long historical tradition of medical writers addressing the "female orgasm," the "female prostate," and "female ejaculation" during the sexual act. But while female moderation in sexuality was recurrently discussed, the practical dietetic literature treated the notion of a liquid sexual economy almost entirely within a male framework.[78] Dietetic texts said that sexual release should be about delight, not lust. Parents produced the healthiest children when they observed moderation, indulging only when there was a natural urge. What counted as sexual moderation was gauged to specific temperaments, stages of life, and momentary conditions. Some writers warned against the release of semen until men reached the age of twenty, others said that the period between ages twenty and thirty-five was the best time for venery, and almost all understood that men of different constitutions required different amounts of venery. Melancholics and cholerics as well as the old and the emaciated needed less; sanguinaries and phlegmatics as well as the middle-aged and the flourishing needed more. There were special circumstances in which sex was considered unhealthy and ought to be avoided: if you were feverish or if you suffered from such conditions as gout, joint disease, or certain types of tumor.[79]

Dietary texts were typically addressed to mature males—constitutionally and metaphorically *hot*—and it was those groups who were judged to be at greatest risk of excess. They were repeatedly warned against immoderate venery and against unnatural modes of sexual release. There was no end of bad bodily consequences of immoderate sexuality: it dissipated natural heat, weakened the body, encouraged the accumulation of vapors, injured the nerves, caused palsies and gout, deranged the senses and the judgment, "and by the commission of uncleanness a rottenness in the *Loins*, and if neglected will penetrate the very bones." And to these bad outcomes were added the injuries of *the French pox*, other diseases occasioned by unnatural and sinful sexual relations, the loss of reputation, and the sense of sin.[80] Immoderate venery was the loss of precious vital fluids, and you might well pay for current pleasure in reduced longevity. Overindulgence was indicated when a body looks old before its time and "causeth a man to have a breef or shorte lyfe."[81] For some constitutions, sexual release was thought to be specially risky. The great architect Inigo Jones knew himself to be a melancholic, and for all those suffering from an excess of black bile, "copulation must be utterly eschewed"; if not, "the best blood of a man is wasted and natural strength enfeebled."[82]

Prudentially considered and with the benefit of age and experience, the hedonistic gratifications of venery might not even be worth it; less pleasure was often experienced than was hoped for. In such matters, dietary texts rarely paid attention to women: they were often portrayed as inducements to excessive venery, threats to the integrity of male fluids. A seventeenth-century popular work offered world-weary deflationary advice to young men in a hurry and in heat: "Consider that there is many a Woman, very desireable to look at, yet if you enjoy them, you will less prize them, and you can find no more pleasure in them, but the evacuation of your own heat and vigor, therefore it is downright folly and madness, to run such great hazard of Soul, body, Estate, and good Name, for a Toy of no value."[83] Immoderate sexuality had dietary causes as well as vital consequences. In the scheme of the things nonnatural, each category was considered to have effects on the others, and incitements to sexual excess included certain regimes of exercise, sleep, and the physical environment of air, waters, and places, but foods and drinks were treated as principal causes of immoderation. Almost all commentators considered that the wrong foods, including those that generated wind and especially those that had marked *heating* qualities, were importantly at fault in exciting venery.[84] In general, excess was considered to excite the sexual urge, improperly pouring more fuel on the fire. In the lists of specific aliments offered in many dietary texts, tendencies to induce venery were prominently mentioned, while there were many instances in which authorities disagreed about effects. Among vegetables, onions and figs were sometimes said to excite "veneryous actes"; cloves could "stir up Venus"; parsnips and carrots could "provoke Lust"; and lobster, oysters, and sparrow could "stimulat *Venus*." And if you wanted to damp the sexual fires, then eating apples and purslane might help.[85]

The wrong foods and drinks in the wrong amounts might increase the urge, but you still had to *decide* what and how much to consume. Dietary texts instructed men to put their body and its urgings under the greatest possible volitional control, to arrange circumstances in which the promptings of venery could be rationally managed. Both health and virtue depended on choice in food as in all the nonnaturals: "Let every man use all honest and lawfull meanes, to suppresse the violent force and fury of his burning lust. . . . [Let him] avoid the imaginations of *Venus*, and such like [so that] he shall not easily be much assailed & tempted with the desire of any carnall appetite; likewise if he earnestly apply himselfe to the study of the holy Scriptures, and morall Phyolosphy, banishing idlenesse, and flie the company of beautifull and amorous women, he shall easily avoid the desire of lustfull conscupiscence. . . . Fly idleness, the greatest occasion of lechery."[86] The complex choreography of traditional dietetics presumed both a rational will—to which medical and moral advice was directed—and physical and environmental circumstances that might compromise the wholesome action of rational will.

Things Nonnatural: Taking Control

Just as traditional dietetics advised how to manage the body's intakes and outgoings, its forms of exercise, its states of rest and waking, and its placement in the environment, it also counseled how to control the emotions, "the passions of the soul." Those passions were understood to be potent and potentially dangerous. You could easily find yourself controlled *by* them—carried away by anger or envy or fear or even by love—and then you had made yourself into the sort of person no rational person would wish to be, someone who was not the master of the passions but rather their slave. To be the servant of the passions was undignified; it was imprudent and was *bad for you.*[87]

The place of the rational will in dietetic advice was both central and problematic. The will was the choosing agent of right conduct, and the will, at the same time, was potentially vulnerable to the influences of the passions, while the passions were in turn vulnerable to bad dietary decisions. You could advise persons under the sway of the passions that they should *resist* them; you could offer such people grounds on which they could extricate themselves from the passions' powers. And in so doing, you worked with a conception of the human agent in which certain aspects of the self could assert themselves over others, in which reason and prudence could act against anger, lust, and the pleasures of the lower senses. Moral and theological writers portrayed Jesus as the ultimate master of the passions. He exposed himself to Satan's temptations and overcame them (Matthew 4:1–4). Jesus did indeed experience the passions of love, grief, and desire, but he was ever *in control*, and in that way he offered sacred warrant for proper civic conduct.[88] You could think of Jesus as a gentleman, with a gentleman's self-control.

The nonnaturals formed an interacting system: the management of any one of the nonnaturals could bear upon the expression of others. Patterns of sleeping and exercise stood, for example, in a reciprocal causal relation with digestion. But commentary on all the nonnaturals invariably noted their interaction with the mind and the emotions, and the vehicles through which they exerted such effects were the humors and spirits. Whatever affected the production and balance of the humors and spirits also affected mental function. "There is great concord," said a seventeenth-century dietetic writer, "betwixt the bodies qualities, and the soules affections." Our bodies are "compacted of the elemental qualities, namely of moysture & drinesse, heat & cold: So among the soules affections are some moist, some dry, some hot, & some are cold." We understand that mirth is hot and moist, sorrow is cold and dry, and so on.[89] Climate, exercise, sleeping, and evacuations all worked on the mind, acting through the humors and spirits, and so too did diet. "*The Passions of the Mind follow the Temperatement of the Body*," wrote a late seventeenth-century popular medical writer, "for such diet as you use, such will be your Blood; and as your Blood is pure, or impure, such will be your Spirits, . . . and consequently all Functions and Operations of Body and Mind shall suitably be altered."[90] Specific foods acted on the mind and body in specific ways. You should know these causal links and should behave accordingly.[91]

Each of the nonnaturals could affect the passions as well as other modes of mental function. It might be tempting here simply to say that traditional dietetics was a form of "psychosomatic" medicine, but the temptation should be resisted.[92] It is not just that the term "psychosomatic" is a nineteenth-century usage; more pertinently, our current appreciation of what is meant by the notion—the causal interaction of mind and body—proceeds from modern ideas about what belongs to the mind and what to the body, and the modern sorting does not capture ancient, medieval, or much early modern sentiment. When the English scholar and cleric Robert Burton felt that he ought to justify meddling in medicine, he pointed out that his subject—melancholy—was "a disease of the soul . . . and as much appertaining to a divine as to a physician, and who knows what an agreement there is betwixt these two professions? A good divine either is or ought to be a good physician."[93] (Indeed, if you want to understand the categories of dietetic medicine, you do not just need a reconfigured notion of the "mind"; you also need a historically appropriate understanding of the "spirits" and of the "soul," and "soul" is not now to be unproblematically translated as "mind." A typical seventeenth-century formulation had it that "the soul and body are so linked and conjoyned, as Partners of each others ill and welfare," but what the writer was saying is not well rendered as "psychosomaticism.")[94] If you think of the humors and their associated qualities as belonging to the body, then you should think of the body as encompassing powers that we now, lacking the category of "soul," consider to belong to the mind. The target of traditional dietetic counsel was the immaterial, rational soul, an entity that could deliberate, weigh, and decide what to do. But that passion-managing soul is missing from credentialed modern thought about body and mind.

The general texture of dietetic advice on managing the passions followed now-familiar form. The passions must be *moderated*. In the early sixteenth century, the humanist Thomas Elyot's book on health said that if the passions "be immoderate, they do not only annoy the bodye and shorten the life, but also they do appayre, and sometyme loose utterly a mans estimation."[95] Protecting the passions from violent manifestations was good for the body: "moderate joy and a chearful spirit doth preserve the body in health, and sound constitution, for it recreates and refreshes the heart and spirits, and whole body; but if joy be excessive, it dissipates and consumes the spirits."[96] *Mens sana in corpore sano*—a healthy mind in a healthy body—as Juvenal said in the first century CE. Endlessly repeated through the early modern period and into the Victorian era, the tag gestured at a state of affairs that everyone was presumed to want and also referred to systems of reciprocal causal interaction between mind and body. A soul under the sway of the passions makes for bad digestion. Food is poorly transformed into blood, excretion is hindered, and noxious fumes are produced, which in turn affect mental function and moral capacity.[97] Under the power of pathological sadness, food loses its savor, sleep is disturbed and beset by bad dreams, and exercise becomes distasteful.

The passions themselves were part of the normal equipment of the human frame, but they were powerfully tidal: we should maintain ourselves not like the Mediterranean but instead like the Caspian Sea, keeping "our selves ever at one heigth without ebbe or refluxe."[98] Yet, unlike such other normal human tendencies as sleeping, evacuating, and eating, the passions were regarded as inherently troublesome, things not just to manage but also whose management might be *worried about*, a worry that was itself encompassed within the passions. Passion management belonged as much to practical ethics as to dietetic medicine, a cohabitation of the good for you and the good that long remained stable in the general culture. A practical text on the training of young noblemen instructed them to control "our unrulie passions" and to quench the "fire of our violent lusts" so as to settle "good order in al our actions."[99] Henry Peacham's guide to being a gentleman urged that we must find a mental "bridle" with which "wee curbe and breake our ranke and unruly Passions."[100] And Richard Brathwait's *The English Gentleman* judged that there was no glory "equall to the command or soveraigntie over our owne *passions*; the conquest whereof makes Man an absolute Commander": "There is no one vertue which doth better adorne or beautifie man, than Temperance or *Moderation*; which indeed is given as an especiall attribute to man, purposely to distinguish him from brute beasts, whose onely delight is injoying the benefit of Sense, without any further ayme."[101] Brathwait's guide to the conduct of a gentleman was soon joined by his companion volume on the English gentlewoman. The management of her passions was, on the whole, treated less problematically, and Brathwait's commendation of self-control was embellished by a hierarchy of moral agents. Discussing the evils of sensuality, Brathwait wrote that "it is this only, which maketh of men, women; of women, beasts; of beasts, monsters."[102]

The passions were called "the accidents of the spirit . . . , strange or sodaine [sudden] insurrections, and rebellious alterations of a tumultuous troubled soule, which withdraw it from the light of reason, to cleave and adhere unto worldly vanities."[103] The passions were unruly subjects in the polity of the self. They needed to be watched and spied upon and, if they asserted themselves, disciplined. They should be the governed, not the governors. Erasmus's 1516 tract on the education of a Christian prince observed that "all slavery is pitiable and dishonorable, but the lowest and most wretched form is slavery to vice and degrading passions. What more abject and disgraceful condition can there be than that in which the prince, who holds imperial authority over free men, is himself a slave to lust, irascibility, avarice, ambition, and all the rest of that malicious category?"[104] The passions, like the other non-naturals, should be disciplined to the golden mean: "Every moderate passion bordereth betwixt two extreames, as liberalitie betwixt avarice and prodigalitie; temperate diet betwixt gluttonie and scarcity; fortitude betwixt desperate boldnes & superfluous feare, called timiditie."[105]

Despite its variety of dietary counsel and when it got down to specifics, dietetic advice amounted to not very much more than direction to "get a grip on yourself." Burton's *Anatomy of Melancholy* showed more than usual awareness that controlling the passions was hard work. Take the example of besetting melancholy, depression, and moroseness. Burton imagined the weak of will to say, "'Tis a natural infirmity, a most powerful adversary; all men are subject to passions. . . . You may advise and give good precepts, as who cannot? But how shall they be put in practice?" Good questions, Burton conceded, but he could do no better than to advise self-control, and he made some concrete suggestions about how you might achieve that: get out in company, listen to soothing music, and reform your diet.[106] Best of all, Burton wrote, was to occupy the mind to keep it from dwelling on melancholic matters: "There is no greater cause of melancholy than idleness, 'no better cure than business.'" Burton represented his sprawling meditations on the nature and causes of melancholy as a cure for his own melancholic disposition, making "an antidote out of that which was the prime cause of my disease" or, as we might now say, to cure psychic upsets by talking them through.[107]

A sixteenth-century dietary text said that situating your house in a pleasant place looking out on views that delighted was an effective way of disciplining dark passions.[108] A seventeenth-century physician recommended intentional acts of "mirth" as a cure of "sadness," and the same sort of advice was both acted on and passed on to others by the philosopher René Descartes, whose last tract was *On the Passions of the Soul*. Descartes said that he had been born with a disposition to sadness but had alleviated his condition through avoiding sad-making scenes and thoughts.[109] The most popular medical text of the late eighteenth century acknowledged limits on the power of self-control while insisting that self-control was both possible and necessary: the ability to make ourselves "good humoured [is] . . . not altogether in our own power; yet our temper of mind, as well as our actions, depend greatly on ourselves."[110] It is up to you to decide whether to brood or not, to keep company with cheerful people or with gloomy ones, to dwell within yourself or to get out and get doing, to listen to merry music and to read uplifting books, or to gloomily devote yourself to depression.[111]

This kind of counsel was considered to be pertinent, practical, and reasonable: you had a rational soul, you could recognize when reasonable things were said, and you could act on reasonable counsels. This was possible and was what you ought to do. It was your fault if you did not, and it was no excuse that your bad way of life had compromised reason. The advice was, again, at once medical and moral. Though injury and epidemic disease might intervene, your bodily ill health was substantially your responsibility; ill health might be taken as a sign that you had lived an immoral life.[112] You might succeed in following rational dietetic advice, or you might fail. Failure was not just bad for your health but was also a sign of a compromised and damaged self. One form of advice on passion management that an early modern physician might offer was then identified as "spiritual": spiritual remedies for spiritual disorders. Physicians might, for instance, "invent and devise some spirituall pageant to fortifie and help the imaginative facultie, which is corrupted and depraved." They might endeavor to "deceive and imprint another conceit, whether it be wise or foolish, in the Patients braine, thereby to put out all former phantasies." If persons deranged by the passions were consumed by jealousy, they might be told stories and given evidence that their jealousy was baseless. Obsessed by love, they might be reminded that there were, so to speak, many more fish in the sea. If gripped by lust, they could be directed to seek serenity by meditating on Holy Scripture and to consider the superiority of God's love to carnal love. Choler, itself a temperament (a surfeit of yellow bile), might also be regarded as a passion of the mind. When the humors were too hot, "they grow to be sulphureous, kindeling of fiery fevers, pleurisies, gall in the stomacke, yealow jaundices, tumours, *Erisipelaes*, itch, and innumerable other maladies, as well externall and internall," the cure of which was "that Christian Vertue *Patience*."[113] But the passions might be managed more mundanely too through the intentional ordering of other nonnaturals. You could be advised to adjust your patterns of sleep, exercise, evacuation, and especially eating and drinking so as to control the emotions, and there was abundant commentary on the dietetic causes and remedies of the specific mental/moral/humoral condition widely known as *melancholia*.

Things Nonnatural: Taking Things In

Food and drink were usually placed second on the list of the six things nonnatural but have been left to last here because they were commonly treated in far greater detail than the others, understandably so because eating and drinking were so tightly integrated into the fabric of everyday life and because what and how you ate were understood as rich indices of personal and collective identity. Medical commentary typically offered long catalogs of edible items, the qualities and consequences of each of them sequentially addressed in the many Renaissance and early modern books called "dietaries."[114] The sense of *diet* as an overall *manner of living* was recognized, but dietetic physicians often wrote books specifically about the qualities and consequences of different forms of food.

How to eat and drink so as to maintain health? The consistent answer was to observe *moderation*, to keep the appetites in check.[115] "Our belly is a troublesome Creditour," it was said, and just as you don't want to be in debt to the moneylenders, so too should you guard against the demands of your stomach ruining your bodily budget.[116] You should oppose gluttony through disciplined and rationally informed moderation. Again, the virtuous mean was located between two extremes, and at the other pole was asceticism. Self-denying ascetics were often identified with religiosity, so there was no question but that the ascetic extreme had its recognized place and cultural value.[117] (*Hunger*, which was the norm for the masses of European society, was almost never mentioned in medical and ethical writings as an ordinary condition. These texts were not meant for the starving multitude, and only *intentional* hunger appeared as an issue.) By the time of the Renaissance and the early modern period, asceticism had little place in civic culture, and writers of practical ethical texts had no significant reasons to single it out for condemnation. When obsessive abstinence was mentioned in such books, it was treated as a vicious extreme. Richard Brathwait counseled the gentleman to "temper your desires, that neither too much restraint may enfeeble them, nor excesse surcharge them." He acknowledged widespread belief that weakening the body strengthened the soul but insisted that dietary abstinence could also hinder "the performance of spirituall exercises."[118] Other practical ethical writers portrayed abstinence and gluttony as viciously conjoined. Gluttons abstain but only to sharpen their appetites for yet more food and drink. The ascetic fasts, but "when his belly is emptie, straight *Gluttony* filleth it, and as soon as it is full, then *Abstinence* empties it."[119] There was, however, notable criticism of those who were overprecise about their diet, for that fastidiousness worked against participation in the normal modes of gentlemanly culture.[120] The extremes of gluttony and abstinence were both deemed dangerous even if excess was more popular and more evident among those classes who could afford it.[121]

Injunctions to moderation in eating and drinking were almost without challenge in medical writing. Thomas Cogan's late sixteenth-century *Haven of Health* noted that "the word 'mediocre,' which Hippocrates applyeth" to all the nonnaturals, "must especially bee applied to meats [meaning food in general], that is to say, that the quantity of meate be such, as may be well digested in the stomacke," though, of course, the position of mediocrity was acknowledged to depend on individual constitution.[122] A seventeenth-century verse rendition of medieval medical wisdom cautioned that "gainst all surfets, vertues Schoole hath taught/To make the gift of Temperance a shield," specifically a sovereign preventive against ill health: "To shew you to shun raw running rheums; / Exceed not much in meat, in drinke, and sleep, / For all excesse is cause of hurtfull fumes."[123] "Esteem temperance and regularity in eating and drinking [is] a great preservative of health," wrote one seventeenth-century English physician.[124] "Too much of any thing is an enemy to Nature," wrote another.[125] And an early modern compendium of prudence judged that "temperance in diet and exercise, will make a man say; a figge, for *Gallen & Paracelsus*."[126] That is to say, a temperate diet—like the apple—keeps the doctor away, and so on and on into the nineteenth century and even beyond.

Dietary texts addressed not just the quantity and qualities of foods but also whether, how, when, and in what order different foods might be combined. Many dietaries vigorously argued against mixing up a *variety* of foods in your stomach. Such variety might index an unwholesome tendency to think too much about the taste and delicacy of foods and might be a sign of excess, but independent physiological grounds were given for its harmfulness. Combining many different sorts of foods at one meal or even during one day made digestion and assimilation difficult. The stomach could not effectively process alimentary jumble, the qualities of one food might work against those of another, or the drain on the spirits from digesting one food might leave another food unprocessed and liable to cause vapors. The quest for variety was a fool's errand, momentarily satisfying a depraved palate but damaging the body.[127] Taking a diversity of foods unnaturally stoked the appetite and encouraged excess. A simple diet was "the best Remedy against Intemperance." It was morally unworthy as well as unhealthy to run after dietary complexity and delicacy. The genuine pleasure of eating did not consist in "the daintiness, and curiosity of Fare, and Multitude of Dishes"; eating was a blameless *social* act—a "Society of Feeding"—"not in our Eating much, but in our Eating together." Eating fueled the body; conviviality made society.[128] Puritanical ethical writers condemned enthusiasm for unnatural and artificial preparations and the eaters who madly sought after them. "Nature affords not Meat delicate enough for their palats, it must be adulterated with the costly mixtures of Art, before it can become *Gentile* nourishment," a Restoration moralist wrote. "Cookery is become a very mysterious Trade, the Kitchin has almost as many intricacies as the Schools."[129] If you pursued variety too much or too long you would wind up ill, unable to enjoy even a normal diet. But an obsession with variety—and especially with the elaborate, the exotically rare, and the luxurious—was also a long-lasting target for moralizers: it was artificial, showed disrespect to the simple foods of one's native land, and might even count as a violation of divinely framed natural law.

Simplicity in eating and drinking was by far the most pervasive advice, but it was not unopposed in the Renaissance and the early modern period. Patterns of court dietary culture indicated that medical and ethical commendation of simple dietary nativism had limited impact. Indeed, there was some medical and ethical sentiment that applauded variety. An important strand of variety approval drew on traditional concepts of humors and qualities, figuring variety as itself a way to securing healthful *balance*. So, there were medical authorities who saw a food with one set of qualities as being *balanced* (or *corrected*) by ingredients whose qualities were opposed, hence the substantive relationship between dietetic medical thinking and the development of *cuisine*. This is the medical sentiment behind the idea of "correcting the seasoning" or combining food with contrasting qualities.[130] In the sixteenth century the French barber-surgeon Ambroise Paré, while acknowledging that there were medical writers with different opinions, wrote that a "variety of meats is good." He maintained that it was right that people acting on a public stage should get used to different sorts of foods, and he reckoned that a yearning for variety was not unnatural, that it might be a reliable sign of what the body required:

> Neither is it convenient that the meat should be simple, and of one kind, but of many sorts, and of divers dishes dressed after different forms, lest nature by the continual and hateful feeding upon the same meat, may at the length loath it, and so neither straitly contain it, nor well digest it; or the stomach accustomed to one meat, taking any loathing thereat, may abhor all other; and as there is no desire of that we do not know, so the dejected appetite cannot be delighted and stirred up with the pleasure of any meat which can be offered. For we must not credit these superstitious or too nice Physitians, who think the digestion is hindred by the much variety of meats.[131]

By the late eighteenth century, there were both political economists and gourmands who defended variety and luxury on grounds other than morality or medicine. Some early economists counted luxury and a taste for the exotic as inducements to worldwide commerce and its benefits, while gourmands approved these as the satisfaction of gustatory tastes that themselves needed no further justification or that contributed to the advance of civilization. The great nineteenth-century French gourmand Jean Anthelme Brillat-Savarin wrote that food connoisseurship "is an impassioned, reasoned, and habitual preference for everything that gratifies the organ of taste." It was good for the body, the mind, and the global order. Regarded "from the point of view of political economy, gourmandism is the common bond which unites the nations of the world," which causes a global flow of capital "incalculable in respect to mobility and magnitude by even the most expert brains."[132]

Views differed on this, and discussions of what it was proper to eat gnawed on moral bones from antiquity through much of the modern period. Caring little for food was sometimes taken as a mark of wisdom, prudence, and sheer common sense. Socrates enjoyed good health, took exercise, and was concerned to maintain manly well-being, but "glory'd in his frugality . . . He who wanted least, came nearest to the Gods." Socrates was accustomed to saying "that other men liv'd, that they might eat; but that he ate only that he might live."[133] Dietary moderation seen as virtuous self-mastery was a key topic in codes of Renaissance and early modern manners and distanced feeding from its carnal realities and made that mastery manifest in ordinary experience.[134] The damage that excess did to gentlemanly reputation was a standard sentiment in Continental Renaissance courtesy texts. It was "well known," Castiglione's *The Courtier* blandly announced, that the ideal "Courtier ought not to profess to be a great eater or drinker."[135] When King James I wrote to instruct his infant son and heir how to conduct himself like a prince, he too warned against "using excesse of meate & drinke; and chiefelie [to] beware of drunkennesse, whiche is a beastlie vice."[136] And Burton's *Anatomy of Melancholy* wagged a warning finger at early seventeenth-century gourmandism. How we now "luxuriate and rage" in food and drink, Burton wrote: a cook used rightly to be accounted "a base knave" but now was "a great man in request; cookery is become an art, a noble science; cooks are gentlemen." Gluttons wear "their brains in their bellies, and guts in their heads."[137] Dietary excess was just undignified. Absolutely everywhere that bookish counsel was offered to early modern gentlemen, the Road to Wellville was signed by the golden mean. Moderation in food and drink made for virtuous people with healthy and long-lived bodies.

Why Gluttony Is Bad

There were all sorts of modifications and qualifications to the seemingly bland counsel of moderation, yet its authority was immense, and contests over its legitimacy occurred only about its proper interpretation. Temperance in all things, the volitional observance of moderation, and the disciplining of unruly appetites were *virtues*. Indeed, in many pagan and Christian schemes, temperance was not just a cardinal virtue—typically along with prudence, fortitude, and justice—but might be regarded as the master virtue without which none of the other virtues could be effectively observed (figure 2.1). An English puritanical courtesy text explained why the mean was golden: "There is no one vertue which doth better adorne or beautifie man, than Temperance or *Moderation*; which indeed is given as an especiall attribute to man, purposely to distinguish him from brute beasts, whose onely delight is injoying the benefit of Sense, without any further ayme. . . . [N]o vertue can subsist without Moderation"—the foundation and root of all the others.[138]

In the early seventeenth century the Flemish Jesuit Leonardus Lessius praised temperance, noting that he wrote as a divine and not as a physician. "Sobrietie"—for Lessius another name for "temperance"—was, as physicians rightly noted, the major condition for securing health and long life, but it was also a fit subject for theologians and moralists, as "it doth wonderfully conduce to the

Figure 2.1. *The Seven Virtues* by Francesco Pesellino (ca. 1450), the virtues and iconic human representations of each. Lists of the virtues varied from antiquity through the Renaissance. Here, from the left: Prudence, Justice, Faith, Charity, Fortitude, Hope, and Temperance. Temperance was often shown pouring wine; she is not abstaining from drink but is taking it in moderation.

attainment of *Wisdome*, to the exercises of *Contemplation*, *Prayer*, and *Devotion*, and to the preservation of *Chastitie*, and other virtues."[139] If Plato's glutton was a beast, the temperate human being was a shadow of the divine. "Temperance," Robert Burton wrote, "is a bridle of gold, and he that can use it aright . . . is liker a god than a man."[140] English divines agreed that gluttony destroyed bodily health, but they laid special stress on dietary excess as both a sin and a handicap to piety. "*Gluttony* is also a *deadly enemy to the mind*, and to *all* the noble employments of Reason," wrote the Reverend Richard Baxter in the 1670s:

> It unfits men for any close and serious studies, and therefore tends to nourish ignorance, and keep men fools: It greatly unfits men for hearing Gods Word, or reading, or praying, or meditating, or any holy work; and makes them have more mind to sleep; or so undisposeth and dulleth them, that they have no life or fitness for their duty; but a clear head not troubled with their drowsie vapours will do more, and get more in an hour, than a full bellied Beast will do in many. So that Gluttony is . . . an enemy to all Religious and manly studies.[141]

Intemperance was a *sin*. In Christian frames, the seven deadly sins (the capital vices or cardinal sins) were wrath, avarice, sloth, pride, lust, envy, and gluttony. Excess in eating and drinking was, as a general matter, morally wrong, but specific reasons were offered for its wrongness. Human beings had to eat. Any form of eating was evidence of what we shared with animals, but both excess in eating and excessive *interest* in eating displayed the willful subjection of your higher and God-given nature to your lower bestial nature. Saint Paul said that gluttons were those "whose god is their belly," practicing an idolatry of the palate and the gut.[142] Early Christian thinkers agreed that gluttony was sinful while debating whether it was the greatest sin, one of the mortal sins, or a source of other vices. At the same time, gluttony was widely recognized as the most difficult vice to control and reject just because you *needed* to eat in order to live. However compelling the other vices were, none were so essential to life. To live was to eat, and in eating you could not nicely distinguish between necessity and pleasure.[143] The one inevitably attended the other.

The Latin term for gluttony was *gula* (gullet, throat, palate).[144] Gluttons were ruled as much by their tongues as their bellies. The tongue had a dangerously dual nature as the organ of both tasting and speaking. Gluttons talked endlessly about food; they ordered their cooks to prepare complicated dishes and had their foods elaborately sauced to arouse their appetites and to make them crave even more. Complex concoctions and especially spices were targets for moralists; they unnaturally stimulated the palate and made the body crave more than was necessary.[145] From antiquity through the early modern period, it was proverbially said that *hunger is the best sauce* and that anyone who demanded contrived and stimulating sauces was going against both the natural and moral orders.[146] Writing about the education of young gentlemen, John Locke urged that the young become accustomed to a "very *plain* and simple" diet: "*Flesh* once a Day, and of one Sort at a Meal, is enough. Beef, Mutton, Veal, &c. without other Sawce than Hunger, is best." People are *made* "*Gorman[d]s* and *Gluttons* by Custom, that were not so by Nature."[147] Gluttony was as imprudent as it was morally odious. By eating too much you selfishly denied food to the poor, and by obsession with luxurious, dainty, exotic, and overly elaborated foods you mocked the proper role of food as sustenance—living to eat, not eating to live—and might wind up ruining your estate and eating up your heirs' inheritance.[148] In a mixed civic-sacred idiom, it was said that gluttons so stuffed themselves with food and drink that they could not perform their proper obligations to themselves, their families and friends, and their prince. Proverbial expressions had it that "the glutton digs his grave with his teeth," that "Intemperance is the Mother of Physicians," and that "By Suppers, and Surfeits more have been killed, than Galen ever cured."[149] Princes (and indeed any other substantial civic actors) should acknowledge an obligation to their health and long life not just for their own sake but also for the welfare of their subjects, their heirs, and others dependent on them.[150]

Gluttony was a vice long before obesity was a disease. Clerics preached against the sin of dietary excess, confident that parishioners were well familiar with the vocabulary of traditional dietetics.[151] Food and drink were understood to fuel carnality; the gut was causally prior to the gonads. The more you ate, the more wanton you were.[152] A Latin tag circulating widely in Renaissance and early modern medical and civic circles was *Sine Cerere et Baccho friget Venus*: Venus grows cold without food and drink (figure 2.2).[153] Chaucer's five-times-married Wife of Bath knew what she was talking about when she cited a punning English proverb: "A liquorish tongue must have a liquorish [i.e., lecherous] tail."[154] In Rabelais's *Pantagruel*, the physician Rondibilis reminded Panurge of that ancient wisdom, adding that "Master Priapus was the son of Bacchus and Venus."[155] Doctors as well as preachers and practical ethical writers reckoned that "Gluttonie and Drunkennes [are] the mother of al vices."[156] Medical understandings of the physiological role of food in the body underwrote that causal link. Thomas Wright's early seventeenth-century treatise on the passions expanded on the traditional saying: "lechery springeth from gluttonie. . . . [G]luttony is the forechamber of lust, and lust is the inner roome of gluttony. . . . [G]reat repasts swim under the froth of lust."[157] Excess could be quantitative, that is, eating too much food, or it might be qualitative, taking food that was too strong, too gross, lingering too long in the stomach and generating the "crudities" and fumes that disturbed a range of bodily functions. The

Figure 2.2. *Sine Cerere et Baccho friget Venus*, a pen painting by the Dutch printmaker Hendrick Goltzius (produced ca. 1600–1603). A voluptuous Venus—her right arm around her son Cupid—is offered grapes by Bacchus (the god of wine) and fruits by Ceres (the goddess of agriculture). The torch held by Cupid indicates how food and wine fuel the sexual flame. There were many paintings on this theme, especially popular in the Low Countries in the late sixteenth and early seventeenth centuries.

passions could not be effectively controlled and the mind could not reason clearly when the fires of desire and rage were stoked by dietary excess. The stronger and fatter the food, the more vehement and hotter the fire.

There were also straightforwardly practical reasons why dietary excess was understood to breed civic inadequacy and vice. Early modern commentators asked why the Spartans used "eat and drink so sparingly," and the reason was that sobriety was necessary for prudent politics: "luxurious and intemperate men were utterly indisposed and unfit for Counsel." So too was the case for military action. If you were habitually drunk, you might not be able to respond to a call to arms, and if you were grossly fat, you couldn't leap on your horse and fight effectively.[158] Those were reasons why you wouldn't want to grow great through dietary excess, but bulkiness was not always considered bad. What we now regard as obesity, a risk factor for disease, was in the past often seen as corpulence, bulk, or largeness of body. Hans Holbein's famous 1537 portrait of Henry VIII depicted not a fat man but a big man. The significance of bodily bulk was variable. If it incapacitated you, bulk was bad, but insofar as it did not, then it was a powerful display of *substance*. All things being equal and in a social order where dietary dearth was the norm, it was good to be big, and corpulence was a visible mark of riches and command.

Shakespeare's Falstaff described himself as "A goodly portly man, i'faith, and a corpulent" (*Henry IV, Part 1*, Act 2, scene 5), continually fuddled with sack and stuffed with beef. Even when Falstaff was the master of his revels, Prince Hal knew just what sort of man his companion really was. The old glutton asked Hal the time of day, and the prince responded, "Thou art so fat-witted, with drinking of old sack, and unbuttoning of thee after supper . . . that thou hast forgotten to demand that truly which thou wouldst truly know" (Act 1, scene 2). But the terms in which his royal companion eventually cast Falstaff aside at his coronation was not an actuarial statement of the health risks to which Sir John's diet made him subject; it was a politic renunciation of a way of life with which a prince could no longer be associated: "Make less thy body hence, and more thy grace; Leave gormandizing; know the grave doth gape for thee thrice wider than for other men. . . . Presume not, that I am the thing I was" (*Henry IV, Part 2*, Act 5, scene 5).

Custom and Common Sense

There is an apparent banality about much traditional dietetic counsel. Although that advice was widely articulated by learned physicians and by reputed ethical writers, the content of dietetic prescriptions can appear as to be little more than folkish common sense. Yet the seemingly banal, proverbial, and commonsensical character of dietetic sentiment is pertinent to its authority. The seeming banality is a way of appreciating the integration of dietetic categories into pervasive practices of knowing about the body; the qualities of the body's aliment and its transactions with the environment; different types of people; different people's qualities, dispositions, and ways of behaving; what counts as prudent and virtuous conduct; and, of course, how to maintain health, manage disease, and perhaps extend life. Be moderate, traditional dietetics recommended: nothing too much. But what did moderation *mean*? How did you go about observing the mean? Dietetic medicine fully acknowledged that advice about right conduct must be geared to circumstance. One size never fitted all; one man's meat is another man's poison. What counted as moderation for one sort of person at one stage of life, one time of year, or in one environment, following one way of life, might not be moderate for another. Dietetic expertise involved knowing general concepts and particular instances and how to bring the one to bear on the other. It was not just mastery of the concepts but also a large store of *experience*—with people, states of health and disease, and circumstances—that counted as dietetic expertise.

At one pole, traditional dietetic culture was attached to the knowledge of experts—physicians and philosophers—and, at the other, touched the knowledge and sensibilities of the laity, appealing to ordinary experience, common judgments, and custom. One way to conceive of custom was as your settled management of the nonnaturals, the historical sum of your interactions with the environment, occupying much of the everyday happenings of your life. Over time, your diet, your manner of sleeping, defecating, and the like could secure your innate nature, ensuring its health, or could even remake the nature with which birth or early life had endowed you. You are what you eat, but eating—and the other nonnaturals—might remake what you are. This was a reason why traditional dietetic counsel emphasized that you should not make abrupt changes in your diet. If you were in reasonable good health, your customary way of life, with some exceptions, stood in little need of expert correction. That it *was* customary and that you evidently functioned in and through it meant not just that it was good for you but also that it represented a powerful mode of self-knowledge. And while some physicians did make claims that they knew what was best for their patients, most acknowledged the power of custom. Indeed, physicians could come to know patients' way of living, their management of the nonnaturals, almost solely through those patients' own testimony.

Many traditional dietetic writers endorsed the adage that "custom is a second nature."[159] What you were used to came to form what you were, and early modern views of the self that circulated in everyday life were, in effect, also *environmental* histories. The edible bits of nature you customarily consumed molded themselves over time to *your* nature. You are the sum of your innate constitution *and* the history of your transactions with the environment, so you should understand that you are a *creature of habit*. Custom, a seventeenth-century English physician wrote, was "that great Imitatrix of Nature," and he thought that people often used the "second nature" saying to justify and excuse their unwholesome way of life. Yet, he recognized the cultural grip of the sentiment and stressed human capacity, through the exercise of the rationally informed will, to *change* customary patterns.[160] If you decided, exercising due care, to reshape your customary patterns of life, you could remake *yourself*.[161]

The surgeon Paré observed that "if Custom (as they say) be another nature, the Physician must have great care of it. . . . For this sometimes by little and little, and insensibly, changes our natural temperament, and in stead thereof gives us a borrowed temper."[162]

A late sixteenth-century English doctor wrote that everyone should take great care in "what quality his meat is of, for custome begetteth another nature, and the whole constitution of body may be changed by Diet."[163] That formula persisted well into the eighteenth century. In the 1730s, the Scottish physician John Arbuthnot observed that "the whole Constitution of the Body may be changed by Diet."[164] To the end of the eighteenth century, editions of William Buchan's hugely popular *Domestic Medicine* repeated that "there is no question but the whole constitution of body may be changed by diet alone."[165] The only people in early modern society who could and should adopt an insouciant disregard to settled habit were those who traveled and had to adapt themselves to a changing alimentary environment: traders, soldiers, diplomats, and, most consequentially, European colonists. For such people, the desired custom was tolerance of the vicissitudes.

Given the power of custom, you should be very wary of radical or abrupt changes in your way of life: such changes might be *dangerous* to your health. This caution was found in the Hippocratic corpus and was repeatedly expressed into the early modern period.[166] The wisdom of the late medieval Salernitan medical school put the sentiment into memorable verse:

> If to any use you have your selfe betaken,
> Of any dyet, make no sudden change,
> A custome is not easily forsaken,
> Yea, though it better were, yet seemes it strange,
> Long use is as a second nature taken,
> With nature, custome walkes in equall range.[167]

In Rabelais's sixteenth-century satire, Gargantua's tutor Ponocrates saw his pupil's "vicious manner of living" and resolved to reform him but was determined to proceed only gradually, "knowing that Nature cannot without great violence endure sudden changes."[168] Paré warned that any who "would presently or sodainly change a Custom which is sometimes ill, into a better, truly he will bring more harm than good." If you had to make changes, you should make them very gradually so as to constitute a new altered nature.

If you have to change, then do it bit by bit.[169] Montaigne said, "I beleeve nothing more certainely then this, that I cannot be offended by the use of things, which I have so long accustomed."[170] "Custome is a second Nature, and no lesse powerfull," he wrote. "What is wanting to [my] Custom, I hold it a defect."[171] Physicians accounted harmful that which was against nature and reasoned similarly about that which went against long-settled custom.[172]

Through the eighteenth century and beyond, this sentiment was shared by laypeople and physicians. The distinguished Dutch Newtonian physician Herman Boerhaave saw no reason to dissent from the traditional wisdom—widely attributed to Celsus—concerning custom: "Custom itself . . . , which is not improperly called a second Nature . . . , makes a surprising Difference with respect to the Effects of Air, Food, Drink, Exercise. . . . Whereof a sudden Change from accustomed to new things, is always and every where very dangerous; even tho' the Change should be from reputed ill Habits, to such as are judged to be good."[173] In these ways, an individual's habitual way of life might be endorsed by medical expertise. And that qualified approval followed from views about how your transactions with the environment—edible or otherwise—made you what you were and how radical disruptions to these habits introduced the risk of the exotic, the foreign, and the abnormal. What you were not used to might harm you.

Proverbial Prudence

The stress on custom in dietetic advice is one way of recognizing its traditionalism. Another way reflects on the *form* in which dietetic counsel was often expressed and its relationship with folkish genres. To modern readers, the practical recommendations and intellectual summaries of much traditional dietetics look like a set of clichés: proverbs, tags, saws, apothegms, aphorisms, or, in common early modern terms, "old wives' tales." Modernizers of natural philosophy such as Francis Bacon and René Descartes—those who reckoned that medicine urgently needed radical philosophical reform and new conceptual foundations—routinely recommended sorting out genuine testimony of vital phenomena from uninstructed lay belief. Sixteenth- and seventeenth-century physicians concerned with defending their distinct expertise against the challenges and skepticism of "folk wisdom" collected the proverbial forms of popular belief, offering these as examples of everything that was wrong about lay thinking.

The sixteenth-century French physician Laurent Joubert compiled a long list of proverbial mistakes in his account of "popular errors."[174] But even the strenuous opposition of some physicians to the clichés and proverbs of popular wisdom gave evidence of their power. And that power is indexed by the central place occupied by these proverbial, apparently folkish forms *within* the body of ancient, medieval, and much early modern expertise itself. The Hippocratic treatise *Aphorisms* is a prominent example of this, collecting short, pithy, summaries of what exists, what to think, and what to do that stick effectively in the mind: "Life is short, and Art is long"; "Those things which require to be evacuated should be evacuated, wherever they most tend, by the proper outlets"; "Neither repletion, nor fasting, nor anything else, is good when more than natural"; "'Little by little' is a safe rule"; and "Desperate diseases require desperate remedies."[175]

From the Middle Ages through the early modern period, the most widely circulated compilation of dietetic wisdom was the so-called Salernitan Verses, the Latin title of which was *Regimen Sanitatis Salernitanum* (*The Salernitan Rule of Health*). The exact identity of the original text is uncertain, but it was understood to represent the accumulated medical knowledge of the great medical school at Salerno, south of Naples, where, according to tradition, Christian, Jewish, and Islamic scholars studied and worked together from about the eleventh century. The text usually comprised seventy ten-line stanzas of hexameter verse, and from its inception through the seventeenth century the *Regimen* went through many versions, acquired many additions, and, with the advent of printing, appeared in at least 240 published editions in almost every European vernacular language (including a Yiddish edition published in Poland), with new printed versions produced even into the mid-nineteenth-century.[176] The *Regimen* was rendered into English in the 1590s and published in 1607 as *The English Mans Doctor, Or the*

Schoole of Salerne. The creative translator was Sir John Harington (1560–1612), who was not a physician (though one of his claims to fame is his invention of an early version of the flush toilet). He had been a courtier (and godson) to Elizabeth I and at the time of composition was tutor to Henry, Prince of Wales, heir to the throne of his father, James I of England.[177] Harington's translation faithfully preserved the original's merger of medical expertise and folkish forms.[178] Many of the counsels of the Salernitan Verses passed into folk wisdom—several of the versified dietetic sentiments appeared in Benjamin Franklin's eighteenth-century *Poor Richard's Almanack*—and some of them remain current.[179] Verse and proverbial formulae are easily archived and retrieved from memory in relatively stable forms, an especially valuable capacity in what was still a culture—folkish, of course, but also learned—that relied heavily on orality and the ability mentally to store and summon to speech the formulae of proper belief and action.[180]

The dietetic content of *Regimen Sanitatis Salernitanum* is wholly unexceptional: it justifies the importance of attending to health; recommends moderation as the way of maintaining health; describes the humors, temperaments, and nonnaturals; discusses the precise conditions of being bled and the right seasons of year for doing so; acknowledges the constitutional heterogeneity of people; and leads readers through the precise qualities of a range of specific items of food and drink: onions, cabbage, cheese, different sorts of wine and beer, and so on. There is nothing at all original or idiosyncratic about the *Regimen*, and that, again, is the point. It is offered as an easy-to-remember synthesis of traditional wisdom, sanctioned by ancient authority and communicated in an aphoristic idiom that itself was accorded authority in a culture only partly shaped by print. *Regimen* embodied tradition in both form and content, in medium and in message. The advice and the summaries of dietetic knowledge in this text—and indeed in very many other similar books—is traditionally commonsensical partly *because* its linguistic expressions make it so. Commonsense sentiments are recognized as such by what they say and how it is said.

Regimen Sanitatis Salernitanum condensed dietetic wisdom about the relationship between rational control of the appetites and political prudence: "A King that cannot rule him in his Dyet, / Will hardly rule his Realme in peace and quiet." The text concisely personified the fundamental principles of dietetic preventive medicine, the use of moderation in the nonnaturals, of passion management, and of looking on the bright side: "Use three Physicions still; first Doctor *Quiet*, / Next Doctor *Merry-man*, and Doctor *Diet*." The text reminded people that they had to adapt their diet to the seasons: "Foure seasons of the yeare there are in all, / The *Summer* and the *Winter*, *Spring* and *Fall*: / In every one of these, the rule of reason / Bids keepe good dyet, / Suting every Season." You shouldn't eat until you feel hunger, which, as proverbs said, is the best sauce: "To keep good dyet, you should never feede / Until you finde your stomacke clean & void / Of former eaten meate, for they doe breede, / *Repletion*, and will cause you soone be cloid, / None other rule but appetite should need." *Regimen* also recognized that you might occasionally drink to excess, offering both a practical remedy and a reminder that excess was, after all, wrong: "If Wine have over night a surfet brought, / A thing we wish to you should happen seld: / Then earely in the morning drinke a draught, / And that a kind of remedy shall yield, / But gainst all surfets, vertue Schoole hath taught / To make the gift of Temperance a shield." And the text sequentially treated the qualities and powers of the various foods—what they did in the body, for whom they were suited, and how and when they should be consumed:

> They that in Physicke will prescribe you food,
> Six things must note we here in order touch,
> First *what it is*, and then *for what 'tis good*,
> And *when*, and *where*, *how often*, and *how much*.[181]

Traditional dietetics told you what the world was like, how you ought to manage your transactions with it, what was prudent and virtuous, what was good and what was good for you. It spoke with the voice of both expertise and common sense.

3

How to Know about Your Food and How to Know about Yourself

Traditional Dietetics: How to Be?

Traditional dietetics offered a vocabulary for recognizing different types of people, knowing how they came to be what they were, and anticipating what they were likely to do. The same vocabulary also provided ways of self-understanding and figured in making sense of how your body and mind worked and in managing your corporeal, emotional, and mental states. And the same vocabulary was a resource for presenting yourself to others as a person of a certain sort. In these ways and from antiquity through the early modern period, traditional dietetics was enfolded in everyday ways of understanding self, others, and the relations between self and others. The categories and concepts of traditional dietetics emerged from a philosophy of nature; they fed into practical views of human nature and gave accounts of how edible nature was transformed into you and how you might choose to manage consumption so as to regulate your nature.

The key dietetic notions for knowing about others and oneself were the qualities, the humors, and the temperaments. Your temperament was the normal tendency of your body. It specified the illnesses to which you were disposed, the regimens that best suited you, and the patterns of your likely behavior and the dispositions of your mind. For people living in the period from late antiquity through about the eighteenth century, it was an important element of competent social and personal knowledge to recognize your own temperamental type and that of others. "Let . . . every man take surveigh of himselfe," a late sixteenth-century book on the temperaments said, "and search out what his nature most desireth, in what state his body standeth, what thing it is that he feeleth himselfe to be holpen [sustained by], and what to be offended withall."[1] But how did you come to know your own temperament, and how did you come to know others' temperaments?

One way was through books. From the invention of printing into the eighteenth century, there were very many books that contained lists of the temperaments, gave accounts of their humoral bases, and described the dispositions belonging to the temperaments and the signs by which you could recognize their types and subtypes. These sorts of books were widely distributed. The literate classes were almost certainly familiar with them, and there is much evidence that the general categories and sentiments in them were also current among the illiterate and those whose reading tended not to include such books. The Salernitan Verses was among the most widely distributed and mnemonically sticky accounts of the temperaments, seamlessly combining descriptions of both appearance and behavioral tendencies. When you *looked* at temperamental types, you *saw* the dominant humor inscribed on body surfaces. The complexion of sanguinary persons was both white and rose red, and they were often redheaded; cholerics tended to be thin and hairy; phlegmatics were short, pudgy, and pale; and melancholics were dark-complexioned and just *looked* depressed. The temperaments that were visible from superficial appearances might also be read from behavioral dispositions. Sanguinaries were full of fun, cheerful, sociable, easygoing, and fond of jokes, card playing, wine, and song. Cholerics were bad-tempered, prone to be proud, ambitious, malicious, aggressive, sarcastic, and mendacious. Phlegmatics were lazy, and they spit a lot. Melancholics were unsociable solitude seekers, suspicious, timid, given to extreme swings of mood, studious, and with depressed sexual appetites.[2]

There was some variation in textual portrayals of temperamental types—both over time and between writers in the same period—but the most notable feature of dietetic accounts of the temperaments was their stability. So, for example, Thomas Elyot's dietetic text added detail to the basic picture given in the Salernitan Verses. The sanguinary's veins and arteries were prominent, with a strong pulse; they digested well, slept easily, and had pleasant dreams. Cholerics had black or dark hair and sallow faces; they slept badly and tended toward constipation. The phlegmatic type had abundant hair, a slow pulse, thin veins, a weak digestion, and dreamed of watery things. Melancholics' hair was thin; their urine was watery, their digestion was poor, and their dreams were often terrible.[3] The textual authority for such accounts tracked back to antiquity. Writing in the mid-sixteenth century, the Dutch physician Levinus Lemnius endorsed Galen's opinion that "there is no surer way . . . certainely to know the humours and juyce in a creature, then by the colour and outward complexion."[4] On the one hand, Renaissance and early modern texts presumed that you could *see* the temperaments; on the other hand, these texts offered the authority of still greater texts to guarantee the reliability of doing so.

Texts offered verbal descriptions of the temperamental types; pictures *showed* them. Albrecht Dürer's stunning engraving the *Melencolia I* of 1514 was one of the most best-known, widely reproduced, and symbolically rich of sixteenth-century engravings: it was a visual depiction of a specific temperamental type (figure 3.1). There is no evidence that Dürer was planning to create prints of the other three types, and *Melencolia I* is sometimes interpreted as a "portrait of the artist," melancholy being the temperament to which the studious and the artistic were inclined.[5] Beyond the sixteenth century, Dürer's *Melencolia I* was commonly brought to mind when this temperamental type was mentioned.[6] Then there were books in which much less vivid pictures were accompanied and glossed by verbal accounts, as in Henry Peacham's syncretic emblem book *Minerva Britanna* (1612), where crude wood engravings were captioned with verse accounts of what the types looked like and how they were disposed to behave (figure 3.2). Melancholy

Figure 3.1. *Melencolia I* by Albrecht Dürer (1514). This great engraving was the most widely circulated Renaissance depiction of a humoral temperament. Its meaning has long been intensely debated, and historians disagree about its proper interpretation. The melancholy figure is solitary, apart from the *putto*, or cherub; is possibly female, possibly androgynous; and sits surrounded by the tools of craft (hand plane, hammer, saw, scales) and the instruments of thought and inquiry (book, magic square, mathematical devices, crucible). The figure is—in various readings—lost in thought, introspective, depressed, forlornly awaiting inspiration, all of which were associated with the humoral condition of melancholy in the Renaissance and the early modern period. Whatever Dürer intended, this image was available as a vivid picture of the melancholic person.

is "Pale visag'd, of complexion cold and drie, / All solitarie, at his studie sits, / Within a wood, devoid of companie." Choler stands "resembling most the fire, / Of swarthie yeallow, and a meager face." Phlegm is shown "Of Bodie grosse, not through excesse of meate, / But of a Dropsie, he had got of yore: / His slothfull hand, in's bosome still he keeps, / Drinkes, spits, or nodding, in the Chimney sleepes." Then look at sanguine "in whose youthfull cheeke, / The *Pestane Rose*, and *Lilly* doe contend: / By nature is benigne, and gentle meeke, / To Musick, and all merriment a frend."[7]

Seeing Human Natures

The possibility of discerning internal states from external bodily signs was often—though not invariably—referred to the dominance in human bodies of one or another of the four bodily humors. Humoral constitution marked the body surfaces, and if you knew how to read those marks, you could reliably *see* the animating causes of individuals' nature (figure 3.2). Since antiquity, it was a commonplace that the face, or more specifically the eyes, were windows to the soul, and the form continued to circulate through the early modern period. Aristotle's *Prior Analytics* noted that "it is possible to infer character from features, if it is granted that the body and the soul are changed together by the natural affections,"[8] a notion further developed in the pseudo-Aristotelian *Physiognomonics* of about 300 BCE.[9] The late Renaissance Neapolitan scholar Giambattista della Porta published a widely distributed and well-illustrated text on physiognomy that juxtaposed human and animal heads, suggesting that similar cross-species' surface appearances revealed similar natural dispositions.[10] "The affections of the mind declare themselves openly in the face and behaviour of man," an English cleric wrote toward the end of the sixteenth century: "As, if we be sorry, our countenance is heavy, sad and cloudy; if we be merry, our face hath a good colour."[11] The English physician Sir Thomas Browne broadly endorsed the reliability of face reading: "Since the Brow speaks often true, since Eyes and Noses have Tongues, and the countenance proclaims the Heart and in-

Figure 3.2. The Four Temperaments. Woodcuts from Henry Peacham, *Minerva Britanna* (1612).

clinations; let observation so far instruct thee in Physiognomical lines, as to be some Rule for thy distinction, and Guide for thy affection unto such a look most like Men. Mankind, methinks, is comprehended in a few Faces," and this, Browne wrote, was as applicable to nations as it is to individuals.[12] Through the early modern period, people found faces reliably informative of what was supposed to lie within, a presumption that evoked practical advice on how to tell what other people were like and what they might do. For example, you might manage your own face to be *less* transparent.

Lady Macbeth told her husband that his face "is as a book where men may read strange matters" (Act 1, scene 5) and counseled him on facial management. So important was face reading as a practical way of discerning inner nature that the reference of "complexion" gradually moved from an internal bodily state—the overall balance of humors—to the appearance of the body's surfaces, for example, the paleness or ruddiness of skin (especially on the face). "Complexion" persisted as a way of designating skin tones and textures even when humoral schemes dropped out of medical and lay culture.

It was the humors that were mainly understood to sign body surfaces, though the mediating role of the spirits might also be invoked. Whether it was the humors or the spirits that were involved, knowing interior states through exterior surfaces might be accounted reliable because over time, habitual expressions and habitual ways of life engraved themselves on the face. A general causal link between body surfaces and inner states persisted in lay attributions. In the mid-twentieth century and in a culture that no longer invoked humors or spirits, George Orwell wrote that "at 50, everyone has the face he deserves."[13] And this was understood to mean that a lifetime's living with your face, using it as an expressive and communicative instrument—smiling, frowning, laughing, and crying—was fixed into the mature face as the end product. Of course, in earlier settings facial displays of smiling, crying, and the like could be seen as animated by humoral constitution and attendant passions, with face marking and face reading both presuming interactions between humors and the historical expressiveness of the face.[14]

Beauty, it was proverbially said, is only skin deep. Surface appearances are no sure guide to inner nature; reading surfaces leads only to superficial knowledge. Yet, the inferences that the proverb warns against were pervasive in the culture where humors, temperaments, and complexions were sense-making resources: skin reading could have deep meaning.[15] There was some variation in directions about just how to recognize people with different temperaments, yet there was robust assent that such inferences from the outer to the inner were possible and, indeed, that surfaces were *part of* constitution itself. This was a sensibility that worked its way into early modern poetry, as in John Donne's *Lady of a Dark Complexion*:

Nor hath dame Nature her black art reveal'd
To outward parts alone, some lie conceal'd.
For as by heads of springs men often know
The nature of the streams which run below,
So your black hair and eyes do give direction
To think the rest to be of like complexion.[16]

There might be little one could do about one's constitution and the inferences that might be drawn from outer appearance, but that same sort of inference also figured in the artificial *management* of body surfaces for the purpose of misleading. The use of *cosmetics* included just this sort of deception, as in the application of red "blusher"—rouge—to create the surface appearance of inner modesty. As John Donne, addressing a "Painted Lady," wrote,

Do I not know these balls of white and red
That on thy cheeks so amorously are spread,
Thy snowy neck, those veins upon thy brow,
Which with their azure wrinkles sweetly bow,
Are artificial and no more thy own.[17]

And giving the cultural screw one more turn, the use of cosmetics could be read as a visible sign of inner *immodesty*, as when in the male gaze a "painted lady" insinuated a prostitute.

Different temperamental types were disposed toward the expression of different passions, and, near the end of the seventeenth century the French painter Charles Le Brun—following Descartes's work on the passions of the soul—produced a treatise about how the passions (admiration, sadness, jealousy, etc.) might be represented by artists, and his text was at the same time a manual for how to *look* at faces and how to *present* a face to others: "Expression is . . . a part that intimates the emotions of the Soul, and renders visible the effects of Passion."[18] Actors as well as painters pointed to representations of the passions and of temperamental types. In the eighteenth century, acting manuals directed players' attention to printed and painted images as a guide to their performances. Actors should, of course, "observe Nature," as Le Brun had done, but they should also examine artistic representations, "not to be a Stranger to Painting and Sculpture, imitating their graces so masterly, as not to fall short of a *Raphael Urbin*, a *Michel Angelo*."[19] So, in the early modern period there was an interpretative circle linking bodies, the performances of embodied people, printed and painted representations of people's natures and emotions, and medical/physiological theories of human natures. You *learned* to be the sort of person you were, and you learned that you were moved by the natural dispositions that certain sorts of person possessed. That is, theories of human nature were resources for the work of *being* a human of a certain sort, and traditional dietetics offered a range of such theories that were *acted upon*.

A systematic and aggressively publicized system of physiognomy emerged in the 1770s, developed by the Swiss theologian Johann Kaspar Lavater (1741–1801). Lavater argued that you could discern the inner character of individuals from their outward cranial and facial appearances and claimed that the methods for doing so were both scientific and disciplined extensions of reliable everyday inference: "All men (this is indisputable), absolutely all men, estimate all things, whatever, by their physiognomy, their exterior temporary superficies" (figure 3.3). This is what physicians do with their patients, what merchants do with customers and trading partners, what lovers do with the objects of their love, and what everyone does with their food: "No food, not a glass of wine, or beer, not a cup of coffee, or tea, comes to table, which is not judged by its physiognomy, its exterior; and of which we do not thence deduce some conclusion respecting its interior, good, or bad, properties." The molding of exterior facial and cranial surfaces to interior states was causal. "There must be a certain native analogy," Lavater wrote, "between external varieties of the countenance and form, and the internal varieties of the mind," with the internal as the cause

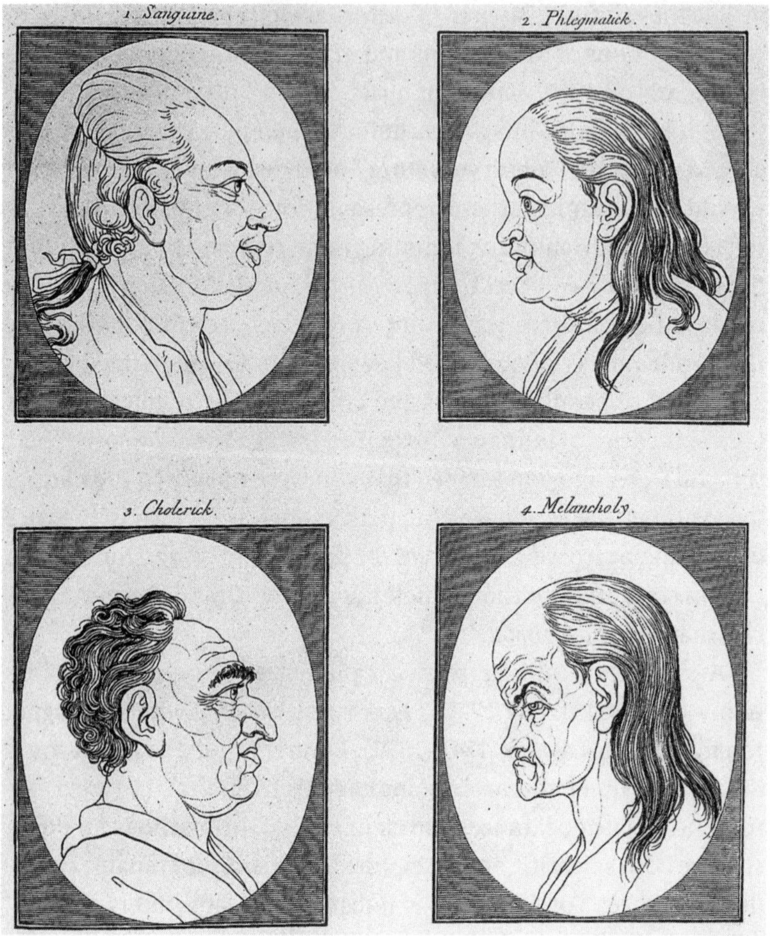

Figure 3.3. Physiognomical Types. Johann Kaspar Lavater published this woodcut—widely reproduced—in his *Physiognomische Fragmente, zur Beförderung der Menschenkenntnisse und Menschenliebe*, 4 vols. (Leipzig: Weidmanns Erben, 1775–1778), Vol. 4. Outer appearances, he claimed, allowed reliable discernment of inner states at the level of both human classes and individuals. Much of Lavater's attention was devoted to knowing individual natures, but the path to individuality might traverse human types, including the species of Galenic temperaments depicted here.

of the external. "Anger renders the [facial] muscles protuberant"; an "active and vivid eye and an active and acute wit are frequently found in the same person," and here too inner nature is said to be the cause of superficial appearances, making body surfaces reliable gauges of inner makeup.[20]

The program of *phrenology* that appeared in the last years of the eighteenth century and gained cultural currency through the 1830s and 1840s was marked by a similar submergence or hollowing out of humoral and temperamental categories. Like physiognomy, phrenology licensed inference from external signs to internal states and dispositions, focusing here on the shape of skull areas as visible marks of underlying mental and emotional powers. In 1830, the Scottish phrenologist George Combe allowed that the four temperaments were "to a certain extent" indices of the brain's constitutional qualities, for each was associated with different degrees of brain activity. The temperamental categories were much the same as those of traditional dietetics, with the phlegmatic temperament now called the "lymphatic" and the melancholic constitution rebranded as "nervous." (Black bile was now said not to exist, and the only form of bile was yellow; hence, the preferred name for the choleric temperament was "bilious.") The phrenologists' temperaments, however, claimed no humoral basis. The inner states underlying the temperaments were identified as organ systems: the nervous temperament arose from active brain and nerves; the sanguine temperament was caused by preponderant activity of heart, blood vessels, and lungs; the bilious proceeded from active muscular and fibrous systems; and the lymphatic was caused by glands and the organs of assimilation. You could still reliably infer inner constitution from such external signs as body shape, facial features, skin, eyes, and behavioral dispositions, but the scheme was detached from the categories of traditional dietetics.[21]

Sensing Alimentary Nature: Qualities

In traditional dietetics, just as human natures had their physiognomy, knowing alimentary nature also had its system of inferring from sensible signs to what those signs signified. At the heart of traditional dietetics was an elaborate physiognomy of food, a way of discerning its qualities and powers from sensory signs. To know the nature of foods and drinks was to take their sensory characteristics as indices of their qualities and powers—what particular foods would do when taken into particular human bodies. There were few concerns more central to dietetic medicine than discerning alimentary natures and communicating them to people who wished to maintain health and alleviate disease. Dietary texts from the Renaissance onward typically incorporated long lists of foods, loosely arranged by category: vegetables; fruits; dairy products; herbs and spices; wines, beers, and ciders; and flesh meat, including different animals, different parts of the animal, the environmental source from which the flesh foods came, their state of development when consumed, and the modes of preparation to which they were subjected. (In sixteenth- and seventeenth-century English usage, "meat" tended generically to designate solid food—especially that forming a major component of a meal, whether or not it was animal in origin—while animal meat was often designated as "flesh.") Each item was then described in terms of its qualities; its effects on bodily organs and systems; its consequences when consumed by different types of people, especially temperamental types; its proper place within a meal; with what other foods or seasonings it ought to be consumed; the best geographical sources; and the time of year most fit to consume it. There was variation among writers about the qualities of specific foods, but contestation floated on the top of substantial overall agreement.

There were several modes in which dietary writers categorized foods: the fundamental *qualities* each food had, the circumstances attending its digestion, its trajectory through the body, and its noticeable effects on the body. The qualities of foods were those associated with the *elements* of ancient natural philosophy. Foods were invariably classified on the traditional matrix of *hot*, *cold*, *moist*, and *dry*. These qualities were possessed by different foods in varying *degrees*, usually expressed from first to third or fourth degree in ascending intensity, where it should be understood that the qualities might have only an analogical relationship to physically ascertainable temperature or wetness. That is, one food could be judged "hot" and another "cold" at the same perceptible temperature or giving the same physical sensation of warmth or coldness. Among vegetables and fruits, for instance, cucumbers were cold and moist, both in the second degree; peaches were cold in the first degree and moist in the second degree; and leeks were hot and dry, both in the third degree. Plums were cold and moist, but they varied among their types: the sweet ones were not as cold as the sour. The age and condition of foods was also a source of qualitative variation: fresh beans were said to be cold and moist in the first degree, while dried beans were cold and dry in the first degree.[22]

Exotic foods, introduced to Europe from the East and later from the Americas, came without authoritative classification from ancient medical sources, and Renaissance and early modern dietary writers were sometimes challenged to fit them into the same qualitative framework as familiar foods. The alimentary characteristics of New World animals (e.g., turkeys) and fruits and vegetables (pineapples, maize, potatoes, tomatoes, etc.) as well as the beverages coffee, tea, and chocolate were noted, but not prominently, in sixteenth- and seventeenth-century dietetic texts, while they were all actively discussed in other sorts of literature. The nutritional value of the potato was quickly recognized in the Old World, and it was widely embraced. By the early seventeenth century, only a few authors knew how to describe the nutritive qualities of the potato using off-the-shelf dietetic vocabulary. One noted that "as to Temperature they are of a middle Nature, but rather warm than cold"; another noted that it was "of a temperate qualitie . . . though somewhat windie. . . . Good for every age and constitution."[23]

The different *parts* of a food might have different qualities. The yolk of an egg was "moderately hot and moist," while the whites were cold and dry. The peel of an orange was hot in the first degree and dry in the second degree, while the juice was cold in the second degree and dry in the first degree. In such cases, the general precept seems to have been that the more sour the juice, the colder its quality, and the sweeter the juice, the hotter, so that might be why lemons were deemed colder than oranges.[24] The spices categorized included both exotics long known in European settings (e.g., various kinds of pepper) and domestic sorts (such as mustard). Cloves were hot and dry in the third degree; pepper was hot, with black pepper being hotter than white pepper and white pepper being hotter than the once common and now rare "long pepper." In certain cases, writers might even specify exactly where in the range of qualitative "degrees" a food was situated: plums, one author observed, were "cold in the beginning of the second [degree], moist in the end of the third." Strongly flavored foods encouraged some writers to reach for extremes, and a late sixteenth-century dietary writer judged that onions were "hoate and drie almost in the fourth degree," while garlic was unambiguously hot and dry in the fourth degree.[25]

Meats were less commonly described through paired qualities than were fruits and vegetables, though many writers counted beef as cold and dry, and some confidently specified that it was cold in the first degree and dry in the second degree.[26] Lamb was said to be moist, though roasting made all meats drier, and the flesh of old beasts was drier than that of young ones.[27] Fish tended to be colder and moister than the flesh of land animals.[28] Wines—though they were, of course, physically liquid—were considered in general to be drying, and they differed in their degrees of heat along several dimensions. Wines that we would now say were higher in alcohol were judged hotter than those lower in alcohol; old wines were agreed to be hotter than young wines, while color and geographical origin also evidently bore upon degrees of heat.[29] The seventeenth-century English writer James Howell said that wines were commonly categorized as "female" (white wines) and "male" (red wines), because commonly the red has more "body and heat" in it.[30] In general, wines coming from hot climates were themselves considered to be hot, though there were exceptions.[31] "Who doth not know," Cogan rhetorically asked in the late sixteenth century, that "Sacke is hotter than white Wine or Claret, and Malmsay or Muskadell hotter than Sacke, and Wine of Madera or Canary to bee hottest of all?"[32] *All* foods could be considered in a medical mode, while there was special attention to the medicinal role of wines. These were considered to powerfully affect the body's fluid balance, its perceived heat and mood.[33]

Sensing Quality: Nutritiousness

When you put food into your mouth you experience the qualities of things—for example, the moistness of a melon—but, after swallowing, foods pass into and through your body. The descriptions of traditional dietetics addressed what happens to foods and what happens to your body when foods are consumed. You have internal bodily sensory experience of foods as they proceed into your stomach, as they are digested, and when they leave your body in the form of waste; those excrements can also be seen, smelled, and touched.[34] Digestion was traditionally referred to as "concoction," and what happened to food in the gut was assimilated to what visibly happened to it during familiar domestic processes of fermentation, putrefaction, or—apparently closer to the vernacular—cooking.[35] Aristotelian sensibilities distinguished between two senses of concoction, both dependent on heat: the first was "mutation of essence, as when food is converted into flesh or blood"—which is the concern here—and the second was "mutation according to quantity or quality, as in fruites that ripen."[36] In the 1630s, the poet Francis Quarles wrote about the parts of the human body: the heart is the "*Great Chamber*," the midriff is "a large partition *Wall*," and the "*Stomacke* is the *Kitchin*."[37] The body was warm; food taken in might be cold. The processes of transforming a variety of differently textured foods into one's own specific warm flesh (bones, fibers, and fluids) and indeed of supplying the body with its vital heat were considered to be continuations of broadly warming, reducing, and dissociating processes that were familiar in the household kitchen. When you cooked food, you took the first step in a transformative process that was carried on in the gut under the direction of vital powers. Many modes of cooking—soup making, for instance—were understood in the vernacular to extract the "good" portions, or nutritious fractions, leaving behind waste materials that might then be discarded. The stomach, Timothie

Bright wrote, was the body's "naturall furnace," breaking aliment down into its parts while preserving the qualities and then transforming those parts into the humors.[38]

You chew a mouthful of food, and both mechanical processes and saliva break it down into a more or less uniform ball of mush, different foods requiring different amounts and different sorts of processing in the mouth before swallowing. That ball of mush—called a *bolus* in modern expert usage but not seeming to have had a special name in the early modern period—proceeds through the gullet to the stomach. Occasionally, the partly digested mass is regurgitated, making visible the results of further concoction. This semifluid and further homogenized mass known as the *chyme* was considered to be the acidified juice of digestion's first processes. The chyme was then transported to the duodenum, where its acidity was modified by bile and pancreatic juices. It was further transformed into a completely fluid and whitish *chyle*, found in the vessels surrounding the small intestine called the *lacteals* (owing to the milky fluids they contained), before it was carried on to the liver and there converted into blood. The condition of visibility of this and subsequent processes during the Renaissance and the early modern period was dependent on skillful dissection, theoretically interpreted.[39] The broad understanding of digestion as cooking and as a transformation of substances and not of qualities probably had some presence in the wider culture, while the specifics—chyme, chyle, lacteals, etc.—were certainly aspects of an elite vocabulary until at least the eighteenth century, when medical texts meant for a more popular audience went into circulation.[40]

The partial visibility of digestive processes meant that there was at least occasional visual access to how foods were transformed in the body. How quickly, effectively, and completely did they pass from mouth to bodily flesh and to excrement? When they became excrement, what was their appearance and odor? And there were also other sensory modes through which you might have access to digestive trajectories. How did foods of various sorts make you *feel* as they passed through gut and bowels? How long did those feelings *persist*? What effects, if any, did foods have on mental function and mood and on physical capacities? These sensory availabilities were pertinent to much dietetic interest in the *nutritiousness* of foods. Foods said to be *nutritious*—or, as it was said, to have "aliment"—were those judged capable of *nourishing*, having the power to support bodily growth and development, sustain life and health, and fuel activity. Foods were deemed different in their nutritiousness; different parts of a given food were not equally nutritious, nor were foodstuffs at different stages of their development. Different ways of preparing a food had different implications for nutritiousness, as did foods in different combinations; different sorts of people had different nutritional requirements and were differently able to cope with concocting specific foods. Nutritiousness had to do with the circumstances of what was eaten and with the circumstances of the eater. Sixteenth- and seventeenth-century dietaries specified that spinach "affords little nourishment," artichokes "afford no good Aliment," oats contain "not much nourishment," old cheese "affords ill nourishment," tame ducks afford "little nourishment" while wild ones yield "good nourishment," and barley yields "not so good nourishment" as wheat. Swine's flesh "nourisheth very plentifully"; blackbirds "make good nourishment"; pheasants "are the best nourishment"; the liver of capons, hens, and geese "nourisheth excellently"; and venison "becomes good nourishment" provided that it is taken with a glass of wine.[41]

Concoction was understood as a heat-mediated process by which the vital powers inherent in foods were made available for *you* and by which the substance of foods might become the substance of *you*. The concocting process, Elyot wrote, "is an alteration in the stomacke of meates and drinks, according to their qualities, whereby they are made like to the substance of the body."[42] Because foods were you *at a remove*, it was medically vital to know which foods were easy and which were hard to concoct. Again, the texts were specific: among vegetables, turnips, carrots, and asparagus were easily concocted, while cabbage and beans were hard; among flesh meats, chicken, capons, and pigeons were easy, while beef was hard especially if it came from old cows or oxen; mutton was easy, and the hearts, kidneys, and brains of land animals were in general difficult to concoct; cheese was hard to concoct, but eggs were easy.[43] Foods said to be "firm" (or "thick") were those that gave abundant nourishment, but some of these were nevertheless not easily digested, requiring "much strength of heat for concoction."[44] Capon, for instance, was firm and easily digested, while pork gave firm nourishment but was hard to digest.[45] The proper concoction of certain foods depended on specific modes of preparation, the order in which they were taken, the presence or absence of other foods in the meal, and their "correction" by seasoning. For instance, beans—notoriously difficult in terms of concoction—needed correction with pepper to be properly digestible. Cold and moist foods were "corrected" by warming and drying spices: apples with cinnamon, melons with prosciutto, and so on. This was the scheme of things—combining textual authority with the everyday experiences of digestion and the deliverances of taste—that informed links between dietetic counsel and *cuisine*, the modes of food preparation that were considered both good for you and good.[46]

The best food, dietetic writers generalized, "yeilds plenty of nourishment, is easie of concoction, not quickly corrupted, nor hath an ill quality, and there remains after it, but few excrements."[47] Here, the notion of *excrements* (or "excrementitious matter") picked out the stuff remaining after the nutritious portions of food had been extracted. Excrements included the wastes eliminated as feces and urine as well as the unconcocted or poorly concocted matter producing stomach gases. Both gave sensory evidence of nonnutritive character, a sign that whatever vital virtue the foods once contained had been extracted. Different foods contained more or less of this excrementitious material. According to one writer, capon "maketh little ordure, and much good nourishment," while another writer was even more enthusiastic: capon was "almost altogether free of excrement." Domesticated geese had more excrements than wild geese.[48] Beans and cabbage were invariably condemned as excrementitious, and other foods might produce much excrement unless they were taken in the right way. Plums and prunes were benign just on the condition that they be eaten before meat; taken after meat, they were "very dangerous" because "they scatter abroad many excrements and that crude." The gas-producing qualities of many foods were prominently remarked on: these were said to be "flatulent," breeding "fumes," "fulliginous vapors," "putrefactions," and "crudities."[49] Digestion was the process by which foods became you, most directly through their substantial contribution to the body's fluids. And because the material body was constructed from the fluids, food ultimately became flesh. Food and drink, a late sixteenth-century physician wrote, "doe alter our bodies, and either temper them or distemper them greatly. And no marvaile, seeing that such as the food is, such is the blood: and such as the blood is, such is the flesh." Dietary writers commented extensively on the capacity of specific foods to make specific "juices," "blood," or "humors," varying in their capacity to make good or bad "*juice*," good or bad "*blood*," this or that "*humor*." Foods that were good made "good bloud" and foods deemed bad "engendered ill bloud."[50] And through foods' contribution to humoral constitution, your diet helped maintain, remake, restore, or alter your *temperament*.

Food by food, drink by drink, dietetic texts told readers about the goodness, badness, and character of the fluids bred from each aliment. The detail and the specificity covered every conceivable thing you might eat and every conceivable way you might eat it. Chicken livers "maketh good juice"; veal "affords good Juyce . . . and yeilds a thicker Juyce then Lamb, or Mutton"; capons, pullets, and chickens are all "of good juice" and "procreate good blood." In general, the flesh of younger beasts made better juice than those of the older, and you could often tell goodness from color: "The more that the fleshe of Beastes doth degenerate from whytenes, the lesse good juice is therein it," supporting the general inclination to approve the flesh of young animals and to warn about things like bull, ox, and hare.[51] Fruits and vegetables were less often parsed in terms of the juices and the humors elaborated from them, but melons were said to make bad juice: cucumbers were similar, but their juice was "not so dangerous"; and sour cherries were to be preferred over sweet ones, as they are not "of so hurtful a Juyce."[52]

It mattered greatly what specific humors were bred from specific foods. "Al fruits," Elyot generalized, "are noyful to man, and do ingender ill humours, by oft times the cause of putrified fevers."[53] But the specific putrefying juice made from foods such as melons and cucumbers was phlegm, and for that reason these were widely accounted very dangerous indeed, encouraging the fluxes accompanying colds, catarrhs, and the dreaded dropsies. Melons were potential killers. Their taste is very pleasant, but they are of "a cold watry moist substance," easily corrupted and becoming "as it were of a poysonous nature . . . whereby many great men are killed." If eaten at all, they should be taken as a first course, "or before meat that they may the easier descend through the belly, and after the eating of them some good Food ought to be eaten, and good Wine to be drank that their corrupting may be hindred." Cucumbers, strawberries, and peaches had similarly risky moisture-breeding qualities; they were less lethal than melons, but their consumption should still be approached with caution.[54] Cold and moist pears needed to be cooked or taken with wine, and the Salernitan Verses made the mode of preparing them a matter of life and or death: "Raw *Peares* a poyson, bak't a medicine be."[55]

Other foods were worried about because they bred choler: eating them risked the derangements marked by the "acrimonies" of excessive heat and dryness: fevers and inflammations of the guts, liver, and kidneys, for example. The "ill humor" produced by old cheese, dried figs, almonds, and pomegranates was often specified as choler, and there were similar reasons to be concerned about garlic, onions, leeks, and mustard.[56] But no humor was the object of more expert dietary concern than melancholy, an imbalance of which was often viewed as the worst of all the "ill humors."[57] Diet, Robert Burton wrote, "is the mother of all diseases, let the father be what he will; and from this alone melancholy and frequent other maladies arise."[58] Among vegetables, cabbages and beans made notably melancholy juice; among fruits, quinces, dates, olives and chestnuts bred melancholy; and among flesh meats, pigeon, hare, geese, duck, goat, and boar and meat from older deer and cattle all made melancholy humors. There was disagreement among commentators as to whether beef was innocuous, though meat from older and fatter beasts was sometimes thought to engender much melancholy. Among organ meats, spleen was notoriously melancholy making, as it was—in beasts as well in human beings—"the receptacle of gross Melancholy blood."[59] Burton's survey of the dietary causes of melancholy covered the usual suspects, though he was specially apprehensive of beef (the wrong sort), pork, goat ("a filthy beast, and rammish"), venison, hare ("a black meat" that "breeds *incubus*, often eaten, and causeth fearful dreams"), most forms of dairy foods (whey excepted and possibly asses' milk), noting that some authorities condemned all fish, while others forbade only tench, eel, lampreys, and any others "bred in muddy and standing waters." Cabbage, he said, "causeth troublesome dreams, and sends up black vapours to the brain," and beans "fill the brain with gross fumes, breed black, thick blood, and also make for bad dreams."[60] Dietetic medicine had something to say about every known food, its goodness for every sort of person and the propriety of every way of preparing it. Medical expertise bid to be present at every mealtime, perched watchfully alongside every dish.

Correct Food, Correcting People

Robert Burton and many other dietary writers identified certain sorts of flesh meat as making melancholy humors, but they did not warn *all* sorts of people against eating them: some people were at greater risk of bad consequences than others. So, for instance, Burton said that beef was safe food for "such as are sound and of a strong constitution," especially for those doing manual labor, but was very likely to make melancholy disorders for the sedentary, the studious, and those of a cold and dry temperament.[61] That was the general pattern in dietary texts. Foods were good or bad according to eaters' temperament and way of life. Just as persons in normal health should *eat what they were*, consuming foods whose qualities matched their own, so too those in a state of unbalanced ill health or tending in that direction should avoid foods whose qualities exacerbated the unbalanced qualities and should seek out those with countervailing qualities. "Every thing we eat and drink" should be regarded as medicine "provided it be *contrary* to a disease."[62] You should know what you were, you should know what foods were, and you should monitor your momentary state of health to understand what you should and should not consume. Diet could *tune* the body and its qualities, on the condition that you *knew enough* about your body and about your food. Some of these things you could know through your own experience; others were transmitted through the channels of medical expertise and those who represented such expertise.

There were happy meats—not many—that suited all temperaments because they themselves were temperate, well-balanced in their qualities. Capon was widely accounted as suited to all constitutions; among organ meats, tongue was good for every temperament; and among fish (generally cold and moist), sole was described as "the Sea-Capon" and similarly commended.[63] But for the rest, wholesomeness depended on a range of considerations. Your own constitution was paramount, including the changes effected by your age and manner of living, but so too were a food's stage of ripeness, its environmental source, its mode of preparation, its accompanying foods, the season of year, and the stage in a meal at which it was consumed. Moist and cold foods—melons and cucumbers, of course, but possibly also trout—could breed phlegm and associated watery ailments, making them dangerous to phlegmatics but acceptable to those of a hot constitution, that is, cholerics and sanguinaries.[64] The heat of onions made them good for phlegmatics and bad for cholerics.[65] Cherries, plums, and damsons combated the heat of choler,[66] while black pepper was good for both melancholics and phlegmatics.[67] Goodness or hurtfulness might also follow from the qualities of a specific *variety* of a food. There were at least three sorts of pomegranate: sweet, sour, and temperate. The sweet pomegranate was hot in the first degree and temperately moist and was good for melancholics but bad for cholerics. The sour pomegranate was cold in the second degree: good for cholerics but bad for phlegmatics.[68] Dietary writers sometimes recognized that people would stubbornly eat foods whose qualities were wrong for their temperament, and the *correction* of cold foods by hot spices was specially commended to phlegmatics. "If any shall eat meats," an early seventeenth-century doctor said, "that are not convenient for his constitution and state of body, by reason of a great desire, that hee hath unto such, hee ought to take them with their correctories; as unto moyst and phlegmatick meats, to adde things of contrary quality and substance: for by this means they will be made more agreeable to the body, and so taken with lesse offence."[69] Wines varying in their *heat* varied in their suitability for different temperaments and stages of life. In the mid-sixteenth century, Andrew Boorde noted that old men—as they grew colder—were less harmed and more helped by stronger wine. In 1650, the English physician Humphrey Brooke cited the common saying "*Vinum est Lac Senum*" (Wine is the old Man's Milk). And in the eighteenth century, the great lexicographer Samuel Johnson offered the formula "Claret for boys,—port for men,—brandy for heroes."[70] Foods might be *corrected* by seasoning, preparation, and combining with other foods, but the ultimate purpose was to use food to *correct* the body, to maintain or restore qualitative balance.[71]

The powers and qualities of foods extended beyond their qualitative effect on the body's humors, and there were few aspects of bodily capacity or condition not thought to be impacted by food and drink. Foods had vital powers, some of which did not clearly emerge from their elemental qualities. They had specific effects on those afflicted by epidemic diseases and infestations of particular organs, emotions, inclinations to sexual activity, and dispositions to sleep and dream. Pepper and ginger were good for tertian and quartan agues (fevers, probably malarial); honey was good for the lungs, with a "cleansing faculty"; parsnips, turnips, carrots, and onions are all aphrodisiacs, some of these also promoting urine and menstruation; leeks and garlic were good remedies against the plague; and asparagus cleansed the kidneys and alleviated obstructions in the liver. One writer said that walnuts make you cough, while another asserted that they "multiply sperme, and provoke sleepe." Cheese "stops the belly," making for constipation, but being an effective way to "close the mouth of the Stomach," it was good to take it after a full meal. Saffron "rejoyceth the heart," and oysters "stimulate Venus." Lettuce was good for the heart and liver; it put you to sleep, and a decoction of lettuce could be taken as a medicine to relieve insomnia.[72] Garlic was accounted as a powerful vital force: it "killeth wormes, provokes urine, excites Venus, opens obstructions, helps the cough, & paines of the breast proceeding of colde, and likewise the winde-collicke. It is also an enemy to all cold poysons, and to the bitings of venomous beasts, a remedie to such as are constrained to take naughty corrupt drinkes or meats, and a Preservative against contagious and pestilent aire, and therefore not unfitly termed, *The Country-mans Treacle*. But if it be often or immoderately eaten, it causeth headach, and hurteth the sight."[73] John Evelyn's seventeenth-century manual on growing greens and preparing salads was intermittently concerned with the effect of various plant materials on eyesight: basil might be "offensive to the Eyes," as were onions "eaten in excess," and there were reasons to be wary of cabbages in general or, perhaps specifically, of cauliflower, while he was skeptical of claims that parsley was "hurtful to Eyes."[74] Whatever fell within the competences of medical expertise—however specific—food was almost always involved its management.

The Qualities of Foods: Who Knew?

In the accounts just given, almost all the descriptions of foods and their qualities come from *books*. The books summarized here are Renaissance and early modern works by medical writers telling readers how to be healthy and by practical ethical authors who used dietetic vocabulary in telling readers how to be virtuous or prudent. In turn, these sorts of books gained their authority by gestures at still other books, writings by Hippocrates, Galen, Celsus, Islamic commentators and synthesizers such as Avicenna and books by Greek thinkers who elaborated natural philosophical frameworks populated by elements and qualities. Dietaries and practical ethical texts were to a great extent books about books, even while all these books retailed views represented as grounded in experience: how to *be*, day by day, year by year, and in the varying circumstances of everyday life. A few of these books (such as the Salernitan Verses) enjoyed very wide distribution—combining the authority of ancient expertise with folkish formulae—but Renaissance and early modern books about foods and their qualities were obviously directed to those sorts of people who enjoyed a great range of choice about how much and what they ate. That is evident in both the hugely repetitive injunctions not to eat *too much* and the range of foodstuffs whose qualities might suit or harm the eater: pomegranates and peacocks, malmsey and mallard.

So, one way people could know that garlic was hot and dry in the fourth degree and was congenial for the old, the phlegmatic, and the melancholic was by reading medical books that said so and that borrowed the authority of other texts, contemporary and ancient. Yet, many of the dietary texts were written by people who were not physicians, while practical ethical tracts on how to be a gentleman, how to form the character of young gentlemen, and how to behave in polite society showed familiarity with the medically warranted characteristics of different foods. But if books were the only way you might know about these sorts of things, the social distribution of that knowledge would be limited, confined to the thin stratum of the then-literate classes. Indeed, there are many elements of these textual descriptions that are hard to imagine as enjoying very wide circulation and recognition: that cabbage injured the eyesight, that egg yolks and egg whites differed in warmth and moisture, and that peacock flesh tended to breed a thick and dry melancholy.

However, there was much about this kind of knowledge that evidently traveled widely in Renaissance and early modern culture. The more general sentiments of traditional dietetics evidently belonged to lay as well as expert knowledge.[75] The prudence of resting after dinner and walking after supper, knowing that desperate conditions called for desperate remedies, that the signs of diseases might be treated by opposites, that custom was second nature, that cheese was "binding," that strong meats didn't suit weak stomachs, and that pepper was hot and melons were cold. Along with much else, these were the sorts of things that were as close to common cultural possessions as anything else in the Renaissance and the early modern period apart from biblical stories, set prayers, and incantations. These sentiments belonged to not just the world of bookish literacy but also the worlds of orality, to the illiterate as well as the learned. Renaissance physicians often complained bitterly about the confidence that ordinary people placed in their knowledge of the powers and wholesomeness of foods, yet texts such as the Salernitan Verses not only incorporated much of this commonsensical knowledge but also deployed the mnemonically powerful forms of rhymes, saws, and proverbs to give their counsel force and to ensure that it stuck in the mind.[76]

At some level, knowledge of the qualities and vital effects of specific foods was widespread in the sixteenth and seventeenth centuries. Laypeople might rely on different authorities to find out whether lettuce put you to sleep and old cheese was binding, and authorities did sometimes disagree. But any such disagreements still used a shared vocabulary to talk about these things and an extensively shared sensibility about how to reason from food qualities to their effects on the human frame. Rabelais's Gargantua took his lessons from his tutor Ponocrates, and Ponocrates took them from the ancient medical writers. They sat down to eat and began to "discourse gaily together, speaking in the first place of the virtues, properties, efficacy, and nature of whatever was served them at table: of the bread, the wine, the water, the salt, the meats, fish, fruit, herbs, and roots, and of their dressing. From this talk Gargantua learned in a very short time all the relevant passages in Pliny, Athenaeus, Dioscorides, Julius Pollox, Galen, Porphyrius, Oppian, Polybius, Heliodorus, Aristotle, Aelian, and others." If in doubt, they called for the books themselves to be brought to the table: "and so well and completely did Gargantua retain in his memory what had been said that there was not a physician living who knew half as much of this as he."[77]

Shakespeare knew these sorts of things, and he knew that his audiences knew them too. He reckoned that they knew about the qualities of beef, sack, mustard, and mutton, or, at the very least, they knew that specific foods possessed specific qualities. He knew that his audience was aware that specific foods had specific effects on body and mind and that these effects might be taken into account in social interaction. Humors, complexions, and the qualities of airs and aliments belonged to a vocabulary Shakespeare pervasively used to tell his audience who was who, what sorts of people they were, and what circumstances and foods suited or did not suit different temperamental types.[78] Shakespeare probably drew on Timothie Bright's *Treatise of Melancholy* (1586) in making up the character of Hamlet, a melancholic temperament that the Globe Theatre audience might have relied on to instantly recognize the sort of person he was: depressive, fearful, and indecisive yet prone to fantastic visions, extremities of mood, and rash action. Bright wrote that southerly winds suited the melancholic temperament, a medical sentiment echoed in the play: "I am but mad north-north-west," Hamlet says. "When the wind is southerly, I know a hawk from a hand-saw," which was to say that he was quite sane (*Hamlet*, Act 2, scene 2).[79] In *The Taming of the Shrew*, Petruchio attempts to remake the shrewish Katharina through dietary management, and indeed, the play centers on the environmental shaping of character. Kate's problem is her choleric nature, and that nature has to be reconstituted, as far as possible, by denying her the hot and dry and encouraging the cold and moist. Petruchio gets his new bride home and cruelly proceeds to damp her fire by starving her body of fuel: "She eat no meat to-day, nor none shall eat." His servants bring in some overdone mutton. Kate thinks it's alright, but she isn't allowed to have it:

> I tell thee, Kate, 'twas burnt and dried away;
> And I expressly am forbid to touch it,
> For it engenders choler, planteth anger;
> And better 'twere that both of us did fast,
> Since, of ourselves, ourselves are choleric,
> Than feed it with such over-roasted flesh. (*The Taming of the Shrew*, Act 4, scene 1)

Kate and Petruchio recognize each other as constitutional kin, and that constitutional knowledge lets Petruchio try to reform Kate into the kind of wife he wants. "What say you to a neat's foot?,"

Petruchio's servant Grumio, acting under instruction, asks her (Act 4, scene 3). Yes, please, she says. But that too is a tease: "I fear it is too choleric a meat. How say you to a fat tripe finely broil'd?" Kate once more takes the bait: "I like it well: good Grumio, fetch it me." Grumio refuses: "I cannot tell; I fear 'tis choleric. What say you to a piece of beef and mustard?'" Kate, on the verge of starving, says, "A dish that I do love to feed upon." But Grumio again denies her; the qualities are all wrong: "The mustard is too hot a little" (Act 4, scene 1).[80]

The effects of different wines on the human body were extensively and minutely treated in dietary writings. Sherry (or "sack") was, for example, "compleatly hot in the third degree . . . and therefore it doth vehemently and quickly heat the bodie."[81] And in *Henry IV, Part 2*, the gluttonous but valiant Sir John Falstaff explains the virtues of sherry and the qualities it imparts to the human frame and disposition. Sherry is a soldier's drink: it suits the military nature and fortifies the military virtues, spreading its qualities through body and mind. Falstaff's speech is a virtuosic tour through many of the categories and causal structures of traditional dietetics:

> A good sherris sack hath a two-fold operation in it. It ascends me into the brain; dries me there all the foolish and dull and curdy vapours which environ it; makes it apprehensive, quick, forgetive, full of nimble fiery and delectable shapes, which, delivered o'er to the voice, the tongue, which is the birth, becomes excellent wit. The second property of your excellent sherris is, the warming of the blood; which, before cold and settled, left the liver white and pale, which is the badge of pusillanimity and cowardice; but the sherris warms it and makes it course from the inwards to the parts extreme: it illumineth the face, which as a beacon gives warning to all the rest of this little kingdom, man, to arm; and then the vital commoners and inland petty spirits muster me all to their captain, the heart, who, great and puffed up with this retinue, doth any deed of courage; and this valour comes of sherris. . . . Hereof comes it that Prince Harry is valiant; for the cold blood he

> did naturally inherit of his father, he hath, like lean, sterile and bare land, manured, husbanded and tilled with excellent endeavour of drinking good and good store of fertile sherris, that he is become very hot and valiant. (*Henry IV, Part 2*, Act 4, scene 3)

Hearing Falstaff's speech, the audience at the Globe recognized what sort of person he was, what qualities sherry possessed and what effects it had on people, and how consuming sherry helped make Falstaff who he was.

Experiencing the Qualities of Foods

There were books telling you about the qualities of foods, but there was also embodied *experience*. One mode of having sensory experience of food qualities occurred at the moment they were taken into the body or even just before. In describing the qualities of foods as hot, cold, moist, and dry, traditional dietetics gestured at the authority of ancient natural philosophy. Yet, each of the four qualities was also apprehended by the senses: by placing an item of food on the body's surfaces—skin as well as tongue—you could *experience* it as hot, moist, etc. The match between sensory experience and textual accounts was neither exact nor superficially obvious: a melon appears to the senses as moist, but onions and leeks are not in the same way self-evidently hot; the heat of black pepper is not the heat of beef taken directly from the roasting spit; the juice of an orange is not obviously cold and dry; and it is not transparently clear why any wine should be described as dry. Nevertheless, the qualities used in traditional dietetics to describe foods were available as *sensory* categories.

Sensory experience of qualities and goodness continues when foods enter the mouth. In ancient and early modern sensibilities, *taste* could be accounted as a reliable probe into wholesomeness. The scheme of things in which taste signaled qualities and vital consequences proceeded from the same philosophy of nature that informed the notions of humors and temperaments.[82] The edible bits of the natural world were made of the same elements as its nonedible parts. And the bits of the natural world that were human bodies were similarly composed, as was the tongue, the tasting organ of the human body. Taste might signal goodness because it was an index of how the qualities of a thing to be eaten related to the qualities of the eating body. Persons in a normal state of health should eat foods whose qualities matched their own complexion: people whose temperament tended to the cold and dry should in normal circumstances consume foods tending to the cold and dry, and so on. A food that was, for instance, cold and moist matched the qualities of a phlegmatic eater, and that match was made manifest to the sense of taste by *pleasantness*, *relish*, and *sapidity*. The term of art traditionally used to express that match of qualities was "agreement," an equivalence of qualities that was directly apprehended by the sense of taste.[83] Foods that were, as it was said, "grateful to the taste" were foods that were, on the whole, *good for you*.[84]

The learned might use a Latin tag to express this sensibility: *quod sapit nutrit*, meaning "what pleases nourishes" or "what tastes good is good for you." (And, conversely, an unpleasant taste signals unwholesomeness or even poison. A bitter taste, for example, was widely taken as an index of the toxic as well as of the medicinal.)[85] Foods tasted good because their qualities matched—agreed with—the qualities of a person's body. With some exceptions, medical writers endorsed the *quod sapit* sentiment. They could acknowledge the power of dietary custom, and they could align themselves with the inclinations of their own patients while maintaining an overall condemnation of gluttonous excess. Tasting preceded digesting, but taste was considered an index of digestibility. In its enthusiasm for spices, a late medieval dietary text joined good taste and healthy digestion: these "are of no small value in a healthy diet because condiments make food *more delectable to the taste and therefore more digestible. For what is more delectable is better for digestion*."[86] Once foods are consumed, the signs of their qualities continue to be available to the senses. You have sensory access to how digestion is going. You sense whether foods are being digested with ease or with difficulty. You sense whether certain foods, taken at certain times, prepared in certain ways, and combined with certain other foods, produce fumes and vapors, and you can see and feel how they come out the other end. Some foods give you a sense of satisfaction or fullness that lasts a long time: they make you feel satiated, replete, or, in medically learned language, *plethoric*. Other foods give little sense of lasting fullness: soon filled, soon empty. You will, though perhaps with less confidence, have sensory experiences giving warrant that certain foods have certain effects on mood, emotion, sexual inclinations, dream life, feelings of vitality, well-being, and so on. "Spirits" now designate alcoholic beverages, but past sensibilities concerned how these drinks worked physiologically *on* the body's spirits. In these ways, dietetic talk of the effects of drink on vital activity and emotions also traces back to sensory evidence. Wine, beer, and distilled beverages are both metaphorically and substantially *spiritual*. Spirits are volatile, aromatic, and inflammable, natural symbols for the stuff of the aetherial heavens. Once consumed, spirits are *ecstatic* substances—they can take you *outside of yourself*—affecting vitality and moving mood. For that reason, spirits strongly touched the principles of vitality and were suitable fuel for contact with the divine. And if you knew that sort of thing, you knew it courtesy of available sensory experience.

You may have sensory experiences of foods that once tasted good and no longer do, or you may find that foods that were once digested well now cause you difficulties. You may find that foods vary in their effects on you according to time of day or time of year. And you may feel that foods you could digest well when engaged in violent physical activity are not well digested when you follow a sedentary way of life. In traditional dietetic culture, the sensory experiences pertinent in these connections might or might not be widely shared, but there was reasonable confidence that *your* experiences reliably testified to the suitability of foods *for you*. The effects of foods on your body and mind were gauges of qualities, and knowing those qualities was a form of self-knowledge.

Transformation and Analogy

Food becomes *you*—the substance of your body and the material aspects of your mind—through a series of transformations: an ox eats plants, and the plant matter becomes animal flesh; a human being eats the flesh of the ox, and the animal flesh becomes human flesh; and a human being eats a plant, and the plant undergoes a double transformation to become human flesh. In traditional medicine, there was no very well worked-out account of how these transformations occurred. Special properties were, of course, attached both to life in general and to human life in particular. Consumed food is concocted in the stomach by the body's natural heat. The elements of food are disaggregated and recombined, excremental matter is separated after blood is formed, the nourishing fluid of blood is transported to various parts of the body where deposits from blood form flesh, and the whole process was guided by the notion of a *nutritive soul*.

The tag "you are what you eat" appeared in the nineteenth century and referred, in one sensibility, to the transformations framed by materialistic nutrition science and, in another, to the cultural and sociological modes in which people were understood to be defined by patterns of consumption. Traditional dietetics offered accounts of both the transformative processes by which food became you and the outcomes of those transformations. You are what you eat through changes in not just material substance but also quality. "Those things that are Aliment," a mid-seventeenth-century English physician wrote, "(*viz.*) which being eaten or drunk are altered by our Naturall Heat, and so prepared by the several Parts destined thereunto, as at length to be Converted into the Habit of the Body it Self."[87] Digestion and assimilation rearranged the alimentary matter, and those vital processes required vital animation, but the processes preserved qualities. The qualities in your food would ultimately determine the qualities of you. Food was the qualitative palette for painting a self-portrait, and your choices in food amounted to your choices about what you were like in health and in disease, in body and in mind.[88] You should pay minute attention to the qualities of what you ate, for food knowledge belonged to self-knowledge, and the choice of foods belonged to the care of the self.

A mode of reasoning pertinent to this awareness and care was *analogical*: reasoning from the observable characteristics of a food source to its consequences for the eater. Inferring the moistness of melons, for example, from the sensory appearance of their watery nature was itself a form of analogical reasoning, yet there were other modes of analogical reasoning, and these had little or no connection to Aristotelian and Galenic categories. The so-called *doctrine of signatures* in strands of sixteenth- and seventeenth-century medicine inferred from the shape of a plant or plant part to the human organ it would affect.[89] A stomach-shaped seed pod said, in effect, "Take me for an upset stomach." Those resemblances were marks of divine intention: God had signed His creations so that we could work out their virtues, what was good for what. Human capacity to recognize and read such signs was part of what it meant to be given dominion over Creation. "Every sort of Food," wrote the English merchant and popular medical writer Thomas Tryon, "hath its operation in the Body, and on the Spirits by way of *Simile*."[90] One form of this inference considered the likeness between the food and the flesh of human beings. Eggs, for example, were commended and were said to be easily digested because of a "a certaine analogie or likeness that they have with mans nature"; mother's milk was spectacularly well suited on those grounds.[91] (The undeniable likeness between human flesh and human eaters informed enormous interest in whether cannibalism was practiced by native peoples encountered by European explorers. That practice was, of course, violently condemned, but there was great curiosity about the physiological consequences of cannibalism.)[92] Analogical reasoning also took in the behavioral and physical characteristics of the beasts that might be used as food. In the mid-sixteenth century, Andrew Boorde folded together analogical and humoral vocabularies in worrying about the eating of venison; deer were shy and anxious creatures, and that timorousness might have humoral consequences: "tymorosyte doth brynge in melancoly humors."[93] A Welsh promoter of colonial settlement in the New World reflected on causal connections between native diet and native character: "one of the chiefest causes of the *Savages* inhumane cruelty proceedes through their devouring of *Wolves* and *Beares flesh*."[94] And in the mid-seventeenth century, the Czech reformer Comenius noted a substantive link between food, the humors, and the moral as well as physical characteristics of eaters: "he that feeds on dry meat, is dry of complexion; he that feeds on moist, is flegmatick, &c . . . because, for the most part a man reteins the qualities of those living creatures on whose flesh he feeds, as he that feeds on beefe is strong; he that feeds on venison, is nimble, &c."[95]

This type of analogical reasoning persisted at least through the early eighteenth century. A French physician strongly influenced by new chemical ways of thinking about the constituents of foods still recorded the views of writers who said that "those who feed on gross Flesh . . . become gross, stupid, and, as a man may say, acquire a Resemblance of Temper and Inclination with those Animals whose Flesh they feed upon. And this was the Reason that in antient Times there were some People who would not eat the Flesh of any Animals, but such as were strong and couragious, in order to acquire the noble Qualities of those Animals for themselves." Even now, he noted, people who live on goats' milk were "lively, active, and nimble," while those who consume the flesh of asses or camels were "usually heavy, and dull of Understanding."[96] Remarks about dry and moist qualities were straightforwardly Galenic: through the humors, the qualities of your food were said to cause the qualities of your body. But while there was some connection between the humors and such characteristics as strength, courage, stupidity, and fleetness of foot, the rich vocabulary of analogical reasoning ranged beyond the moist and dry, the warm and cold.[97]

Traditional dietetics reckoned that we eat what we are. As it was then said, we have ourselves on our plates.[98] We should understand that and should act accordingly. We are uniquely placed—through both reason and sensory self-knowledge—to know what foods agree or disagree with us. Our choice of what we eat is ultimately self-making. Much of dietetic counsel was concerned with self-making for individuals, but it also addressed self-making at collective levels. How did Spanish food make Spanish people? How did English food make English people? Traditional dietetics offered answers to such questions.

4

You Are What You Eat: Types of People and Types of Food

Who Are What They Eat?

In *Taming of the Shrew*, Petruchio and Katharina recognize that they are temperamentally similar. One is a man, the other a woman; one is powerful, the other vulnerable. Petruchio proposes to remake Kate through coercion, to be sure, but also through an imposed dietary regime. Shakespeare expected the audience to know what it was to be choleric, melancholic, phlegmatic, and sanguinary, and he expected them to understand what people of these different constitutions were like, what they might do, how they came to be the sorts of people they were, and what sorts of interactions with the environment—edible and nonedible—were customary and appropriate for such people. Traditional dietetics offered a vocabulary through which people might understand much about themselves and about those with whom they routinely dealt. Practical ethical writers and playwrights used these understandings to present their characters as characters of a certain sort. Hamlet, Hal, Falstaff, Cassius, and Caliban were offered up as unique individuals but also as *types*, and those types were significantly configured through the vocabulary of traditional dietetics.

What worked for the recognition of individuals—encountered on the stage or in the street—worked also for the recognition of *collective* identities. The categories of male and female were, of course, basic to systems of social identity, sorting, and power, but for all that, sex roles were not central features of the descriptions and recommendations in dietary writing. The notion that women were colder and moister than men was widespread, though as the Petruchio-Kate passages suggest, Renaissance and early modern culture was capable of recognizing women and men whose constitutions differed from any humoral generalization.[1] The general silence about women and their constitution in the dietetic literature certainly had to do with male authorship, maleness as the norm in a patriarchal society, and a focus on men as both public actors and the major agents of *choice* in matters nonnatural. However, women and their conditions were not absent from the general medical literature. Doctors had long been engaged with those conditions and vital processes that were particular to women, notably including menstruation, pregnancy, nursing, and also the pathologies of the womb and breasts. And in these connections dietary management often figured, along with a range of other preventive and therapeutic measures.

A seventeenth-century treatment of the female jaundice called chlorosis (or "green-sickness") referred the condition to an obstruction of the veins involving the womb and other organs. And this obstruction had something to do with an "evil Diet" that caused a loss of "Natural heat." Failure to menstruate was said to afflict women with a dry constitution, and the "external Causes" of the condition included exposure to cold winds and water as well as eating too little or too much meat, salt, or spice.[2] The only marginal presence of women in the dietetic literature continued into the eighteenth century, though again, dietary causes and therapies were frequently cited when medical men addressed specifically female conditions. William Buchan's ubiquitous *Domestic Medicine* (1769) had a section on "diseases of women" and, treating chlorosis, he blamed "unwholesome food," lack of exercise, inadequate exposure to fresh air, and insufficient cheerfulness. Much of this was said to be women's own fault: "Fond of all manner of trash, they often eat every out-of-the-way thing they can get, till their blood and humours are quite vitiated." Proper management of the nonnaturals was also indicated in pregnancy, menopause, and barrenness, and here Buchan insisted that the dietetic management recommended is "applicable to men as well as to women."[3] Later in life, offering medical advice to mothers, Buchan reckoned that women's care for "personal beauty" was a positive inducement to proper dietetic management. Female vanity was part of nature's design for health: vanity is "a powerful check on excesses of every kind, and is the strongest incitement to cleanliness, temperance, moderate exercise, and habitual good-humour. . . . Beauty . . . is nothing more than visible health,—the outward mirror of the state of things within,—the certain effect of good air, cheerfulness, temperance, and exercise." A pervasive cause of female ill health identified by the doctors was the sedentary nature of their lives—here neglecting the arduousness of much female labor, both domestic and nondomestic—and advice to get out and get moving was common.[4] There was nothing here about a distinct female constitution or a dietetic regime appropriate to any such supposed constitution.

While the constitutional specificity of men and women was only a marginal concern of dietetic writing, the situation was different when the subject was the identity of different races, nations, and occupational roles. Dietetic vocabulary was drawn upon to give accounts of the natures of those living a holy or civic life, laborers and courtiers, and soldiers and scholars, and it was a resource in descriptions of the dangers and opportunities faced by colonists and travelers to foreign lands. From antiquity through the early modern period, dietetic texts ascribed racial and national characteristics to transactions with the environment, notably the foods and drinks that different people preferred and customarily consumed. What foods, for example, suited the English, the Scots, the Welsh, the French, and the Spanish? In England, what suited people from the West Country and what suited Essex man? If you knew what the different races and nations ate and drank, you might understand something of their dispositions, for their diet worked to make them what they were.

This sensibility persisted into the eighteenth century, as, for example, in Montesquieu's 1748 *Spirit of Laws* and its account of differing collective ways of life. The Muslims of Arabia did not drink wine. They lived in a hot climate, and their blood lost water through perspiration. Water could reconstitute the lost fluid, while spirituous drinks would further congeal the blood. In cold climates the opposite obtained, and this is why northern people tended to take strong drink. It was not a matter of Islam but rather of the environment and the nature of indigenous people. Montesquieu claimed that the Arabs drank water and shunned wine *before* Muhammad, and so too did the Carthaginians, as their climate was much the same as Arabia's. The Scandinavians are drunkards, and the Arabs abstain: that is *who they are* at a constitutional level. "It is very natural, that, where wine is contrary to the climate, and consequently to health, the excess of it should be more severely punished than in countries where intoxication produces very few bad effects to the person, fewer to the society, and where it does not make people frantic and wild, but only stupid and heavy. Hence those laws which inflicted a double punishment for crimes committed in drunkenness were applicable only to a personal, and not to a national, ebriety. A German drinks through custom, and a Spaniard by choice."[5] Local aliment and local natures were points on a causal loop. In many Mediterranean countries, Robert Burton wrote,

> they live most on roots, raw herbs, camel's milk, and it agrees well with them: which to a stranger will cause much grievance. In Wales . . . they live most on white meats; . . . in Holland on fish, roots, butter; and so at this day in Greece . . . they had much rather feed on fish than flesh. With us, . . . we feed on flesh most part, . . . as all northern countries do; and it would be very offensive to us to live after their diet, or they to live after ours. We drink beer, they wine; they use oil, we butter; we in the north are great eaters, they most sparing in those hotter countries; and yet they and we following our own customs are well pleased.[6]

The causal nexus linking constitution and aliment might reach to the county level. Cider and perry are common beverages in Worcestershire and Gloucestershire, and they are "cold and windy" drinks, yet as Burton observed, "in some shires of England, [and] Normandy in France . . . , 'tis their common drink, and they are no whit offended with it."[7] A late sixteenth-century English East Anglian writer described the characteristic and appropriate diets for people from Yorkshire, Lancashire, Essex, Kent, Middlesex, and Wales: "For the Northeren-man, White-meates, Beefe, Mutton, Venison: for the Southerne man, Fruites, Hearbes, Fowle, Fish, Spice, and Sauce." Essex men ate veal, and Welshmen, of course, ate leeks and cheese.[8] Montaigne, always a supporter of custom over confident claims to rational expertise, noted the force of custom in shaping national tastes: "A Spaniard can not well brooke to feede after our fashion, nor we endure to drinke as the Swizzers."[9] And he noted that people had been found in the New World "to whom our usuall flesh and other meats were mortall and venomous."[10] That relationship between diet, climate, and constitution—even apart from the question of what was locally available—continued to be acknowledged into the nineteenth century and beyond. In the 1820s, a prominent English physician conceded that vegetables were not easy to obtain in northern regions but insisted nevertheless on constitutional *suitability* as the reason for the association between flesh-eating and the north: "An animal diet is probably better fitted for producing the vigour and hardihood of frame, which is requisite to brave the rigour of an arctic climate."[11]

There were a few English dietary writers who disapproved of wine for Englishmen, but they made exceptions for those whose stomachs were disordered, since wine was understood to be an aid to digestion. Canary wine or sherry was commended as a powerful cordial, but some warned against the routine use of all rich and sweet wines as "being not at all agreeable to our Northern Constitutions." They heated and disordered the blood, making "*Men* too *Effeminate* and *Women* too *Salacious*" and setting "the *Gate of Venus open*."[12] An ascetically inclined late eighteenth-century physician advised that if "the stomach be weak and cold, the constitution languid, weak, cold, and relaxed, and the blood poor and watery, then a glass or two of wine will be of service; but people in health require no wine; it ought only to be used as spices are." Wine suited the natures of those who lived where the vine naturally flourished—indeed, that too was part of the divine plan—but it was in no way necessary for any normally healthy person living in England's cold and wet climate: "Wine was never designed for common use. In warm countries it is very necessary, nor can health be preserved without it. For the heat of the weather exhausts the strength, weakens the inside, and hurts digestion; therefore we see providence has provided for their wants by giving plenty of grapes, which are the produce of warm climates only."[13]

European travelers and especially colonists were anxious about the possible effects of exotic environments on their own constitutions. Would the climates of new lands help them or harm them? The English in North America, for example, were concerned to find temperate zones congenial to their constitutions and similar to that of their supremely temperate homeland. One colonist was pleased to observe a climatic similarity between old England and New England, both of which were situated in "a golden meane betwixt two extreames: I meane the temperate Zones," and another enthusiastically concurred, noting that New England agreed with "our *English* bodies."[14] An early English settler in the Massachusetts Bay Colony waxed poetic about "the Temper of the Aire": "there is hardly a more healthfull place to be found in the World that agreeth better with our English Bodyes. . . . For here is an extraordinarie cleere and dry Aire that is of a most healing nature to all such as are of a Cold, Melancholy, Flegmatick, Reumaticke temper of Body. . . . [A] Sup of *New-Englands* Aire is better then a whole draft of old *Englands* Ale."[15] Massachusetts was powerful medicine. Hippocratic sensibilities about airs, waters, and places also directed practical attention to diet and particularly the effects of eating unfamiliar native foods. What would happen to travelers and colonists in the East or West Indies or in the Americas if they ate as the natives did? Would their constitutions *change*, and if so in what ways? Would such changes be good or bad for health? If the natives were what they were through long-term interactions between climate, diet, and bodies, would European colonists go constitutionally *native*?

Traditional modes of dietetic reasoning strongly suggested that substantial change through novel diet was likely. Could colonists in the New World digest native foods? Would such foods be palatable? On Hispaniola, Christopher Columbus and his men ate some of what the "Indians" gave them, but they were wary of other local foods: "They eat many such things as would not only make any Spaniard vomit but would poison him if he tried them." Some, indeed, found that Indian foods "disagreed with them very badly, since they were not used to them."[16] In the early North American colonies, the English closely monitored the suitability of local aliment, gauging such things through the acceptability of English foods to the Native American constitution and paying attention to their own bodily changes in over time.[17] What the English called "Indian corn" (maize) was the basic grain in the diet of native populations, and it was of special interest to traders and colonists. Early opinion about maize was unfavorable, and the English herbalist John Gerard was scornful: The "barbarous Indians" ate it and thought it good, but they were ignorant of better fare and were "constrained to make a virtue of necessitie": "It nourisheth but little, and is of hard and evill digestion, a more convenient foode for swine than for men."[18]

Over time, the colonists—Spanish, French, and British—came to terms with maize: they grew it and consumed it, though much was apparently used to feed slaves and animals.[19] Over time, the colonists fought back against Gerard's dismissal. The son of one of the founders of the Massachusetts Bay Colony wrote to Robert Boyle that with the benefit of "much Experience," maize was indeed found wholesome and that colonists' adaptation to American environmental conditions was assisted by adjusting to native foods, and, as for its digestibility, experience proved Gerard wrong.[20] Gerard's opinion continued to circulate into the eighteenth century, to the considerable annoyance of Benjamin Franklin. In 1766, Franklin's criticisms of the punitive British Stamp Act and Sugar Act were joined to a celebration of maize. The English seemed to think that the colonists could not do without tea for breakfast, since alternative breakfasts made from "Indian corn" were neither tasty nor easily digestible. But the English didn't know what they were talking about: "take it for *all in all*," Franklin wrote, Indian corn "is one of the most agreeable and wholesome grains in the world"; its roasted ears "are a delicacy beyond expression," and a "*johny* or *hoecake*, hot from the fire, is better than a Yorkshire muffin."[21] The English countered by asking why if Indian corn was so wholesome the "white men give it to their slaves," reserving more familiar grains, such as wheat, for their own tables. Franklin admitted that slaves were given maize but that this was evidence of nutritiousness: "[Maize] keeps them healthy, strong and hearty, and fit to go through all the labour we require of them. Our slaves, Sir, cost us money, and we buy them to make money by their labour. If they are sick, they are not only unprofitable, but expensive. Where then was your *English good sense*, when you imagined we gave the slaves our Indian corn, because we knew it to be *unwholesome*?"[22] In the East Indies, an early seventeenth-century botanist insisted that the goodness or noxiousness of foods had to be referred to the local environmental context, not that of Europe. Dutch aquatic fowl, for example, were accounted unwholesome because they fed on slime and weeds, but the opposite was the case for their East Indian equivalents, since these lived in free-running rivers and fed on better fare. Europeans struggled intellectually to assimilate exotic aliments to traditional qualitative and descriptive categories, sometimes succeeding (as with the potato) and sometimes failing (as with the pineapple, where many commentators confessed stunned inability to describe its exceptional taste and other characteristics).[23] The well-being of European colonists depended on knowing themselves and knowing the *local* qualities of *local* foods and drinks in relation to local conditions.[24]

There were contests over the proper diet for the colonial body. Some European commentators recommended that colonists observe and follow native dietary usages, adjusting their appetites to what nature locally provided, confident that what was initially disgusting and functionally upsetting would eventually come to taste delicious, prove innocuous, and protect against local illnesses.[25] But other Europeans were anxious about the relationship between the colonial constitution and native aliment and waters. Spanish colonists in Latin America were seriously concerned about the causal link between food and human constitution. The right foods—those that the colonists were accustomed to, notably wheat bread and wine—would, it was thought, protect the colonial body from the physiological risks of the New World environment, while eating local foodstuffs might transform it into the inferior native body. Native foods were responsible for native humors and temperaments. The colonial enterprise was understood to depend on maintaining the constitutional difference between colonists and colonized.[26]

Beef Eaters

The English were widely identified as beef eaters, and the roast beef of Olde England was a notable instance of analogical reasoning, from the qualities of aliment to the qualities of people.[27] The historical frames for understanding why the English were beef eaters and how beef eating accounted for their racial characteristics run across the cultural spectrum. English beefiness is partly intelligible from within the basic Galenic system. Many dietetic writers judged that beef is a "cold" meat, suiting the "cold" English nature, so explaining why it was not suited to those—such as students and scholars—suffering from the distempers of melancholy and why the hot-tempered Italians had notably little taste for it.[28] Joined up with the Galenic vocabulary of qualities, analogical reasoning testified even more powerfully to the effects of beef eating on the bodies and minds of those who ate it. Whatever characteristics were imputed to cattle—and specifically to their varieties (oxen, bulls, calves, cows)—were available to understand the consequences of eating their flesh.[29]

Beef *defined* Englishness, and its generous and unadorned presence on the English table was routinely contrasted with elaborate and effete foreign fare. Englishmen reflected on and often celebrated that carnal diet: a Restoration survey of "the present state of England" noted that "the English are great *Flesh-eaters*," this despite the abundance of fish in the surrounding seas. Great feasts of beef were then not so common as in the past, when the brother of Henry III had 13,000 dishes of meat at his marriage feast and Richard II at Christmastime "spent daily 26 oxen," together with 100 sheep and the appropriate trimmings.[30] Foreigners remarked on not just the English obsession with beef but also their disdain for more complex Continental cuisine. An Italian diplomat in the late sixteenth century recalled "a speech of Sir *Roger Williams* to an idle *Spaniard*, boasting of his country citrons, orenges, olives, and such like: Why (saith he) in *England* wee have good surloines of beefe, and daintie capons to eat with your sauce, with all meat worthy the name of sustenance; but you have sauce and no sustenance."[31] That dietary nationalism resonated into the eighteenth and nineteenth centuries. Robert Burns—no Englishman, of course, and writing in praise of the great Scots haggis—scorned artificial French concoctions, then increasing in popularity throughout Great Britain: "Is there that owre his French ragout, / Or olio that wad staw a sow, / Or fricassee wad mak her spew / Wi' perfect sconner, / Looks down wi' sneering, scornfu' view / On sic a dinner?"[32] And Benjamin Franklin, writing in defense of colonial corn eating, noted how the English celebration of roast beef figured in bullish patriotism. The Englishman, he wrote, "shews in nothing more his great veneration for good eating, and how much he is always thinking of his belly, than in his making it the constant topic of his contempt for other nations, that *they do not eat so well as himself*. The *roast beef of Old England* he is always exulting in, as if no other country had beef to roast; reproaching, on every occasion, the Welsh with their leeks, the Irish with their potatoes, and the Scotch with their oatmeal."[33] In early nineteenth-century England, a prominent diet doctor—ironically named Paris—joined a chorus of complaint that natural appetites had been destroyed by the artificial contrivances of fancy French chefs. Headaches, flatulence, hypochondriasis, and many other illness "have arisen from the too prevailing fashion of loading our tables with that host of French *entremets*, and *hors-dœuvres*, which have so unfortunately usurped the roast beef of old England."[34] Beef eating was also physiologically efficient: an

Englishman could eat one pound of beef, while the Dutchman—to obtain the same nourishment—would have to take in two pounds of cabbages or turnips and yet not be "near so active as the English Man."[35] There are indications that in England beef was more associated with country than city modes of life. In an early seventeenth-century dialogue between a "North-Country-Man" and a "Citizen," the northerner said that city fare was typified by stuffed capons, while "our Beefe and Bacon feeds us strong in the Countrey," and it was understood that country people and those who did hard labor had stronger stomachs and were better able to digest beef and other "gross" meats.[36]

How much beef did Englishmen actually eat in the early modern period? Historians disagree on matters of quantity but not on the cultural significance attached to beef eating, certainly by the early eighteenth century.[37] At least at that point, beef seems to have been normal fare for Englishmen—for those who could afford uncured "butcher's meat," it seems to have been the preferred form—so one account of why Englishmen ate beef was just that this is what English people *did*. The well-off apparently did so in great quantity, and the lower orders ate beef when means allowed, though all forms of meat eating were limited by expense.[38] Foreign visitors often found the English attachment to beef noteworthy. An Italian merchant visiting London in the 1560s was astounded: "It is almost impossible to believe that they could eat so much meat," noting that the English preferred roast beef to veal but that all kinds of roast meat were abundantly available at London inns.[39] Touring England in 1698, the Frenchman Henri Misson wrote that "it is a common Practice, even among People of good Substance, to have a huge Piece of Roast-Beef on *Sundays*, of which they stuff till they can swallow no more, and eat the rest cold, without any other Victuals, the other six Days of the Week."[40] The Finnish naturalist Peter Kalm, visiting England in the mid-eighteenth century, noted how "the English nation differed somewhat particularly from others in this; that butcher's meat formed with them the greater part of the meal, and the principal dishes. . . . I do not believe that any Englishman, who is his own master, has ever eaten a dinner without meat."[41] The normal gave you the normative, and in this case it had long done so. In the 1580s, a dietary writer noted that "biefe of all flesh is most usuall among English men"; it is "plentifull . . . throughout this land,"[42] and its presence on the plate defined hospitality and generosity; it was, so to speak, the fatted calf. Beef was good for you, and beef was good to offer to guests.

Samuel Pepys's Restoration *Diary* recorded his consumption of a fair amount of beef, more as his wealth increased through the 1660s, but it is clear that this remained a food for special occasions and especially to be remarked on: "a fine piece of rost beef," "a most brave chine of beef," "some good ribbs of beef roasted," "a fine collation of collar of beef," and "a rare piece of roast beef."[43] Into the early nineteenth century and afterward, roast beef was celebrated as "the pride and glory of this happy island" and as the cause of the people's "manly" character.[44] Cattle flourished in England, and the English were proud of the abundance of the beasts and the quality of their meat: "The cattell which we breed are commonlie such, as for greatnesse of bone, sweetnesse of flesh, and other benefits to be reaped by the same, give place unto none other."[45] The country specialized in raising cattle suited for the table, and as England suited the nature of the beast, eating its flesh suited the nature of Englishmen: "how well [beef] doth agree with the nature of English men, the common consent of all our nation doth sufficiently prove. Yea that it bringeth more strong nourishment than other meates, may plainly be perceived, by the difference of strength in those that commonly feede of biefe, and them that are fedde with other fine meates."[46] True, people living in hot climates tended to eat less meat, and Englishmen dwelling in such places were well advised to follow local examples, but that was because, among other reasons, "*Flesh* in hot countries is nothing so firm, good and wholsom as in cold."[47]

Beef eating agreed with English natures and was understood to help *make* English natures. Because custom is second nature and because habitual transactions with the environment could remake your constitution, routine eating of beef transmitted into English human natures the natures of the beasts themselves. And here it was sometimes thought important that Englishmen should eat the flesh of *English* cattle: no imported beasts, as their qualities would be different and therefore less suitable.[48] (Physiological qualities merged into assessment of commercial quality, and that is one reason why the characteristics of beef from other lands were so closely monitored.) Cattle—think now of oxen and bulls rather than cows and heifers—were strong and obstinate, "stubborn as an ox." (The flesh of bulls was indeed eaten, and its effects were recorded by proverb collectors—"He lookes so big as if he had eaten bull-beefe"—while some early seventeenth-century physicians evidently believed that bull blood was poisonous.[49] You had to be a rough and vulgar sort of person to eat that sort of meat: "Bulls beefe is of a ranke and unpleasant taste, of a thick grosse and corrupt juyce, and of a very hard digestion. I commend it to poore hard labourers, and to them that desire to looke big, and to live basely.")[50] Beef eaters were "strong, vigorous, and hale," and those who advised against consuming beef might be thinking either of meat from very old animals or of eaters who were lacking in strength and vigor.[51]

Beef, Blood, and Bloody-Mindedness

In *Henry IV, Part 1* (Act 2, scene 2), Poins asks after "the martlemas," that is, John Falstaff; in *Part 2* (Act 3, scene 3), Prince Hal addresses Falstaff as "my sweet beef"; and in Act 2, scene 4, he is "that roasted Manningtree ox with the pudding in his belly."[52] These expressions gesture, by way of beefiness, at the abundance of Falstaff's flesh as well as the stolidity, valor, and carnality of his character. The feast of Martlemas (or Martinmas), on November 11, was the traditional autumnal time for the slaughter of beasts, when they were at their fattest and their meat was at its sweetest and before their flesh was preserved by salting or smoking, hence the common expression "Martlemas beef." Falstaff was a great eater of beef, and it was a diet understood to appeal to those who delighted in their food and to help make them brave.[53] It was common for traditional dietary texts to identify rich food and strong drink as causes of licentiousness.[54] The caustic *The Truth of Our Times* by the English practical ethical writer Henry Peacham told the story of a mother overjoyed to see her foppish and spoiled son turn to eating beef, "which she protested hee never did before in his life, and now she verily believed hee would prove a souldier."[55] Just as it was thought that the routine consumption of hares' flesh would make men fearful, beef eating rendered the English "couragious and undaunted in perills."[56] In Shakespeare's *Henry V* (Act 3, scene 7), on the eve of Agincourt the French forces were talking about the state of mind and likely qualities of their foe. The discussion settled on diet. If the English had any sense they would run away, so puny were their forces. True, England "breeds very valiant creatures; their mastiffs are of unmatchable courage." Yes, indeed, a French general agreed, brave but too stupid to know when they're overmatched, "leaving their wits with their wives." They take "great meals of beef and iron and steel," which they eat like wolves and which makes them "fight like devils." But there was no reason to worry: "these English are shrewdly out of beef," and that means that "to-morrow they will have only stomachs to eat and none to fight." An army, as Napoleon said, marches on its stomach, and stomachs in Henry's army were not just empty but also empty of their fighting fare.[57] Daniel Defoe's celebration of the *true-born Englishman* agreed that he was beefily bold beyond other races:

> Eager to fight, and lavish of their Blood;
> And equally of *Fear and Forecast* void.
> . . . .
> The Climate makes them Terrible and Bold;
> And *English* Beef their Courage does uphold.[58]

In a chivalric culture, it was good to be brave, fierce, stolid, and appropriately violent. But causal idioms relating food to human nature could also be used to oppose chivalric virtues and to suggest a dietary remedy. In the late seventeenth century, the pacifist Thomas Tryon reckoned meat eating, and specifically the eating of bloody meat, to be a cause of a violent society. It wasn't here, as some modern vegetarian writers have thought, that killing animals or eating animals slaughtered on your behalf made you insensitive to violence, though that might indeed happen. Rather, it was that the *qualities* of the beast were carried in its blood and were transferred to human beings through bestial blood consumed. "Blood . . . doth not only contain the Spirits, but the very Humour, Dispositions and Inclinations of the Creature," and so, by eating bloody meat you become bloody-minded. Different forms of aliment had the power, Tryon said, of awakening their "similes" in the human body. Fruits and vegetables are "of a clean Simple Nature and Opperation, which being well prepared and temperately eaten, have onely power to waken their Similes in the Body and Senses." By contrast, animals—some sorts more than others—"are endued with all kind of Beastial Passions, as Anger, Revenge, Covetousness, Love and Hate, which dispositions and Passions of the Flesh, but especially the Blood, doth retain after such Animals are Killed."[59] The dispositions of all animals were "Bestial and revengeful," and, "after Deaths painful stroke" these passions come to "center in the Blood and Spirits," where they may be transferred to the people who eat them.[60] That brutalizing effect might obtain even if one was only exposed to the material effluvia from abattoirs, from which places "fiery wrathful Spirits do evaporate themselves into the Air, being continually breathed into the [bodies]" of people congregating near such places. This was why butchers were "more fierce and cruel, sooner moved to wrath than others."[61] During the English Civil Wars, it was observed "that one Foot-Regiment of Butchers, behaved themselves more stoutly than any other," though they had never before killed anything but beasts.[62] Butchers were "surly," "merciless," and "cruel": the moral effects of their bloody way of life made them that way, but so too did the physical effects of the blood of the beasts they slaughtered.[63]

This was a causal inference that persisted well into the nineteenth century and even when the powers of foods were being attributed to a range of newly identified chemical substances. In the 1820s, the German chemist Frederick Accum discussed radically different national food preferences, finding it a "striking fact, that the English soldiers and sailors surpass all those of other nations in bravery and hardihood," and attributed this to "the effect of a considerable proportion of animal food." The Irish were potato eaters and therefore "more indolent and sluggish" than English beef eaters, but when they were recruited into the British army they became as valiant as the English, the change owing solely to a new diet.[64] Lecturing to a public audience at London's Science Museum in the early 1860s, the physician Edwin Lankester represented as a well-known matter of fact that "those races who have partaken of animal food are the most vigorous, the most moral, and the most intellectual races of mankind." In a rare outbreak of philosemitism, he said that the ancient Jews—though their prohibitions of pork and animal blood were well known—"partook largely of meat, and were amongst the most vigorous people of their day."[65] Sometime later, the German chemist Justus von Liebig, the greatest nineteenth-century propagandist for the nutritional value of protein, wrote that meat eating had marked effects on the nervous system: "It is essentially their food which makes carnivorous animals, in general, bolder and more combative than the herbivorous animals, which are their prey."[66] In the 1880s, an English physician observed that "the conquering Anglo-Saxon,—the master and too often the exterminator of aborigines whose lands he covets—is a meat-eating man *par excellence*." A carnivorous diet and tendencies toward bravery and domination were, the doctor said, "linked in some subtile manner with [Anglo-Saxon] masterfulness," though he offered no fleshed-out scientific account of why that should be.[67] Around the same time, Friedrich Nietzsche, writing to his mother from Venice, observed that a vegetarian diet induced "irritation and gloom" among German intellectuals attracted to it: "Just look at the 'meat-eating' English," Nietzsche said, "till now, that was the race which knew best how to found colonies. *Phlegm* and roast beef—that was till now the recipe for such 'undertakings.'"[68] And yet "even the English diet, which in comparison with German, and indeed with French alimentation, seems to me to constitute a 'return to Nature,'—that is to say, to cannibalism, is profoundly opposed to my own instincts. It seems to me to give the intellect heavy feet, in fact, Englishwomen's feet. . . . The best cooking is that of Piedmont."[69]

Criticism of national characteristics as well as their celebration also proceeded by way of diet, and in the late seventeenth century Tryon's views inverted those of people who celebrated the beef-eating heroes of Agincourt. There were Englishmen who condemned English partiality for not only beef eating but also taking their meat blood red. Defoe was embarrassed by that national preference. "When Strangers see us feeding thus," he wrote, they understandably take us "to be, if not Canibals, yet a sort of People that have a canine Appetite." Civilized foreigners say that we "*Devour* our Meat, but do not Eat it; *viz. Devour* it as the Beasts of Prey do their Meat with the Blood running between our Teeth."[70] Bloody meat was causally connected to bloody manners. That sensibility also survived into the Enlightenment, with the French agreeing that it was bloody beef eating that made the English what they were, that is, "rosbifs." It was Rousseau's view that this bloody diet made people fierce and cruel—"the English are noted for their cruelty"[71]—and a similar opinion was expressed in La Mettrie's *L'homme machine* (1747): "Raw meat gives a fierceness to animals; and man would become fierce by the same nourishment. This is so true, that the *English*, who eat not their meat so well roasted and boiled as we, but red and bloody, seem to partake of this fierceness more or less, which arises in part from such food."[72] Even the Scots, who reckoned that they were also pretty fierce, agreed. This, for example, was the opinion of the popular medical writer William Buchan, discerning a causal link between the violence of the English nature and the beefiness of their diet. No nation, he wrote, eats more animal flesh than the English: "There is no doubt," Buchan continued, "but [that flesh eating] induces a ferocity of temper unknown to men whose food is chiefly taken from the vegetable kingdom."[73]

English beef eating was also implicated in causal stories about characteristic English pathologies. Beef was one of those foods that made notoriously "ill juise" and "grosse bloude," engendering melancholy.[74] (Veal was widely judged innocuous or wholesome.)[75] This sort of thing was common knowledge, circulating in both lay and expert cultures. The Salernitan Verses listed beef prominently among the foods "that breed ill bloud, and Melancholy, / If sicke you be, to feed on them were folly."[76] Henry Fielding's 1731 patriotic ballad "The Roast Beef of Old England" celebrated the mental benefits of beef: "When mighty Roast Beef was the Englishman's food, / It ennobled our brains and enriched our blood," but there were serious doubts about the matter among the doctors. The famous English Malady, described by the Scots physician George Cheyne in the early eighteenth century, was a mental and moral condition said to be partly caused by the poor digestibility of red meat.[77] The city-dwelling and sedentary classes were advised to avoid beef, and it was understood that country people and those who did physical labor had stronger stomachs, better able handle it.[78] Beef eating, especially eating the flesh of older beasts, might cause melancholy through its Galenic coldness or might do so through difficulty in digestion. These causal stories inform another Shakespearean digression on food and human character. This is in *Twelfth Night* (Act 1, scene 3) when Sir Andrew Aguecheek explains to Sir Toby Belch that "I have no more wit than a Christian or an ordinary man has: but I am a great eater of beef and I believe that does harm to my wit."[79] This too was evidently a contemporary commonplace: a 1601 English tract on the passions noted that "Superfluity of meat, causeth dulnesse of minde . . . blockishnesse, and dulnesse of wit. . . . A fat belly engendereth not a subtile wit."[80] Into the late eighteenth century and beyond, popular medical writers warned that a diet heavily based on animal food held special risks for scholars. Buchan cautioned that the sedentary and the studious ought to observe a mainly vegetable diet, for taking much meat induces "a putrid diathesis" in the system, and this in turn "renders men dull, and unfit for the pursuits of science."[81]

Knowing and Not Eating

The idea that beef eating (or, more generally, eating animal flesh) made you stupid was present across cultures and over time. In Diogenes Laërtius's (third century BCE) life of Diogenes the Cynic, the philosopher had suggested that athletes weren't very sharp, and, "being asked why the Gamesters were men of no Sense, he said, Because they were built up of Beef and Bacon."[82] Centuries later, Plutarch wrote that meat eating "by clogging and cloying [our bodies], [renders our] very minds and intellects gross. For it is well known to most, that wine and much flesh-eating make the body indeed strong and lusty, but the mind weak and feeble."[83] In an exchange of insults in Dryden's Restoration play *Troilus and Cressida*, Thersites calls Ajax "beef-witted."[84] That is, the beef eater is precisely John Bull, described by Washington Irving as "a plain, downright, matter-of-fact fellow, with much less of poetry about him than rich prose. There is little of romance in his nature, but a vast deal of a strong natural feeling. He excels in humor [where humor tended to mean quirkiness] more than in wit; . . . melancholy rather than morose."[85]

If melancholy was considered a temperament characteristic of the English (and sometimes of northern races in general), it was even more strongly linked to philosophers, scholars, and all those living the life of the mind. The philosopher dealt with the eternal, the transcendent, and the abstract and lived an abstracted life, abstracted from society and its civic expectations, the mind abstracted from the body, its appetites and demands.[86] Melancholy was understood at once as an excess of black bile and as a consequent complex of moodiness, anxiety, hypochondriasis, depression, sighing, anxiety, irritability, sleeplessness, bad dreams, groundless fears and sorrows, an unhealthy preference for solitude, eccentric behavior, and suicidal thoughts. It was said to be "a Disease more terrible than Death" and was regarded as the characteristic occupational disease of philosophers and "hard students," the scholar's "inseparable companion."[87]

Melancholy, withdrawal, fragile bodies, and dietary asceticism were a durable cultural ensemble, made compellingly visible, for example, in the great Dürer engraving *Melencolia* (see figure 3.1). The display of a melancholic temperament, the embrace of an ascetic way of life, and withdrawal from society were meaningful gestures, whether this manner of living belonged to pagan or Christian, sacred or secular, cultures. Ascetic melancholia was recognized as a condition for seeking abstracted knowledge, and those who wanted to be seen as living the life of the mind could draw on ascetic repertoires as a way of presenting themselves as authentic truth seekers. In antiquity, Aristotle asked why men of genius tended toward melancholy, and Seneca asked why God afflicted the wisest men with ill health.[88] Stoic philosophers, content with water and plain bread, able to miss their dinner without complaint or even without noticing, were celebrated for the simplicity of their diet.[89] Pythagoras and his followers were famous for their abstemiousness. Legend had it that they routinely performed "an exercise of temperance": "There being prepared and set before them all sorts of delicate food, they looked upon it a good while, and after that their appetites were fully provoked by the sight thereof, they commanded it to be taken off, and given to the servants."[90] The Greek seeker after truth was recurrently said to eat only enough to keep life going. To eat more than a bare minimum or to yearn after delicacies was to compromise the philosopher's ideal self-sufficiency. The maxim was "eat to live; do not live to eat." The condition for truth was an austere dietetics.

After Jesus wandered in the desert for forty days and nights, "he hungered" (Matthew 4:1–2; Luke 4:1–2). Satan's first temptation was not power but food: "If thou art the Son of God, command that these stones become bread. But he answered and said, It is written [quoting Deuteronomy 8:3] Man shall not live by bread alone" (Matthew 4:3–4; Luke 4:3–4); "Is not the life more than the food and the body more than the raiment?" (Matthew 6:25). When the disciples wondered that the rabbi did not eat, "he said unto them, I have meat to eat that ye know not" (John 4:30–32). For the faithful, Jesus himself was "the bread of life: he that cometh to me shall not hunger, and he that believeth on me shall never thirst" (John 6:35). Paul lectured the Corinthians: "Meats for the belly, and the belly for meats; but God shall bring to nought both it and them" (I Corinthians 6: 13).[91] In the early Christian period, ascetic desert hermits performed spectacular feats as hunger artists, acting out more and more theatrical displays of dietary self-denial in the search for spiritual perfection. For Saint Anthony, as historian Peter Brown has written, "the most bitter struggle of the desert ascetic was presented not so much as a struggle with his sexuality as with his belly." In early Christianity, it was food, not sexuality, that was the stage for ascetic self-denial.[92] The Desert Fathers regarded eating as a matter of both shame and spiritual danger: "The body prospers in the measure in which the soul is weakened and the soul prospers in the measure in which the body is weakened."[93] Another legend tells of a friend, concerned for the health of the hermit Abba Macarius, bringing him a bunch of grapes. Macarius was unwilling to indulge himself and sent them to another hermit, who then passed them on to still another until at last they came back to Macarius, uneaten.[94] Here the religious life of the mind appears not just disembodied but also specifically disemboweled.

The ascetics of late antiquity tended to conceive of the human body as an "autarkic" system. In ideal conditions and, tellingly, before Adam's original sin—it was food, after all, that brought him down—the body was thought capable of running "on its own heat" and needed just enough food to maintain that heat. It was only "the twisted will of fallen men" that gorged the body with surplus food, and it was this dietary surfeit that produced the excess energy manifested in "physical appetite, in anger, and in the sexual urge." The passions, including that of sexuality, were in part epiphenomena of dietetics: food before sex. Brown writes that "in reducing the intake to which he had become accustomed, the ascetic slowly remade his body. He turned it into an exactly calibrated instrument. Its drastic physical changes, after years of ascetic discipline, registered with satisfying precision the essential, preliminary stages of the long return of the human person, body and soul together, to an original, natural and uncorrupted state."[95] Through the late Middle Ages and beyond, asceticism remained close to the center of ideas about the authenticity of knowledge, both sacred and secular, about the texture of the life of the mind and about the respective virtues and vices of the contest between the active and the contemplative lives.[96] Was pathological melancholy an occupational disease necessarily associated with the ascetic and disengaged life of the mind? Was the presentation of melancholy how you recognized a "hard student"? And if scholarly melancholy was in need of treatment, what were its dietary causes and—if any existed—its dietetic remedies?

Eating and Not Knowing

People living the life of the mind tended toward melancholy, but how was it thought that this way of life produced pathology? And how, if at all, might the disorder be managed if not cured? Traditional dietetics addressed both questions. Your natural constitution was, of course, one of the causes of melancholy. A melancholic temperament disposed you to melancholic disorders, though not all constitutional melancholics wound up "infected with this miserable passion."[97] But the causes of melancholy also took in management of all the nonnaturals. Scholars did not get out enough, took insufficient exercise, remained studiously awake when they ought to be sleeping, and lived and worked in close and confined quarters where the air might be stale or putrid. Their abstruse studies rarely evoked pleasant emotions, their general poverty bred disillusionment, and they lived a solitary life, which (as Burton wrote) was both "cause and symptom" of the disorder.[98] And although this was usually treated with circumspection, the celibate way of life associated with the scholarly role blocked sexual release. Melancholy might be bred through "immoderate or no use at all of Venus," with some writers thinking that the absence of "carnal copulation" and the retention of the "natural seed" overlong "sends up poisoned vapours to the brain and heart," making the celibate "heavy and dull."[99] (Dietetic texts were overwhelmingly concerned with males, but melancholia was also recognized among women, where it was sometimes associated with a "cold distemper" of the "matrix," or womb.)[100] Although there were bodily and mental consequences of the celibate way of life, celibacy was an institutional requirement for many scholars. Their diet should therefore take sexual circumstances into account. "A scholer must be of a spare and moderate diet," it was said by an early seventeenth-century English encyclopedist, "for how ever *Venus* cannot florish without the helpe of *Ceres* and *Bacchus* [food and wine], yet will Minerva and the Muses live gloriously by the pleasant waters of *Hellicon*."[101] Again, rich and abundant food was understood to feed the sexual fires: the scholar had to control the urgings of his stomach to control the urgings of his sexuality.

Melancholics also put themselves at risk by eating the wrong sorts of foods.[102] As melancholy was an excess of cold and dry black bile, whatever foods increased the qualities of that humor were hazardous for the constitutionally melancholic. Meats that were cold and dry were generally warned against, while the warm and moist were commended. There were long lists of foods that melancholics should avoid: beef was notably bad (though proper preparation might mitigate the peril), but so too were hare, mutton, goat, boar, venison, fish from standing waters and large sea fish, many water fowl, and most offal. Among vegetables making melancholy blood, there were cabbages, beans, and garlic, while brown bread, hard cheese, and thick or darkly colored wines were also dangerous.[103] Modes of preparation might be crucial: an early seventeenth-century dietary, for example, commended soft-boiled eggs while cautioning that fried eggs send up "bad vapours to the brain and heart."[104] Here as elsewhere, analogical reasoning inferred from the sensed qualities of the foods to their likely effects on the humors, the body and the mind. "Melancholicke juyce," it was said, "is nothing else then the dryer and thicker part of blood, altogether like unto Dregs and Lees that settleth in the bottome of the Vessell."[105] So, crude, dense, and dark foods elaborated their qualities in the body, making bad "juice" and its pathological consequences. Dark foods produced black bile, hares were "mad" ("as a March hare"), eel and other slimy fish generated slimy juice, and beef, in its virtuous mode, enhanced the stolid and chivalric virtues that stood in contrast with those belonging to the contemplative and studious life.

All meats "as are hard of digestion breed melancholy," Burton wrote.[106] The intermediate causes between food and disorder were the fumes and vapors produced by difficult and incomplete concoction: "Indigestions . . . are the *Principal* and *Adequate* Cause of *Vapours*," wrote an early eighteenth-century physician.[107] Crude foods generated "crudities" in the stomach, a "vicious concoction of thinges received, they not being wholly or perfectly altered."[108] Stomach gas was a bodily annoyance and a social inconvenience, but it was considered to have specific mental morbidities as well: effects on the retention of thoughts, on dreaming, emotion, and judgment.[109] The wrong sorts of foods impaired the memory: the poet William Basse wrote that "crude and grosse Flesh, unripe fruits, greene herbes, and all other things, cold by nature or vaporous, . . . send up grosse humors into the braine," making us forget what we ought to remember.[110] Many difficult-to-digest foods—vegetables such as leeks, beans, and cucumbers; dairy products such as hard cheese; and the flesh of red or fallow deer—were condemned in the dietary literature for their tendency to cause turbulent sleep.[111] Treating the causes and circumstances of habitual trains of thought, the Scottish Common Sense philosopher Thomas Reid implicated diet: "Crudities and indigestion are said to give uneasy dreams, and probably have a like effect upon the waking thoughts."[112] Cabbage, in Galenic terms, was cold and dry; it was "hard of concoction, affords little nourishment, and that thick and melancholy, from whence fuliginous [smoky] vapours fly into the head and produce turbulent sleep."[113] The vapors were also a cause of ill-considered and inconsequential speech, hence the notion of "vaporing," and of ill-considered political and religious judgment, so forming a link in a causal chain connecting corrupt ideology to the corruptions of bad diet and bad digestion. Unsteadiness in controlling the limbs—"staggers"—and unsteadiness in right religious conviction—"inconstancie"—both tracked back to the stomach, whose digestive troubles afflicted the brain with "vapours, grosse and tough humours, or windy exhalations, either lodging in the braine, or sent thither from the stomack, turning about the animall spirits: hence the braine staggers with giddinesse. This *spirituall* Inconstancie ariseth from like causes. If it be in religion, it proceeds from cloudy imaginations, fancies, fictions, and forced dreames, which keepe the mind from a sober and peacefull consideratenesse. Multitude of opinions, like foggy vapours, mist the intellectuall faculty, and like reverberated blastes whirle about the spirits."[114]

Dietary choice in low things had consequences for such high things as the cultural and political orders, so connecting cabbages and kingship. Scholarly melancholy—with its fearfulness, depression, and uncontrollable trains of thought—was at the end of a causal series that started with bad diet and proceeded through bad digestion to bad dreams and bad thoughts.

The medical and moral categories engaged with this condition were roughly—not tightly—molded to the contours of gender. The conditions called "the vapors," "hysteria," and "hysterick fits" were generally attached to women, while men were said to suffer from a range of related disorders variously known as "hypochrondriasis," "lowness of spirits," "the English Malady," and "the spleen," the organ though which some physicians thought the vapors passed or where they were elaborated (as *Flatus Hypochondriacus*). There were, however, eighteenth-century instances of men afflicted with "the vapors" and of women tormented by "the spleen."[115] Suffering from the vapors was serious, nasty in its consequences and proliferating in its bodily and mental manifestations. One early eighteenth-century physician announced that it was "a Disease which more generally afflicts Humane Kind, than any other whatsoever; and *Proteus*-like, transforms it self into the shape and representation of almost all Distempers."[116]

The spleen (or the vapors) was both a cause of ailments and itself a disease—even a grievous one in need of heroic treatment—but it might also be taken as a mark of refinement and sensitivity, and there was an eighteenth-century epidemic of the condition among the fashionable and the literati. (Historians have emphasized the significance of *nervous sensitivity* as a recognized cause of both disease and intellectual refinement, but digestion figured largely too.)[117] As the "*English* Malady," it might also serve as an ambivalent advertisement for the advanced state of a national culture and way of life.[118] Descriptions of the condition were unstable. In the 1730s, the fashionable Scottish physician George Cheyne noted that "the *Spleen* or *Vapours* . . . is of so loose a Signification," ranging from farting to grief, from swelling in the stomach to fidgeting, "that it is a common Subterfuge for meer Ignorance of the Nature of Distempers." Cheyne's preferred designation of the "English Malady" was no less protean, and what he called "the *nervous* or Flying *Gout*" was a condition that overlapped with both the spleen and vapors.[119]

This story also implicated analogical reasoning. Food that was hard to digest became putrid, breeding crudities in the stomach. These in turn gave rise to vapors and fumes, where the category of vapors bore a family resemblance to those malodorous airs and "miasmas" arising from swamps, foul water, rotting organic matter, coal fires, sewers, slaughterhouses of crowded cities, and those sulfurous exhalations from the bowels of Earth that were associated with Hell. (That is one reason why vapors—like their parallel, positively valued category of spirits—appeared so often in a theological context. The vapors emanating from both the gut and the subterranean realm both reeked of sulfur—the stench of evil.) The trajectories and consequences of stomach gas might also vary. Belching and farting were, of course, audible signs of internal gas, so sufferers had direct experience of indigestion and its consequent vapors. This gas had bad consequences in the body, and that was a reason for recommending its expulsion or at least for warning against too great an effort at keeping it in. The Salernitan Verses, recall, recommended that you not strive too hard to keep gas in; one of Sir Thomas More's verse epigrams warned that "wind, if you keep it too long in the stomach, kills you; on the other hand, it can save your life if it is properly let out," and Montaigne ruefully reflected on his inability to let his wind out whenever he wanted: "And would to God I knew it but by Histories, how that many times our belly, being restrained thereof, brings us even to the gates of a pining and languishing death: And that the Emperour, who gave us free leave to vent at all times, and every where, had also given us the power to doe it."[120]

There were medical reasons to let your wind go free. If you didn't, the vapors might take a different and dangerous path within the body. Just as steams and exhalations ascended from stagnant marshy waters, infecting the pure air with their putrid stinks, fumes elaborated from poor digestion were also said to ascend from the stomach to the spleen and then on to the brain. (Descartes thought that sleeping on your left side exerted pressure on the spleen, allowing its vapors to rise to the brain and causing insomnia, so the preventive position was evident.)[121] Reaching the brain, the vapors caused a range of bodily ills: headache, eye troubles, catarrh, drooling, colds, agues, scurvy, vertigo, and even epilepsy.[122] The Salernitan Verses noted that one of the reliable marks of melancholy was gas: "bitter belches from the stomacke coming."[123] The vapors "bemist and darken" so that their pathological effects included the mental and emotional.[124] "Those vapours and fumes, which cloud and overshadow the clearnesse of the Brain," Lessius wrote in the early seventeenth century, "*are chiefly caused by the meat taken down into the stomack.*"[125] The gloomy and cloudy thoughts that marked the melancholic state were the direct result of dark vapors and might be the indirect result of poor dietary choices.

There were dietary causes of the vapors, and there were dietary remedies. In the seventeenth century, novel beverages such as coffee were both celebrated and warned against for their physiological effects. Those who commended coffee claimed that it "dries up Crudities in the Stomach" and "supresses Vapours," and a range of other medicinals and aromatics were also said to "dissipate the Cloudy Vapours disturbing the Brain."[126] A tincture of saffron clears "the Cloudy Vapours."[127] Lettuce repressed the vapors,[128] and so too did the ague-specific cinchona (or Jesuit's bark); extracts of aloes, anise, and fennel seeds; a concoction of mustard seeds, horseradish, and herbs; and certain sorts of mineral water.[129] Opium and proprietary pills were taken for the vapors, and, as always, blood might be let.[130] An abstemious diet suppressed vapors while also reducing the mental damage that the vapors caused. A mid-seventeenth-century English divine wrote that "the fewer vapours the belly sends to the brains, besides what are necessary, the clearer is the skie in that upper region: the best rule therefore for such as feast plentifully, is to fast frequently."[131] There were few conditions that did not benefit from dietary moderation, and the vapors was no exception. Opinion was split on whether the effect of wine on the vapors was good or bad, and some maintained that white wine was less vapor-inducing than red and that ale was worse than any wine.[132] Many ascribed drunkenness and its mental effects to those vapors—the spirituous nature of wine and other inebriating drinks being an index of their vapor-inducing character—while others celebrated the positive effects of certain sorts of wine. Recall that one of Falstaff's recommendations of sherry was its remedial effect on the vapors: "A good sherris-sack hath a two-fold operation in it. It ascends me into the brain; dries me there the foolish and dull and crudy vapours which environ it; apprehensive, quick, forgetive, full of nimble, fiery, and delectable shapes; which delivered o'er to the voice, the which is the birth, becomes excellent wit" (*Henry IV, Part 2*, Act 4, scene 3).

The Scholarly Stomach

The culture describing the relations between food, vapors, and disease took specific forms when dealing with the role of the *scholar* and of others living the contemplative life of the mind. Students and scholars, it was persistently said, had weak stomachs. They digested their food poorly, and indigestion in turn produced the dark and cloudy vapors that made for scholarly melancholy. But *why* did they have weak stomachs and poor digestion? One answer figured in a tract on the diseases of the learned, and scholars were among many groups treated in the first systematic account of what became known as "occupational diseases." In *De Morbis Artificum Diatriba* (*Diseases of Workers*), written in 1700 and widely translated, the Italian physician Bernardino Ramazzini (1633–1714) added his views to centuries of accumulated commentary on what the learned were like in body and mind. The chapter on "the diseases of learned men" was the last in a text dealing with, among others, miners, bakers, wrestlers, masons, and (oddly) Jews. Ramazzini claimed that one major cause of scholarly diseases was fundamental to the sedentary life: scholars and other students just spent too much time *sitting*. This prolonged posture mechanically compressed the stomach, interfering with proper concoction and initiating the causal chain that proceeded from "crudities" to "vapors" to constipation and on to the protean marks of melancholy.[133]

The vocabulary of foodstuffs, qualities, and humors was used to conceive of the body as a *chemical* system: foods undergoing alteration of state as they were (successfully or incompletely) concocted. Taken into the body, foods experienced the sorts of transformations with which medieval and early modern chemists were familiar, transferring their qualities from food to the eater's body and mind. But the vocabulary of *spirits*—those material but very fine entities that powered the body's functions—also understood the human body as an *energetic* system, a distribution of powers animating various functions. And Ramazzini was one in a long line of writers accounting for learned indigestion and their other characteristic pathologies by way of the spirits and what at various times they were called upon to do. "There's no hard Student but what complains of his Stomach," Ramazzini wrote, repeating a commonplace that went back through the Renaissance to antiquity.[134] The neo-Platonist philosopher Marsilio Ficino wrote that learned men suffered from both melancholy and weak stomachs, explaining this conjunction through body's spiritual economy. Digestion was powered by the natural spirits, and while digestive work continued, little of the spirits' power was available for any other bodily function.[135] In the late sixteenth century, an English physician wrote that whoever "shall begin long and difficult contemplation, shall of force draw the spirits from the stomake to the head, and so leave the stomack destitute: whereby the head shall be filled with vapors, and the meat in the stomacke for want of heate, shall be undigested or corrupted."[136] This was one reason for the proverbial recommendation that after dinner—the day's main meal—you should "rest a while."[137] In 1614, Lessius recognized that "the exercises & employments of the minde do very much hinder and disturb the concoction," and they did this because mental activity called up the "whole force" of the body's spirits, making them unavailable for other purposes. Deep thinking and meditation "withdraw not only the animall, but the vitall and natural spirits themselves from their proper services." Scholars need less food than those who work with the body, and they *should* have less food because their bodies' powers are unavailable to process it.[138]

Any sort of food that occasioned vapors and fumes might have an indirect effect on the spirits, and that is how a late seventeenth-century English physician explained the mental "dullness which is so common during the digestion of Victuals; for seeing that cannot be performed without the elevation of Vapours into the Brain, which entangle the Spirits, and make a kind of Obstruction in the Orifices of the Nerves."[139] You need the spirits' energy to power digestion, and you need the spirits to fuel serious mental activity, and these two functions could not be effectively performed at once. You could not think deep thoughts and feed at the same time; if you attempted to do so, you would suffer pathological consequences:

> Nature in contemplation is directed wholly to the brain and heart and deserts the stomach and liver. For this reason foods, especially the more fatty or harsh foods, are poorly digested, and as a result the blood is rendered cold, thick, and black. Finally, with too little exercise, superfluities are not carried off and the thick, dense, clinging, dusky vapors do not exhale. All these things characteristically make the spirit melancholy and the soul sad and fearful. . . . But of all learned people, those especially are oppressed by the black bile, who, being sedulously devoted to the study of philosophy, recall their mind from the body and corporeal things and apply it to incorporeal things. The cause is, first, that the more difficult the work, the greater concentration of mind it requires; and second, that the more they apply their mind to incorporeal truth, the more they are compelled to disjoin it from the body. Hence their body is rendered as if it were half-alive and often melancholic.[140]

Ficino's opinion was a medical and cultural commonplace. In the early seventeenth century, Burton's *Anatomy of Melancholy* agreed that scholarly contemplation "dries the brain and extinguisheth natural heat; for whilst the spirits are intent to meditation above in the head, the stomach and liver are left destitute, and thence come black blood and crudities by defect of concoction."[141]

Gluttony made the problem of powering both stomach and mind much worse. Ficino wrote that "excessive food recalls all the power of nature first of all to the stomach to digest it. This renders nature unable to exert itself at the same time in the head and for reflection."[142] A little later than Ramazzini and two centuries after Ficino, the Anglo-Dutch physician Bernard Mandeville concurred: the stomach—and no other internal organ—"is influenced in a more than ordinary manner by that Part of us which thinks." That was because of the spirits' various functions. Whenever their "Power is tired or exhausted by the Labour of the Brain, the Stomach suffers."[143] "We say commonly, that Melancholick Persons are Ingenious," Ramazzini wrote, "but we have more Reason to say that Ingenious People turn Melancholick, the more spirituous Part of their Blood being consum'd in the Exercise of the Mind, and only the earthy drossy Part left behind."[144] All focused and rigorous thinking posed problems for digestion, but special dangers were attached to what was called "speculative," "abstruse," "difficult," or "deep" thinking, and there was some agreement that the worst of these studies was mathematics, which made demands on the spiritual economy unmatched by any other intellectual activity. The mathematical mind, Ramazzini observed, "must of Necessity be disjoyn'd from the Senses, and cut off in a manner from all Commerce with the Body, in order to contemplate and demonstrate the abstrusest things that lie most remote from matter; this set of Men, I say, are almost all of 'em Stupid, Slothful, Lethargick and perfect Strangers to human Conversation, or the Business of the World."[145] The Swiss physician Samuel-Auguste David Tissot (1728–1797) confirmed both expert opinion and "daily experience" of "the power of the soul upon the stomach": "Those most addicted to reflection, perform their digestion . . . with greater difficulty; on the contrary, those who think little, generally digest well." He cited a Portuguese physician to the effect that "a vitiated stomach attends learned people, as surely as a shadow follows the body." The causal connection between deep thinking and poor digestion was a "wasting of the spirits" by "keeping the nerves too long in a state of action."[146]

A major dietary problem attending the life of the mind was one of timing. It was the simultaneity of deep thinking and abundant eating that was most dangerous. There were better and worse times for thinking. Some dietary writers specified that the period immediately after dinner was a poor time for study, as was the time just before going to bed, since that would interfere with sleep. Morning was widely agreed to be best of all: the mind was fresh, the air was good, and the stomach not yet burdened with food. But worst of all was mixing philosophy with dinner. Lémery wrote that "we should also during the Time of digesting our Victuals, forbear too serious Applications of Mind; and in a Word any Thing, that is apt to cause violent Distractions in the Animal Spirits, and to hinder the natural Heat from continuing the Work it hath begun."[147] The table was not a fit setting for inquiry, scholarly pedantry, or disputation. Shakespeare noted the bad bodily effects of tension at table: "unquiet meals make ill digestions" (*The Comedy of Errors*, Act 5, scene 1). And King James I's advice to his son was to keep "an open and cheerefull countenance" at table, perhaps requiring others to read "pleasant histories" to him and entertaining "pleasant, quicke, but honest discourses."[148] The genre approved on both civic and medical grounds was known as "table-talk": this was conversation, the sort of thing that made only light demands on the spiritual powers needed for concoction and that solidified rather than strained social bonds. That sort of conjunction of pleasant speech with dining was commended by Montaigne and possibly sanctioned by his physicians: "Concerning familiar table-talke, I rather acquaint my selfe with, and follow a merry conceited humour, than a wise man."[149]

These causal loops between eating, the workings of the stomach, and the operations of the mind featured in explaining scholarly gloominess. They offered scripts for self-presentation to those pretending to be sacred or secular knowers, were resources in the display of mental refinement and sensitivity, provided a basis for commending ascetic or temperate dietary regimes, and were mobilized in attempts to remedy learned melancholy. Melancholy and its dietary framing were widely associated with high social standing, partly by way of links between that standing and the world of learning. The historian Michael MacDonald has written that melancholy was "à la mode in Jacobean England, and the rage for this fashionable affliction popularized medical ideas about emotional distress." The educated classes knew all about melancholia from such texts as Timothie Bright's *Treatise of Melancholy* and Robert Burton's *Anatomy of Melancholy* and from characters in plays such as *Hamlet* (the Prince) and *As You Like It* (the gloomy Jaques). "Melancholy and gentility became boon companions" in Jacobean England, MacDonald wrote, and courtiers often liked to have their portraits painted as melancholy lovers. Examining a contemporary physician's casebook, MacDonald noted that 40 percent of diagnosed melancholics were peers, knights, and ladies, while seventeenth-century commentators presumed that ordinary folk were just too stupid to be melancholic. When their minds were judged disturbed, the common people were much more likely to be categorized as "mopish" and "troubled in mind."[150]

The English were reflexively aware of their collective identity, and they understood the role of food in coming to be who they were. The learned also knew who they were, what sorts of life suited them, and how to prevent or at least manage the distempers to which they were subject, and they too understood the constitutive role of food and the other elements of traditional dietetics. That these sorts of understandings were widespread in the culture, enduring over a long period of time, is clear. What, however, can and should be questioned is the *authority* enjoyed by dietetic expertise. Traditional dietetic language was built into self-understandings and into everyday practices of being and interacting. Were there, for all that, conditions in which expert authority might be challenged or set aside? *Could* laypeople talk back to the doctors? And could laypeople dispute or reject the claims and categories of traditional dietetics?

5

Talking Back to the Doctors: Medical Expertise and Ordinary Life

Knowing Yourself and Managing Yourself

The authority and the cultural reach of traditional dietetics were immense. Its general injunctions, such as observing moderation and paying careful attention to the things nonnatural, were without significant challenge. Many of the more specific descriptions and recommendations—for example, the degrees of cold and dryness of cabbage and whether veal was more wholesome than beef—were widely ignored and sometimes disputed, but the general categories and modes of reasoning remained stable in Renaissance and early modern culture. Even when deliverances of dietetic expertise were disputed, dissent still used its categories and patterns of inference. Dietetic vocabulary and sensibilities spread across the social and cultural landscape, and the integration of traditional dietetics into a range of everyday practices and beliefs made dietetics durable and robust.

For all that, the authority of traditional dietetics was never absolute. Who was an expert, or the best expert, on your state of health? Who had the most reliable knowledge of your unique body? Who could advise on its management and on what grounds? Just because dietetics was so enfolded in the practices of everyday life, the acknowledged obligations of that life might conflict with expertly delivered advice on how to maintain health and live long.

Recall that one of the inscriptions on Apollo's temple at Delphi was "know yourself," and recall also the ancient dictum, continually repeated through the early modern period, that everyone should be their own physician: "There is no man nor woman the which have any respect to them self, that can be a better Phesycion for theyr owne savegarde, than theyr owne self can be."[1]

The tag "every man his own doctor" was ambiguous. One sense was that you *ought to be* your own doctor, not because you disdained the basic form and content of dietetic expertise but because you should have internalized its ordering principles and were aware of its practical regimes. It was an element of prudence closely to monitor yourself. How did you respond to each of the nonnaturals, hour by hour, day by day, through the seasons of the year and the arc of aging? You learned something about the type of person you were through your response to dietary regimes, and you learned something about the components of proper diet through knowing the type of person you were.

To be your own doctor might not be to reject medicine but rather to extend medical sensibilities, to build them into your everyday beliefs and actions. To be your own doctor, in that sense, was something that physicians might urge on you as an extension of their authority into your everyday life management. Or you might announce that you were your own doctor—with or without the benefits of physicians' promptings—because you, *of course*, knew yourself better than anyone else possibly could. You might well concede the quality of your doctor's theoretical knowledge and wide experience of others' bodies; you might acknowledge the wisdom contained in the dietary books the doctor had on hand and might even have written. You knew best what agreed with you, you knew where the shoe pinched, and although you might be able to *tell* your physician about your personal experiences with your body and how it responded to various regimes, the doctor could never know them with the richness and security that you did. Physicians could secure adequate knowledge about the unique patient that was you: your constitution, your history, and the changes in way of life that you had experienced and what effects these changes had on your body. But if your physicians came to know these things, they knew them courtesy of you—by way of your self-knowledge and your testimony.[2]

There were indeed Renaissance and early modern skeptics about physicians' expert knowledge and its power. Some lay and educated people reckoned that doctors couldn't do much to save you if "your time was up," that fraud and self-interest were epidemic among physicians, that some doctors were learned idiots, or that bookish knowledge didn't have much relevance to your specific illness. And there were many books by doctors that assembled and condemned such "popular errors," including presumptions by the uneducated that folk remedies were superior to anything a physician might recommend.[3] That said, criticism of *physicians* was not the same thing as rejection of the categories of physicians' knowledge. If, for example, you thought that the doctors were wrong about a food's degree of warmth and moistness or if you asserted that a food they said was hard of concoction actually agreed with you, you were dissenting from the specific while entrenching the general. And after all, that ability to speak intelligibly about your individual body and to assess how it fared under dietetic regimens was a big part of what it was to obey medical advice to *be your own doctor*. So, the notion that you should be your own physician might count as an extension of expert medical authority into the day-by-day management of your body *and* a limit to that authority. It was often both. The physician directed you to regulate your own body and offered you the fundamental principles of that self-management, but it was that same extension of authority into one's bodily management that allowed patients—some of the time and in some circumstances—to talk back to their doctors, even to reject the idea that physicians' knowledge was legitimate and that their counsel was effective. What was at issue was the legitimacy and value of different kinds of *experience*.

A Temperate Individual

A short book by the Venetian engineer, architect, and (supposed) nobleman Luigi Cornaro (ca. 1467–1566)—his first name sometimes given as Alvise or Anglicized as Lewis—was among the most influential and certainly one of the most enduring dietetic texts produced in the period from the Renaissance through the eighteenth century. Begun in 1550 and originally published in 1558, Cornaro's book was titled *Discorsi della Vita Sobria* (*On a Temperate Life*), and part of its interest arose from the fact that its author was not a physician at all.[4] (How long Cornaro lived is relevant to the authority of his story. His death occurred in 1566, but some sources say he was born in 1467—which would mean he died just short of a hundred years old—while others place his birth in 1475, 1482, or 1484. Those who were most favorably disposed to Cornaro's diet often referred to him as a centenarian.) The book went through several editions in Cornaro's own lifetime; it was known and remarked on by a wide range of commentators through the early modern period and beyond. At the end of the nineteenth century, seeking to repair his own health in Italy, Friedrich Nietzsche wrote about Cornaro's book. "Everybody knows the book of the famous Cornaro," Nietzsche declared in *Twilight of the Idols*. "Few books have been so widely read, and to this day many thousand copies of it are still printed annually in England." But Nietzsche, who thought about diet a lot, detested Cornaro's way: "I do not doubt that there is scarcely a single book (the Bible of course excepted) that has worked more mischief."[5] *De Vita Sobria* survived Nietzsche's criticism and remains in print, considered as both a historical object and a practical guide to health and longevity, making it almost certainly the most widely distributed and long-lived account of an individual's dietetic practice ever produced.[6] Centuries after it was written, *De Vita Sobria* was said to "have obtained a celebrity beyond almost any publication of the sort."[7] Enthusiasts—including some medical people—continued to commend Cornaro's tract as sound dietetic advice into the twentieth century. His version of temperate dietetics was ardently endorsed by Horace Fletcher, famous for his thorough-chewing discipline, and, through Fletcher, by the Seventh-day Adventist dietary program launched by the great American cereal entrepreneur J. H. Kellogg. In 1901, the British Medical Association was treated to a lecture titled "Was Luigi

Cornaro Right?" by the English physician and Fletcher ally Ernest Van Someren, and in 1903 Kellogg collaborated with Fletcher in encouraging a new edition of Cornaro's text.[8]

Cornaro appropriated the tags "know yourself" and "every man his own physician" to great effect: "No man can be a perfect Physician to another, but to himself onely. The reason whereof is this, Every one by long experience may know the qualities of his own nature, and what hidden properties it hath, what meat and drink agrees best with it: which things in others cannot known without such observation, as is not easily to be made upon others."[9] *De Vita Sobria* was an extended assessment of his own medical condition and the results of his attempts to manage it. Reaching his midthirties and having (he said) followed his own tastes and having led a life of excess, Cornaro was told by physicians that he did not have much longer to live: he had immediately to mend his ways and adopt a temperate diet.[10] He knew very well that this was the sort of advice that doctors generally gave, and he accepted that they were right to do so: moderation could indeed cure disease, and if you kept to a moderate diet, you would have your best chance of preventing further illness.[11] Cornaro embraced rigorous temperance, and the improvement in his condition was rapid: "within a few dayes . . . I was exceedingly helped, [and] within lesse then one yeare . . . I was perfectly cured of all my infirmities."[12] Impressed with the effects of his new temperate regime, Cornaro decided to devote the rest of his life to refining its dictates, both for himself and, through his book, for all those who might benefit from his hard-won personal experience. Further editions of his book continued to appear as Cornaro reached extreme and healthy old age, with each edition securing more authority for a temperate diet.[13]

Cornaro recalled his own initial reluctance to take medical advice to adopt a diet suitable for invalids, thinking that foods that tasted good to him were indeed likely to be good for him, and he misled his doctors about what he was actually consuming. But he eventually came to realize both the force and the error of the widespread view of "that which savours, is good or nourisheth."[14] This was, he said, simply a proverb that gluttons used to justify stuffing themselves with the foods they liked, and Cornaro subjected the old saw to systematic self-experimentation. He had a taste for certain kinds of wine, melons, lettuce, fish, pork, sausages, and cake, but he learned from experience that when he consumed these things his health took a plunge, and when he eliminated them from his diet he flourished. He knew what agreed and what disagreed with him. Taste was not a guarantee of healthfulness, and, through reflection and experience, Cornaro had become a dietetic expert. He followed his physicians' general counsel but learned to fine-tune it and adapt it to his individual constitution and circumstances, noting carefully what was well or ill-digested and which foods led to health and vigor and which to disease and malaise.[15]

Over the years, Cornaro accumulated immense experience with dietary regulation and its effects on his body and mind. His food included bread, eggs, some flesh meat (not much), and broth (probably made from capon) and no fruit or fish—even though he considered that there was no real *danger* in them—because they did not *agree with* him.[16] And notably, he precisely *weighed* his food, giving moderation a secure quantitative meaning. The total weight of all the solid food consumed in a day was "twelve ounces exactly weighed" and "the measure of my drink" was "foureteen ounces."[17] He was now convinced that he knew more about temperance and a wholesome diet than the physicians who had originally advised him. Moderation now had its measure, and it was Cornaro's version of that measure that came under pressure when both his friends and (evidently long-suffering) physicians eventually became concerned that he was in fact practicing not moderation but rather harmful asceticism; at one point he was down to just a single egg as his daily intake. Cornaro countered that his meager dietary custom had now become second nature, and for that reason change was risky.[18] But he ultimately gave in to their urgings, slightly increasing the quantity of his meat and drink. He upped his now-accustomed twelve ounces of meat, bread, soup, and eggs to sixteen ounces. This, Cornaro announced, proved to be a huge mistake. He should have listened to his own body and not to his doctors' expert recommendations. Just ten days after elevating his diet, Cornaro fell desperately ill in both body and mind. From his normal "cheerfull and merrie" self, he became "melancholie and cholerick"; he developed a pain in his side and a "terrible fever" that lasted more than a month. He could not sleep, and "all gave me [up] for dead." Reasserting his own expertise, Cornaro resumed the low diet to which he had become accustomed and "by the grace of God cured my self."[19] A peppy old man at eighty-three, Cornaro announced that "I am continually in health, and I am so nimble, that I can easily get on horseback without the advantage of the ground, and sometimes I go up high stairs and hills on foot."[20] The lesson was not just about the health-giving benefits of what was widely taken to be an abstemious and not just a temperate diet; Cornaro also advertised the importance of vigilant self-knowledge. The doctors thought that he was taking temperance to a dangerous ascetic extreme; Cornaro knew better. His experience taught him "how great is the power of order and disorder; whereof the one kept me well for many yeares; the other, though it was but a little excesse, in a few dayes had so soon overthrown me."[21] That tiny detour from the temperance that Cornaro had permitted himself proved perilous. Doctors were *not* useless; they were "greatly to be esteemed for the knowing & curing of diseases" afflicting those who *fail* to follow a temperate diet.[22] But temperance had to be tested in individual cases, and a prudent individual knew best. Cornaro was his own doctor.

Was Cornaro an Authority?

Discussion of Cornaro's text and its presentation of dietetic discipline continued through the following centuries. It is hard to find a major seventeenth- or eighteenth-century thinker on these subjects who did *not* refer to Cornaro.[23] Some were wholly enthusiastic about both the temperate regime and the capacity for dietetic self-knowledge. Physicians and clerics were among the most fervent, praising the disciplining of appetite, the rejection of gluttony and debauchery, and the joining up of health and virtue.[24] Leonardus Lessius was so enthusiastic that he translated *De Vita Sobria* into Latin and appended a tract titled *Hygiasticon*, his own rules for right living but closely modeled on those of the Venetian. *Hygiasticon* commented on existing evaluations of Cornaro's way and defended the Italian's dietary expertise. (Lessius anticipated that some might think it odd that a cleric such as himself had written a medical text, but this "transgression" was not uncommon at the time, and the Jesuit pointed out that "it is the divine vertue of Temperance, which is chiefly in question." The business of temperance was "not altogether Physicall, but in great part appertains to Divinitie and morall Philosophie." Expertise in dietetic medicine was of a piece with moral authority.)[25] The two texts were both translated into English and appeared, bound together, in 1634, the preface reporting great success among those in England who adopted Cornaro's regime. Lessius was sure that Cornaro's version of temperance was powerful medicine and endorsed the precise weighing of food. If there was any ambiguity about what moderation meant, it was all spelled out in Cornaro's tract: so many ounces of meat, so many of drink, together with accounts of the Italian milk- or broth-soaked bread gruel called *panada* or *panatella*.[26] Moderation might mean different things for people of different constitutions of different ages and engaged in different ways of life, living in different climates, and at different times of the year, but Cornaro's measure was nevertheless reasonable and robust advice for all.[27] Lessius treated Cornaro as an expert not because the Venetian was thought to command medical *theory* but because of the depth and quality of his personal *experience*, supported by traditional wisdom about the prudence of a spare diet and especially because Cornaro's success was evident. Some literati in eighteenth-century polite society also commended Cornaro, though usually without the theological gloss. Writing in *The Spectator* for 1711, Joseph Addison celebrated Cornaro's temperate diet as a sure way to a long and healthy life, not at all the commonly dour and joyless sort of austerity-preaching. *De Vita Sobria* was, Addison thought, written "with such a spirit of cheerfulness, religion, and good sense, as are the natural concomitants of temperance and sobriety. The mixture of the old man in it is rather a recommendation than a discredit to it."[28]

Physicians approved Cornaro too. After all, the doctors were invested in crying up the power of temperance and warning against the diseases and the shortened life caused by excess. Cornaro had been his own physician, and now his regime could be a valued resource for physicians themselves. The most celebrated diet doctor of the first of part of the eighteenth century, the Scot George Cheyne (1672–1743), who was himself often accused of prescribing too-great austerity, was wholly on board with not only Cornaro's abstemiousness but also his practice of precise weighing. Cheyne had it that Cornaro practically starved himself, applauding his apparent insight that it was the quantity rather than the quality of foods that was most important.[29] (Those inspired by attempts to mathematize medicine in the late seventeenth and early eighteenth centuries were greatly attracted by the possibility of getting rid of many features of Galenic qualities in favor of a mechanical and quantitative orientation to food and body function.)[30] The celebrated Swiss physician (and antimasturbation campaigner) Samuel-Auguste Tissot similarly commended the quantification of aliment, holding it mainly responsible for Cornaro's long life.[31] At the century's end, a much-reprinted tract on diet and longevity by the German "macrobiotic" physician Christoph Hufeland agreed that Cornaro had succeeded in achieving long life by "the simplest and strictest regimen" and, again, by meticulous attention to the exact quantity of aliment.[32] Editions of William Buchan's hugely popular *Domestic Medicine* asserted that Cornaro had "experimentally and incontestably proved" the power of proper regimen to cure diseases, even those "which had resisted the force of medicine." Cornaro's way was said to be *better* than many doctors' doses, purges, bleedings, and lax dietary regimes, and the Venetian gentleman was commended especially by those physicians who continued to urge the importance of rigorous diet over more aggressive therapies.[33] In the middle of the nineteenth century the French physiologist Pierre Flourens, writing on human longevity, could point to no "better authority" than that of old Cornaro: "The whole *marvel* of his regimen was moderation," and Flourens celebrated the efficacy of Cornaro's exact quantification.[34]

There was a certain amount of resistance too from both physicians and the laity. One possible reservation—that a moderate regimen was not actually effective in ensuring health and longevity—was rarely expressed. Most commentators through the seventeenth century and beyond agreed that dietary excess *did* shorten life and breed disease. It was said that there were many examples of pious laypeople and philosophers—from biblical times through antiquity to the present—whose long lives were attributable to the strict observance of temperance or even asceticism. Cornaro's own longevity was one of the more widely cited proofs of this principle. Nevertheless, there was some skepticism about a causal link between the ascetic way and long life. Francis Bacon, who was among many early modern philosophers greatly concerned with longevity and its causes, conceded that "a *Pythagoricall*, or *Monasticall Diet*, according to *Rules* . . . seemeth to be very effectuall for long Life," but not invariably so: "amongst those that live freely, and after the common sort, such as have *good Stomacks*, and *feed* more *plentifully*, are often the Longest liv'd."[35]

If physicians and practical ethical writers didn't want their readers to be gluttons and drunkards, neither did they approve asceticism, conceived as the opposite extreme and, for that reason, constituting a vice.[36] Burton's *Anatomy of Melancholy* condemned excess at length but more briefly noted the mischief wrought on bodies and minds by those going to the other extreme: "too ceremonious and strict diet, being over precise, cockney-like, and curious in their observation of meats." Remarkably, Burton wrote, such people *weighed* their food—"just so many ounces at dinner . . . , so much at supper, not a little more, nor a little less, of such meat, and at such hours, a diet-drink in the morning, cock-broth, china-broth, at dinner plum-broth, a chicken, a rabbit, rib of a rack of mutton, wing of a capon, the merry-thought of a hen, etc." This was no way for healthy people to live: "to sounder bodies this is too nice and most absurd." Such precision was uncivil and unsound, but there were other dangers: monks and anchorites were well known to have driven themselves mad "through immoderate fasting,"[37] and an abstinence that was too severe might make for holiness while harming health. Around 1600, an English physician warned that the exact measuring of your food wasn't necessary for your bodily health and was certainly bad for your mental health: "The prescribing of meat by drams or ounces, driveth many fears into a weak mans mind, taketh al alacrity from the heart, maketh a man jealous of his owne fingers, daring to eat nothing with chearfulness, because he ever suspecteth that he eateth too much."[38]

The nineteenth-century Scottish physician Andrew Combe—brother of the phrenologist George—lumped together Cornaro and George Cheyne, and their fault was, Combe said, the attempt to prescribe a standard and rigorously ascetic diet for *everyone*. "Cornaro, Cheyne, and others, have, most absurdly, attempted to determine a standard quantity of food for all mankind, and have fixed it at the lowest possible limit," but this was just wrong; it ignored variations of constitution and of way of life.[39] Disapproval of "Cornaro's plan of adhering rigidly to a weighed and measured diet" continued into the early twentieth century.[40] In 1909 an American general practitioner, writing in a popular periodical, translated Cornaro's diet into the newly popular language of calories—calculating that it amounted to only 1,200 calories—and found it inadequate to sustain a normal life. It was, the physician said, a "starvationist diet," and "no normal, average, unspiritual individual has even been able to live upon any such diet as 1500 calories a day without impairing either his health or his working power."[41] And twenty-first-century enthusiasm for longevity by way of radical "caloric restriction" diets sometimes explicitly positions itself in a lineage tracing back to Cornaro.[42]

The Rule of Number

Weighing and measuring, aiming for precision and exactitude, and turning from the qualitative to the quantitative all marked emerging scientific practices of the seventeenth century. The "moderns" of the Scientific Revolution liked biblical warrant for mathematical approaches to natural phenomena, and a favorite text was the apocryphal Book of Wisdom, or Wisdom of Solomon: "Thou hast ordered all things in measure and number and weight" (11.20). So, how did quantification stand with respect to the human body, its everyday management and nourishing, and the prevention and cure of its ailments? Since antiquity, it was standard practice to weigh and measure all sorts of traded goods. In long-distance commerce, there were problems agreeing on quantities because of widely varying measuring standards, both between countries and within a given country. But it was not routine, even in books of recipes for food preparations, to precisely specify quantities or, even in medical texts, to specify measured quantities of food that a person should consume.[43] (That is one reason why Cornaro's regimen stood out and attracted attention.) With the exception of trade, in matters of everyday life the qualitative usually took precedence over the quantitative. It was generally presumed that you ate what you were given or until you were no longer hungry or that you consumed the amount you usually took or that was suitable for the occasion. Quantities in such contexts *might* be specified as, for example, a "quart," a "hogs-head," a "bottle," or a "bushel" or simply as "a goodly capon," a "brave chine of beef," or a "brace of partridges." These directions were presumed to be just what you needed to know; the rest could be assumed. Being exact about quantities in a dietetic setting might be deprecated as unmanly and uncivil fastidiousness.

But what about knowing how much of *you* there was, how the mass of your body was known and represented? George Cheyne was greatly concerned about his own health, since he too suffered from the so-called English Malady—depression, anxiety—about which he wrote so influentially. And he was concerned about his bulk, not (as we might now say) as a "risk factor" for disease but instead largely because of its many inconveniences: a mechanical inability to get in and out of his coach and a shortness of breath when climbing stairs or walking just a hundred paces.[44] (Other considerations bearing on body bulk and on why you might want to reduce it included the possibility—not the certainty—that it was socially unattractive and the greater possibility that it would be read as a sign of the sin of gluttony, a lack of bodily control.) Cheyne recorded his *weight* and changes in his weight at various times. At different points in his life, he was spectacularly fat. Sometime before he composed the autobiographical essay included in *The English Malady* of 1733, he recorded that his feeding had become so out of control that "upon my last weighing I exceeded 32 Stone" [448 pounds], and he accounted himself "overgrown beyond any one I believe in Europe."[45] It was not, however, common for people before the early eighteenth century to know their weight in numerical terms.[46] The first lay fad for weighing oneself probably appeared in the mid-eighteenth century. Weighing scales became part of the public furniture of Paris in the 1740s, and enthusiasm for personal weighing in civic spaces developed in London around the same time. There were weighing scales in London coffeehouses from the 1760s, and the London wine merchant Berry Brothers was equipped with large scales used for their goods: several registers survived at least into the early twentieth century "containing the signatures of celebrated personages from the year 1765 onwards, who, whilst giving orders for wines, improved the occasion by ascertaining their correct weight."[47] (The routine availability of weighing scales in such public places as agricultural fairs, train stations, and drugstores emerged from the mid-nineteenth century, and the ubiquity of domestic bathroom scales dates from about the 1920s.)[48]

Until at least the eighteenth century, knowledge of body bulk was overwhelmingly qualitative, visually available, and interpreted in the frame of constitutions, conditions, and capabilities and was not mediated by weighing instruments and assigned a numerical value. People were, for instance, big, small, lean, fat, corpulent, or plump. Nor did they assess their state of health and well-being through scalar measures, norms, and quantitatively established liabilities for disease, though many people, including physicians, had notions of what sorts of diseases the fat and ruddy or the lean and pallid might tend to suffer and even how close they might be to illness or death. The complexions, after all, encompassed body shape, texture, and, of course, the appearance of the body's surfaces, including the emotional and affective expressiveness of the face.[49] Body mass, in that qualitative sense, belonged to traditional dietetics and was a meaningful sign of your constitution, your state of health, and your liability to certain sorts of ill health.

One of the iconic images of early seventeenth-century science is "Santorio's chair." Like Cornaro, the physician Santorio Santorio (1561–1636) was a Venetian; he was a friend of Galileo, and Galilean impulses were evident in much of Santorio's medical work devoted to quantifying physiological states.[50] Santorio's "weighing chair" was his most celebrated invention, and the image of the device he included in an early edition of his *Medical Statics* (originally published in Latin in 1614) became one of the most widely distributed images of quantitative discipline in modern science and medicine (figure 5.1).[51] The chair was an instrument designed to quantify the relationship between food, excrement, and bodily weight. For about thirty years, Santorio weighed himself: he first weighed himself as he sat in the chair suspended from a steelyard, and then he weighed the food he was going to consume, and finally he weighed his urine and feces. Eight pounds went in, three pounds eventually came out, and his body weighed the same the next day. This unaccounted-for five pounds he ascribed to "insensible perspiration"—-not the visible sweaty sort that came out through the pores of the skin—and he went on to prescribe how that perspiration might be effectively managed to maintain a steady weight, a numerical balance he associated with health.[52] Santorio presented himself as a hero of a modernizing, quantifying impulse: the perspiration of which people were sensorily unaware was conclusively established by rigorous quantification of inputs and outputs, and, crucially, the insensible loss from the body was greater than that of the visible excreta. Using Santorio's chair, you *might* establish your absolute weight, though its primary use was to determine weight *change*.

Santorio was aware that some people might find the device and the practices associated with it absurd, so he recommended that the chair be set up in a "secret place," in a room above the customary dining place, because it might be considered "unsightly" and because there are silly people "to whom all things that are unusual seem ridiculous."[53] He was right: opinion was split over whether

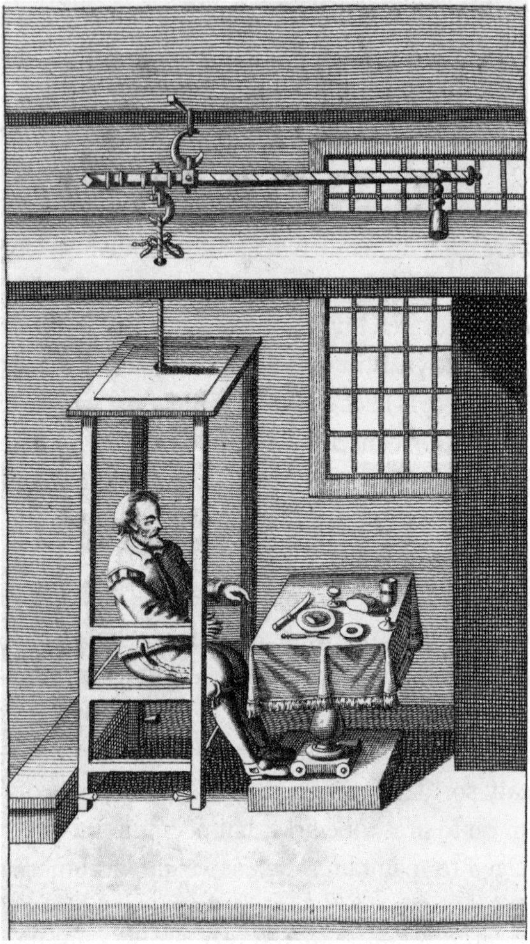

Figure 5.1. Santorio's weighing chair. This engraving served as a frontispiece to Santorio's *De Statica Medicina* (1614) and was much reproduced into the eighteenth century. The physician sits in a weighing chair of his own devising. The chair hangs from a steelyard suspended from beams above his dining room. The chair was itself poised just a little above the floor, and a desired weight was attained when a specified quantity of food ingested caused the chair to descend to floor level. The image became a powerful and pervasive emblem of quantified medicine.

the chair was useful or bizarre. Into the eighteenth century, many physicians did admire Santorio's discipline and recommended frequent weighing as a way of maintaining health. An eighteenth-century British editor of the *Medical Statics* reminded readers of the old saying that "a man at Forty is either a Fool or a Physician," and even though he understood that some laypeople would bridle at its discipline, he identified the use of Santorio's chair as a defining mark of medical prudence.[54] But some doctors had reservations, and many nonmedical writers found the practice ludicrous. James Mackenzie reckoned that there was no need for weighing: your "natural undepraved appetite" was a wholly reliable guide to how much food you should consume, "for nature never craves more, nor is easy with less, than what is proper for her."[55] And the Scottish physician James Keill offered as a medical aphorism a quantitative version of *Quod sapit*: "The Rule for eating to every Body, is a natural Appetite; and by this Monitor may every one be advertised of the Quantity proper to be taken in, without weighing. . . . The natural Appetite is certainly the best Guide, both in the Quantities and Qualities of the Meats and Drinks to be taken in."[56]

Joseph Addison's *The Spectator* spoofed Santorio's obsessive quantitative self-monitoring. In 1711, the periodical supposed that it had received a letter—in fact, a fabrication—from a correspondent who described himself as "one of that sickly tribe who are commonly known by the name of valetudinarians." He had attempted to perfect his health by the close observance of medical rule, but the effect was quite the opposite. He became an anxious pulse feeler and fancied himself afflicted with whatever disease he was then reading about. Resolving finally to achieve complete health, he took up Santorio's tract and read the description of "a certain mathematical chair," with which the learned world was "very well acquainted." Using the chair for three years—not Santorio's thirty—the correspondent established that when in "full health," his body was "precisely two hundred weight," varying about a pound and a half either way depending on his intake. He carefully adjusted his consumption when the increase or decrease slipped beyond those limits: "I do not dine and sup by the clock, but by my chair." Santorio's chair did not, however, work its promised medical magic: "I find myself in a sick and languishing Condition. My complexion is grown very sallow, my pulse low, and my body hydropical." Nothing diminished in his enthusiasm for the newest medical discipline, the valetudinarian begged *The Spectator* "to give me more certain rules to walk by than those I have already observed." Addison basically told his correspondent to man up; he had, so to speak, nothing to fear but fear itself: "The fear of death often proves mortal, and sets people on methods to save their lives, which infallibly destroy them." Meticulous observation of numerical rules was both ineffective and undignified. It was reasonable and right to take care of your health, but people who subjected themselves to mathematical rule, who "tempered [their] health by ounces and by scruples," were doing an injury to both themselves and their society.[57] There were all sorts of dietary disciplines pushed by the doctors and their moralistic friends, but many polite writers found the discipline of numbers too much. It implied not only that you could not *control* yourself but also that you could not *know* yourself, that some scientific instrument knew better than you when enough was enough.

Henry Peacham was one of several seventeenth-century practical ethical writers who warned against going too far in attempts to avoid dietary excess: "Neither desire I, you should be so abstemious, as not to remember a friend with an hearty draught, since wine was created to make the heart merry, *for what is the life of man if it want wine*? Moderately taken it preserveth health, comforteth and dispereth the naturall heate over all the whole body, allayes cholericke humours, expelling the same with the sweate, &c., tempereth Melancholly. And, as one saith, hath in it self a drawing vertue to procure friendship."[58] Buchan's ubiquitous *Domestic Medicine* repeated the standard commendation of dietary moderation while judging that "mankind were never intended to weigh and measure their food. Nature teaches every creature when it has enough of food, and a very small degree of reason is sufficient for the choice of it." There was just no practical need for Cornaro's preciseness.[59] Even Puritan clerics, working themselves into pious frenzy about gluttony as "self-murder" and generally approving the example of Cornaro and his followers, felt obliged to concede that they may have overdone it. Richard Baxter's *Christian Directory* (1673) recommended "Cornario's Treatise of himself" as well as the tract of Lessius, "though yet I perswade none without necessity to their exceeding strictness."[60] The more prevalent criticisms of Cornaro and of physicians' recommendations of a highly regulated temperate dietetics were on grounds that were not—in contemporary terms—strictly medical at all. That is to say, people who accepted that a rigorous regimen of moderation *did* make for health and longevity might nevertheless reject it for other reasons. That was indeed talking back to the experts, not denying that what they recommended *could* secure the stated goals of health and long life but instead disputing that those goals constituted adequate counsel about *how it was right to live*.

"To Live Physically Is to Live Miserably": Dietetic Discipline and the Civic Life

A person living an active life—taking a normal part in society, in its governance, and in the institutions of extending its power and securing its commodities—might push back against many of the dietetic disciplines intended to attain perfect health and extend life. A proverb had it that "to live physically is to live miserably." A medically regulated life was unpleasant, the sort of thing suitable for a sacred or secular ascetic but not for those accepting their role in ordinary society. Securing health by denying yourself all forms of pleasure and of spontaneous action wasn't worth the mental, social, and moral cost. Asceticism did have a recognized value, but it wasn't for everyone. Medical advice, however well founded, shouldn't keep otherwise healthy people away from the active sphere. If occasional feasting and liberal drinking were central to the public life—and in early modern gentlemanly society they definitely were—then it couldn't possibly be a point of prudence or of practical morality to embrace medical counsel withdrawing public actors from those scenes in which public action occurred and in which social solidarity was made, maintained, and occasionally subverted. The sort of self-indulgent discipline that might be acceptable for sequestered scholars, monks, and gentlemen retired from the world was not proper or permissible for the civic actor. La Rochefoucauld put it aphoristically: "To keep well by too strict a regimen is a tedious disease in itself."[61]

Written toward the end of the sixteenth century, Montaigne's essay *Of Experience* was in large part addressed to questions of what doctors knew, how their knowledge varied from one apparent expert to another, how any physician's understandings of your body stood with respect to your self-knowledge, and, most pertinently, whether you should in all cases follow their advice. (His conclusions were eloquently his own—close to the heart of his twin projects to know himself and to know what it was possible for *anyone* to securely know about anything else—but his sentiments were widely shared in early modern gentlemanly culture.) Montaigne was a skeptic with respect to medicine and many other systems of knowledge and morality, but one thing he did not doubt was that physicians could do *something* and that you should, of course, call upon their services and take their advice if you had the resources to do so and when you were genuinely ill. One of the inconveniences of dietetic medicine, however, was its preventive aspect, its orientation to the regulation of many aspects of your life when you were *not* actually ill. In those circumstances of ordinary life, was it right to "live physically" when your body and mind were able to perform all their ordinary functions? The answer from Montaigne, and indeed from other early modern civic actors, was no. The early modern public actor—the gentleman, the courtier, the politician, the diplomat, the merchant, the soldier—came down firmly on the action side of the ancient debate between the active and the contemplative lives, that is, whether it was the place of virtue to be fully engaged with civil society or to withdraw from it. A strict dietetics of moderation had to be tempered by other ethical and prudential concerns, and indeed, the *meaning* of moderation might even be interpreted so as to align the notion of temperance with these other considerations. That is why it was said of Francis Bacon—a great admirer of Montaigne—that "he did, indeed live Physically, but not miserably."[62]

Of Experience was an extended display of Montaigne's well-known skepticism about external expertise and a vigorous defense of the moral and practical integrity that such skepticism assisted. In form, it is an essay about the general superiority of prudence to systemic pretension, of self-knowledge to experts' knowledge of you. In substantial content, much of the essay is a tract about the management of the Galenic nonnaturals. If you were a prudent person, then over the years your dietary routine had been informed by the patterns of your body's responses to food and drink, and in turn, your body had grown accustomed to that dietary routine: Montaigne knew his body, and he knew from long experience that it could suffer no harm from its habitual regimes. But in ordinary circumstances, habit, having become second nature, was to be respected, and Montaigne thought that you should be careful about making abrupt changes in dietary routine unless they were absolutely necessary. Still, Montaigne—soldier and public actor that he had once been—insisted that he was never a slave to habit: "I have no fashion, but hath varied according to accidents"; "the best of my corporall complexions, is, that I am flexible." The vicissitudes of life were one thing; the sudden and sometimes restrictive changes that flowed from adopting your doctors' expert systems were quite another.[63]

The sense that legitimate social obligations could trump dietetic rule—even if observing that rule was not conducive to health—was widespread in civic life. It was an appreciation linked to the sense that dietetic medicine must be adapted to individuals' varying ways of life. The Italian physician Guglielmo Gratarolo—a contemporary of Montaigne—took the "one size doesn't fit all" sensibility from Galen: "even as a shoemaker cannot make one shoe to serve every mannes foote, so neither can a Physician describe and appoint any one generall order and dietarie for all manner of persons. . . . Now, whereas there is both diversities in bodies, and also diverse trades of livinge, it cannot be that any one absolute way should be appointed to serve everie nature in everie facultie generally." There are people whose lives aren't under their own complete control—those of the servant class and courtiers who serve princes (women were not mentioned here)—and such people just could not follow whatever dietary rule they favored and that they believed secured their health. No doctor could legitimately prescribe a special regime to suit servants: dietary texts were intended for the independent, those who "live on them selves frely and are not enthralled." Choice pertained to them and not to the unfree and the dependent. To choose whether or not to submit yourself to dietary rule and, if you did, to choose in what ways and circumstances was a substantial aspect of what it meant to be a free actor.[64]

The Rule of No Rule

These sorts of sensibility were central to practical ethical notions of how governors, courtiers, noblemen, and genteel civic actors should conduct their lives. In early eighteenth-century Spain, the Benedictine monk Feijóo y Montenegro wrote that "the rule of Celsus to accustom yourself to eat like the generality of people . . . seems to me very reasonable. . . . It is part of good education, though many rich persons fail in it, to make their children from time to time eat of everything," so they become adaptable to change in situation or fortune.[65] The Scottish courtier and ethical writer James Cleland similarly warned against living to rigid dietary rule: "Bee neither uncivillike a grosse Cynicke, nor affectuatly niggard, like a dainty Dame, but eate in a manly, round, and honest fashion. Use most to eate of reasonable grosse, and common meates, as well for making your body strong, and durable for travell at all occasions, either in peace or in warre, as that yee may be the hartier received by your mean friends in their houses, when their cheere may suffice you."[66] In the early seventeenth century, the French theologian Pierre Charron generalized the injunction to dietary adaptability: "It is a good thing for a man not to accustome himselfe to a delicate diet, lest when he shall happen to be deprived therof, his bodie grow out of order, and his spirit languish and faint."[67] It was a point of policy as well as of courtesy to take the foods and drinks offered you by your followers and allies. You will need to rely on them in peace and in war, and you must not offend them by declining their generosity or being squeamish in accepting it.

Montaigne fell in with much civic opinion: if you put your conduct under the care of physicians, they would squeeze the joy as well as the spontaneity out of your life. Forbidding this and forbidding that, the doctors unman you and ultimately undo you: "If they doe no other good, at least they doe this, that betimes they prepare their patients unto death, by little undermining and cutting off the use of life."[68] Dietetic medicine—in both its Christian and secular affiliations—reckoned that a man who submitted himself to animal appetites was less than a man. But a similar sentiment also applied to people who made themselves slaves to the medical system, who were governed by their doctors and by medical regimens assuring health and longevity. Those who were ruled by others or by a book of rules were not free actors; they lacked the integrity central to gentlemanly identity. That sentiment was classical, and Montaigne knew very well that it was: Plutarch, for example, wrote in the first century CE that "a diet which is very exact and precisely according to rule puts one's body both in fear and danger; it hinders the gallantry of our soul itself, makes it suspicious of every thing to do with any thing, no less in pleasures than in labors; so that it dares not undertake any thing boldly and courageously."[69] Montaigne approved that general moral sentiment and applied it specifically to dietetics. Change was physiologically good and civically obligatory; it was better and more possible for vigorous youth than feeble age: "There is no course of life so weake and sottish, as that which is managed by Order, Methode and Discipline. . . . The most contrarie quality in an honest man," Montaigne declared, "is nice-delicateness, and to bee tied to one certaine particular fashion." Not to do, not to eat, or not to drink what was going wherever you were was shameful: "Let such men keepe [to] their kitchin." And in a chivalric age, that sort of fastidiousness unfitted a man for the life of a soldier. Attend, of course, to authentic medical expertise, but do not give up your freedom of action. Montaigne said that he knew of and pitied people who, while healthy, had made themselves prisoners of their doctors. Just as you shouldn't been a slave to your appetites, you shouldn't be a slave to your physicians.[70]

Montaigne was not the only influential early modern critic of people who subjected themselves to rigorous dietetic rule, whether self-imposed or dispensed by doctors. Lessius noted that "many authors have written about preservation of health, but they charge men with so many rules, and exact so much observation and caution about the qualitie and quantitie of meats and drinks" and about the other nonnaturals "as they bring men into a labyrinth of care in the observation, and unto perfect slaverie in the endeavouring to perform what they do in the matter enjoyn." But that sort of regime was doomed by the realities of civic life, legitimately limiting the effectiveness and the scope of dietary advice.[71] Some of these critics were themselves physicians, and their usages indicate the cultural power of lay rejections of "living physically." Burton's *Anatomy of Melancholy* agreed that "it is a hard thing to observe a strict diet" and reluctantly noted the popularity of the saying "*et qui medicè vivit, miserè vivit,*" that "the physic is more troublesome than the disease."[72] In 1650, the English doctor Humphrey Brooke dismissed the practice of precisely measuring dietary intake. Different people had different bodily requirements and tolerances, and not everyone whose health might benefit from a rigid regimen could or should live that way: we should leave "the strictness of *Lessius* and *Cornaro* to Speculative and Monastick men, as somewhat above us."[73] And George Cheyne—notorious for the rigid "lowering" regimes he prescribed for many of his patients—nevertheless conceded the fundamental wisdom of declining to "live physically": "The Reflection is not more common than just, That he who lives *physically* must live miserably. The Truth is, too great Nicety and Exactness about every minute Circumstance that may impair our Health, is such a Yoke and Slavery, as no Man of a generous free Spirit would submit to. 'Tis, as a *Poet* expresses it, *to die for fear of Dying*."[74]

The Rule of Celsus and the Place of the Golden Mean

Cornaro claimed that his regimen was moderate, so helping himself to the practical and moral prizes long attached to the idea of the golden mean. But moderation never had an unambiguous meaning; it was open to a range of interpretations, and while many commentators celebrated Cornaro as a pattern of temperance, others reckoned that his way of life was in fact an ascetic extreme or that if it was a mode of moderation, it represented only one contestable *interpretation* of what genuine moderation might mean. Laypeople understood and widely approved medical directions to position themselves at the dietary golden mean, but they also understood that it was not always an easy matter to locate the point precisely equidistant from vicious extremes. After all, for ordinary actors, day-to-day practical life was at issue, not physiological theory or points of ethical principle, and virtuous mediocrity here was a matter not of logical exactness but of the more or less, the reasonably close to some notion of the mean, the right thing to do for *you* in *these specific circumstances*. Since you might often miss the mark, on what side should you err? And a related question is how much it mattered if you sometimes fell away from some conception of perfect moderation. Advocacy of the dietary golden mean was formulaic from antiquity through the early modern period, but many commentators in lay society thought that the formula offered little reliable advice for the management of life-as-it-was-actually-lived. Adherence to the golden mean still counted as prudent, but you needed to rightly understand on specific occasions just where the mean was located, and you needed to understand the consequences of departing from it.

There was a popular way of coping with this problem in civic circles, and it drew its force from ancient medical authority. This was the so-called Rule of Celsus (introduced in chapter 1), one version of which was the advice not to "live physically" when you were in fact healthy, and a related version pointed to problems attending any decision to live according to the principle of *moderation*. The rule was a widely cited medical principle associated with the problem of locating the golden mean. It offered an interpretation of the mean that appealed to laypeople as much as it might offend medical advocates of a strictly rule-governed way of life. Into the seventeenth century and beyond, Celsus rivaled Galen as a source of medical wisdom. Francis Bacon notably appropriated Celsus as authority for a proper reading of dietary moderation. "*Celsus* could never have spoken it as a *Physician*, had he not been a Wise Man withall; when he giveth it, for one of the great precepts of Health and Lasting; That a Man doe vary, and enterchange Contraries; But with an Inclination to the more benigne Extreme: Use Fasting and full Eating, but rather full Eating; Watching and Sleep, but rather Sleep, Sitting, and Exercise, but rather Exercise; and the like."[75] And similarly in his philosophical work on longevity, "where the Extremes are Hurtfull, there the Meane is best; But where the Extremes are Helpfull, there the Meane is Nothing. . . . Meane while, [the advice] of *Celsus*, who was not only a Learned Physician, but a wise Man, is not to be omitted. Who adviseth Interchanging, and Alternation of the Diet, but still with an inclination to the more Benigne: As that a Man should sometimes accustome himselfe to watching, sometimes to sleepe; But to sleepe oftenest: Againe, that hee should sometimes give himselfe to Fasting, sometimes to Feasting; But to Feasting oftenest."[76] There was no practical reason, Bacon said, to embrace asceticism. There was no evidence that it did you much good, and there were substantial problems associated with asceticism that the doctors and their moralist allies rarely considered. The Rule of Celsus retained its authority as a guide to conduct while showing itself open to interpretation. What *was* the rule for following the Rule of Celsus? And the answer given by many commentators greatly expanded definitions of what conduct came under its domain and what did not.

Bacon disputed the idea that dietary extremes, regardless of circumstances, were necessarily vicious and damaging. Sometimes, moving toward one extreme might be in itself more beneficial than moving toward the other: eating a lot was better for you than fasting, and sleeping a lot was better for you than staying long awake. And when this was manifestly the case, then the point of prudence was momentarily shifted toward—though not reaching—the beneficial extreme. For Burton, there was Hippocratic warrant for veering toward dietary plenty: "'They more offend in too sparing diet, and are worse damnified, than they that feed liberally and are ready to surfeit.'"[77] (It was not that the mean was a bad place to be; it was instead a matter of the side on which one might occasionally and safely err.) Bacon's essay on regimen fell in with Montaigne's sentiments: "In *Sicknesse*, respect *Health* principally; And in *Health*, *Action*."[78] Bacon practiced what he preached. His seventeenth-century biographer noted that Bacon followed "rather a plentifull, and liberall, Diet, as his Stomack would bear it, than a Restrained."[79] Bacon not only wrote but also ate like a lord chancellor, a great courtier, a philosopher who didn't live in contemplative isolation but rather for much of his life on the most public of stages.

Here and elsewhere, the sensibility inscribed in the Rule of Celsus was a powerful resource, enabling civic actors to talk back to medical experts and to do so in terms that physicians were obliged to acknowledge (if not necessarily to approve), using concepts and sentiments that authentically belonged to medical tradition. Writing in the first part of the eighteenth century, Feijóo y Montenegro noted what he called "the rule of Galen," which was always to get up from the table with a little appetite remaining, but preferred the Rule of Celsus as "more complaisant": "as he prescribes at times that we should exceed a little in our diet, and always that we should make the stomach digest as much as it can.—He advises us [that] we ought always to eat as much as we can, so the stomach does but digest it."[80] Civility *agreed with* Celsus.

The Meaning of Moderation: Access to Excess

John Aubrey, writing in *Brief Lives* about his friend Thomas Hobbes, said that the philosopher had on the whole lived a temperate life. Aubrey noted something that Hobbes had once told him; "that he did beleeve he had been in excesse in his life, a hundred times; which, considering his great age, did not amount to above once a yeare. When he did drinke, he would drinke to excesse to have the benefitt of Vomiting, which he did easily; . . . but he never was, nor could not endure to be, habitually a good fellow, i.e. to drinke every day wine with company."[81] In Hobbes's long life—he died at age ninety-one—that would probably amount to being drunk and puking maybe once or twice a year. That story was told about a unique individual, but it is a "portable" one, related and discussed by many people over very many years, and its persistence is a sign of focused contest over the meaning of moderation and over the authority of dietetic advice. Did moderation mean that you should *never* drink to excess? Did it mean that you should drink no more than a glass or two a day? Did it mean that you should not drink intoxicating beverages at all? Or, as the telling of the "Hobbesian" story was usually meant to recommend, did it mean that you could, even should, get drunk occasionally but not make a *habit* of it? Occasional drunkenness was one of the set pieces through which the proper meaning of moderation was continually debated. The prudence of moderation was *never* itself contested; what was at issue was the legitimate interpretation of what moderation *meant* and how moderation should figure in the course of ordinary social life.

Ancient medical expertise was cited for moderation as only occasional excess, and in this connection a dictum of Avicenna was repeatedly invoked to support the notion that it was good to get drunk from time to time and that its virtue had to do with the system cleansing of the subsequent vomiting.[82] Montaigne's approval of occasional excess encompassed both food and drink: "If he believe me, he shall often give himselfe unto all manner of excesse: otherwise the least disorder wil utterly overthrow him; and so make him unfit and unwelcome in all conversations."[83] You should get used to occasional excess, and if you didn't, then even one lapse might cause you serious injury. And since normal civic life *did* involve at least periodic excess—feasts and endless toasts—it was good to get your body used to handling those sorts of thing.[84] Much of therapeutic medicine was physician-directed taking of purgatives (to empty the bowels) and emetics (to encourage vomiting), and in lay sensibilities, the puking that followed alcoholic excess was thought to have similar purifying powers. Indeed, some sixteenth- and seventeenth-century physicians, even those of a modernizing disposition, recognized the judiciousness of drinking too much from time to time. In the 1580s, Thomas Cogan's *Haven of Health* summarized contemporary debate among physicians—all citing different ancient authority—that it was good to get drunk and to have the advantage of vomiting once a month, though he was unsure whether this was the advice of Hippocrates or "some *Arabian*

Physitian."[85] Even the physician Santorio—criticized by many for his overly fastidious weight-maintenance regimen—nevertheless wrote that "they who use a regular Diet, want the Benefit of those who Debauch once or twice a Month."[86] But for the most part, Avicenna's commendation of occasional drunkenness was more opposed than embraced by Renaissance and early modern physicians and moralists.[87] It was a lay belief and practice that the doctors were generally worried about and that they opposed in the name of moderation.

In the 1570s, Gratarolo noted the unsoundness of Avicenna's view—"alledging the opinion of diverse others"—"that to be dronk once in everie month is holsome and healthfull to the bodie." It was, he said, a practice especially harmful to those who "have weake and feeble braines."[88] The early seventeenth-century English doctor Tobias Venner fretted about the grip of lay opinion—supposedly based on the authority of "the ancient Physitions"—that "being drunke once or twise in a moneth" was good for you, by "inducing sleepe, . . . provoke[ing] vomiting, urine, and sweat," and expelling "the evill humors" from the stomach. But this was self-indulgent and dangerous, nonsense: "all drunkennesse is evill, and hurtfull to the true health of the bodie."[89] The English physician Sir Thomas Browne merged moralism and medicine, noting "that 'tis good to be drunk once a moneth, is a common flattery of sensuality, supporting it self upon physick, and the healthfull effects of inebriation. This indeed seems plainly affirmed by Avicenna, a Physitian of great authority, and whose religion prohibiting Wine could lesse extentuate ebriety." While Browne thought it was quite all right to drink to the point of merriness, he could not see that drunkenness was anything but unwholesome.[90] In the 1660s, the pious experimental philosopher Robert Boyle, much concerned with medical reform, similarly cited the opinion of those who maintained that "a constant Diet wants the help of those that once or twice in a Moneth do exceed," but he offered no support for any such view, nor did his virtuous and regular pattern of life suggest that he accepted it.[91]

Both medical opposition to intermittent drunkenness and physicians' recognition that this was a perniciously popular belief—

apparently supported by some ancient medical authority—continued into the eighteenth century and beyond. In the 1750s, it was illegitimate citation of Hippocratic warrant that irritated James Mackenzie: "Some have pretended that Hippocrates . . . advises people to get drunk on certain occasions. Others have gone farther, and recommended the getting drunk once or twice every month as conducive to health; and have quoted Hippocrates to justify their intemperance. But such opinions have no sort of foundation."[92] Avicenna's license continued to circulate in the culture. In the early eighteenth century, a pseudonymously published text, *The Praise of Drunkenness*, quoted "the Sentiment of the most able Physicians" of antiquity, approving occasional inebriation in bad verse—"That what to Health conduceth best, / Is Fuddling once a Month at least"—while the star turn was, again, Avicenna, who "absolutely approves getting Drunk once or twice every Month, and alledges it for physical Reasons."[93] The generality of physicians continued to oppose drunkenness—intermittent or habitual—but they were well aware that other meanings of moderation were loose in lay culture, that these meanings drew on other versions of ancient authority, and that they constituted important ways of talking back to expertise or of contesting its ownership.

Sense, Common Sense, and Expert Knowledge

Contest about dietary excess and about the meanings of moderation was one site where the experts talked back to lay skepticism. But opposing all sorts of lay talk-back was a well-developed and pervasive learned genre. A prominent popular error corrector was the French physician Laurent Joubert (1529–1582), and his 1578 *Erreurs Populaires* was a sprawling compilation of the silly things that laypeople—and especially the *common* people—believed. The "ungrateful common people" were, Joubert said, always likely to give themselves and their commonsense beliefs and practices the credit that rightfully belonged to medical expertise, just as they "forget rather easily the benefits they receive [from physicians' care] and retain in memory the most insignificant mistakes." They attribute to "nature's doing" or to "good soups" what was really done by physicians' expertise, uniquely effective dietetic advice, and well-judged therapeutic interventions.[94] Joubert rejected the idea that there were legitimate gentlemanly values more important than looking after your health or that submission to expert medical system conflicted with those supposed values. It was, the physician wrote, very prudent to live young as if you were old, for then you really would attain vigorous old age. Common sense itself conceded this, and the proverb here was "young old, old young." And if princes and public actors objected that they were busy and that it was better for them "to be loose and not observe any rules, schedules, or system," then Joubert settled for the best that he could get, and more than the laity were usually accustomed to giving. People should nevertheless observe "the strictest codes and rules [they] can possibly manage to apply, as much as . . . circumstances will allow."[95]

It was true that there were natural (or divine) limits to how long you would live, but it was within your competence—and after taking expert advice—to avoid *shortening* your allotted years, taking "food and drink in proper substance, quality, quantity, time, and order" and "to abstain from what, by consuming and taxing our natural heat, such as excessive work, spices, staying up at night, worries, and diverse emotions, but above all excessive carnal copulation, especially at improper times, and other similar things that one can and should avoid, according to the orders and prescriptions of medicine."[96] Folkish adages opposing sound dietetic advice and rejecting physicians' expertise circulated perniciously among the people, including the gentry and aristocracy, and they needed to be refuted. So Joubert collected those sayings and held them up to ridicule: "Against those who say that we will live until we die in spite of physicians"; "against those who say: 'Piss clear and make a fig at physicians'" and others who say that "he who pisses, sleeps, and wags well has no need of Doctor Bell"; against those who say that there are "more old drunkards than there are old physicians"; and especially "against those who say that to live on medication is to live in misery."[97]

Expertise continually countered lay skepticism, but neither side ever achieved clear victory. In the 1760s, the Swiss physician Tissot produced one of the most widely distributed attempts to instruct the common people, especially country people, how to avoid and cure the most common diseases. This was a text framed by despair about national depopulation and about the persistent popularity of erroneous beliefs contributing to ill health. The people lay down in cold places when they were hot, congregated in poorly ventilated rooms, piled animal dung near their houses, and drank wine when suffering from fever when they should take a light infusion of elder flowers or some sweet whey. Why did the common people think they knew better than the doctors? Why did they stubbornly persist in error? Why did they prefer mountebanks and quacks to genuine experts? Was it that the quacks cannily indulged their patient's "Tast and Humor" while the genuine physician "disgusts" by prescribing strict regimen?[98] The battle continued, and it remains unresolved. Yet, whatever skepticism the laity had about the quality, propriety, and effectiveness of expert counsel never seriously undermined experts' standing and the authority of dietetic ways of talking about health, disease, and the management of everyday life. Lay skepticism was a problem for expertise, though it was generally a manageable one. The physician's role persisted, and so too did a degree of skepticism about the quality of physicians' advice.[99]

The experience of taste is subjective, external experts cannot access your experience, and those experts can know about it only through your testimony. Taste is at once "soft"—varying, private, inaccessible, difficult to formally describe—and "hard," belonging to the experiencing subject and strongly motivating patterns of consumption. Both medical and lay culture possessed a sensibility that rendered the experiences of taste consequential and reliable, subjective indicators of the objective characteristics of both food and the consuming body. The notion that long expressed the causal connections between taste and bodily consequence was *agreement*, used to describe foods that went well with certain bodies and the pleasant or innocuous sensory experiences accompanying their consumption. Agreement appeared as both description and prescription: both dietetic expertise and lay wisdom held that you *should* eat those foods that *agreed with* you. “My rule,” Feijóo y Montenegro announced, “is that every person should follow what agrees with him best, as the most secure method. Indeed, with regard to eating and drinking, let our own experience be our guide in every thing.”[100] Agreement was framed by traditional dietetic ideas about the qualities of foods and the qualities of human bodies. Coming to *know* agreement was something you might do by reading any of the Renaissance and early modern dietary books detailing the characteristics of foods and the bodies for which they were suitable.

The Latin tag *quod sapit nutrit* (and its many ordinary language versions)—if it tastes good, it’s good for you—might have the aura of ancient authority.[101] Some writers attributed the formally cited version to Hippocrates and others to Avicenna, and there was Aristotelian warrant for the general sentiment: “It is by taste that one distinguishes in food the pleasant from the unpleasant, so as to flee from the latter and pursue the former.”[102] But the tag was probably proverbial, with no definite authorship. This is the connection in which it is useful to remember that *taste* once carried the general sense of *testing* or *trying*, as it did for Shakespeare: “I hope, for my brother’s justification, he wrote this but as an essay, or taste of my virtue” (*King Lear*, Act 1, scene 2). (Now, the tasting-trying sense obtains only in the case of cooking, as when the cook *tastes* in order to *test* whether things are right.) So, the *quod sapit* sentiment characterized both lay and expert medical sensibilities. It was a notion that experts used to give advice, but it was also a resource used to oppose expertise. It could be used to assert the sufficiency of self-knowledge; it warranted inference from sensory appeal to wholesomeness, from subjective experience to objective vital consequences. Agreement figured in experts’ advice to patients and in patients talking back to experts.

Traditional dietetic culture picked out the *sensible* aspect of edible natures. The seventeenth-century distinction between “primary and secondary qualities” deflated the power of subjective experience and altered appreciation of how the senses might know the world.[103] The findings of all the senses were rendered suspect by this distinction, while the information delivered through the contact senses of touch, smell, and especially taste was long discounted in philosophical traditions tracking back to Aristotle. But so long as the natural philosophical categories of traditional dietetics retained their cultural force, the deliverances of taste could, in principle, testify to what things—including food—were *like* and what powers they might exert when taken into the body. Yet, through the seventeenth century, there were commentators utterly unaffected by ancient or modern devaluations of the sense of taste. Writing in 1700, Thomas Tryon celebrated the sense of taste as a wholly trustworthy guardian of bodily health. Taste, he wrote, "may be justly stiled the Prince, King or compleat Judge over Life and Death." And that is why God placed the organ of taste in the mouth, ensuring that nothing passed further into the body that gave gustatory signs of unwholesomeness or danger: "The Palate, Taste and the Vessels or Servants thereunto belonging, are the proper and natural Judges, of what is good and agreeable."[104] The experiences of taste and of agreement belonged to *you*. Their medical legitimacy remained widely accepted by physicians while also offering a vocabulary that allowed laypeople—if they wanted—to assert the primacy of their experience.[105]

Quod sapit nutrit was at once philosophical generalization and practical maxim. It spoke about taste and the qualities of things and offered guidance to practical action based on your own accumulated sensory experience. *Quod sapit* sensibilities informed much medical enthusiasm for sweet things, though like many other exotic foods and spices, sugar had long been treated as a medicine and sold by apothecaries.[106] In 1620, an English physician announced that "sugar agreeth with all ages, and all complexions," though it was recognized that an excess of sweet things, like all other forms of excess, was bad for you, and there was worry that the attractiveness of the sweet might be an inducement to surfeit.[107] Shakespeare channeled a popular qualification to the wholesomeness of the sweet: in *Richard II* (Act 1, scene 3), John of Gaunt cautioned that "things sweet to taste prove in digestion sour." However, the Scottish chemist and physician William Cullen insisted that the agreeableness of taste was a wholly reliable guide to nutritiousness: "In general, the more sweet substances are all nutritious," while "those of an acrid, bitter, nauseous nature are improper. Every body, *en gros*, will allow the truth of this."[108]

In the 1670s, the French Cartesian philosopher Nicholas Malebranche defended the reliability of the senses, including taste, as guides to prudent dietary practice. A healthy person with senses in good order did not need a physician, Malebranche said. He was well aware of the common objection that "if we followed our senses, [we would] often eat poison," but he didn't believe it: "as to poisons, I do not think that our senses ever lead us to eat them; and I believe that if, by chance, our eyes excite us to taste something poisonous, we would not find it to have the kind of taste that would make us swallow it." Disgust—the sense of *bad* or *repellent taste*—was functional.[109] The alimentary exotic generally revolted, and it was good that it did so: the initially disagreeable taste of unfamiliar foods would cause us to take none or very little of them, and that too was prudential. In fact, if our tastes ever lead us astray, that is because our bodies are not in their natural state, since our senses have been corrupted by bad diet or because foods have been prepared in an artificial way, disguising their true nature. Beware of the elaborate dishes prepared by skilled cooks: "if cooks have found the art of making us eat old shoes in their stews, we must also make use of our reason and distrust these bogus meats that are not in the state that God created them." It was good to eat simply: we then taste things one by one and in their natural states.[110] Having celebrated the reliability of taste to pick out what was and was not good, Tryon understood how common it was for taste to be corrupted and its proper function lost. A taste for "improper and unequal Foods and unnatural Drinks becomes as it were natural by Life, Habit and Custom," and then "this Noble Sense" loses its ability to discern "the try Nature and Taste of Things, not knowing nor distinguishing their Friends from their Enemies."[111]

In the 1760s, the reliability of native taste and its pervasive corruption by custom and convention was a centerpiece of Rousseau's radical proposals for the education of children. In the state of nature, taste and appetite were sound guides to health. A taste for simple things was natural to us. Our first food was mild milk, and our initial experience of strongly flavored foods and drinks was unpleasant: tastes for these things had to be learned. There were indeed reasons why people might want to acquire tastes to which they were not born—for example, if their profession required them to travel to lands where people ate differently—but like Malebranche, Rousseau was unimpressed with his native cuisine: "the French are the only people who do not know what good food is, since they require such a special art to make their dishes eatable."[112] Eighteenth-century aesthetic philosophers addressed the relationship between judgment and taste, debating the emerging metaphorical extension of palate taste to aesthetic taste.[113] In doing so, they could call on a standard of taste that was grounded in both physiology and theology and held out the possibility of moral critique: "The taste of the palate," wrote Thomas Reid, "may be accounted most just and perfect, when we relish the things that are fit for the nourishment of the body, and are disgusted with things of a contrary nature." That sort of taste was "the manifest intention of Nature" and, so assured, we could search for the pathologies—mental derangement and bad habits—that had corrupted natural taste.[114] Taste was not *for* connoisseurship; it was the capacity that enabled us to observe divinely ordained natural law.

The topics that philosophers discussed in a theoretical mode were at the same time folded into the dietary and medical practices of everyday life, structuring relations between external dietary expertise and your own knowledge of particular foods and your own particular body. Montaigne recognized taste and his own body's responses to the tastes of food as sound bases for practical action, at least as sound as those supplied by the doctors. He considered that his appetite was a quite reliable guide to what was good for him. If he liked it, it probably liked him. He did not believe that he had ever been harmed by foods and drinks to which he was accustomed and that were agreeable to his taste, and he did not think that following his appetites had ever been injurious: "I never received harme by any action that was very pleasing unto me. . . . *It is in the hands of custome to give our life what forme it pleaseth*: in that it can do all in all. It is the drinke of *Circes*, diversifieth our nature as she thinkes good."[115] Montaigne was no fool: when his body began speaking to him in unaccustomed ways, he listened. When sharp sauces started to disagree with him, he went off them, and his taste followed suit; when he was ill wine lost its savor and he gave it up.[116] Many people recognized the connection between illness and wine losing its appeal. Malebranche later used just that example, writing almost a century later that "a man in fever, for example, finds that wine is bitter, and wine is also then harmful to them. The same man finds it pleasant-tasting when he is in health, and wine is then good for him."[117] And the fashionable diet doctor George Cheyne stated, as an example of the "*infinitely wise* Contrivance of Nature," that ill people in general experienced a "*Loathing* and *Inappetency*, or at least a Difficulty in Digestion," for all those foods whose consumption (or overconsumption) had caused the illness.[118]

The modernizing early eighteenth-century Dutch physician Herman Boerhaave offered a new kind of explanation for "why things agreeable to the Taste are also speedy Restoratives."[119] Feijóo y Montenegro followed Malebranche in arguing that "that we had better govern ourselves by our own sense, in the preservation of our health, than by all the laws of physic." Taste and smell were, on the whole, reliable guides to the fitness of foods, and "a strong desire . . . is a sign that the stomach has within it some ferment proper to dissolve the matter for which it so eagerly wishes." We may and should "follow the will of our appetite in the choice of what we eat and drink. Certain it is, that nature has made a union between our palate and our stomach, consonant to the habit of our bodies, and that, what is agreeable to the one, will be amicable to the other. The Almighty has given us senses to be as watchmen for our preservation, and that of taste alone will inform us what is conformable to our present constitution or otherwise. Experience shews, that the stomach never embraces with affection what the palate receives with disgust."[120] Into the nineteenth century and despite substantial conceptual changes in medicine and physiology, some medical and much lay sentiment continued to favor a version of *quod sapit*.[121]

That said, the *quod sapit* sentiment was intermittently qualified and suspected. In the sixteenth century, Cornaro understood the lay popularity of the notion that what tastes good is good for you, but he was keen to dismiss it. Before reforming his diet, he recalled that he used to follow "my Appetite, and did eat of meats pleasing to my taste, . . . telling my Physicians nothing thereof, as is the custome of sick people." He poured special scorn on "that Proverb, wherewith Gluttons use to defend themselves, to wit, *That which savours, is good and nourisheth*." It was a sentiment as demonstrably and hypocritically false as it was destructively self-serving.[122] And a mid-eighteenth-century English doctor (and poet), while agreeing that you should in general follow your natural appetites, lectured his patients not to let those appetites become perverted by the unnatural dishes elaborated by cooks, the effect of which was to corrupt the reliability of taste as a gauge of goodness.[123] The English philosopher David Hartley warned against foods having an "acrid" taste and anything initially "disagreeable" to the palate even though, like coffee, it might be "made grateful by custom." So, your first impression of an agreeable taste, not your habituated impression, was to be accounted natural and reliable.[124] Current custom—or the artificial stuff devised by fashionable cooks—had corrupted taste. Once, in humankind's natural state, taste had been a reliable guide to goodness; now, it might no longer be.[125]

Physicians from the Renaissance through the early modern period knew what they were up against when their patients (or those who *ought* to have been their patients) talked back to them by asserting the legitimacy of taste and appetite as reliable indices of wholesomeness. Some physicians fought back; others agreed with some version of *quod sapit*, seeking to harness it as a resource in the interplay between medical expertise and self-knowledge. But the principle was powerful, and not just because Hippocrates, Celsus, or Avicenna might be cited in its support. Traditional dietetics did indeed validate many forms of self-knowledge, and its categories linked the sensible aspects of foods to their ultimate qualities and bodily powers. To the extent that there was genuine and substantial discontent with traditional medicine—its efficacy and the legitimacy of the principles informing medical practice—there was a constituency for something more than the sinuous choreography that recognized appropriate roles for *both* self-knowledge and external expertise. That "something more" presented itself as a revolution in the conceptions of nature and of the human body upon which dietetic medicine had long depended.

6

Dietetics Modernized? Mathematics and Mechanism

Medicine Doesn't Work

Despite the authority and cultural reach of traditional dietetics, resistance happened. Moral and practical objections were the most pervasive forms of talk-back: it was considered wrong to subject yourself to rigid medical rule, other legitimate concerns of ordinary life made it difficult or impossible to do as the doctors advised, and self-knowledge was more pertinent than physicians' knowledge. But it was also possible to dispute claims of efficacy. In the Renaissance and the early modern period, there were indeed doubts about the power of traditional dietetic medicine to do what it claimed to do: prevent disease, maintain health, and prolong life. Much of that skepticism over efficacy, however, needs historical qualification. It was *not* then universally expected that physicians' ministrations could cure any and all diseases or that medical failures were evidence of inadequate knowledge and technique. Some diseases, it was accepted, just could not be cured, and sometimes physicians explained that they could do little because, for instance, they had been summoned too late in the course of disease. The category of epidemic disease was usually acknowledged to be beyond physicians' ability to prevent or do much to cure. And, of course, there were the afflictions of old age. Even with the emergence of vigorous criticism of traditional dietetic notions in the late seventeenth and early eighteenth centuries, physicians took for granted that they could not cure all patients and that a rational explanation of what was wrong (diagnosis) and how the disease would proceed (prognosis) was all that could be legitimately expected. As a modernizing English physician wrote in the 1720s, if the physician "should be so unfortunate as to lose his Patient (and who can prevent Death in all Diseases?) yet, if he timely prognosticate the real Danger attending the Disease, he will come off with Honour, and have the Satisfaction, of having done his Duty."[1] There were periodic promises that dietetic measures *could* ensure long life—Luigi Cornaro was a notably visible case—but neither dietetic physicians nor their later critics were commonly held to blame when their patients died aged three score and ten (or even less). It was common to think that your longevity was in God's hands but that you might *shorten* your allotted years through your freely chosen bad way of living.[2]

Yet the learned did offer systematic criticisms of medical efficacy, and some of these were particularly sharp in the modernizing movements of the sixteenth and seventeenth centuries. Here it was maintained that a fundamental reform of the content and methods of natural knowledge held out real promise of curing disease and extending life and, conversely, that the foundations of traditional medicine in the ancient philosophies of Aristotle and Galen guaranteed that potent cures or life extension could never be achieved. Medicine was currently impotent, but it could be made effective on the condition of revolutionary change in its intellectual foundations. Skeptical laypeople often inferred medicine's powerlessness from what they saw as pervasive disagreement among the doctors about the qualities of different foods; which foods, taken in which orders, suited which temperaments; regimens of drugging and bleeding; diagnoses and prognoses; the underlying physiology of the body; and, in general, the right thing to do in particular cases. You could always—it was thought—get a "second opinion" or a third or any number you liked. Some commentators thought that this expert variation was in the nature of things. Others reckoned that disagreement was a mark of professional hubris—doctors advertising novel and potent therapies in a competitive medical marketplace. An English cleric, writing around 1700, said that some physicians notoriously publicized the infallibility of their remedies, giving their patients to "think no Disease incurable." But therapies that worked in principle very often failed in practice, and doctors themselves differed in their assessments of what *counted* as effectiveness. Critics recommended that physicians would be well advised to be more modest about what they could deliver.[3]

From the lay point of view, medical dissensus might mean that you had to take care to select the right doctor, or it might be taken as a sign that *none* of the candidates for reliable knowledge and effective practice was very good. This was notably Montaigne's opinion: "If your Physitian thinke it not good that you sleepe, that you drinke wine, or eate such and such meates: Care you not for that; I will finde you another that shall not be of his opinion." You might as well do what you liked or observe no expertly sanctioned rules at all. The curative power of nature was, in any case, probably more effective than the educated art of any doctor: "A man must give sicknesses their passage: And I finde that they stay least with me, because I allow them their swinge, and let them doe what they list. And contrary to common received rules, I have without ayde or art ridde my selfe of some, that are deemed the most obstinately lingering, and unremoovably obstinate. *Let nature worke*. Let hir have hir will: She knoweth what she hath to doe, and understands hir selfe better than we do." Here, Montaigne knew that there were physicians who agreed with him. There was a long-standing position among medical experts, going back to Hippocrates, that insisted the physician should do little more than observe and, if possible, assist the *vis medicatrix naturae* (the healing power of nature, the tendency of the body, when disordered, to restore itself naturally to its normal state).[4] Although Montaigne did call in the physicians when he judged himself actually ill, he was largely skeptical about the doctors' curative powers: "The Arts that promise to keepe our body and minde in good health, promise much unto us; but therewith there is none performeth lesse what they promise."[5]

Robert Burton was one among very many laymen who "meddled" in medicine, because the melancholic conditions he dealt with were not exclusively the preserves of physicians and because even some medical authorities he cited were exasperated by expert disagreements. Burton indexed the opinion of the Danish "chemical" (and anti-Galenic) doctor Petrus Severinus on the brawling state of medical opinion: "Unhappy men as we are, we spend our days in unprofitable questions and disputations."[6] In the early eighteenth century, Feijóo y Montenegro offered his own views on preserving health, arguing against "the too great confidence placed by the world upon the professors of physic." They were so occupied by their busy attendance on the ill that they could not properly reflect on the grounds of their knowledge. They prescribed indiscriminately, and their remedies were often harmful. Still, Feijóo warned against unreasonable expectations about what *any* form of medicine could do: "The state of imperfection is all that can be expected in physic, as practiced by learned and prudent men; while ignorant empirics have driven it into that of error and corruption." The best physicians admitted their uncertainties and doubts, their powerlessness in the face of many diseases, and the lack of progress that had been made in prevention and cure. Even some authors of recent medical treatises pointed to the "uselessness" of many writings that were "full of obscurity, uncertainty, and falsehood."[7] Medicine was a mess. Some laypeople thought this, and so too did some physicians.

Better Medicine through Better Natural Philosophy

It was not the laity but rather the natural philosophers and chemists whose criticisms of existing medicine had the greatest effect on what eventually happened to traditional dietetics. There were many aspects to philosophers' vigorous rejection of ancient natural knowledge—and it bore upon many different intellectual and practical activities—but one promise was that the reform of knowledge would deliver better *medicine*: healthier bodies and longer life. New natural knowledge would allow a better description of the ultimate nature of the world—of the matter that made up food, bodies, and everything else—and better accounts of how the body worked, that is, how its solids and liquids were formed, how its vital functions were performed, how the body normally worked, and how it went wrong. Just as the ancient Greek world of elements and qualities was a foundation of traditional medical practice, prominently including dietetics, a new philosophy of nature might make possible a new and more powerful medicine. In the late sixteenth and early seventeenth centuries, Francis Bacon made wide-ranging criticisms of the state of medical science and technique. "Medicine," he wrote, "is a science which hath been . . . more professed than laboured, and yet more laboured than advanced; the labour having been, in my judgment, rather in a circle than in progression." Unless medicine was refounded on a purified register of natural fact and a proper philosophy of nature, there would never be more effective treatment of disease or the prolongation of life: "the science of medicine, if it be destituted and forsaken by natural philosophy, it is not much better than an empirical practice," working by trial and error and, for the most part, doing no good. Medical practice lacked both evidential discipline and philosophical system. But were it to achieve rigorous discipline and rational system, people might be free of disease and live much longer.[8] The "prolongation of life," Bacon wrote, is the "most noble" of all the parts of medicine, and he wrote extensively about how vast extension of human life might be achieved.[9]

But there were soon new ideas about how the human body worked that could be celebrated as advances over ancient knowledge. William Harvey's discovery in 1628 of the circulation of the blood was advertised as a paradigmatic achievement of contemporary anatomy and physiology, a concrete example of how ancient views of the body and its workings had been superseded.[10] Harvey did not offer his achievement as part of a new systematic understanding of nature, but it was soon recruited by modernizing natural philosophers. In France, Descartes's vision of the practical medical goods that would certainly be delivered to those possessed of right philosophical reason was more clear, coherent, and ambitious than that of any other seventeenth-century intellectual modernizer. The *Discourse on Method* of 1637 announced that a renovated philosophy would make us, "as it were, the lords and masters of nature." This mastery would, of course, be desirable "for the invention of innumerable devices which would facilitate our enjoyment of the fruits of the earth and all the goods we find there, but also, and most importantly, for the maintenance of health." Medicine "as currently practised does not contain much of significant use," and most honest medical practitioners would freely acknowledge the poor match between what we now know and what remains to be known. However, if medicine were to be refounded on proper philosophical principles, "we might free ourselves from innumerable diseases, both of the body and the mind, and perhaps even from the infirmity of old age." The essay concluded with Descartes's resolution "to devote the rest of my life to nothing other than trying to acquire some knowledge of nature from which we may derive rules of medicine which are more reliable than those we have had up till now."[11] Descartes hinted to friends and associates that he was hot on the trail of a powerful new form of scientific expertise, ensuring that death might be indefinitely postponed or even banished. It was not a matter of "knowing yourself"; it was about knowing the real nature of matter and its combinations and modes of motion.

The preface to Descartes's *Principles of Philosophy* (1647) introduced the celebrated metaphor of philosophy as a tree: "The roots are metaphysics, the trunk is physics, and the branches emerging from the trunk are all the other sciences, which may be reduced to three principal ones, namely medicine, mechanics and morals."[12] This is how things should be and, following Cartesian reform, how they would be. In *The Passions of the Soul* (1649), Descartes (or at least a close associate) wrote that medicine, "as it's practiced today by the most learned and prudent in the art," does *not* depend on physics: "they are content to follow the maxims or rules which long experience has taught, and are not so scornful of human life as to rest their judgments, on which it often depends, on the uncertain reasonings of Scholastic Philosophy."[13] Descartes's mechanistic physiology was formally elaborated in a number of works: the

Discourse, the *Treatise of Man* (comp. 1629–1633), the *Description of the Human Body*, and *The Passions of the Soul*. Descartes had read Harvey's tract on circulation, and while Descartes disputed aspects of the English physician's views, he developed the notion of the heart as a pump whose mechanical action was fundamental to vital function.[14] The heart powered the animal machine. Its motion was the source of all other bodily motions, and its natural heat was the natural source of those motions. The heat of the heart rarefied the blood, "and this alone . . . is the cause of the heart's movement," and in the *Discourse on Method*, Descartes offered this account of the motion of the heart as *the* exemplar of his mechanical philosophy.[15]

Having sketched his physiological system in the *Description of the Human Body*, Descartes declared that "this will enable us to make better use both of the body and of soul and to cure or prevent the maladies of both." And given the significance of cardiac theory in his physiology, Descartes insisted that without accurate knowledge of "the true cause of the heart's motion"—which he considered he now had—"it is impossible to know anything which relates to the theory of medicine."[16] The body was a hydraulic machine, with vital heat as its motive force, and the passage of particles through its various pipes and canals described how the machine worked.[17] So, for example, digestion occurred mechanically in the stomach "by the force of certain liquids, which, gliding among the food particles, separate, shake, and heat them just as common water does the particles of quicklime, or aqua fortis [nitric acid] those of metals. . . . And the food is ordinarily of such a nature that it can be broken down and heated quite of itself just as new hay is if shut up in the barn before it is dry." The agitation and heating of the particles of food break them up, with the coarsest particles being excreted through the rectum and the finest and most agitated flowing through pores in the vein that takes them to the liver, where the particles went depending on their size and states of motion.[18]

Like Descartes, Robert Boyle was much concerned with medical therapeutics, and his advocacy of pharmaceutical "specifics"—remedies acting on a particular disease or organ of the body rather than systemically on the humors—was framed in terms of his version of corpuscular natural philosophy. "The Physitian borrows his Principles of the Naturalist," Boyle wrote, so the doctor should borrow only the best theory. Until we have a better natural philosophy, "'tis hard to arrive at a more comprehensive Theory of the various possible causes of Diseases." Boyle insisted that "a deeper insight into Nature may enable Men to apply the Physiological Discoveries made by it (though some more immediately, and some less directly) to the Advancement and Improvement of Physick."[19] The "Dietetical part" of medicine would be advanced; the quality, the nutritiousness, and even the taste of foods and drinks would be improved; the maintenance of health would be more effective; the causes of disease would be better understood; the mechanisms of drug action would be comprehended; and new drugs would be discovered.[20] Mechanical natural philosophy issued a new promissory note.

Even following the widespread adoption of new philosophical foundations in the late seventeenth and early eighteenth centuries, modernizers continued to reflect on the past poverty of medicine. The Scot Archibald Pitcairne, briefly professor of medicine at the University of Leiden, wrote about "why the *Art of Physic* has so long baffled the Endeavours and Studies of so many Learned Persons," and the hope he held out for better medicine resided in the new mechanical and mathematical natural philosophy. So much had recently been achieved in the sciences, but in medicine "we who have shook off the weight of Stupidity" have achieved little. Medicine could never be a certain science if its philosophical foundations were uncertain, and so it needed the degree of intellectual assurance that, he declared, could only come from the mathematicians and their allies. Nothing should be allowed as a medical principle that could be called into question by "persons who are the least entangled with Prejudice."[21] Aphorisms, empirical experience, and the discredited principles of ancient natural philosophy were not up to the task; what medicine needed and deserved was the certainty of demonstration. Pitcairne pointed to both medical dissensus and lay skepticism as reasons to start all over again. "An Art which of all others promises Safety and Health to Mankind, ought not in reason to be involved in the Conjectures and Dreams of Disputants," Pitcairne wrote, "for no Man of common Prudence would entrust his Life to Him, whose Reasoning seems false to the Generality, and probable to very few." The doctors ought to adopt the intellectual standards of the astronomers and the mathematicians, whose principles and findings, he said, allowed no dissent. Medical dissensus was not then universal. The broad principles of dietetics and especially the importance of managing the nonnaturals in preventive medicine were still widely accepted, but much else had long been disputed, notably what to do in particular cases. What was increasingly called into question during the seventeenth and early eighteenth centuries was whether the traditional vocabulary of elements, qualities, humors, and complexions had much value.[22]

Medical reform and the hope for more powerful medical practice were close to the heart of seventeenth-century programs to modernize natural philosophy. Life and death issues mattered. They mattered to ordinary people and to the learned too, since even Francis Bacon, Robert Boyle, and René Descartes had vulnerable and, as it disappointingly turned out, mortal bodies about which they cared a great deal, viewed at times as instruments to their extraordinary souls and at other times just as everybody else views their bodies, wishing that they would be less painful, more serviceable, and more durable. That is one reason why the ability to prevent and cure disease, alleviate suffering, and extend human life was recurrently used as a public test of the truth and power of philosophic and scientific systems and why the learned too might share in that public assessment. There were, however, reasons particular to the learned classes, and more particular to the great philosophical modernizers of the seventeenth century, as to why medicine mattered, even to those who were not themselves physicians. In general terms, medicine was bound to matter because of the close but contested relationship that historically obtained between natural philosophy and the science of medicine and, through the science, the practice of medicine. For Aristotelians and Galenists, knowledge of the hidden makeup and workings of the human body was wholly integrated within an overall philosophy of nature. The humors and the temperaments were just parts of the same system of elements and qualities that allowed philosophers to explain what the natural world was made of and why inanimate bodies moved as they did. For Aristotle, Galen, and their followers, "medicine was the philosophy of the body," or, put another way, medicine was philosophy in action, put to the test. And so it remained for both the ancients and their modernizing critics.[23]

Nature's Little Machines and How We Know about Them

The reformed natural philosophy that promised to deliver better medicine depended on changes in understandings of natural reality and the conditions on which we could know it. In ancient Greek philosophy, the qualities of things in the world were significantly available to the senses. Things that possessed the quality of hotness were so sensed by the body, those that were moist were sensed as moistness, and so on. Beyond the traditional four qualities associated with the four elements, there was in general assurance that our sensory experience of things might access the nature of things: the sweetness and redness *sensed as* belonging to a rose, for instance, did indeed *belong to* the rose. The human senses were in that way imperfectly but broadly reliable philosophical instruments; the categories of things sensed corresponded to and were caused by the qualitative categories belonging to things.

A challenge to this story about the status of our sense-based knowledge was, at the same time, a challenge to the story of about how things in the world were ultimately to be *described*, that is, what they were *like*, at the most fundamental level. In ancient philosophical schemes, sensations of things matched the nature of things; in the proposed new scheme there was no such match, and a distinction was made between what were called "primary" and "secondary" qualities. That distinction had ancient roots, but it became central to the modernizing tendencies of seventeenth-century natural philosophy. Its earliest clear seventeenth-century articulation was by Galileo in the 1620s, though it was repeated and developed by Boyle, Locke, Descartes, and other modernizing philosophers. In *The Assayer* of 1623, Galileo noted that people commonly have sensory experience of objects they call "hot." As a report of subjective sensation, there is nothing wrong with saying that "this pot is hot." Where people go wrong, Galileo said, is in supposing that "heat is a real phenomenon, or property, or quality, which actually resides in the material by which we feel ourselves warmed." Though we cannot conceive of an object without thinking that it has a certain shape, size, place, and motion, Galileo noted that we can easily think of objects that are not red, sweet, or hot. These latter are qualities that present themselves to our senses when we encounter a particular object, not qualities that belong to the object in itself: "Hence I think that tastes, odors, colors, and so on are no more than mere names so far as the object in which we place them is concerned, and that they reside only in the consciousness."[24]

Primary qualities, by contrast, are those that really belong to the object *itself*: the size, shape, "texture" (or spatial arrangement), and the motions of its ultimate material bits. They are called primary because no object or its constituents can be described without reference to these qualities. Secondary qualities—rigidity, yellowness, bitterness, warmth, moistness, and so on—are *derived from* an object's primary qualities and therefore belong to not the object but rather to the processing work done by the senses. The primary causes (and is held to explain) the secondary. You might *think* that the rose is red or that the plum is sweet but, in terms of the ultimate realities of its ultimate bits, there is no quality of redness *in* the rose or sweetness *in* the plum. The fundamental constituents of things are in themselves neither rigid, nor yellow, sweet, or hot, but their size, shape, arrangement, and motions may produce these *subjective* effects in us. All the experienced diversity of natural objects is to be accounted for by the mechanically simple and primitive qualities that necessarily belong to all bodies as bodies, and not to roses, rigid iron bars, or plums as types of bodies. As Locke put it at the end of the seventeenth century, "There is nothing like our *Ideas* [of bodies], existing in the Bodies themselves." The ideas we have of color or warmth or sounds are but the effects *on us* of "the certain Bulk, Figure, and Motion of the insensible Parts" of bodies.[25] And what applies to color or heat or sound applies, Locke insisted, also to taste and smell: if the eye and the ear and the tactile sense do not truthfully testify to ultimate reality, neither do the sensations belonging to the tongue and the nose. The distinction between primary and secondary qualities drove a wedge between the domain of philosophical knowledge and that of commonsense experience, including the sensory experience you might have of foods and drinks.

Subjective experience was severed from accounts of what objectively existed. Our actual sensory experience, it was now insisted, offered no reliable guide to how the world *really was*. A gulf now officially existed between how things *seemed* and how things *were*. Gulfs of that general sort had long existed. The Aristotelian world, for instance, presumed the existence of entities called "substantial forms"—invisible organizing principles that gave this dog its "dogness" and this particular plum its "plum-ness"—and these "forms" were roundly criticized by seventeenth-century modernizers as occult, unintelligible, dreamt into existence. Yet, occult as the forms might be, they were knowable by the same means that allowed people to recognize specific dogs and roses when they were encountered: forms were *abstractions* from particular sensed things. Modernizing natural philosophers contested the idea that ordinary sensory experience accurately testified to the nature of things—truth and appearance were fundamentally different—but as they offered accounts of what ultimate reality was like, they also offered assurance that this reality was knowable by *reason*.

That world was ultimately made up of small bits of stuff, characterized by geometry—their size, shape, and arrangement into certain configurations—and by their state of motion. That is to say, you should think of nature as you thought of artificially designed *machines*. Machines were made up of inanimate parts—of certain sizes, shapes, and arrangements—and could be put in motion by a range of external forces. Machines could produce patterned effects, *suggesting* to the untutored mind that they possessed capacities such as reason, but they had no such capacities in themselves, and all such appearances resided in human agency: the intelligence that arranged machine parts to do work and to produce intelligence-imitating effects. You could think of the solar system as great clockwork, and you could think of animal bodies as machines such as the intricately crafted medieval and Renaissance automata—roaring lions, singing and wing-flapping birds—that adorned aristocratic gardens and collections of curiosities. Descartes did not invent the notion that you should consider animals to be machines—that existed in medieval thought—but he offered a mechanical explanation of vital function as a model of intelligibility.[26] That thinking should be continued down in physical scale: the so-called mechanical philosophy of the seventeenth century argued the legitimacy, intelligibility, and fruitfulness of conceiving of nature's littlest bits—the parts out of which the visible and tangible parts of nature are composed—as *micromachines*. Those ultimate bits—variously called particles, atoms, and especially corpuscles—possessed only the *primary* qualities given by geometry and mechanics but not the *secondary* qualities which they occasioned in the senses. In these ways, the distinction between primary and secondary qualities challenged existing accounts of how you knew the world, while micromechanism constituted a new language to put in place of the four elements and associated four qualities.

How did you know that the world was so constituted? And what was the quality of such knowledge? Locke did not think that we could ever understand exactly how micromechanical states of affairs produced the visual sensation of redness or the gustatory sensation of sweetness.[27] The security of the micromechanical account of nature was based not on sensory access to the world of the very small but rather disciplined rational inference. A few philosophers, such as the mechanic and experimentalist Robert Hooke, hoped that new and improved microscopes would eventually make visible "the figures of the compounding Particles of matter": the microscope had made visible hitherto invisible domains of the very small, and the microscope revealed granularity in what unaided vision perceived as smooth or unitary. Perhaps microscopes more powerful than any yet available would carry vision into the domain of ultimate granularity, the fundamental corpuscles, but most accepted that the corpuscular world would forever remain inaccessible to human vision. *Analogy* between the visible and the invisible worlds did more work here than was done by existing microscopic technology.[28]

There were modernizing natural philosophers confident in their knowledge of what ultimate particles were like: some were said to be triangular while others were spherical, and some were large while others were small, but the intelligibility of these claims rested on grounds other than direct or instrumentally aided sensory access. Different seventeenth- and early eighteenth-century "mechanical philosophers" offered different versions of the micromechanical world, how certain you could be of the structure and processes of that invisible microworld, and how micromechanical structures and processes produced the vast range of phenomena that ancient philosophy supposedly explained. Boyle gave a hugely influential mechanical account of air pressure while doubting that the micromechanisms causing that pressure could ever be known with anything approaching certainty. In France, Descartes—while formally aligning himself with the view that our knowledge of micromechanical realities could be no more than *probable*—went on to produce explicitly hypothetical but highly detailed and specific accounts of the micromechanisms explaining a wide range of natural phenomena—gravitation, magnetism, the nature of light and vision, everything about animals other than human beings, and much about human beings—including reflex action, emotion, and digestion.

Since sensory access to the micromechanical world was not possible, specifying the geometrical and kinetic properties of ultimate particles lacked empirical warrant. Analogical inference figured in traditional dietetic reasoning about the nature of things, and analogical inference was also significant in the credibility of mechanical philosophy. For example, the French philosopher Jacques Rohault, a follower of Descartes, drew attention to our familiarity with ordinary workmen's tools and how the different shapes and sizes of these tools allowed them to perform different functions. Rohault also invited readers to find it "reasonable to think that the most imperceptible Parts of Matter, seeing that they have every one a certain Figure, are also capable of producing certain Effects in Proportion to their Bigness, like those we see produced by the grossest Bodies."[29] The notion that the sensory experience of heat was produced by the percussive movements of the ultimate particles of matter also helped itself to an analogy from the familiar sensory world: if, for example, you rapidly rub your hands together, you will feel an increase in warmth.[30] When Boyle hypothesized about how air pressure might be ascribed to the properties of the corpuscles of air, one speculation was that they were like little springs, exerting a resisting force when they were compressed, and another was that they might be shaped and arranged like the fleece of wool. Other speculations about the micromechanical domain helped themselves to analogies from the visible ordinary worlds of billiard balls, balances, gear wheels, screws, and, of course, the properties of human-made machines, while still others invoked familiar macroworld phenomena of magnetic attraction and chemical affinity. The opportunities and resources for philosophical speculation about micromechanism were enormous, and they were aggressively exploited by modernizing natural philosophers and also by those concerned with the reformulation of medical knowledge and medical practice.

Invisible Particles and Their Medical Powers

The inspiration for new medical ways of thinking about food and its physiological consequences came in part from the mechanical and mathematical natural philosophies of Descartes and Newton and in part from traditions of chemical thought that mechanical philosophers sometimes drew upon and sometimes opposed.[31] In work done in the late 1620s and early 1630s, Descartes speculated widely on how corpuscles' geometrical and kinetic properties might account for a wide range of natural phenomena, including those of the human body: "We see clocks, artificial fountains, mills, and other such machines which, although only man-made, have the power to move of their own accord in many different ways," Descartes wrote, and we ought to conceive of the human body and its vital functions in the same terms, qualified only by the facts that all animal machines were divinely designed and that unlike man-made automata, the human machine was ultimately animated by an immaterial, rational soul that alone escaped the mechanical frame.[32] Later, Boyle announced that for humors and qualities, you should just substitute the notions of corpuscular textures and mechanisms: "I think the Physitian . . . is to look on his Patients Body, as an Engine, that is out of Order, but yet is so constituted, that, by his Concurrence with the Endeavours, or rather Tendencies, of the Parts of the *Automaton* itself, it may be brought to a better State." Following Cartesian sensibilities, Boyle wrote that human bodies are "Living *Automata*" and that same notion proved popular among many physicians.[33] Automata could be comprehensively understood by their component parts, so the physician's task was to comprehend and, if necessary, adjust the mechanical components of ailing human bodies to restore the human machine to normal function. John Locke—a physician as well as a philosopher—reckoned that we might, in principle, come to understand the medical powers of drugs and foods if only we could come to know their micromechanical properties:

> I doubt not but if we could discover the Figure, Size, Connexion, and Motion of the minute constituent parts of any two Bodies, we should know without Trial several of their Operations one upon another, as we do now the Properties of a Square, or a Triangle; and we should be able to tell before Hand, that R[h]ubarb would purge, Hemlock kill, and Opium make a Man sleep; as well as a Watch-maker does that a little piece of Paper, laid on the Ballance, will keep the Watch from going till it be removed; or that some small part of it, being rubb'd by a File, the Machin would quite lose its Motion, and the Watch go no more.[34]

In the early years of the eighteenth century, George Cheyne pronounced the mechanical revolution in medical thought and practice essentially complete. The human body just *is* a hydraulic machine, and the function of medicine was effectively to adjust its hydraulics: "The Wiser part of Mankind are now perswaded, That this Machine [our body] We carry about is nothing but an Infinity of Branching and Winding Canals, fill'd with Liquors of different Natures." If you want to know how to restore a diseased human body, think of what you do with a machine that is out of order: you find and isolate the defective functional part, identify that part as the efficient cause of derangement, and repair the part.[35] Digestion, reflex action, the production and effects of the emotions, and so on were all to be accounted for in mechanical terms. (Cheyne sometimes invoked a notion of animal spirits compatible with a hydraulic mechanical model, sometimes postulating variously configured corpuscles making up the body's substance.) So, apart from reformulated notions of the spirits, the vocabulary of traditional medicine—elements, qualities, humors, complexions—was to be purged and replaced by a new set of concepts for conceiving of vital and medical phenomena. The body was a machine; its invisible bits and the invisible bits of its aliment were quality-less corpuscles whose mathematical characteristics accounted for vital functions. Some natural philosophers exempted a range of vital phenomena from mechanical explanation, but Descartes's full-blooded (though formally hypothetical) mechanical account of bodily function enjoyed enormous appeal among the learned.

In the mid-eighteenth century, the French materialist physician Julien Offray de La Mettrie (1709–1751), in his *Man a Machine*, praised Descartes for being the first "who perfectly demonstrated animals to be mere machines" but announced that he was going further than Descartes. La Mettrie attributed self-organizing powers to matter and declined any exemption for human beings: "The human body is a machine that winds up its own springs," and "organized matter is endowed with a moving principle." There was no need for an immaterial soul powering the body. "He is the best physician, and most worthy our confidence," La Mettrie wrote, "who has the greatest knowledge in the physical and mechanical constitution of the human body; who does not trouble himself about the soul, nor all that train of perplexities which this chimera is apt to raise in foolish ignorant brains, but is seriously busied with pure nature."[36] In fact, Descartes did almost nothing to bridge the gap between the theory of medicine and its practical realization in hygiene and therapeutics. He rarely spelled out exactly how it was that certain dietetic practices or therapeutic interventions *worked* in mechanical and corpuscular terms or how proper philosophical foundations would actually lead to new and uniquely effective medical practice. With a few exceptions—notably concerned with how the passions acted on the soul—Descartes's medical mechanism was largely an *argument in principle*, following a model of what intelligible explanations look like.[37]

Descartes was one major source for the philosophical reformulation of medical thinking; another emerged from some of the work of Isaac Newton. Newton's *Mathematical Principles of Natural Philosophy* (1687) had nothing about the human body, its functions, or the science or practice of medicine, but throughout his career Newton was greatly interested in chemical (indeed alchemical) processes, and these interests overlapped with physiological concerns. A short essay composed in 1692 dealt with acids, and here as well as in speculations contained in later editions of the *Opticks* (originally published in 1703), he hypothesized about how some elementary properties of matter might bear upon both chemical and vital phenomena. In the earlier paper "On the Nature of Acids," Newton suggested that short-range attractive forces could account for the cohesion of bodies and their volatility and flammability, and he offered some speculations about the size of fundamental particles.[38] The later sprawling speculations in the *Opticks*—carefully identified not as certain knowledge but instead as "queries"—dealt with how "a great Part of the Phænomena of Nature" *might* be produced through supposing that the smallest particles of matter possessed "certain Powers, Virtues, and Forces, by which they act at a distance." Speculations about short-range attractive forces between particles were extended to vital phenomena: these forces might be the key to how "the Parts of Animals preserve their several Forms, and assimilate their Nourishment; the soft and moist Nourishment easily changing its Texture by a gentle Heat and Motion." It was "the Business of experimental Philosophy to find out" just what these attractive forces were and how they worked in nature.[39]

There were physicians who were dazzled by the new agenda elaborated by modernizing natural philosophers and especially by what they saw as the fertility of mechanical and mathematical modes for a revolution in physiology and medicine. Here, they considered, were views of nature—its constituents, properties, and powers—that could replace Aristotle and Galen and fulfill dreams of a powerful medical practice that could prevent disease, cure those diseases that could not be prevented, and especially extend life.[40] The already-acknowledged predictive and explanatory power of these new frameworks lay partly in their applications to problems in natural knowledge, for example, in astronomy, mechanics, optics, and pneumatics (the study of the properties of air). But immense enthusiasm soon developed for generalizing mathematical and mechanical modes in the study of many other sorts of natural phenomena, notably including those of human life and its medical management. What emerged in the late seventeenth and early eighteenth centuries was no single coherent mechanical and mathematical vision for the reform of medicine but rather a disparate *range* of programs contending for legitimacy as the most potent application of the new philosophies to medical problems.

Some modernizing critics of Aristotelian and Galenic medical frameworks opted strongly for mathematics. The influence of Santorio Santorio's 1614 *Medical Statics*, his array of instruments to quantify physiological processes, and especially his celebrated weighing chair, remained influential well into the eighteenth century, though Santorio's methods were not grounded in a micromechanical view of the body. It was Santorio's admirers who saw the possibility of joining broadly quantifying impulses to micromechanical theory. One of the most radical and most coherent programs for a new medicine was elaborated by Archibald Pitcairne, Newton's Scottish follower. What was required, Pitcairne reckoned, was a medicine securely founded on mathematics, hence the compound designation "iatromathematics" (*iatros* is Greek for "healer"). The ambition was to reform medicine on secure modern mathematical and mechanical foundations, the two modes so often occupying the same intellectual terrain that iatromathematics and iatromechanism were often interchangeable terms.[41] But agendas for mathematizing and mechanizing physiology and medicine had other sources than Descartes and Newton. From the middle of the seventeenth century, the Italian Galileans Giovanni Borelli and Lorenzo Bellini were concerned, respectively, with the mathematical modeling of muscular motion and secretion. Boyle's pneumatic research bore importantly on respiration, digestion, and animal heat, and the English physician Thomas Willis, while specializing in the anatomy of the brain and the nervous system, combined mechanical and chemical frameworks in a search for medical understanding.[42] But Pitcairne's agenda made a distinction between mechanism and mathematics: Pitcairne was skeptical of identifying the physical *causes* of vital processes and enthusiastic about the potential of pure mathematical modeling.

Medicine could be a secure science, Pitcairne maintained, only when it rested on mathematical foundations and took on the certainty of mathematics. His fellow iatromathematician Richard Mead announced that soon "mathematical learning will be the distinguishing mark of a physician from a quack; and that he, who wants this necessary qualification, will be as ridiculous as one without Greek or Latin."[43] Mathematicians could identify the laws that regulated the behavior of physical bodies even if they could not necessarily discern the causes of that behavior. Physicians ought to embrace the same distinction. They had to know *that* therapies worked, and they had to know the laws of therapeutic operation. They need not speculate about *how* therapies operated, though such knowledge might be desirable in principle. And here the analogy with Newton's law of gravitation was exact: you could write down the laws of attraction between any two objects, and that sort of knowledge was mathematically certain, but you need not know how it was that gravity physically operated. Pitcairne was then familiar with a range of speculations about the physical bases of vital phenomena—about fermentations; the relationship between the body's pores, its "sieves" and "strainers" (glands) and the particles of its liquids; and the geometrical properties of aliment—and he judged that knowing any of these supposed physical causes and structures was formally unnecessary to the theory, much less the practice, of medicine. Unwarranted speculations of this type undermined the recognized certainty of medicine.

George Cheyne, Pitcairne's protégé and fellow Scot, made a stunning start to his medical career with *A New Theory of Continu'd Fevers* (1701), based on what he took to be Newtonian methods and concepts. In a text liberally sprinkled with geometrical diagrams and equations, largely concerning the flow of liquids in the body, Cheyne presumed that "the General and most effectual cause of all Fevers"—and, as he later wrote, of many other diseases—is the "Obstruction or Dilatation . . . of the *Glands*." Blocked flow in the body's canals resulted in the "Stagnation of the Fluids" so as to "vitiate their nature." This blockage was the proximate cause of fever, together with many of its manifest symptoms: headache, elevated temperature, thirst, difficulty in respiration, etc. The free flow of liquids—blood and the nervous fluid—defined health; physical restrictions of that flow made for disease.[44] Hydraulic theory was to inform therapeutic practice, and much of Cheyne's subsequent care of his patients involved the management of diet, drugging, and bloodletting to maintain or enhance fluid flow.

At about the same time, the English physician Jeremiah Wainewright joined the iatromathematical chorus, blandly confident that "what Improvements there have been, or are likely to be made in the Theory of *Medicine*, are only under the conduct of *Arithmetick* and *Geometry*." Mathematical principles underpinned the workings of the "curious *Machine*" that is the human body, the laws of which are "govern'd by the Laws of *Gravitation*, *Impulse* and *Reaction*," as established by Newton and other modern natural philosophers. Wainewright understood that some physicians were in principle dismissive of "Theory," but he insisted that effective medicine depended on a mastery of "the known Laws of Motion as aply'd to *Mechanics*, and *Hydrostatics*, with the application of 'em to the alterations made in Human Bodies."[45] Hippocratic stress on the medical importance of airs was only to be understood with reference to Boyle's findings about air pressure and the springiness of its corpuscles, and it was this discovery that explained the pathological response of the body to weather changes. The same explanatory power was brought to bear on the distribution of certain kinds of disease. Agues—usually taken as fevers in general or, specifically, the intermittent fever that came to be called malaria—abound in the Cambridgeshire fens and in the marshy parts of Essex, because the saturated earth fills "the *Air* with Vapours, whereby its *Elasticity* is weakned, the Fibres of the Body relaxed, and the Pores of the Skin obstructed."[46]

The English iatromechanist Nicholas Robinson declared that "certainty" in medicine was now available and that it depended on mastering "the general Laws of Matter, its Extension, Figure, and Motion, its Gravitation, Attraction, and Repulsion, as also its Density and Division," all such things known by courtesy of the then still-living "Ornament of the *English* Nation, Sir Isaac Newton."[47] The human body was a machine, with its animating "Springs, Wheels, and Pullies." That had become a standard iatromechanical gesture, but Robinson reckoned that *all* the principles of natural philosophy could be encompassed by the Newtonian law of attraction and its hypothesized pair, the law of repulsion. Attraction was a philosophical way of accounting for the coherence of all material bodies; repulsion was the principle of motion and change.[48] There was only a loose connection between the advertised certainty of these general laws and the explanation of any specific disease, and Robinson's medical management centered on achieving some notion of physically based "balance" between the body's solids (fibrous nerves, muscles, ligaments, vessels, and bones) and its fluids (blood and chyle).[49]

Not all medical modernizers of the late seventeenth and early eighteenth centuries, even those enthralled by the achievements of Descartes and Newton, were quite so enthusiastic about mechanical and mathematical methods. If mathematics and mechanics constituted two aspects of the modern rejection of ancient authority, then rigorous observation, the control of testimony, and experiment were other ways of opposing Aristotle, Galen, and their still-numerous followers. In the Netherlands, the immensely influential medical professor Herman Boerhaave (1668–1738) was greatly impressed by the natural philosophies of Boyle and Newton. Boerhaave offered a view of the fluids passing through the body according to mechanical laws and argued that "qualities arising from the magnitude, figure, and motion of bodies" must be considered in medicine.[50] But at the same time, Boerhaave admired Hippocrates and despised adherence to metaphysical systems, especially the deductive forms associated with the Cartesians.[51] Medicine should free itself of the dead hand of ancient authority, but it should do so without enslaving itself to mechanical systems. The Anglo-Dutch doctor Bernard Mandeville (1670–1733) fell in with this sentiment: "I know it is a received Opinion now-a-days, that a Man of Sense who understands Anatomy, and something of Mechanick Rules, ought to penetrate into the Manner of every Operation that is performed in a Human Body, it being a mere Machine." It was rational to think that the particles of different types of food had different geometrical characteristics and that these characteristics had specific effects in the body, but the security and the utility of mechanical explanations in medicine were being oversold.[52] For many modernizing physicians, the ambitions of the new systematizers had to be disciplined by the close observation of cases and the accumulation of large stocks of empirical knowledge, and here the empirical and theoretically modest work of the seventeenth-century English physician Thomas Sydenham (1624–1689)—sometimes called "the English Hippocrates"—was also important to what counted as modernized medicine.[53] Mathematical and mechanical models might count as the leading edge of medical progress, but their proponents still had to confront the authority of empirical experience.

Eating Machines

The modernizing physicians of the early eighteenth century wanted to replace the vocabulary of traditional dietetics with mathematical, mechanical, and chemical categories, a replacement that they considered would guarantee better medical practice. The proposed successor vocabulary was a pastiche: there was no single and coherent mode of iatromechanism, iatromathematics, or iatrochemistry that took the place of much-criticized Galenism and traditional dietetics. Different physicians proposed different ways of conceiving the makeup of foods and their role in the body; individual practitioners often moved back and forth between "old" and "new" categories, and versions of the "new" or even of what counted as mechanical or mathematical modes differed markedly. Nevertheless, the reform of medicine was represented as rational, and much of its power was considered to flow from *some* version of micromechanism.

Deferential gestures to the philosophical authority of Descartes and Newton were routine, but the resources taken from the modern philosophical masters varied, as did their extension into the domains of medical theory and practice. The Cartesian Rohault speculated about how the tastes of different foods were produced by the geometrical properties of their ultimate particles. "The Form of a Body which causes Taste, consists in *the Bigness, Figure and Motion of its Particles*." Foods taste as they do because of the effects on the sensory apparatus of the tongue of the shape, size, and motions of their fundamental particles, which were in themselves devoid of qualities. Just as the primary-secondary distinction announced that the ultimate particles of a rose were neither red nor fragrant in themselves, those of a melon were neither moist nor cold. There was no match between the qualities experienced and the qualities in the foods consumed. Reality was not appearance. An ultimate particle, for example, might be small, triangular, and rapidly moving, but it was in itself neither bitter nor sweet or indeed cold or moist. A sweet taste in ripe fruits, Rohault wrote, arose from the effect on the palate of "broken, blunted, and . . . small" particles, and their surfaces were shaped such that they would "tickle" the tongue in specific ways; fruits that were *too* ripe occasioned a disagreeable bitter taste because their particles were so broken, blunted, and diminished in size that they physically contacted the palate in a different manner; acid or sour tastes were caused by particles that were "long and stiff . . . , which in some measure resemble small Needles," and so on.[54]

George Cheyne was among the most fashionable British diet doctors of the first part of the eighteenth century and also one of the leading proponents of micromechanical frameworks for explaining how the body worked in health and disease. The human body, Cheyne announced, is "nothing but a *Compages* or Contexture of Pipes, an *hydraulic Machin*." The alimentary tract "is, at it were, a *Common* Sewer, [which] may be foul'd or clean'd in various Manners." The "elasticity" of the body's solids—its muscles and nerves, for example—was a durable paternal inheritance, not easy (though not impossible) to alter over time. But the fluid parts—"the juices"—came from the mother, and these were readily changed by way of life and especially by patterns of consuming food and drink.[55] The main task of the dietetic physician was to manage the juices: if these could be mended by expert dietetic advice, "*they* will in time . . . rectify and confirm the *Solids* into their proper Situation and *Tone*." Following the sensibilities of Boerhaave, Cheyne asserted that in health, the fluids enjoyed free passage through these canals, while many sorts of diseases resulted from obstructed flow. The expert physician could mathematically model blood flow and could demonstrate precise causal links between the particulate configurations of foods and their effects on fluid flow. This expertise was what allowed the learned doctor to give good advice: "*Art* can do nothing but remove Impediments, resolve Obstructions, cut off and tear away *Excrescences* and Superfluities, and reduce Nature to its primitive Order; and this only can be done by a proper and specific *Regimen* in Quantity and Quality." It was dietary regimen, prescribed and managed by medical expertise, that alone could maintain health and cure disease.[56]

The "*Grand Secret*" of health and long life was simple: it was "to keep the Blood and Juices in a due State of Thinness and *Fluidity*, whereby they may be able to make those Rounds and Circulations through the animal Fibres, wherein Life and Health consist, with the fewest Rubs and least Resistance that may be." But if through age and especially through improper diet the fluids become "*viscid*, *thick* and *glewy*," the circulation slows, producing first disease and then death.[57] Cheyne's tract on the melancholic-depressive condition called the "English Malady" began with a causal explanation of "Chronical Distempers" in general, the paramount cause being a "*Glewiness*, *Sizyness*, *Viscidity*, or *Grossness* in the Fluids."[58] The "*best* Blood" was the "thinnest and most fluid Blood," as it "most easily circulates thro' the *capillary* Vessels, which is the most solid Foundation of good Health and Long Life."[59] You were meant to understand that Cheyne's advice to eat this and not to eat that, to take certain drugs, and to adopt specific regimens of exercise proceeded from his deep and systematic knowledge of the invisible world and that the quality of this expertise was vouched for by its derivation from chemical traditions but especially from Newtonian natural philosophy and mathematics.

Cheyne described several ways to keep the blood and other bodily juices thin, sweet, and flowing. Here elements of the "non-naturals" in traditional dietetics continued as organizing principles. Exercise (of the right sort) was important; so were air, emotional management, bloodletting, and the judicious use of drugs, but "it is *Diet* alone, proper and specific *Diet*, in *Quantity, Quality* and *Order*, that continued in till the Juices are sufficiently thinn'd, to make the Functions regular and easy, which is the sole *universal Remedy*."[60] "A *thin*, *fluid*, spare and lean Diet" made for thin and free-flowing juices.[61] People prone to depression and anxiety should "religiously . . . study and practice the *lightest* and the *least* of Food." Anything heavier or more coarse would constrict "tender delicat *Fibres*, *Nerves*, or *Membranes*," which would hinder their "easy, pleasant, and natural *Play* and *Vibrations*, which is the immediat Cause of *Lowness* and *Anxiety*."[62] Analogical reasoning assured Cheyne that he could identify aliments that assisted free flow or that obstructed or irritated the vessels. At a visible level, Cheyne could observe how "Milk, Oil, Emulsion, and such like soft Liquors" run freely "through Leathern Tubes or Pipes (for such Animal Veins and Arteries are indeed) for many Years, without wearing or destroying them," while other sorts of liquids—"Brine, inflammable or urinous Spirits, *Aqua fortis*, or *Regia*, and the like acrimonious and burning Fluids, corrode, destroy, and consume them in a very short Time."[63] The intelligibility of micromechanical accounts of the invisible world was buttressed by analogical inference: the visible world of medium-sized objects was evidence of the invisible (or barely visible) world of the body's pipes and the fluids flowing through it.

Cheyne went on—in the then-familiar micromechanical idiom—to describe the geometrical properties of the invisible particles in foods, drinks, and drugs that affected flow, how the size and shape of particles either damaged or maintained health. The ultimate particles of thin and fluid aliments were—possibly, probably, or certainly—themselves thin and fluid, while gross and sharply flavored foods and drinks were made up of large and angular particles, likely to scrape the vessels and to deposit an obstructing crust.[64] Cheyne proceeded to speculate—Newton-style, with propositions, scholia, and corollaries—about the specific shapes of specific sorts of aliment, about the attractive forces between them that accounted for the strength of their coherence: "Possibly the Figures of Acids, may be *triangular Prisms*"—elsewhere they were supposed to be "thin double wedged-like Particles"—and this geometry allowed them, wedge-like, to separate the particles of other bodies.[65] "Rich Foods, high Sauces, Aromaticks, Delicacies, fine Flavours, and rich and generous Wines" all owe their deliciousness and appeal to the presence in them of large amounts of salts and sulfurous compounds, all of which irritate, corrode, and deposit plaque-like concretions on the vessels.[66] Animal flesh was a problem for Cheyne, for in many cases he encouraged patients to adopt "low diets"—sometimes exclusively vegetarian—while at other times he insisted that animal food was well suited to fallen post-Edenic humankind. Animal flesh was said to be easily digested and nourishing.[67] But in cases where digestive weakness was diagnosed, or where Cheyne considered that habitual consumption of strong foods and drinks had damaged the vessels, he urged patients to immediately give up such things and shift toward the "lowest" possible diet: "Those *Substances* that consist of the *grossest* Parts and hardest of *Digestion*; the constituent *Particles* coming into more *Contacts*, and consequently adhering more firmly." People who had a proper care for their health should avoid "indulging in strong high Foods, which the concoctive Powers cannot break and divide into Parts *small* enough to run into red Globules, or circulate through the *small* Vessels."[68]

Cheyne greatly approved of water drinking and worried about stronger stuff. "*Pure Water*," he wrote, is "the only *Beverage* designed and fitted by Nature for *long Life*, *Health* and *Serenity*." To drink water as your sole dietary liquid "is the only Preservative, I am certain, known or knowable to Art." If you started with a course of exclusive water drinking when young and persisted with it, you "would live probably till towards an *hundred* Years of Age." Water is the "true and universal *Panacea*, and the *Philosopher's Stone*."[69] Strong spirituous drinks contained sharp and angular particles that damaged the vessels, and these ought to be avoided.[70] Despite the confident promise of longevity, this was radical advice, seriously unpopular in free-toping Georgian England.[71] Cheyne appreciated that shunning wine, ale, and other spirituous drinks could be justified only on the strongest philosophical grounds. The ultimate particles of water were fine and smooth, and it followed that this was the "*sole* Fluid that will pass through the smallest animal *Tubes* without Resistance." The micromechanical structure of water was what made it such a good solvent of vascular obstructions and such an effective vehicle for keeping the juices flowing.[72] Iatromechanical expertise in advising regimens of food and drink amounted to privileged knowledge of the micromechanical world. Cheyne did not confine himself to dietetic advice: there was much drugging as well, but micromechanism figured largely here too. He offered detailed accounts of the properties and medical virtues of mercury and, to a lesser extent, of antimony and sulfurous compounds. The particles of mercury are the smallest of any known fluid; they are the most perfectly spherical, the heaviest, and the most powerfully attractive and repulsive. Mercury, of all substances, is the most easily raised by heat and possesses the greatest momentum and is the most able "to pass through all *animal* Substances, which are lax and porous." Physiologically, this meant that mercury was uniquely suited to break up viscid and gluey accretions bunging up the vessels. And from these physiological capacities, the specific medical uses of mercury could be rationally deduced: mercury was the most potent medicine against scurvy, palsies, gout, and indeed all the "*chronical* Distempers caus'd by Excesses."[73]

The application to medicine of the new vocabulary of seventeenth-century mechanical and mathematical philosophy had many of the appearances of revolutionary change. Thinking of the world in mechanical and mathematical terms was meant to fundamentally change how you conceived of food and its bodily career, and this shaped the conditions in which you could know about the qualities of foods, indeed, whether you *should* think of foods in terms of their *inherent qualities*. The mathematization and mechanization of medical thinking put in place new relations between expert and lay knowledge. Iatromathematics and iatromechanism had the capacity to reconfigure the categories in which you thought about food and what you thought about your self, notably including your mental and emotional makeup. Instead of the four elements, four qualities, and four temperaments, mechanism informed a program for talking about foods not in terms of their *qualities* but rather of their *constituents*, describing ill health not in terms of imbalances of humors and associated qualities but instead through the mechanical effects of particles as they passed through and became part of your body. The food you ate was properly talked about not in the language of qualities that might literally feed the qualities of the human body and mind but rather in a quality-less geometrical and kinetic language of machines and machine parts. That was the basis of the intelligibility claims associated with mechanical explanation.

For physicians and laypeople who actively embraced versions of micromechanism, there were far-reaching consequences for how the constituents of aliment were *known* and how different foods and drinks were understood to work on the body. Several sensibilities central to traditional dietetics lost their legitimacy. You no longer knew the qualities of foods through the senses as you once could, and neither could you "know yourself"—a key recommendation of traditional dietetics—in the same way: what you had to now know about the constituents of your diet was to be known *courtesy of experts* who described the structures and processes of the micromechanical world. Unless you were yourself adept in those domains, you had to locate knowledgeable sources and then trust their knowledge of a world that was inaccessible to your senses. The in-principle elimination of the vocabulary of elements, humors, and temperaments meant, among other things, that the recognition of character—in oneself and in others—could no longer proceed in the ways that it once had, and this meant that your sources of knowledge of aliment and its effects were now wholly indirect.

If the natural world was no longer constituted by the elements fire, air, water, and earth, if the qualities of hot, cold, moist, and dry no longer described what things were like at an ultimate level, and if the body's nature was not to be accounted for by the role and relations of four humors, a preponderancy of each one elaborating a different temperament of body and mind, then people were meant to describe *themselves* and others in new ways, with the role of food in sustaining, disturbing, and possibly changing people's nature also changing. Iatromechanism, taken cask strength, disrupted the traditional relationship between the instrumental and the moral aspects of dietetic medicine: the link between what was *good for you* and what was *good*. Physicians might continue to oppose gluttony—now more likely in terms of health effects than as sin—but medical advice now began to appear as quite distinct from moral counsel. Medical expertise continued to be an important aspect of the general culture, but in its iatromechanical mode, its cultural reach was restricted. Together with the other volitionally managed nonnaturals, you remained the product of what you ate, but iatromechanism, along with its proposals for replacing the language of traditional dietetics, was intended to alter ideas about who *you* were and the role of *what you ate* in making you who you were.

7

Dietetics Revolutionized? Common Sense, Common Life, and Chemical Expertise

Eating for Experts

The iatromechanical physicians were intent on revolution. The house of traditional dietetics was to be pulled down and an entirely new one constructed on radically different and incomparably more secure philosophical foundations. What is less evident is the extent to which iatromechanical ambitions were realized in practice. What did modernized medicine *do* that was new, uniquely informed by new medical theory, and judged to be more effective? How far indeed did the vocabulary of modernized medicine spread across the *professional* landscape? What was its reach into the *lay* culture that had been so marked by the categories and sensibilities of traditional dietetics? And what then *happened* over time to the vocabulary as well as the practices of traditional dietetics? Substantial changes were indeed occurring in ways of knowing about food, the body, and the consuming person. But for whom did these things change, and on what occasions was change manifest? How did changes in conceptual vocabulary relate to changes in practice and in judgment of what it was right to do? And what remained stable across real and substantial changes?

Iatromechanical and iatromathematical programs were much in fashion in the first part of the eighteenth century, especially in the capital cities and the spa resorts to which ambitious physicians and their polite clients flocked. Elite patients expected their physicians to be learned, and now—for some, not all—that display of learning was no longer mastery of Aristotle and Galen, Latin and Greek, but rather a show of Cartesian and Newtonian philosophy. This was how the "rational physician" was distinguished from the "mere empiric" in what has been called the "medical marketplace" of the eighteenth century.[1] But while elements of mechanical philosophy and mathematics were incorporated into education in medical theory—notably in such leading centers as Leiden and Edinburgh—the most full-blooded forms of the new medical theories were largely abandoned by the latter part of the century. In the 1740s, even the Royal Society's *Philosophical Transactions* was publishing papers that were frankly critical of mechanical explanations of vital processes. Iatromechanism had seemed to overpromise and underdeliver, and other scientific frameworks were then being offered as candidate replacements for officially discredited Galenism.[2] Notable English physicians were speaking out against modern iatromechanical theory. The reputation of Hippocrates was recuperated, Thomas Sydenham's stress on systematic observation was celebrated, skepticism was expressed about modernizers' claims to greater efficacy, and statistical methods—not new theory—were looked to as a source of explanatory and therapeutic progress. Many of the modernizers had stressed theoretically informed drugging, but it was now often said that they were wrong to do so: there ought instead to be a return to careful, cautious, and case-specific dietary management.[3]

In Germany, the physician and chemist Georg Ernst Stahl (1659–1734) reacted against physiological and medical materialism, insisting that there was an immaterial animating power unique to living things. In England and Scotland too, "vitalism" was resurgent, with advanced medical thinkers identifying the explanatory limits of materialism and mechanism: much about living things might be explained with reference to matter and motion alone, but much required the role of forces special to life.[4] Physicians well disposed to Newtonianism appealed to a "vital Force," which was "the Sum of all those Powers in an Animal Body which converts its Aliment into Fluids of its own Nature."[5] Indeed, several of the mechanical natural philosophers from whom iatromechanical physicians drew for inspiration—notably Boyle and Newton—elaborated schemes that importantly included "active principles" and provided for agencies in living things that were neither material nor mechanical. (Well into the nineteenth century, some versions of "spirits" continued to do significant work in accounts of the human body, while their always ambiguous status between the material and immaterial allowed for a range of interpretations of what sort of entities spirits were and how their activity might relate to diet.) The same attractive forces to which Newton referred in the *Opticks* could be invoked to distinguish physiological from merely mechanical processes.[6] For Newton especially, those active principles were not part of an ad hoc attempt to patch up regrettable failures of mechanical explanation; they were celebrated within a philosophy of nature that centrally included a role for God and spiritual agencies.[7] The materialisms of René Descartes (and Thomas Hobbes) were different from Newton's mechanism, and even the aggressively materialist *Man a Machine* by La Mettrie offered the notion of organized matter as self-moving, a concept that in other hands could look quite like a vital power.[8]

What, then, about everyday medical practice? What about the medical and specifically the dietetic advice given to patients by early and mid-eighteenth-century physicians who were persuaded of the explanatory power of mechanical accounts of the human body and its workings? In what ways did such practical advice differ from that of traditional dietetics? Here, the relationship between new theory and dietetic practice is problematic. On the one hand, the intended abandonment of traditional dietetics loosened once-strong links between medical expertise and lay moral and prudential sensibilities. Traditional dietetics had joined up the notions of *what was good for you* and *what was good* and did so through the notions of moderation in dietary qualities, balance among the humors, and temperance as a sacred and secular virtue. But what if foods were *not* rightly characterized through their qualities, and what if the state of the quality-carrying humors was *not* the correct way to characterize health, disease, and the self? What happened to practice when the relationship between eating and being was so radically reconfigured?

Descartes His Own Doctor

No modernizing natural philosopher promised more of medical reform than Descartes. While not himself a physician, Descartes was much occupied by anatomical and physiological inquiries, and he set himself the goal of improving the powers of medicine to prevent disease and extend human life. He had little doubt that he would achieve these goals once his grand philosophical system had been completed. Meanwhile, Descartes reckoned that he had learned enough about practical medicine to treat his own illnesses and to take measures that, he believed, would greatly extend his life, perhaps even to several hundred years. He was consulted on medical matters by friends and associates and was generous in the advice he offered. And yet, despite the continuing search for a perfected philosophical system that would allow for a perfected medicine, Descartes's practical medical counsel was little different from that proffered by contemporary dietetic physicians.[9] Visiting the sickly young mathematician Blaise Pascal in 1647, Descartes recommended bed rest and soup. Consulted by Princess Elizabeth of Bohemia at about the same time in connection with her skin condition, Descartes told her not to be bled until the spring equinox—which was quite standard medical advice—and if the condition persisted to take "some gentle purgatives or refreshing broths which contain nothing but known kitchen herbs" and not to eat food that was "too salty or spicy."[10]

The medical regimen that Descartes prescribed for himself was also little different from dietetic tradition. His personal dietetics was temperate in most respects, and he was skeptical about a range of heroic medical therapies that purported to do better than the healing powers of nature. Descartes reckoned that wise people should know themselves, and he considered that he *did* know himself well enough to act as his own physician. This meant that you learned, over a long period of time, which foods agreed with you and which did not and in what order and quantity they should be taken. You learned what manner of living, forms of exercise, patterns of sleeping, modes of emotional expression and restraint, and types of intellectual stimulation were best for your individual constitution. If you needed medicine, then you might call on a stock of personal experience to judge what was required for your specific condition. That is to say, learning to be a prudent person meant learning how best to manage the nonnaturals not as a general and theoretical matter but instead as a particular, personal, and practical matter, as a regimen for the ordering of everyday life.

"Our own experience," Descartes told a visiting young scholar, teaches us "whether a food agrees with us or not, and hence we can always learn for the future whether or not we should have the same food again, and whether we should eat it in the same way and in the same order." Even when we are ill, we should let our appetites be our guide. This was Descartes sounding like Montaigne and like many civic gentlemen talking back to the purported dietary experts. All prudent people knew their bodies and their bodies' requirements through experience, and it was only academically trained physicians who might counsel otherwise: "perhaps if doctors would only allow people the food and drink they frequently desire when they are ill, they would often be restored to health far more satisfactorily than they are by means of all those unpleasant medicines."[11] While Descartes himself was dietetically temperate, he did not make a fetish of his food. He was neither gluttonous nor abstemious. He was restrained in his diet, yet he was not one to fail in his obligations as a gentleman. As his earliest biographer wrote,

> His course of Diet was always uniform. Sobriety was natural to him. He drunk little Wine, and was sometimes a whole Month together without drinking a drop yet seeming very jocund and pleasant at table, his frugality not burthensome to his Company. He was neither nice nor difficult in the choice of his Victuals, and he has accustomed his Palat to every thing that was not prejudicial to the health of the Body.[12]

Descartes was no ascetic: "He knew Nature must be supply'd." He took moderate exercise, counted bodily health "the greatest Blessing in this life next to Virtue," and systematically explored the dependence of his mind on his body and his body on his mind. "His judgment was, that it was good always to keep the Stomach and other *Viscera* a doing, as we do to Horses," and he commended "Roots and Fruits, which he believed more proper to prolong the Life of Man, than the Flesh of Animals."[13] Why shouldn't that which was good for the animal machine be good also for the human machine? So, he said, "the best way of prolonging life, and the best method of keeping to a healthy diet, is to live and eat like animals, i.e. eat as much as we enjoy and relish, but no more."[14] His "two grand Remedies" were "his spare regular Diet, and moderation in his Exercise."[15]

In his quest for health, Descartes also stressed the control of the passions. By far the most pervasive medical counsel that he offered was the injunction to *cheer up*. He managed the passions by subjecting them to rational control. Many colleagues and friends who consulted him got the advice to cheer up; Descartes said that he took it himself, to great effect. In 1644, Princess Elizabeth consulted Descartes about her upset stomach. Descartes recommended moderate diet and exercise but specially drew her attention to the influence of the soul on the body. *Believe* that you are basically healthy, and you *will* in general be healthy. Next year, Descartes was plainly concerned about Elizabeth's continuing health worries. He told her what the remedy was, assured her that he had taken it himself and that it worked, and, as her philosophical master, told her *why* it worked in terms particular to his modernized system. The therapy was "so far as possible to distract our imagination and senses from [sources of distress], and when obliged by prudence to consider them to do so with our intellect alone." Descartes himself had "found by experience" that cheering up had "cured an illness almost exactly the same as hers, and perhaps even more dangerous." This was the "dry cough and pale colour" he had inherited from his mother. But early on he recognized and reflected upon his own "inclination to look at things from the most favourable angle and to make my principal happiness depend upon myself alone, and I believe that this inclination caused the indisposition, which was almost part of my nature, gradually to disappear completely."[16] Acting as his own physician and moral counselor, Descartes had cured himself of a condition that, for all practical purposes, was part of his innate constitution. He had changed his nature by acts of will. His custom, based on a rationally informed regimen, had become second nature.[17]

Descartes understood the passions as *systemic*. Having a sensation of sadness involves the soul and all the organs of the body that might be pertinent to the emotional experience: the heaviness or heat of the heart, the sensation of sluggishness arising from the composition of the blood, and the condition of the spleen. Because of the systemic interaction between the soul and the body's solids and fluids, you could cure yourself by the rational management of not only thought but also the other nonnaturals. Diet mattered: aliment was the basis for all bodily constituents, including the spirits.[18] Airs and waters mattered, for example, as they bore upon the composition and consistency of the blood: when the blood was in a fine and thin condition, joy was a likely consequence, and when it was too thick and sluggish its movement in the heart tended to produce sadness. Taking the waters at spas (and very judicious phlebotomy) might achieve that thinness, while dry air rendered the blood "more subtle."[19] The management of sleep mattered: sleep allows the blood to refresh the brain substance, making it more responsive to the movement of spirits.[20] And the active control of the passions mattered very much. Descartes started out by defining the passions as those things that were *not* subject to volitional management, but he wound up treating them in a therapeutic context in just the same way as the traditional dietetic doctors: something about which you could give medical advice, expecting that the prudent person would rationally act upon it.

So, if Descartes were your doctor, this is the sort of advice you would get. First of all, once you have obtained a sufficient stock of experience with your own body, reflect on and trust that experience, and do not be led by medical experts whose knowledge of your body is manifestly inferior to your own. Be your own doctor.

Within the terms of that general restriction, some robust sorts of advice included these: observe dietary moderation; be neither an ascetic nor a glutton; go for a high-fiber diet, more vegetables than meat; avoid very spicy and salty foods; and don't drink to excess. A variety of foods is good. Soup is very good. On the whole, let your appetites be your guide. Your body is probably telling you something; listen to it. Take exercise (and, by implication, venery) in moderation. Get plenty of bed rest, and do whatever you know works best for you to secure a dreamless sleep. Don't get up from your bed too suddenly, and don't try to think too hard for long periods or at times that might interfere with your sleep. Above all, cheer up: avoid thinking about things that make you distressed, dwell on pleasant objects and memories, and look on the bright side of life.

When the rationalist push came to shove, when new theory was mobilized to inform new medical practice, *this* is what Descartes's revolution came down to. He said that he would reform philosophy on foundations wholly new, rejecting everything that he had been taught in the schools, and said that practical effects on human health and longevity were among the principal aims of his philosophical career. Yet, the medical advice he offered and took was *wholly recognizable* as belonging to long-standing and approved dietetic traditions. In the main, it was the kind of advice you might give if you were a physician embracing traditional dietetics. Even the philosophical skepticism about many sorts of medical expertise resonates with the sort of civic skepticism that was broadly expressed in gentlemanly circles and that indeed was elaborated by the great antirationalist Montaigne. Dietetic regimen proved more durable than dietetic theory.

George Cheyne and the "New Thing" in Dietetic Practice

Expertise in the micromechanical world was supposed to trump personal sensory experience, merely empirical knowledge, and ordinary prudence. Micromechanical expertise drove a wedge between the good for you and the good, constituted reliable dietary knowledge as something the uninstructed laity did not possess and might only come to have courtesy of learned experts, and stripped sensory experience of foods of its practical pertinence. If you reckoned that you were your own doctor, then—at any age—you were a fool. This new expertise licensed iatromechanical practitioners to advise patients to do things that might be dietetically and therapeutically new, something more radical than the powerful yet superficially bland injunction to follow moderation and balance. Of course, while moderation and balance were always central to the preventive aspect of traditional dietetics, the Hippocratic Corpus recognized that "desperate diseases call for desperate remedies," and in the early eighteenth century it was common for physicians to direct their patients to take extreme measures, just on the condition that the doctors believed patients to be in imminent danger and could persuade those patients that death was near. When patients "talked back," sometimes citing the Rule of Celsus, they either had grounds for believing themselves *not* to be desperately ill or, like Montaigne, accounted it morally improper to subject themselves to extreme medical measures when, so to speak, "their time had come." But if claims to micromechanical expertise were credited, then physicians asserting such authority might reckon that they possessed new and more legitimate grounds for telling patients *when* they were in great danger—when radical interventions were necessary—even when patients might feel themselves to be well or at least not terminally ill.[21]

Almost a century after Descartes's death—the great philosopher did not live, as he had hoped, for several centuries, dying of a cold in Sweden aged fifty—the dietary physician George Cheyne also confronted the relationship between new understandings of how the human body worked and the sorts of practical medical advice he offered patients.[22] Cheyne was primarily known for prescribing regimes of eating and drinking, but like many iatromechanical physicians of the eighteenth century, his supposed expertise in the micromechanical realm also informed elaborate drugging practices, with therapeutic action often specifically referring to the geometry and behavior of drugs' ultimate particles. The "diet doctor" Cheyne routinely ordered bleeding and prescribed a wide range of purgative, emetic, carminative, and febrifuge drugs—rhubarb, ipecac, hiera picra, hellebore, extracts of spruce, mercury and its compounds, cinchona, elixir of vitriol, spirit of niter, tinctures of steel and soot—as well as specifically sourced mineral waters whose mode of action was said to flow from the unique geometry of their ultimate constituents.[23]

The diet for which Cheyne was most famous—or even notorious—was a severe "lowering" regime, especially suited to valetudinarians, the sedentary, the studious, and the otherwise fine-nerved who, he warned, were risking their lives by persisting with what *they* accounted a normal course of food and drink: "I advise . . . all Gentlemen of a *sedentary Life*, and of *learned* Professions, to use as much *Abstinence* as possibly they can," even descending, if necessary, to a regime of asses' milk and seeds. And Cheyne laid it down as the first aphorism in his *Essay on Health and Long Life* that the "most infallible" means to "preserve *Life, Health* and *Serenity*" was to take "the *lightest* and the *least* of Meat and Drink a Man can be tolerably easy under."[24] (Cheyne pointed to Luigi Cornaro as a model of how to do this, but the doctor also drew liberally on widely distributed beliefs about the causal relationships between extreme longevity and the austere dietetics of ancient philosophers and early Christian desert fathers.)[25]

Many people (including some of his fashionable patients) swore by Cheyne's lowering diet, announcing that it had saved their lives; others swore at it, considering it bizarre, unbalanced, rigidly doctrinaire, theoretically incoherent, unnecessary, impossible to maintain, and probably ineffective. It was regarded as expertise overreaching itself. Even with the success of micromechanism in much of learned cultural life, the charge of extremism against any sort of medical counsel was a serious matter. New theory might simply lay a gleaming veneer of modern natural philosophy on top of traditional practice or might offer formal justifications of radically new therapeutic and dietary advice. Cheyne encountered both sorts of criticisms from fellow physicians, but much of that criticism targeted what was considered bad medical practice and specifically bad advice about food and drink. Some critics said that Cheyne laid too much emphasis on his unique regimen of eating and not enough on the other nonnaturals. Indeed, while the nonnaturals continued to organize much of Cheyne's thought and practice, his work represents a substantial move toward the emerging notion of "diet" as just food management. He had gone so far as to lay down the exact quantities of food and drink suitable for everybody. Not enough attention had been paid to patients' individuality. Cheyne was accused of formulaic and global prescriptions and of approving dangerously radical departures from customary ways of life. He did not, it was said, appreciate how custom might constitute second nature. His medical critics said that in citing the authority of micromechanical expertise, Cheyne had departed from the proven traditional wisdom of moderation and balance, and this was probably the most consequential charge he faced.[26]

Even after dietetics was "revolutionized" by modernizing physicians following Descartes and Newton and even after doctors such as Cheyne used the authority of micromechanism to sweep away much of the theoretical foundations of traditional dietetics, the counsel of moderation and balance retained immense authority. When accused by his critics of therapeutic and dietetic radicalism, Cheyne sometimes acknowledged that this was a fair charge, but more often he denied it. Moderation was still a great prize for both modernizers and their more traditional colleagues and patients. Asserting that his dietary regime was, after all, moderate, Cheyne at times condensed all of his counsel to one simple rule, the sort of thing you might encounter in the essays of Montaigne or Bacon or even in proverbial common sense: "If any Man has eat or drank so much, as renders him unfit for the *Duties* and *Studies* of his Profession (after an hour's sitting quiet to carry on Digestion) he has overdone." There was no better advice than moderation in all things: "If Men would but observe the *golden Mean* in all their *Passions*, *Appetites* and *Desires*" and if they "follow'd the uncorrupted *Dictates* of *Nature*," they would enjoy better health, have "their *Pleasures* more *exquisite*; live with less *Pain*, and die with less *Horror*."[27] Cheyne complained that his critics unfairly represented him as an extreme ascetic. He did indeed commend dietary restraint, but the charge of extremism was unjust. He knew very well that "*total Abstinence*" was seriously damaging, and so he recognized that there were *two* extremes: the too much and the too little. Accordingly, Cheyne presented himself as falling in with traditional appreciations of both the bodily and moral virtues of the middle way.[28] Living under Dr. Cheyne's physic was not at all to live "miserably." Cheyne repeated his claim that *some* patients were deemed to be in such desperate danger that desperate dietetic measures were needed, and he could point to this as an indication that he was no slave to theory and was sensitive to individual cases.

Cheyne was accused of being a rigid vegetarian—a regimen that most of his compatriots regarded as absurd or even impossible—and this too he denied. In our original unfallen state, we might indeed have consumed no animal flesh, but fallen we now are, and our present physical condition is as degraded as the fertility of Earth itself. Taking our bodies as they are and not as they once were, "*animal Food*, their Juices," and—reminding readers of Cheyne's micromechanical expertise—"*integral Particles*, are the fittest and most proper" to restore the fibers of our bodies and constitute our blood. After all, was it not clear that the "Particles" of animal flesh were already "*form'd, figur'd* and adjusted" to be more similar to the human body than were the particles of vegetables?[29] The great virtue of the golden mean might even apply at the micromechanical level: "There seems to be a *Medium*, or golden Mean, between the extreme Minuteness of some *Particles* of Matter, and the too great Coarseness and Largeness of *others*, that is most proper for the Nourishment of animal Bodies, and best suited to our present Situation."[30] Cheyne's defense against the accusation that his regime was extreme was to reaffirm the wisdom of moderation and to claim it for himself, all the way down to the microworld about which only new forms of scientific expertise might claim legitimate authority.

The More Things Change

The stability, authority, and cultural reach of traditional dietetics were indeed shaken during the course of the eighteenth century.

The challenges posed by mechanical and mathematical frameworks were confident and aggressive: a new and powerful medical practice would, it was promised, follow from the extension of Newtonian or Cartesian philosophies to the understanding of vital processes. This represents a moment in the burst of enthusiasm for mechanical and mathematical natural philosophies that was later damped by confrontation with the complexities of human bodies and by the resistance of sick bodies to the therapies said to follow from those philosophies. It is the judgment of history—and especially the judgment of *historians*—that iatromechanism and iatromathematics "failed." That judgment is irrelevant to present purposes; what is pertinent is the judgment of eighteenth- and early nineteenth-century medical practitioners, and they too considered that micromechanism had not fulfilled its promises for either medical theory or medical practice. But still more apposite is an engagement with what modernizing challenges to traditional dietetics actually looked like *in practice*. What measures to prevent and cure illness were proposed? What conceptual language did modernizing physicians use to frame their accounts of how the body worked in health and disease?

Descartes advocated dietary moderation, and Cheyne insisted that even his radical lowering regime should be understood as a mode of moderation and as an embrace of Hippocratic adaptation of medical therapies to desperate medical realities. Descartes's *Discourse on Method* famously announced that everything taught in the Schools was to be rejected, that the house of existing philosophy was to be torn and reconstructed on foundations wholly new. Custom held sway when certain knowledge was not available, but when philosophical certainty was available, custom must be rejected. Yet, Descartes admitted that it might be some time before the work of radical renovation was completed and that in the meantime he would follow the judgments of the most judicious people in whose company he was living. Until the promised secure philosophical foundations were in place, custom would have to be good enough to be getting on with. And when there was diversity of judgment, when "many opinions were equally well accepted," Descartes said that he "would choose only the most moderate, both because these are always the easiest to act upon and probably the best (excess being usually bad)." Moderation was not perhaps golden, but it was good enough to regulate judgment and to order the affairs of life as it was actually lived.[31]

Under attack from iatromechanical modernizers, the counsel of moderation in medicine bent but did not break; traditional dietetic categories were challenged but did not disappear; and new dietetic and therapeutic practices were advocated, but they coexisted with increasingly disconnected bits and pieces of tradition. Much academic and professional expertise had rejected the concepts of traditional dietetics, yet the result was not so much a global and total shift in beliefs, vocabulary, and practices as it was a fragmentation, decentering, and social dispersal of what once had been a widely shared and extraordinarily stable culture. From sometime in the eighteenth century, official experts and the laity began to speak different languages about the nature of food, the body, and the relationships between what was *taken into* the body and what people *were*, between the body and the mind, between the *good for you* and the *good*. Much thought and practice about food and its effects did shift remarkably through the eighteenth and early nineteenth centuries, but much remained the same. Some advertised changes proved to be largely illusory, just as some appearances of stability mask consequential change. What happened during that period to the culture of traditional dietetics, and what effects did those happenings have on ideas about our food, our knowledge, and our selves?

Descartes and Cheyne—a founder and a follower, so to speak—represent just two perspicuous examples of change with continuity, of fitful accommodation of theory and practice. But the medical pastiche of the new and the traditional was a general phenomenon. One of the key figures among the British iatromechanicians of the early eighteenth century was the fashionable physician Richard Mead (1673–1754), who had studied briefly with Archibald Pitcairne at Leiden and attended Isaac Newton and whose career culminated in immense wealth, a position as royal physician, and becoming an object of satire in *Tristram Shandy*.[32] In 1702, Mead sought to show that the effects of poisons were "owing to something more than the bare qualities of heat or cold," and he aimed to "discover the footsteps of mechanism" in how poisons worked on the body. This was a familiar sort of iatromechanical performance, meant to show how far the effects of poisons could be assimilated "to the known laws of motion" and to "geometrical reasoning." The replacement of Galenic philosophy by mechanical categories was promised to yield a new understanding of medical phenomena and to inform different and more powerful therapies. Mead looked forward to a time when physicians would become wholly familiar with the language of mathematics and mechanics, using its vocabulary as a matter of course.[33]

Yet, years later, when Mead addressed regimen, he acknowledged that what he was about to say "seems to be rendered almost superfluous by the precepts of Celsus":

> For such is the natural constitution of the body of man, that it can easily bear some changes and irregularities without much injury: had it been otherwise, we should be almost constantly put out of order by every slight cause. . . . [So] diseases from inanition are generally more dangerous than from repletion. . . . Upon the same account also, though temperance be beneficial to all men, the ancient physicians advised persons in good health, and their own masters, to indulge a little now and then, by eating and drinking more plentifully than usual. . . . [A]ll changes in the way of living should be made by degrees. It is also beneficial to vary the scenes of life; to be sometimes in the country, sometimes in town; to go to sea, to hunt, to be at rest now and then, but more frequently to use exercise. . . . But a mean is to be observed in all these things.[34]

When theory was called on to advise on the texture of everyday life, Mead's mechanism seems—like that of Descartes—to have had nothing better to recommend than prudential common sense: moderation and traditional dietetic regimen tempered by the circumstantial disruptions central to the ancient Rule of Celsus.

Balance and Moderation in Modernized Medicine

Santorio's *Medical Statics* appeared in 1614, long before the natural philosophical work of Descartes and Newton.[35] *Medical Statics* did not offer a micromechanical account of the body and its workings, but it was celebrated well into the Enlightenment for its program of quantifying bodily states and processes, so rendering medicine rational, experimental, and intellectually *certain*. Eighteenth-century editors decorated Santorio's text with ringing endorsements of mathematical methods and offered lengthy "explanations" of the truth of Santorio's aphorisms by way of more recent mathematically and mechanically informed views of the body. Yet, editions of Santorio appearing even during the height of enthusiasm for iatromechanism and iatromathematics retained a notable feature of the book's original organization: the core of the text was a serial trip through the six things nonnatural, with moderation in the management of each of the nonnaturals newly supported by a quantifying idiom. Santorio's aphorisms identified foods that were good or bad for you according to their contribution to the "insensible perspiration," but what *counted* as good and bad, together with the counsel of moderation, was not much different from that found in late sixteenth- and early seventeenth-century dietary works contemporary with the Italian physician. Excess was bad because it increased bodily weight beyond what was normal for you; fasting was bad if it reduced the body's weight "below its natural Standard." Meats that were once judged bad for you because they produced "crudities" were now bad for you because they hindered perspiration. Big meals were more hurtful to the sedentary than to those who led an active life because exercise encouraged excretion, while sitting led to constipation. The amount of food you should consume is that which could be "perfectly digested." At one time you could only know that state by way of (literal) gut feelings: the audible and olfactory evidence of gas and the visual inspection of excreta. Now, in Santorio's program you could know these things with very great certainty by the expert instrumental weighing of inputs and outputs.[36]

The goal for Santorio—and for his many eighteenth-century admirers—was indeed *balance*, and while balance could now be defined through the maintenance of steady bodily weight, varying around a numerical norm, the authority of the notion of balance still traded on long-standing moral and even religious sentiments. Balance was a measure of justice and virtue. From antiquity through the Renaissance, human *bodies* were *not* routinely weighed, but human souls and human virtues (metaphorically) *were*. Psychostatics historically preceded bodily and medical statics.[37] In the book of Daniel (5:27), the writing on the wall warned about the afterlife: "Thou art weighed in the balances, and art found wanting." Job (31:6) asked to "be weighed in an even balance, that God may know mine integrity," and Psalms (62:9) said that "surely men of low degree are vanity, and men of high degree are a lie: to be laid in the balance, they are altogether lighter than vanity." In law courts, the scales of justice are held by a blindfolded figure, an icon of disinterested legal virtue. Balance was equipoise of weights, and dietary moderation was poised between the extremes of the nonnaturals. So, a quantitative medical idiom could be advertised as a modernizing alternative to the qualitative vocabulary of traditional dietetics, but its authority was supported by the recognized wisdom accumulated in medical and ethical counsels of moderation.

Balance and its virtues did not disappear from the culture, nor did the nonnaturals and other core categories of traditional dietetics. Cultural branding is one thing—and branding is indeed pertinent to establishing distinction—but substance is another: there are differences between a new label on the can and new contents in the can. Early in the eighteenth century, the iatromechanical physician Jeremiah Wainewright did not aim to expel the nonnaturals from modernized medicine but instead meant to give properly detailed mechanical accounts of how their management worked. Announcing allegiance to both Santorio's quantifying program and Newton's mathematical natural philosophy, Wainewright declared that the key state of the body accounting for health and disease was the elasticity of its fibrous solid bits: nerves, muscles, vessels, etc. And it was to the fibers' states of tension or relaxation that dietary advice was geared.[38] Indigestible foods were those that caused contraction of the stomach's fibers; foods that were compounded, as opposed to simple foods, were hard to digest because of the "different degrees of Cohesion" of their parts; hunger put the fibers in a contracted state, while feeding relaxed them; digestion required contractile work; and sleep relaxed the fibers. Practical advice was said to be informed by that sort of philosophical understanding. The standard counsel of traditional dietetics against eating too soon before going to bed was here philosophically explained: since sleep relaxes the fibers, digestion could not effectively happen, and sleep and digestion—either or both—would be disturbed. In these and other respects, iatromechanism characteristically put old prudential counsel into new philosophical containers. Wainewright supported the position associated with Celsus, Montaigne, and Bacon that occasional excess was more healthful than long-maintained asceticism, saying that a reliable index of the right quantity of food consumed was a sense of repletion. He recommended that a healthy person rise from the table with some appetite remaining. He thought that a proper test of moderate consumption was the ability to perform physical and mental tasks as one normally did, and he reckoned that gluttony and other modes of excess were harmful, provided that individuals' varying temperaments and ways of life were taken into account.[39]

Wainewright was not alone in offering mechanical accounts of the nonnaturals. Joining the traditional authority and cultural reach of the nonnaturals to the new authority and cultural prestige of mechanical explanations was a common move among advanced physicians. The North of England doctor John Burton produced a similar text in 1738, again stressing the role of the fibers, their states of tension and relaxation. The details of how mechanism worked were treated only sketchily, and Burton pulled back from specifically *mathematical* ambitions in accounting for bodily functions, announcing that he was "taking great Freedom . . . in decrying Mathematics *as useless in the Cure of Diseases*." Claiming the pertinence of mathematics in medical therapy was "a Cheat, and as errant a Piece of Quackery, as a Stage and a Merry-Andrew."[40]

The nonnaturals serially organized Burton's text in much the same way as the dietary texts of the centuries before. The physiological effects of air were now glossed with the vocabulary of Boyle's work on its pressure and the elasticity of its corpuscles. Moderation was endorsed, and gluttony was condemned, occasionally with reference to dietary excess "stretching the Fibres, and compressing the Vessels." Burton joined with generations of past dietetic physicians in suspecting that a vegetarian diet was dangerous, and perhaps impossible, this owing to the material makeup of plant and animal food and the demands made on the powers of digestion and assimilation. Digesting vegetables required a double transformation from plant matter to animal flesh and then to specifically human flesh. Vegetable food requires "more Labour of the Stomach and other Viscera to reduce it to be fit for Nourishment."[41] Different wines had different physiological effects; water drinking was good; you should walk a bit after supper; and excessive venery had bad effects on the tension of your fibers. At the same time, gestures were made toward the Rule of Celsus and its warnings against too rigid an insistence on regularity. The suitability of particular aliments occasionally referred to the characteristics of their ultimate particles but not in any detailed or systematic way. Patients were cautioned against making sudden changes in their way of life. The mind and body were understood to be causally connected, but the manner of that connection, while proceeding through the tension of the fibers, was a mystery, and Burton was content that it should remain so.[42]

Stress on the medical significance of the body's solids as opposed to the fluids was a major vehicle for the advertised rejection of humoral theory. Nicholas Robinson's *New Theory of Physick and Diseases Founded on the Principles of Newtonian Philosophy* was published during Newton's lifetime and offered one of the more detailed accounts of how the body functioned, what regimens maintained health or caused disease, and how those regimens worked. Again, it was the state of the fibers—their contraction and relaxation—that was central to the story. The way to wellness in Robinson's story was *balance*, a notion key to traditional dietetics, though now respecified in terms of the tension of fibers and its effects on the body's fluids. The "Balance of Nature, or the Standard of Health," Robinson wrote, is the state obtaining "when Solids and Fluids are so exactly balanc'd, that they answer to each other's Motions without the least Resistance." Blood flows freely when the solids are in their natural state of tension; it is obstructed when they are not.[43] People are individuated through the state of their fibers, whether brought about by innate characteristics or, over time, by their way of life. Some people, for example, found that hearty beef or pork meals agreed with them while others were unable to digest even a chicken wing, and this was due to "the different Structure and Springiness of the Solids, and the Power they have to overcome the Fluids." Strong meat and strong wines caused the fibers to abnormally contract, and these should be used with caution by those who found them hard to digest. The state of the fibers was, however, inaccessible to the senses, while experience with the effects of foods on the body were readily knowable. It was only by experience that patient and physician could come to know what agreed with a particular person and therefore to discern something about the state of their fibers.[44]

The temperaments were identified and named, though they no longer significantly organized medical accounts; the nonnaturals were still named and sequentially treated, though their importance was no longer tied to humoral constitution. A person could, for example, be talked about as sanguine or bilious without specific reference to the humoral vocabulary of now-discredited traditional dietetics.[45] The idea that food—"the most necessary of the Non-naturals"—had *qualities*, and that these qualities accounted for bodily consequence was frankly denied.[46] Stories about the fundamental realities of body and food changed, while practical counsel retained much that had belonged to traditional dietetics: moderation remained good advice—"A Temperate Diet is best"—and so too for all the nonnaturals. Excess is the major cause of disease, but if there had to be a departure from some ideal standard of the temperate, it was better to veer toward the full rather than the spare diet.[47]

For all the fashionability of mechanism and mathematics, some eighteenth-century medical writing explicitly opposed modernizing moves, while others took slight notice, and here too the humors, temperaments, and nonnaturals persisted. A prominent example was *The History of Health* (1758) by the English Midlands physician James Mackenzie. Moderation in all things remained the way to well-being, and excess was always dangerous: "It is very injurious to health to take in more food than the constitution will bear." The task of the physician was to precisely adjust diet to patients' individual constitutions and ways of life, and patients were the most knowledgeable about what did and did not agree with them. At the same time, excess in abstinence had its own risks, as did radical departures from routine. The scholarly and the sedentary needed to be especially mindful of their aliment, their exercise, and the risks of solitude: Luigi Cornaro continued to be held up as an example of temperate body management and its benefits for longevity.[48] From 1762 to 1770 there were six editions of *Rules for the Preservation of Health* by the English Quaker physician John Fothergill, all of which were structured by the usual list of the nonnaturals, whose management, it was said, should be governed by temperance. A person should choose food by "consult[ing] his own constitution, and eat[ing] only what perfectly agrees with him." Fothergill made no mention of micromechanism or of chemical findings.[49]

So, by the mid to late eighteenth century, modernizing physicians, advertising their affiliation to mechanical philosophies, succeeded in substantially eliminating some of the vocabulary of traditional dietetics, changing the reference and meaning of other items, and signaling new cultural connections between the physical sciences and the art of medicine. Change was happening, but the result was not the wholesale reconstruction of the fabric of medical knowledge and medical practice that had been so enthusiastically promised. Physicians increasingly talked a newish language that nevertheless incorporated elements of the old; they counseled new ways of living, eating, sleeping, exercising, and so on, but the nonnaturals continued to be treated as central to medical management, and moderation remained good advice. The modernizers refounded the physician's expert knowledge, but patients were still regarded as possessing pertinent self-knowledge. The iatromechanical revolution did prompt reform, but the promised sweeping away of the ancient and its replacement by the new produced what might be better described as a pastiche of old and new.

The new encyclopedias that served as the Enlightenment's arsenals of universal knowledge also came to terms with change and tradition in dietetic thinking. Through the eighteenth century and beyond, there was an overall decline in encyclopedias' explicit engagement with key concepts in traditional dietetics, but the falloff was patchy, and the traditional categories never completely disappeared. Ephraim Chambers's *Cyclopædia* of 1728 described "diet" as "a sovereign Remedy against all Diseases arising from Repletion" while marking the beginnings of a change from "diet" understood as regimen to a more restrictive sense of "diet" as a disciplined and usually reduced food intake. "Regimen," however, was still defined as "a Rule or Course of living," though food and drink had pride of place. While the achievements of the moderns were celebrated and Galenism was generally rubbished, the nonnaturals retained their place as "the Causes and Effects of Diseases."[50] The great French Enlightenment *Encyclopédie* retained the category of the nonnaturals in its articles on "hygiene" and "diet." *Encyclopédie* recognized that normalcy had to be differently defined for individuals with different constitutions and construed "diet" in its traditional sense of "an ordered manner of living." Health was to be preserved by an appropriate management of the nonnaturals.[51] And the Edinburgh-based *Encyclopædia Britannica* of the early 1770s listed the six things nonnatural as core medical categories—"so called because by their abuse they become the causes of disease"—and said that diet "comprehends the whole regimen, or rule of life, with regard to the six non-naturals," which were then listed in their traditional order. The *Encyclopædia Britannica* had an entry on "temperament," which listed the traditional tetrad, with the traditional humoral bases, but this was flagged as a legacy of the past. There were matter-of-fact references to a "sanguineous temperament" and a "bilious constitution"; "melancholy" appeared quite frequently, but this was now to be understood as a protean pathological condition without a specific causal basis in the excess of black bile.[52] In the 1820s and 1830s, the English physiologist John Bostock acknowledged that the four Galenic temperaments continued to circulate in both vernacular and expert usage, while "the [humoral] hypothesis on which it was founded is universally discarded." A variety of physiological and anatomical theories sought to refer traditional categories to modern theory, such as the vital properties of the animal system—irritation, sensation, and so on—and Bostock accepted that Galenic temperamental categories needed little modification to align them with empirical observations of what different people were constitutionally like.[53]

The decline of traditional dietetic categories was slow. Indeed, it isn't straightforward to say that these categories *disappeared*, and a widened cultural focus suggests that its elements were relabeled and dispersed. A text on regimen from the early nineteenth century, for example, omitted explicit reference to the nonnaturals while maintaining the customary sequential treatment of food and drink, exercise, and sleep.[54] Into the 1820s, there were expert medical texts discussing the new chemistry of aliment that alluded to the nonnaturals as though fellow practitioners still knew very well what they were and found them pertinent to the prevention and treatment of disease.[55] The nonnaturals were, however, absent from a 1830s text on digestion by the eminent Scottish physician Andrew Combe. He almost completely dropped the language of humors but continued to list "the four principal constitutions" (temperaments) of the body while changing aspects of their nomenclature ("lymphatic" for phlegmatic and "nervous" for melancholic), referring each of these to the state of the pertinent physiological system: muscular, lymphatic, nervous, vascular. At the same time, Combe prescribed to persons of each constitution individualized regimes of food and drink that are recognizable from the dietetic tradition: "To be serviceable, *the food must be adapted to the age, constitution, state of health, and mode of life, of the individual, and to the climate and season of the year*."[56] But Combe recognized that traditional attention to the requirements of specific constitutions had already become "entirely overlooked" and that much of this individualized knowledge had been purged from medical expertise.

Dietetics and Lay Knowledge in the Age of Reason

When it is said that ideas, sensibilities, and practices have *disappeared*, what is often meant is that they no longer figure in some arenas of elite culture. The presumed disappeared things may, however, persist in lay circles or may even be features of elite culture when the elites are, so to speak, *not at work*, when they are behaving and talking in ordinary nonexpert modes. But there hasn't been great historical interest in exploring the conditions, modes, and extent of "disappearance" except for intermittent expressions of irritation about lay "backwardness" and about the persistence of error and fable among the common people. Fine recent historical scholarship has addressed the "withdrawal" of high from low culture at a specific historical moment, but high and low as well as expert and ordinary are always distinct and are always involved with each other.[57]

Books explicitly intended for a popular readership are a problematic category. They mean nothing to the illiterate except when their contents might be orally related; the literate poor can afford few books, and it is always possible that books advertised for a popular readership wind up serving the elite. Nevertheless, there is solid evidence that some medical texts meant for "the people" did have wide circulation, more in the eighteenth and nineteenth centuries than earlier, and over time, more and more of such books were produced and enjoyed success in the marketplace.[58] In the mid-eighteenth century, there were medical texts meant for ordinary people that covered traditional dietetic territory but were critical of many aspects of elite medicine. Both doctors and moralists supplied a popular market for practical medical counsel: how to cure and manage a range of diseases, how to maintain yourself in health, and how to prepare medicaments. Many of these sorts of books urged that *you could be your own physician* and offered instruction about how to do that. However, the scope of the pertinent *you* was vastly expanded from what it once had been.

In 1747 John Wesley (1703–1791), the founder of English Methodism, produced *Primitive Physick*, whose medical sentiments matched his efforts to revive a purified Christianity. Wesley lamented the decline of a folkish and pristinely religious medical practice based on common sense, prudence, and solid experience. Common people used to know what herbs were good for what ailment; no professional training was needed to understand God's natural laws for the regulation of body, mind, and moral conduct. This once-common knowledge was being eroded through the rise of scientific pseudoexpertise and ought to be retrieved and its value asserted. Confronted with tradition, "Men of a Philosophical Turn, were not satisfied with this," and they began to require causal and, specifically, mechanical explanations of how the body worked in health and disease, building "Physick upon Hypotheses: to form Theories of Diseases and their Cure." Traditional remedies were forgotten, and new-fangled ones intruded by the vaulting ambition of theorizers. And these innovations were more difficult to apply, "as being more Remote from Common Observation," until at last medicine "became an abstruse Science, quite out of the reach of ordinary Men." The currently fashionable language of medicine embraced obscurity not because practice really required esoteric concepts but instead because professionals wanted protection from lay interference, bidding to win patients in the medical marketplace by ostentatious displays of modern learning.[59] The rules for maintaining health and curing most diseases were clear, simple, and wholly in accord with common sense, God's laws of nature, and Christian morality: observe moderation and occasionally abstinence; consume the foods that agreed with you; and be wary of novel highly spiced and elaborately compounded dishes. Those in sedentary occupations should eat less, drink water rather than spirituous beverages, be early to bed and early to rise, and take exercise in moderation. They should seek out fresh air, take steps to avoid or remedy constipation, avoid strong emotions, and know that the love of God "is the Sovereign Remedy of all Miseries."[60]

The bulk of Wesley's text was an alphabetical listing of common ailments and their proper treatments: for "Hypochrondriac and Hysteric Disorders," take a cold bath and an ounce of mercury every morning; for diabetes, drink wine boiled with ginger; to prevent the plague, eat marigold flowers; and for baldness, rub the scalp first with onions and then with honey.[61] Surprisingly, Wesley celebrated the Newtonian modernizing doctor George Cheyne as a paragon of commonsense medical advice, and even though Wesley elsewhere displayed general familiarity with modern iatromechanical modes, *Primitive Physick* set aside Cheyne's micromechanical explanations and his enthusiasm for a regimen of seed and asses' milk.[62] Wesley's book had enduring appeal, evidence of an enduring embrace of much traditional dietetic counsel and sensibilities. Twenty-three editions appeared in the author's lifetime. There have been at least thirty-eight British and twenty-four American editions, and the last edition appeared in 1859.[63] It has been said that the book was once "found in almost every English household, especially in those of the poor, usually beside the Bible," selling more copies than any other medical handbook of the eighteenth century.[64]

Soon there was major competition in the market for books allowing ordinary people to "be their own doctors." The most prominent representatives of this popular genre were *Advice to the People* (1761) by the Swiss physician Samuel-Auguste David Tissot, followed by *Domestic Medicine; or the Family Physician* (1769) by the Scottish physician William Buchan (1729–1805) and, later, the most widely distributed American "frontier" self-help text, *Domestic Medicine, or Poor Man's Friend* (1830) by John Gunn (1797?–1863). Of these, Buchan's was the most broadly distributed, influential, and longest lasting. His text was explicitly intended for ordinary people, but it has been plausibly said that its actual audience was probably "a cross-section of the literate, servant-employing, and self-consciously improving middle-orders."[65] Tissot was oriented toward acute diseases, while Buchan took in chronic illnesses as well because these were the most common and especially because they were the diseases whose cure, he said, "chiefly depends on a proper regimen."[66] Buchan's book did not lecture readers on the conceptual foundations of dietetics, such as elements, qualities, and temperaments; however, much of the vocabulary of traditional dietetics was pervasive, and the role of regimen was repeatedly invoked. *Domestic Medicine* went through 142 editions and remained in print until 1870.[67] The substance of the editions varied, especially after Buchan's death when his name became a marketable "brand." The text was then open to all sorts of revisions while continuing to advertise Buchan as the author. *Domestic Medicine* was a product of the English-speaking world, but its sentiments became globalized, testifying to the cultural reach of dietetic categories. Buchan's book was translated into French and Italian (several editions of each), Spanish, Portuguese, German, Swedish, and Russian. In Brazil, the Portuguese translation—adapted to take account of local conditions—dominated the market for popular medical books and remained in print until the mid-nineteenth century. Through the Dutch translation, *Domestic Medicine* was rendered into Japanese.[68]

Regimen and moderation retained pride of place in the self-help books. In Tissot's *Advice to the People*, "regimen," like "diet," was evidently a notion in transition: at one point Tissot treated "Regimen or Way of Living" while elsewhere writing about people "put into a regimen," or "on a diet," which, in context, usually meant just a special-purpose way of eating.[69] The management of airs, food and drink, evacuations, sleeping, exercise, and the emotions had a prominent place, and moderation in these things was commended. "The most certain Preservative, and the most attainable by every Man," Tissot wrote, "is to avoid all Excess, and especially Excess in eating and drinking," though he made no reference to anything called the "nonnaturals," and neither Galen's name nor ancient medical authority was invoked.[70] The six nonnaturals retained their organizing place in Buchan's *Domestic Medicine*, including nineteenth-century editions produced after his death. Buchan warned against "all extremes" and observed that "moderation [was] the only rule with respect to the quantity of food"; he thought that it was dangerous to make sudden and radical changes in your way of life; and he recommended adjusting patterns of eating to your particular constitution and circumstances. Food was self-making: "The whole constitution of the body may be changed by diet." Buchan endorsed a version of the Rule of Celsus, saying that "living too much by rule might make even the smallest deviation dangerous. It may therefore be prudent to vary a little, sometimes taking more, sometimes less than the usual quantity of meat and drink, provided always that regard be had to moderation." The traditional causal connection between "flatulent food," vapors, and mental disorders was affirmed, and a light diet was commended. With respect to the passions, Buchan accepted that "how mind acts upon matter" might remain forever a mystery, but we know as a fact that there is a reciprocal interaction between "the mental and corporeal parts, and that whatever disorders the one will likewise hurt the other."[71] In these and many other respects, the most widely distributed Anglophone popular medical text of the period from the 1760s into the mid-nineteenth century was marked by significant continuities with the categories and sentiments of traditional dietetics.[72]

There were other traditional dietetic concepts that populated *Domestic Medicine* while their reference changed, sometimes subtly, often substantially. It was, Buchan wrote, important to keep "the humours sound and sweet," to guard against "gross," "acrid," "depraved," "vitiated," "viscid," and "putrid" humors," though there were no gestures toward the canonical four humors of Galenic medicine, and Buchan's usages appear more as generalized reference to notions such as "the body's fluids" as opposed to its "fibers" and "solids."[73] There were, however, no attempts to say what the humors *were* or to show their fundamental importance in the body's economy. In the original Buchan edition, the humors included blood, phlegm, and yellow and black bile, but there were also references to such things as "scorbutic humours," not explicated in the text but evidently considered as the basis of skin conditions and the watery secretions emanating from the gums and other bodily orifices.[74] By the 1802 edition—in which Buchan's authorial presence was diluted—the "constitutions" were described and enumerated, but the author used the notion as little more than a way to designate the type of illnesses and disorders to which different people were liable. Commonly recognized constitutions included several that were apparently similar to the Galenic tetrad, but now the list had expanded, and the idea of "constitution" had broken free of its traditional causal basis in a dominant humor. The 1802 text listed at least a dozen "constitutions" including the phlegmatic, plethoric, bilious, costive, lax, flatulent, gouty, rheumatic, scrophulous, hot and cold, and consumptive,[75] while the invocation of the Galenic temperaments and their "hybrids," such as the phlegmatic-bilious and the phlegmatic-sanguine, appeared in medical texts into the end of the nineteenth century and the early twentieth century. Some medical writers relabeled the melancholic complexion as the "nervous," the phlegmatic as the "lymphatic," and the bilious as the "hysteric," and others maintained the traditional temperamental categories while distancing them from a basis in the "humors."[76]

The popular medical genre flourished throughout the nineteenth century, notably in the United States, where skepticism about orthodox ("allopathic") medicine had deep roots in democratic, antielite culture. Chemical therapies and bleeding were widely rejected; botanic cures were embraced; purified, simplified, and "natural" diets were advertised and widely accepted; strong links developed between popular medical reform and evangelical religion; and the moral place of food and drink was reaffirmed. Moderation *remained* good for you and good, and the prudence of its counsel enjoyed new force and authority in American popular medicine. John Gunn's 1830 *Domestic Medicine* treated each of the six nonnaturals and advised moderation in all things, though it neither used the term "nonnaturals" nor offered the traditional sequential listing. Differing human constitutions were mentioned matter-of-factly, without focused attention on their underlying bases, and different sorts of food were commended or warned against without reference to Galenic "qualities" or to chemical constituents.[77]

Disconnected bits and pieces of traditional dietetic vocabulary persisted in all sorts of medical texts through the nineteenth century and even beyond, and not just in writings intended for the laity. The four humors and four temperaments can be found in late eighteenth-century texts by some of the most progressively oriented medical professors. At the great medical center of Edinburgh University, the physician and chemist William Cullen described the "sanguineous," "melancholic," "choleric," and "phlegmatic" temperaments while ascribing them partly to the state of the body's "solids" as well as fluids and qualifying them as "the *ancient temperaments*." Cullen, however, stressed that people's actual temperaments "are much more *various*, and very far from being reduced to their genera and species."[78] Elite physicians continued to describe the condition called "melancholy"; some people could be described as "good-humored" in both learned and lay speech; others might say about their favored foods that "if I like it, it likes me" or that a disliked food "doesn't agree with me"; and there were physicians who carried on saying that you were your own best doctor and that your own experience with what foods agreed with you was the best guide. There is nothing wrong and much that is right-headed about the historical search for continuities and "survivals." But there were changes in thinking about food and its effects, and those changes amounted to more than the rebranding of traditional notions and sentiments. Traditional dietetic culture was indeed deeply rooted, and it would be facile to presume that something so pervasive and so folded into the practices of the self and of the moral order should simply go away, to be replaced by new and distinct expert knowledge. The death of traditional dietetics was not sudden, and the proper way of talking about what happened might not be "death" at all but rather some other notion such as "submergence," with detached bits and pieces of its ideas and practices continuing to circulate in lay culture while being replaced in expert culture, another moment in the withdrawal of elite culture from lay culture.[79]

The period from about the middle of the eighteenth to the middle of the nineteenth century witnessed change in the distribution and authority of traditional dietetic culture, and that change can be—with qualifications—rightly described as decline and fall. And it is also right to look for ideas and vocabulary that supplanted those of traditional dietetics. What new conditions appeared for describing food, the human body, the self, and the conditions of knowing about food and its consequences when consumed? What happened to the role of qualities in talking about food and the people who consumed it? How did this new vocabulary index new relationships between ordinary sensory knowledge and expert knowledge, between self-knowledge and the knowledge that came from external expertise? And what happened to the traditional substantial link between the moral and the medical, the *good for you* and the *good*? What happened to the relationships between medical and moral counsel?

Attraction and Repulsion: Chemistry, Mechanism, and Common Sense

The new vocabulary belonged to the discipline of *chemistry*. The chemical categories that eventually became central to the understanding of food and its vital consequences were not inventions of the seventeenth-century Scientific Revolution, and their relationship to the mechanical philosophies of Boyle, Descartes, and Newton is problematic. Pertinent origins can be traced at least as far back as the sixteenth-century work of the Swiss-German alchemist and medical man Paracelsus (1493–1541).[80] While earlier histories of science went to great lengths to distinguish (legitimate and protoscientific) chemistry from the pseudoscience of alchemy (mystical and concerned almost solely with transmutation of lead into gold and finding the elixir of eternal life), it has long been accepted that the alchemists accumulated a huge store of empirical knowledge. They knew much about how substances underwent changes of qualitative state, knowledge that was of great interest to modernizing natural philosophers and physicians. The alchemists' vocabulary and their stock of empirical knowledge played a substantial role in the new mechanical philosophies, and many of the modernizing practitioners—notably Boyle and Newton—were skilled chemists with an intense interest in questions to do with material transformations. And some of that interest bore upon vital phenomena, including digestion; the assimilation of aliment; the nature, maintenance, and diminution of animal heat; respiration, perspiration and transpiration; and growth, development, and death.

While Paracelsus was strongly oriented to medical problems, his philosophy of matter was fully general. There were, Paracelsus held, three principles comprising all things: the so-called *tria prima* of salt, sulfur, and mercury. In some alchemical thinking, the *tria prima* were conceived as principles rather than as the familiar material substances going by those names, and in this way they had a standing similar to that the Aristotelian elements in relation to common air, earth, water, and fire. But in other articulations, it was said that the *tria prima* could be extracted from such familiar everyday substances as sulfur and marine salt.[81] In Paracelsian thought, salt was the principle of solidity, crystallization, and permanence; sulfur was the principle of combustion and evaporation; and mercury was the principle of fluidity, volatility, and change.[82] By the early seventeenth century, some chemists expanded the *tria prima* to include such additional fundamental principles as "phlegm"—or "oil"—and "earth," and, by the eighteenth century the list of "chemists' elements" (or "principles") offered by the Dutch physician Boerhaave included earth, air, and fire but also alcohol (spirits of wine), mercury, and the "Spiritus Rector" (or essential oil of natural bodies).[83] If physicians properly understood these principles and their operation, it was claimed that the prevention and cure of diseases would be vastly more effective.

Through the sixteenth and seventeenth centuries, these categories filtered into natural philosophical and medical thinking, strongly allied to growing discontent with Aristotle and Galen, while the alchemists' generally low social standing posed legitimacy problems. Yet, alchemists' influence was not much diminished by criticism of their unintelligibility. Robert Boyle, notably, attacked both the obscurity of alchemical notions and the specific idea that all bodies are indeed composed of these three (or five) substances.[84] Alchemy and the mechanical philosophy made apparently strange bedfellows. The alchemists tended to think of the ultimate constituents of material substances possessing distinct qualities, while the mechanical philosophers aimed, in principle, for an account of ultimate constituents as quality-less, characterized solely by their geometry and states of motion. In practice, however, mechanists incorporated aspects of alchemical language, its appeal partly flowing from widespread expert and lay familiarity with the physical and sensory qualities of many chemical substances, while quality-less corpuscles (particle or atoms) were wholly theoretical entities.[85]

For these reasons, and despite modernizing natural philosophers' often-expressed contempt for the alchemists, much of the mechanical philosophy that so impressed the iatromechanists of the late seventeenth and early eighteenth centuries was already a pastiche of different views of matter. These views encompassed corpuscular concepts (in which the apparent qualities of things were the effects on the mind of quality-less particles) and alchemical thinking (in which the apparent qualities of things was testimony to their ultimate nature).[86] Among mechanical philosophers, Boyle and Newton talked about corpuscles having only geometrical and kinetic properties, but they also wrote straightforwardly about substances such as sulfur and mercury and about what properties and powers these substances possessed. They sometimes speculated, for example, about how corpuscles of sulfur and mercury might be constituted and arranged so as to produce the properties and powers that belonged to known chemical substances. But they also dealt with discrete substances and their properties just as they were, trusting that readers would have general knowledge of what such substances as mercury and sulfur *were like*. This pastiche-like quality was a pervasive feature of much "modern" natural philosophical thought from about the middle of the seventeenth century.[87] In theory, iatromechanism and iatromathematics were bodies of thought quite distinct from iatrochemistry; in practice, chemical language was present even in tracts that censured the chemists.

Iatromechanism and its related modes proved to be passing fashions, and in terms of medical thinking and practice, they came to little. But with chemical categories, the story was different. The *tria prima* belonged to the vocabulary of alchemists from the late medieval period into the seventeenth century; they were central to alchemists' overall philosophy of the world and its workings, of human beings and their place in the world. But the substances known as salt, sulfur, and mercury were, at the same time, more or less familiar in the everyday world: salt, of course; sulfur, since antiquity, as brimstone, in gunpowder, and as a component of various ointments and fumigants; and mercurial substances, as both widely used medicaments and a felting agent in hat manufacture. You did not have to be an alchemist, a physician, or even a learned person to know about things such as salt, sulfur, and mercury. And the same cultural spread pertains to a range of other things and properties that were known to chemists and had more or less stable chemical meanings in the seventeenth and eighteenth centuries. Householders and artisans were familiar with the properties of acidic and alkaline substances (aqua fortis, later nitric acid; muriatic acid, later hydrochloric acid; oil of vitriol, later sulfuric acid; and lye, later sodium or potassium hydroxide), with spirits of wine (later ethyl alcohol), sugar, soda, potash, niter (or saltpeter), and so on.[88] Many sorts of people seem to have understood that soaps were a mixture of oils and salts, that salts were soluble in water and other liquids, that some salts might have laxative effects, and that there were "chalybeate" (or iron-containing) mineral waters that were supposed to dry out the humors. So, when chemical categories appeared pervasively in thinking about food and the body, they counted as alternatives to the Galenic categories of traditional dietetics. And in many cases, chemical categories were also familiar in the world of things and their properties widely encountered by medical men and by many of their patients. Reference to entities such as salt, sulfur, mercury, acids, and alkalis never belonged exclusively to the expert culture of chemists.

The distribution of this "common culture," so to speak, was one of the advantages of chemical categories in medical thought and practice. Like the vocabulary of traditional dietetics, chemical terms might be as familiar to patients as they were to physicians. Chemical categories were, of course, central to the medical thinking of followers of Paracelsus, but they eventually broke free of those specific ties, and by the late seventeenth and early eighteenth centuries, chemical language pervaded much writing about food and the body, sometimes more or less uneasily cohabiting with mechanical categories. Medical views on the nature of food—its assimilation; role in powering the body; transformation into muscle, bone, and nerve; suitability for different sorts of people; and role in health and disease—were all increasingly expressed in chemical terms.

Foodstuffs and Their Chemical Constituents

It was common in eighteenth-century chemistry to treat substances in relation to their membership in the three great "kingdoms" of nature: the vegetable, the animal, and the mineral, two out of the three having a substantial connection with foodstuffs. There was also a general tendency to move away from Galenic accounts of the qualities of foods. The French chemist Louis Lémery, writing in 1702, went through a list of alimentary items—the sort of thing familiar from the texts of traditional dietetics—promising an alternative to Galenic categories and also giving "by Chymical and Mechanical Reasons" an account of the qualities of different foods.[89] But Lémery's serial descriptions of fruits, vegetables, meats, and drinks and their suitability for different sorts of people was, in fact, a hybrid of humoral and chemical categories, only lightly seasoned by mechanism: some foods were better for choleric people, while other foods were better for phlegmatics. All food, Lémery wrote, consists of four sorts of principles,

> *viz*. Oil, Salt, Earth, and Water, and so the Difference that is between one Sort of Food and another, consists exactly in the Conjunction and different Proportion there is between these same Principles. . . . Food proves more or less agreeable to the Taste, as its Parts are more or less subtil, and apt to pass lightly over the nervous Fibres of the Tongue. . . . Food is more or less nourishing, according as it abounds more in those Parts that are oily, balsamick, and apt to stick to the solid Parts, and according as there is

> more Resemblance between the Contextures of its Parts, and that of our Bodies.[90]

Beef "contains much Oil, volatile Salt, and Earth" and agrees with "young bilious People . . . and are pretty much used to Exercise or Labour." Cucumbers "are ill of Digestion, and produce gross and phlegmatic Humours"; "they contain a little Oil, much Phlegm, and an indifferent Measure of essential Salt," proper for "young Persons of an hot and bilious Constitution."[91]

George Cheyne's medical writings persistently ran together chemical and mechanical vocabularies.[92] All aliment, he wrote, contained in various combinations the principles of sulfur ("from whence Spirit and Activity"), salt (with its "hard angular" and highly attractive particles), water ("from whence alone Fluidity"), and earth ("the base and *Substratum* of these others"). For Cheyne, it was "past all Doubt in Philosophy, and in philosophical *Chem[istr]y*," that animal foods were richer in the first two principles, while vegetable foods were richer in air, water, and earth.[93] The physician John Arbuthnot wrote about "the constituent Parts" of animal flesh used for food. These he explained in terms of the principles of all natural bodies according to the vocabulary of the "Chemists." These were "Water, Earth, Oil, Salt, [and] Spirit," of which, Arbuthnot assumed, "every one has some general Notion." He admitted, however, that his readers might not be familiar with their various forms and that these details might occasion "some Confusion in the Minds of such as are ignorant of Chemistry."[94] What chemists knew about all natural bodies applied also to the forms of aliment. Animals differed in their chemical composition and therefore in their physiological effects. Fish, for example, had more oil and salt than did terrestrial animals, and this could be confirmed because certain dried fish tasted of sal ammoniac. And in general, aliment differed in its acidity ("acescence") and alkalinity ("alkalescence"). Carnivorous animals tended to be more alkaline than herbivores; mushrooms "contain an Oil of a volatile Salt"; and fish, "being highly alkalescent, wants to be qualified [its properties corrected] by Salt and Vinegar," the original of the traditional fish-and-chips seasoning. And Arbuthnot explained how acidity, alkalinity, and the constituent parts of different foods worked their effects on the body and also what foods—with what chemical properties—were good for different sorts of people. Some people had acid constitutions; some were alkaline. If you became too acidic, then alkaline foods were recommended and vice versa.[95] The traditional catalog of foods and their qualities was chemically reformulated: beef, for instance, is alkalescent; olives are antiacid "by their Oil"; and garlic and leeks "abound with a pungent volatile salt and Oil."[96] In Prussia, the physician and chemist Friedrich Hoffmann (1660–1742) declared that acidity made foods difficult to digest and that alkalinity was good, as in general were foods high in oily, mucilaginous, and sulfurous substances. A healthful diet sought to match the chemical characteristics of food with those of the chyle and blood.[97]

This sort of chemical language explaining the properties and dietary consequences of foods circulated widely through the eighteenth century. Medical writers who held no brief for mechanism found chemical notions congenial in talking about food and its effects. The English physician John Burton described both bodily constitutions and aliment as acid or alkaline. The acidic was cooling and the alkaline was heating, and on that basis physicians could advise what foods were proper for individual temperaments and conditions.[98] At Edinburgh, William Cullen's course on materia medica described the properties and nutritiousness of foods in terms of acidity and alkalinity as well as acerbity, sweetness, and texture, while other expert physicians listed such properties as "viscidity," "glewiness," and "solubility."[99] By the late eighteenth century, chemically inclined physicians were subjecting alimentary constituents—such things as "farinaceous matter" (or "mucilage"), sugar, and vegetable acids—to experimental analysis in the laboratory, working on the assumption that the processes of digestion in the human body were substantially the same as those that the chemist could produce experimentally.[100]

Tasting Chemicals

The hybridity of eighteenth-century ways of talking about food and the body is also indexed by the role of *taste*. What could the sense of taste testify about the nature of foods? Acidity and alkalinity were chemical properties with chemical meanings that were widely recognized among the laity, while such substances as sulfur, mercury, and, of course, earth, water, and oil were familiar in both lay and expert culture. But there were other categories used by chemically inclined eighteenth-century physicians to describe foods and their bodily effects that were familiar to ordinary people because—like the four qualities of traditional dietetics—they were also *sensory categories*. Physicians had long found the taste and odor of bodily fluids—urine and blood, for example—useful guides to patients' states of health or disease, and some seventeenth-century physicians specifically related those sensory characteristics to chemical constituents in bodily fluids that were ultimately derived from food.[101] In much eighteenth-century medical writing, acerbity and acrimony might describe a pathological condition of the body's fluids, but these were also qualities of aliment that could be tasted and sometimes smelled. The *taste* of acerbity—bitter, sharp, or irritating—testified to foods that had an acerbic principle *in them*, just as the sensory experience of warmth or moistness in traditional dietetics testified to foods that *possessed* the qualities of warmth and moistness. In Holland, Boerhaave causally linked the experience of taste to basic chemical principles. The causes of tastes were the pertinent chemical principles, which, when rendered soluble, penetrated and excited the tongue's taste buds and set their nervous fibers into motion. That motion was "conveyed from them to the common Sensory, where it excites an Idea in the Mind of Saltness, Acidity, Alcaly, Sweetness, Vinosity, Spirituousness, Bitterness, Spiciness, Heat or Pungency, Roughness, or else a Taste compounded of these." Boerhaave thought that taste could testify as to whether some things were physiologically good for you, and this was because taste (and sometimes smell) were, again, reliable indices of alimentary constituents.[102] Arbuthnot causally connected taste to chemical and physiological properties—"Tastes are the Indexes of the different Qualities of Plants as well as of all Sorts of Aliment: Different Tastes proceed from different Mixtures of Water, Earth, Oil and Salt"—and also listed the different sorts of aliment in terms of their chemical composition. The different sorts of aliment were commended or warned against according to their chemical properties, many of which were evident to the senses, including taste and smell. Wheat bread, containing bran, and rye bread were more wholesome for some constitutions because they were more "Acescent." Cucumbers and melons were considered dangerous—as they had long been in traditional dietetics—but now because they contained "a great deal of cooling viscid Juice, combined with a nitrous Salt, which sometimes makes them offensive to the Stomach."[103]

For Cullen too and many other eighteenth-century medical writers, taste continued to be a more or less reliable index of both the nutritiousness and the safety of foods. There were four basic "qualities" in foods such as fruits, each of which was available to the consumer's senses and also possessed a chemical meaning: acerbity or bitterness (not good for you), acidity (qualifiedly good in a moderate degree), sweetness (very good and "perfectly innocent"), and texture (how long things took to be digested, their wholesomeness depending on your constitution and way of life).[104] You could know what vegetables, for instance, were good for the human body by taste: vegetables that tasted mild, bland, or "agreeable" signaled that they were nourishing; those that tasted acrid or bitter indicated that they were not. True, we do occasionally consume such bitter vegetables as celery and endive, but our well-justified custom with such foods is first to blanch or boil them to remove the bitter principle. And Cullen was one of many physicians—especially in his native country—to express unrestrained enthusiasm for the goodness of sugar. He acknowledged medical disagreement about sugar but came down unambiguously in its favor. Sugar is good for you; it doesn't rot your teeth, as some allege, and "the mischiefs of what is called in Scotland eating of sweeties, are wrongly imputed to sugar." Sugar, especially in its most refined forms, "affords a pure and copious nourishment," and while Cullen did not use the old tag *quod sapit nutrit*, here and elsewhere he significantly treated pleasantness of taste as a gauge of nutritiousness.[105]

To the end of the eighteenth century, the language of chemistry infiltrated medical thought far more successfully than that of micromechanism. "Modernity," so to speak, *happened* to thought about food and the body, but it did not happen in the ways promised by enthusiasts for the mechanical philosophies of Descartes and Newton. Chemical vocabulary describing aliment was different from that of traditional dietetics but enjoyed some of the reach and authority that had attached to Galenic culture. First, some chemical language was to a degree "owned" by professional experts while also being intelligible to many laypersons, and so it was part of a culture that was still—to a degree—"held in common." Although chemists and chemically minded physicians might have special operational senses of what it meant that a substance was acidic or alkaline and what exactly constituted the principles of mercury, salt, and sulfur, they could presume that there were patients and readers who broadly recognized such language and what it referred to. Second, chemical categories were often the sorts of things whose meanings might be in many cases available to the senses. Acidity, alkalinity, sweetness, saltiness, bitterness, and the like could be *tasted*, and spirituous things could also be *smelled*, so to talk about aliment in these terms and to use this language to account for the role of foods in bodily function *made sense*.[106] Chemical vocabulary provided a "new way of talking" about food and the body, yet it also had a degree of familiarity in the culture and accessibility to the patient and the reader.

The significance of expert chemistry for talking about food and its effects on the body grew enormously in the latter part of the eighteenth century and beyond, and much of that significance had to do with how chemistry itself changed. What if the notion of an "element" no longer included *any* of the four traditional elements: fire, air, earth, and water? What if water itself was reconceived as a "compound" of two "elements," the fundamental particles of "hydrogen" and "oxygen" that no one had ever heard of before the late eighteenth century, that were given their names by expert chemists who defined their properties and the operations for identifying them, and that had no presence in ordinary sensory experience? What if air also ceased to be an element and became a complex and varying mixture of different *gases*? What if fire was also no longer an element but rather one chemical manifestation of combustion, conceived as a specific mode of oxygenation? What if hydrogen and oxygen were joined as elements by an array of fundamental entities that were similarly defined by chemical expertise and that had been similarly absent from common speech about natural kinds—nitrogen and chlorine, for example? And what if common substances familiar to chemists, artisans, and others turned out to be compounds of newly named elements: magnesium (from the mineral magnesia), potassium (from potash), sodium (from lye or caustic soda), and carbon (from charcoal)? Finally, given this new world of chemical elements and compounds, how should people—both expert and lay—think about the nutritive aspects of food, digestion, and assimilation? What bearing did these new accounts have for the idea of the qualities of food? Were foodstuffs indeed properly thought of in terms of *qualities* or in terms of nutritive *constituents* with newly defined chemical identities? Who had the authority to tell these new stories and to circulate them in the culture? And what did these new ways of thinking about food signify for self-understanding?

8

Nutrition Science: Constituents and Their Powers

Cookery . . . is, strictly speaking, a branch of chemistry. . . . A kitchen is, in fact, a chemical laboratory.
—Frederick Accum, *Culinary Chemistry, Exhibiting the Scientific Principles of Cookery*

A living body considered as an object of chemical research, is a laboratory, within which a number of chemical operations are conducted; of these operations, one chief object is to produce all those phenomena, which taken collectively are denominated *Life*; while another chief object is to develop gradually the corporeal machine or Laboratory itself, from its existence in the condition of an atom, as it were, to its utmost state of perfection.
—Jöns Jacob Berzelius, *Traité de Chimie*, Vol. 5

Nutrients: The Naming of the Parts

In traditional dietetics, it was proverbially said that "the stomach is the body's kitchen"; by the early nineteenth century, the homely kitchen was giving analogical way to the chemist's laboratory. The kitchen offered a way of understanding what happened to consumed food that was accessible to essentially anyone; invocations of the laboratory, however, pointed to operations occurring in a set-apart expert space. Chemical concepts were changing rapidly in the late eighteenth and early nineteenth centuries, and their relationship to lay appreciations of the characteristics and powers of food remained problematic for much of the nineteenth century and beyond. If the vocabulary of traditional dietetics did not precisely die in the nineteenth century, medical expertise nevertheless used its categories less and less, while an ever-greater role was assumed by a new vocabulary being produced by academic chemists and physiologists. Early in this historical process, participants recognized that distinct ways of talking about food and nutrition were emerging in different expert communities: "The chemist investigates the composition of an aliment, and arranges it according to the proximate principles which predominate in its composition, . . . while the empirical practitioner distributes the various kinds of food in an order which answers to his notions of their relative nutritive value, or to the supposed facility with which they are digested in the stomach."[1] In turn, the new expert language emerging from the laboratory increasingly infiltrated and informed the understandings of ordinary people: their ways of appreciating what food was, what happened when it was consumed, and how foods should be selected to maintain or restore health. Finally, that new language was accompanied by instructions telling laypeople to whom they might turn for reliable knowledge of food and its constituents. Knowledge about food and its bodily career was being *externalized*. Experts knew about such things; self-knowledge was increasingly pushed to the side, or its shape was remolded to fit the contours of expertise.

Descriptions of foods as acid or alkaline, oily or saline, persisted through the nineteenth century, and so too did a range of other categories that had circulated in the past: acridity, viscidity or glueyness, and so on. Notions such as these remained a feature of writings intended for both experts and laypeople. The "revolution" in chemistry ascribed to the research of such chemists as Joseph Priestley, Carl Scheele, and Antoine Lavoisier (on combustion) as well as Henry Cavendish (on the compound nature of water) and

John Dalton (on chemical elements and their combinations) coexisted with and partly redefined earlier ways of describing aliment. And these persistent ways of talking referred for the most part to the operations of expert chemists, physiologists, and physicians while also remaining intelligible within lay culture. But by the end of the eighteenth century, at least some alimentary substances began to be designated through characteristics whose recognition depended *wholly* on chemical expertise. Traditional dietetics always made sense within some overall theories of nature, the body, and vital processes. Yet, historians often point to the late eighteenth and early nineteenth centuries as the moment at which dietetics "became scientific," because this is when its vocabulary came to include notions that twentieth- and twenty-first-century moderns recognize as *our own*. And this is when talk of elements, humors, temperaments, and qualities were given new meanings, detached from their reference in the culture of traditional dietetics, sometimes ejected from expert accounts of food and its powers.[2]

Consider, to take one example, the career of the category of the albuminous substance. On the one hand, very many ordinary people were familiar with albumen. (The word was derived from the Latin for egg white.) It may not have been standard currency among the uneducated, but it was widely used by European physicians, chemists, and cooks, and while albumen retained its primary association with the eggs of fowl, some chemists came to extend its reference—analogically or literally—to fractions of human blood, the endosperm of plant seeds, or other natural substances that in some way *resembled* egg whites or behaved under certain conditions as egg whites behaved, for instance, being white or transparent, having a certain viscid character, being soluble in water but not in alcohol or ether, coagulating when heated below the boiling point of water or when exposed to acid or alcohol or, experimentally, when placed in an electric field.[3] Describing a nutritive portion of human food as albuminous was the sort of thing that late eighteenth- and early nineteenth-century chemists did, but the usage did not *necessarily* depend on special chemical expertise: many people—lay as well as expert—were familiar with albumen. They were also familiar with "jelly" derived from meat or bones (which was rebranded and commercialized as "gelatine" and subjected to chemical analysis in late eighteenth- and early nineteenth-century France).[4] By the 1830s and 1840s, albumin and gelatine were joined with other previously known alimentary substances—for example, "fibrin" and "casein"—in a new class of chemical substances eventually recognized as one of the fundamental "nutrients."[5]

The language of chemical *constituents* was on the way to displacing that of *qualities*. Nutrition researchers sought to characterize the different sorts of aliment in specifically chemical terms and to distinguish them in terms of their ultimate elemental constitutions. These inquiries occupied the attentions of leading chemists in the late eighteenth and early nineteenth centuries, notably the Frenchmen Joseph-Louis Gay-Lussac, Louis-Jacques Thénard, Claude Berthollet, Antoine François de Fourcroy, and François Magendie; the Swede Jöns Jacob Berzelius; the Scot Thomas Thomson; and the Englishman John Bostock, all of whom reckoned that a basic set of alimentary principles had distinct chemical identities and that these should be defined in terms of the new language of elements and compounds. That chemical work was rapidly assimilated by physicians, only some of whom had substantial chemical expertise. For the English physician John Ayrton Paris (1785–1856) writing in 1826, the "albuminous" principle (found in eggs and in other "animal matter") was one of nine classes of "Nutrientia" that had been securely established by chemical research, others including "fibrinous aliments" (found in the flesh of mature animals), "gelatinous aliments" (in the flesh of younger animals), "caseous aliments" (in cheese and other dairy products), "farinaceous aliments" (in grains), and "mucilaginous aliments" (in carrots and other root vegetables). In the first part of the nineteenth century, this sort of work often serially listed foods and their physiological effects. So, in form, this was similar to the dietary texts of the past, but now "nutritive" characteristics were no longer associated with Galenic qualities—warmth, moisture, and so on—but rather with distinct and identifiable chemical constituents. "The nutritive qualities of a substance depend upon its composition," Paris noted.[6] Subsequent chemists broke down the different sorts of aliment into their nutritive fractions with ever-greater analytic and quantitative specificity: the muscular parts of beef, mutton, pork, and chicken are rich in albumin and fibrin and have about a fourth as much of gelatine; carrots have about 9.5 percent of sugar and 0.3 percent of mucilage; and butter has butyric, capric, caproic, caseous, and caseic acids.[7]

There were nutritional categories identified by chemists that might also be recognized by laypeople, while determining their elemental constitutions belonged solely to the chemists. By the 1830s and 1840s, chemists were producing organized tables of foodstuffs, with each type of aliment analyzed into its elemental constituents: carbon, hydrogen, oxygen, and nitrogen, of course, but also phosphorus, sulfur, iron, potassium, and so on. And this sort of knowledge was increasingly circulated and standardized in research agendas aimed at accounting for nutritiousness by way of chemical composition, with chemical expertise now providing a language that could define good eating.[8] One chemical element in particular came to assume very great nutritional significance. Paris was aware that there was one recently identified element—then called "azot" (or "azote") and later increasingly known as "nitrogen"—and he gestured at nutritional controversies swirling around this element. Was azote in animals derived from food they consumed or from atmospheric air they inhaled? Was it indeed true that people consuming no (or very little) animal food—for example, slaves who were said to "live a long time without eating any thing but sugar"—could survive or even thrive? Were azote-containing substances such as albumin found—or found to a significant extent—in plant food?[9] Some chemists considered albuminous, nitrogen-containing foods to be restricted to animal sources; others referred to "vegetable albumin" (or gluten) contained in the endosperm of cereal seeds.[10] So, the search for the specific chemical makeup of the nutritive parts of foods also belonged to long-standing arguments about good diet and especially, to debates over the relative nutritiousness of animal and plant foods.

Just after Paris published his *Treatise on Diet*, the English chemist William Prout (1785–1850) sought "to determine the exact composition of the most simple and best defined organic compounds" that were used for human food. These were allotted to "three great classes," which he called the saccharine, the oily, and the albuminous. The new categories counted as "the three great staminal principles from which all organized bodies are essentially constituted" and are responsible for their nutritive character.[11] There was continuing debate over how substances sharing a number of egg albumin's physical and chemical properties should be categorized: some said that gelatine was quite different from albumin while others held that only substances derived from animals should count as albuminous, and a settled chemical identity for the "albuminous principle" was not then available. Prout said that plant matter generally contained carbon, hydrogen, and oxygen, as did animal matter, and he underlined the special significance of azote in animal aliments, "to which they appear to owe many of their peculiar properties."[12] Prout's contribution to the great series of Bridgewater Treatises—the last of eight commissioned volumes whose purpose was to establish the existence and attributes of God from the evidence of Nature—was a survey of divine wisdom in the domains of chemistry, meteorology, and the processes of digestion. In this widely distributed form, Prout's text was an important early exercise in popularizing the new vocabulary for talking about the nutritive constituents of food.[13] The recently established repertoire of chemical elements—oxygen, nitrogen, hydrogen, potassium, etc.—were surveyed and classified according to their roles in supporting combustion or in forming acids and alkalis. Nitrogen was not accessible to the senses, being colorless, tasteless, and odorless, so it could be known only courtesy of chemical expertise. Yet, nitrogen had unique significance in vital phenomena and was celebrated as "the characteristic element of animal substances."[14] You have no sensory experience of consuming elemental nitrogen or of its role in forming the fabric of your body, and indeed, as a reader of a natural theological tract in the 1830s, you may never have previously *heard* of nitrogen or azote, but now you had, and now you knew who possessed special expertise in knowing these things.

Knowledge of the chemical makeup of foodstuffs, including those collected together as "albuminous," was advanced in major ways by the Dutch chemist Gerrit Jan Mulder (1802–1880) and the German Justus von Liebig (1803–1873). In the late 1830s, Mulder published analytic studies of a series of albuminous substances—fibrin, casein, gelatine, and gluten—with the aim of determining their elemental composition and their proper chemical classification. That work soon came to the attention of the Swedish chemist Berzelius (1779–1848), who suggested that they be called by the collective name "protein," and Mulder agreed. (Berzelius liked this gesture to the Greek πρώτειος—"primary"—because he considered that this was the *fundamental* stuff of animal nutrition.)[15] Mulder believed that the physiological significance of protein to vital phenomena was immense: "Without it no life appears possible on our planet. Through its means the chief phenomena of life are produced." Protein is found to some extent in all plant foods, and animals can accumulate it by eating plant matter and use it to form the fabric of their bodies, though Mulder was not sure whether protein might also be *formed* within the animal body. He claimed that all proteins had identical proportions of carbon, hydrogen, oxygen, and nitrogen, so suggesting that there was a root chemical substance—or "protein organic radical"—that was identical in all proteins. Chemical and physical differences among proteins arose, Mulder reckoned, from the different amounts of sulfur and phosphorus each contained, and he canvased several candidate compositions for the protein radical, one of which characterized it as $C_{40}H_{31}N_5O_{12}$. Add two atoms of sulfur and one of phosphorus to this basic radical, and you have serum albumin; add to the radical one atom of sulfur and one of phosphorus, and you have fibrin; and add just one atom of sulfur, and you have casein. It was these small differences between basically similar proteins that accounted for their distinct physiological properties. Proteins were indeed protean, small compositional changes giving rise to very different physiological roles.[16] It was a compelling vision, holding out real hope that organic analysis would yield a thorough understanding of nutrition.[17] The vision commanded public attention. By the 1860s, a series of public lectures at the Science Museum in London noted that protein had "produced more discussion than any other substance in the whole range of chemical inquiry."[18]

The Power of Protein

Working in Stockholm, Berzelius was one major interlocutor of Mulder's early work; another was Justus von Liebig at the small German university of Giessen, where his laboratory was becoming the world's most productive factory for the analysis of naturally occurring organic substances.[19] It was through Liebig, rather than Mulder, that the general culture was most effectively made aware of protein and of the practical goods promised by the chemistry of nutrition. Mulder and Liebig were later caught up in a nasty controversy over the identity of the protein radical,[20] but the German chemist originally supported Mulder's claims about the very similar proportion of carbon, hydrogen, oxygen, and nitrogen in all proteins, and Liebig came to devote increasing attention to the analysis of organic compounds of practical interest in both agriculture and nutrition. Like almost all early protein researchers, Liebig and Mulder also worked on the chemistry of other types of aliment, including sugars, starches, and fats. And what Liebig especially contributed was a set of aggressively publicized claims about the different physiological functions performed by the different categories of alimentary substances.[21]

The animal body, Liebig noted, was fueled and kept warm by now well-understood chemical processes. Some forms of aliment underwent combustion and so supported respiration. That is, these substances combined with oxygen, and the body was powered and heated by what was qualitatively the same sort of process as the familiar burning of wood or coal in a fireplace. Combustion of flammable materials proceeded quickly while combustion of food in the body proceeded more slowly—like the rusting of iron—but chemically speaking, the processes were identical. In traditional dietetics, the body's warmth was causally connected to the qualities of warmth contained in food. Foods that were, for instance, "warm in the third degree"—garlic, mustard, and distilled spirits—might make people of a choleric or sanguinary temperament tip over into pathological overheating. There was an accepted analogy between the body's "vital flame," fueled by food, and the flame of a candle, fueled by wax or tallow. What Liebig and others were proposing, however, was *no analogy*: "The animal body acts . . . as a furnace, which we supply with fuel." It was not the *quality* of warmth in food that warmed the body; in the conceptual world of nineteenth-century chemistry, body warmth was the result of the carbon and hydrogen in food combining with oxygen breathed in from the air. In this respect, there was nothing either special or mysterious about food and life: "The animal body," Liebig wrote, "is a heated mass, which bears the same relation to surrounding objects as any other heated mass." Food, among other things, is fuel: "In the animal body the food is the fuel; with a proper supply of oxygen we obtain the heat given out during its oxidation or combustion."[22] So far as vital activity is concerned, the end results of consuming food were heat and its oxidation products: the carbon in food becomes carbonic acid (CO_2), the hydrogen becomes water (H_2O), and the substances that are not used to power the body or build its fabric are regarded as *nonnutritious*, expelled as excrement. So far as the basic chemistry of fuel is concerned, all food is on a level, and all food comes to the same end. Different foods are mixtures of a standard palette of chemical constituents. While certain constituents and certain mixtures of constituents have different nutritional powers, there are no *qualitative* differences to be considered.

The foods that fuel the body and support respiration are the fats and oils, the starches, and the sugars, that is, the substances later designated as *carbohydrates*.[23] These substances do not produce blood, and in Liebig's scheme, only those forms of food that can become blood—and, by way of blood, produce and maintain the fabric of the animal body—were properly designated as "nourishment."[24] (Nutritiousness here was understood to be about *making* the body, not about *fueling* the body.) It was a singular usage, soon to be disputed by other researchers, though it was consequential for the practical impact of Liebig's animal chemistry. But not all food serves as fuel or, at least, does so in normal circumstances. The forms of aliment properly called nourishing were chemically distinguished by the presence of nitrogen. Blood and the animal tissues made from blood have nitrogen; starches, sugars, and fats do not. It was understood that the animal body does not manufacture nitrogen from atmospheric air and that its only source for making, growing, and maintaining its physical fabric was nitrogen-containing food, that is, in effect, alimentary protein. The fundamental distinction drawn between aliments was that dividing the "*nitrogenized* and *non-nitrogenized*." The "nitrogenized constituents" of animal food have a chemical composition identical with that of blood, and so too do the nitrogenized compounds found in plant matter, such as gluten.[25] The nitrogenized foods (proteins) were called the "*plastic elements of nutrition*," and the carbohydrates were the "*elements of respiration*."[26] Liebig offered a rhetorically compelling, if disturbing, image as a way of grasping the scientific facts of the matter: "In a chemical sense, . . . it may be said that a carnivorous animal, in supporting the vital process, consumes itself. That which serves for its nutrition is identical with those parts of its organization which are to be renewed."[27] The notion of *agreement* was here displaced from its reference in traditional dietetics—the match between sensory experience and the *qualities* in food—and was redefined as the match between the chemical *constituents* in food and those of the blood.

Protein made the body's fabric, but it did more than that. Liebig claimed that the energy driving muscle contraction was derived from breaking down portions of their proteinaceous fabric. The cause of what Liebig called the "waste of matter" is oxidation. Muscular contraction was powered by the breakdown of muscle substance—that is, protein—and this was evidenced by the presence of nitrogen-containing urea (or uric acid) in the urine.[28] So, there were chemical techniques that could reliably assess and quantitatively measure bodily *work*: you need not rely on observation or on self-reporting. Protein was a prerequisite for work. The more physical work you did, the more protein you needed, and Liebig offered a *chemical* argument for the nutritiousness of beef (and other animal flesh), replacing the analogical and qualitative presumptions of traditional dietetics. The more work you did, the more nitrogenous matter appeared in your urine. This made protein the master nutrient in the sense of body-building, of course, but also in the sense of the capacity for work. Similarly, you could define *health* not in terms of the traditional balance of humors (and their qualities) but instead in terms of a measurable *chemical* balance "among all the causes of waste and of supply."[29] Liebig surrounded protein with the aura of scientific authority. Protein was a master molecule, and Liebig was the high priest of the church of protein.[30]

Food Chemistry in the Culture

The new food chemistry, and especially Liebig's ideas about protein, spread rapidly through the general culture. The language of food composition was effectively disseminated, and practical lessons were taught about proper food choice and preparation. In large part, this spread was owing to Liebig's own entrepreneurial activities. While all of his findings appeared in the technical literature, notably in the journal that he controlled (the *Annalen der Chemie und Pharmacie*), Liebig did not miss a beat in diffusing the messages of alimentary, animal, and agricultural chemistry in the popular literature. *Familiar Letters on Chemistry*, *Animal Chemistry*, and *The Principles of Agricultural Chemistry* were bestsellers in all the major European and North American markets. Liebig enriched himself from his writing for a wide global audience and from consultancy, demanding and receiving significant sums for endorsing the wholesomeness of British beers, which had been suspected of toxic contamination.[31] Even those critical of some of his claims about food chemistry acknowledged his celebrity. Liebig had many followers and some critics, but the critics too acknowledged how much he had done to affect both expert and lay thinking about food—its constituents and powers. The encyclopedias, including those pointed at working-class readers, adopted Liebigian theories of nutrition, and the vocabulary of proteins and carbohydrates was well on its way to becoming part of the vernacular by the middle of the nineteenth century.[32] In the 1830s the *Penny Cyclopædia*, launched by the British Society for the Diffusion of Useful Knowledge, introduced the literate lower orders to the chemistry of foods, and supplement volumes appearing in the late 1850s channeled the views of writers who had further popularized Liebig's own popularizations.[33] Herbert Spencer gave prominent attention to the protein theory in his *Principles of Biology*,[34] and his friend George Henry Lewes—philosopher, critic, scientific dabbler, and partner of George Eliot—called Liebig's claims pathbreaking: they have, Lewes said, "agitated" general thinking about nutrition and have "made the tour of Europe, and become stereotyped in lectures and text-books."[35]

The wide influence of Liebig's way of describing nutrition was indexed as much by limited critical pushback as it was by general enthusiasm. Lewes objected specifically to protein worship: it was dogmatically wrong, he said, to deny a tissue-forming role to fats, starches, minerals, and salts and to deny to protein a role in fueling the body.[36] More generally, Lewes bridled at the cultural aggressiveness of defining nutritiousness and body function solely in chemical terms. The chemists, notably including Liebig, had vastly advanced our knowledge of food constituents, but they were not justified in presuming that to know alimentary substances was to know chemical composition. Chemistry is not physiology, Lewes insisted, and chemistry can never answer questions about what food *does* in the body. "Food is a physiological question, inasmuch as it relates to an organism," but the subject "has fallen into the hands of the chemists," with results that are partly positive and partly the pathologies of intellectual imperialism. "Scarcely a single alimentary problem has been solved" by the chemists, Lewes wrote. Physiological questions dealt with the fate of foods in the body, and while chemical knowledge was undeniably relevant to such questions, it could not answer them: "what ordinarily takes place in the laboratory will not at all take place in the organism," and the workings of the stomach are infinitely more complex than those that take place in the chemical laboratory.[37] Chemists could understand the stuff that went in and the stuff that came out, but Lewes judged that they could offer no more than speculation about the processes through which food was transformed into you.

The animal body took in the chemicals contained in food, and the organism worked on and with those chemicals to perform its functions. Understanding vital function, Lewes wrote, belongs to *physiology*. The chemist asks "what are the chemical constituents of nutritive substances?"; the physiologist asks "what are the substances which will nourish the organism?" An organism may die from hunger surrounded by nutritive substances that it will not eat or that it cannot assimilate.[38] Physiologists (and physicians) understood that organisms *vary* with respect to their capacities and powers: what nourishes one individual body will not nourish another body of the same species. Chemistry was tone-deaf to these variations. The chemists, for example, had little or nothing to say about *taste*, about "'why one man's meat is another man's poison.'" And here Lewes fell in with a version of the old tag *quod sapit nutrit* (what tastes good is good for you). Some people just *cannot* eat foods that others relish. Mutton was much liked by some and caused violent vomiting in others; some could not abide cherries or strawberries, which were considered delicacies by their friends; and some rejected the very idea of eating an egg. These things were facts about variation in preferences, not delusions: parents should not insist on their children eating things that disgusted them, even if the chemists insisted that they contained the right nutrients in right amounts. For adults as for children, likes and dislikes should be considered reliable signs that foods were either wholesome or unwholesome *for that individual in these specific circumstances*. The chemists may call such tastes a "caprice," but they are wrong. Tastes should be taken as signs of a particular physiological state of the organism, "which we should do wrong to disregard. And whenever a refusal is constant, it indicates a positive unfitness in the Food." "Only gross ignorance of Physiology, an ignorance unhappily too widely spread," Lewes wrote, "can argue that because a certain article is wholesome to many, it must necessarily be wholesome to all. Each individual organism is specifically different from every other." If you look at almost any work of modern expertise, "you will meet with expositions of the theory of Food, and the nutritive value of various aliments, which are so precise and so unhesitating in their formulæ, that you will scarcely listen to me with patience when I assert that the precision is fallacious, and the doctrines demonstrably erroneous."[39]

The chemist is a generalizer and, so far as specific living bodies are concerned, an abstracter. In Lewes's view, the chemist speaks of nutrient substances and their role *in general* and in the *abstract*, while the physiologist and the physician want to know about food in the organism and food in this particular body in health and disease. And ordinary eaters want to know whether *this* piece of roast beef is good for *them*, just now, at this stage of their life, considered in light of their normal patterns of living. Chemical abstractions were good to think about but not a satisfactory guide about what to eat. Lewes was as disengaged from Galenic medicine as any Victorian intellectual but blamed the views of chemical experts for losing the focus on individual identity that had once been characteristic of traditional dietetics. Chemistry could only legitimately claim to identify the elemental materials out of which organic substances were made, not how they were constructed. "This being so," Lewes went on, "it is clear that every attempt to explain chemically the nutritive value of any aliment by an enumeration of its constituents, must belong to . . . 'the physiology of probabilities.'"[40] It followed from a proper appreciation of the limits of chemical expertise and the quality of its knowledge that we should not look to the chemical laboratory to tell us what to eat: "The chemical point of view is incapable of directing us to a single rule in practice."[41]

Liebig and other chemists were, however, strongly oriented toward concrete practice, nutritional practice, of course, but also the practices of producing food from the land and preparing it in the kitchen. Through the 1840s and 1850s, Liebig was greatly concerned with agricultural technology. He used chemical analysis to devise new artificial fertilizers to increase productivity by replenishing the soil with essential minerals. The nitrogen-phosphorus-potassium (NPK) formula that is the basis for much of the modern synthetic fertilizer industry traces back to Liebig's chemical analyses, together with the skepticism about the use of animal manures that eventually made Liebig a hate figure for the organic agriculture movement emerging in the twentieth century.[42] There were violent controversies, notably involving English agriculturists hostile to Liebig's prescriptions and critical of his neglect of practical field trials. Liebig was commercially invested in the rightness of his theories: he took out patents on his fertilizer formulae and was financially involved in several fertilizer manufacturing companies.[43] Lewes criticized Liebig for failing to close the gap between chemistry and the dynamic process of the animal organism, but Liebig maintained a long-lasting interest in practical human nutrition. And based on organic analysis, he offered specific instructions on how to prepare the most nutritious foods.

Liebig Makes Soup

Soup was Liebig's most celebrated food preparation. Soup was much discussed in traditional dietetics, barley soup and potage at length but also the broths produced from animal flesh, notably those of fowl and cattle. Soups were good for you; they were commended for ease of digestion, were ideal for the ill and the aged, and were reckoned to be restorative. The fowl broth concerned in traditional dietetic culture was often specifically that of the cock (cockerel, rooster) and sometimes, even more specifically, of the cock's comb. So, the restorative quality widely ascribed to chicken soup had long implicated analogical reasoning. True, the flesh of domestic fowl was widely said to be easy to digest, but the cock and its comb possessed the "masculine principle," a source of strength and power. More traditional European societies still commend soup by way of the idea that "the goodness" of the flesh is released through long boiling. Soup remains good for you because, it is thought, the virtues of its animal ingredients have been thoroughly extracted from the flesh, making them digestible and physiologically available.[44] In late eighteenth- and early nineteenth-century France, the new institution of the *restaurant* appeared, a place where anyone could come to have food prepared for them, chosen off a menu (or "carte") and where they could dine at personal tables with their companions and not at a common table, picking their food from common dishes. But the "restaurant" was a dish before it was a place, specifically it was a form of *restorative* beef broth.[45]

Meat soups of all sorts—broths, bouillons, and "beef teas"—had long been considered highly nourishing, and in the late seventeenth century the medically trained French mechanic Denis Papin devised a "digester" (or pressure cooker) that aimed to improve the process of extracting the "essence" of meats and even bones, transforming cheap and tough cuts into convenient and nutritious forms.[46] But during the late eighteenth and early nineteenth centuries, the academic chemists got involved, giving accounts of the specific *constituents* of these soups and extracts that were responsible for nutritive value and working to isolate and concentrate those constituents. Liebig gave new chemical authority to long-standing lay traditions of preparing liquid meat extracts, and he maintained that major nutritive parts of meat are contained not in its fibers but rather in the fluid parts.[47] One such constituent rose to cultural prominence as an outcome of early nineteenth-century French controversies over the nutritive value of *gelatine*. In this setting, gelatine was considered as the semisolidified tremulous stuff that resulted from long-boiled meat or bone broth. The matter was potentially of enormous interest, for if you could produce, concentrate, and stabilize the nutritive portions of meat, you might have a portable and cheap nutritional powerhouse capable, in theory, of supplying the military and efficiently feeding the impoverished. Liebig later agreed with the earlier French chemists that gelatine was not that substance, and, as he later claimed, it was not even a protein. Trials of gelatine cubes on French naval voyages produced poor results, and later animal experiments cast further doubt that bone gelatine was a nutritious substance.[48]

An alternative meat constituent emerged from chemists' inquiries in the first decade of the nineteenth century. This was osmazome, a specific water- and alcohol-soluble substance named in 1806 by the chemist Louis-Joseph Thénard (1777–1857), from the Greek words for "smell" (or "aroma") and "soup" and alternatively referred to, for example, as "extractive Matter of Broth" and "animal extractive matter."[49] Osmazome was identified as both a highly nutritive extract of meat and an aromatic substance, giving cooked meats—beef but also other animal flesh, especially dark meats and those from mature animals—their characteristic savory odor. Meats smelled "meaty": some enthusiasts said that the odor of the extracted osmazome was "the essence of flavour. . . . It abounds in the browned savoury crust of roast meats, which owe, indeed, all their piquancy and relish to the osmazome developed on the spit."[50] For these reasons, osmazome became central to reframing long-standing historical sensibilities about the relationship between nutritiousness and *taste*. In traditional dietetics, taste had a substantial role in identifying the "qualities" of foods and informing the notion of "agreement" between food and one's body, but that relationship had been marginalized by emerging understandings of chemical constituents.[51] Yet, in osmazome, taste retained a role as a sign that nutritiousness was on offer, nutritiousness carried by a specific chemical *constituent* that expert chemists had identified.

The chemists worked to extract osmazome and characterize it. Some thought that osmazome was just fibrin under another name;[52] some, including Liebig, identified it with the nitrogenous organic acid known as creatine.[53] (Through the century, chemists sought to analyze osmazome in the same way they had settled the elemental constituents of, say, casein and gelatine, but no consensus was reached, and analytic attempts were eventually given up.)[54] But the gourmets were mainly content to sing the praises of osmazome and to applaud the chemists who had identified it, and no gourmet did so more lyrically than the French lawyer Jean Anthelme Brillat-Savarin (1755–1826), whose book *The Physiology of Taste* (1825) is one of the great philosophical and literary reflections on gourmandism. Brillat-Savarin celebrated the pleasures of the table as the summit of the civilizing arts and sciences. Both gourmands and ordinary eaters owe a debt to science: "Since the time when analytical chemistry became an exact science, a great deal has been discovered about the dual nature of the elements composing our bodies and the substances which Nature seems to have created to repair the losses we suffer." In Brillat-Savarin's aphorism "Tell me what you eat: I will tell you what you are," he gave that sensibility a specific reference to analytic chemistry, noting that "man is largely composed of the same substances as the animals he eats. . . . The most detailed and praiseworthy researches have been carried out . . . , and it has been found possible to break down both the human body and the foodstuffs by which it is strengthened, first into their secondary parts, and then into their basic elements."[55]

Osmazome provided the most telling concrete evidence of the debt that modern pleasure owed to modern science. Brillat-Savarin displayed his currency with the latest chemical findings. He maintained that good cooks had *always* understood osmazome and the conditions that produced it; they just hadn't known it in formal chemical terms. Here Brillat-Savarin appears as one of the first advocates of what came to be called *molecular gastronomy*.[56] "The most signal service rendered by chemistry to the sciences of food is the discovery, or rather the precise comprehension of osmazome," Brillat-Savarin declared. He explained what this substance was, how it was prepared, and which sorts of animal flesh had more of the sapid substance: "It is in osmazôme that the principal merit of a good soup resides; the brown of roast meat is due to the caramelization of osmazôme; and it is osmazôme that gives its rich flavor to venison and game." Because osmazome is more concentrated in dark meat (and in the white meat of larger animals), "this is why the true connoisseur has always preferred the fleshy thigh. . . . [T]he instinct of taste anticipated science."[57] The French led the way in celebrating osmazome in particular and the contribution of science to gourmandism in general. Some of the English pushed back: osmazome suited the French; plain roast beef agreed with English natures. In the essay "Gastronomy and Civilization," written in the middle of the century, the novelist Thomas Love Peacock—himself a considerable epicure and recipe writer—observed that "we have brought chemistry into our kitchens, not as a handmaid but as a poisoner: she would have taught us the principles of assimilation, affinity, and harmony, and would have instructed us in the laws of preparation, arrangement, and the true theory of the application of heat, but we desired her to conjure bread with muriatic acid and soda, and separate osmazome from gelatine and albumen. . . . We introduce foreign manners without object and made ridiculous by misapplication."[58] Through the late nineteenth and early twentieth centuries—and even as the chemical identity of osmazome became more problematic—the substance maintained its place in texts of practical cookery and domestic management. Eliza Acton's *Modern Cookery* explained to British cooks how to prepare bouillon, "the Common Soup or Beef-Broth of France," giving practical methods for best extracting the "osmazome (*which is the most savoury part of the meat*)," and she endorsed Liebig's way of sealing in the nutritive meat juices by either immersing the meat quickly in boiling water or searing it at high heat. The title page of *Modern Cookery* editions from the mid-1850s advertised its "carefully tested recipes in which the Principles of Baron Liebig and Other Eminent Writers Have Been as Much as Possible Applied and Explained." For people who had not already read Liebig's *Familiar Letters on Chemistry*, this best-selling Victorian cookbook was a major vehicle for introducing the chemist's way of describing the nutritive constituents of food to the Anglophone vernacular. This was advertised as *scientific* cookery, explicitly adopting methods based on chemists' understandings of nutritive constituents and how these might be retained or concentrated.[59]

The leading and most long-lived British manual on cooking and running a home—commonly known as *Mrs. Beeton's Book of Household Management*—copied some of the Liebig material from Eliza Acton, calling Liebig "the highest authority on all matters concerned with the chemistry of food."[60] Isabella Beeton gave middle-class women concise lessons on the chemical constituents of food as a foundation for the theory and practice of cooking, including soup making. Meat, they were instructed, contained fibers, fat, gelatine, albumin, and osmazome, the latter of "which gives flavor and perfume to the stock." The aroma of osmazome was so much more pronounced in roasted meats that she counseled the cook to add leftover bits to the stockpot. Beeton described the relative concentrations of osmazome in the flesh of young versus older animals and in dark meat versus white.[61] Good cooking, good eating, and good nutrition involved awareness and management of these chemical substances. In 1906, a Boston magazine of household management celebrated the preservation of osmazome as "the essential quality which makes the whole difference between good and bad (or unintelligent) cookery."[62] Editions of *Mrs. Beeton's Book of Household Management* and other popular cookbooks into the post-Edwardian era instructed cooks to mind their osmazome. And even as the term "osmazome" dropped out of the vocabulary of chemists in the twentieth century, it maintained a tenuous presence in the vernacular as an explanation of hearty meat flavors, to be replaced only recently by the term "umami," chemically identified with the amino acid glutamate and something cooks can add by the use of products such as monosodium glutamate (MSG).[63]

Liebig did not pretend to be a connoisseur or to have expertise in domestic management, but he offered a detailed recipe for the preparation of beef soup, together with accounts of the nutritious principles generally found in meat and how they might be most effectively preserved. In the matter of beef soup and in other areas of food production and preparation, the political economy of nutrition occupied much of Liebig's attention. And here *taste* continued to figure in identifying nutritious constituents. Informed by the earlier French disputes over gelatine, albumin, and their physiological powers, Liebig had been concerned for some time with preparations of meat that would maximize its nutritive properties and their availability. And in the late 1840s, he published extended discussions about how best to prepare meat dishes. If you boil meat, you extract water-soluble substances, including what Liebig called "nutritive salts," while the meat itself retains insoluble substances, including fibrin and some phosphate salts. Practical advice followed from chemical analysis. Whenever custom dictates that you eat the boiled meat and discard the broth, you are losing a good part of the nutritive chemicals, and the editor of the English edition of Liebig's text on the chemistry of food, the Edinburgh chemist William Gregory, specifically lamented the English practice of throwing away the stock. The result was tasteless meat, of course, but this lack of taste importantly testified to a lack of nutritional value. If you were going to subject meat to long boiling, then you must eat the broth too. "The addition of stock to any dish not only improves the flavor," Gregory noted, "but often restores the soluble matter removed in previous operations . . . and thus renders it much more wholesome and nutritious than it would otherwise be. . . . A good cook judges of almost every thing by the taste," and Gregory endorsed Liebig's chemical explanations of why this was so. Chemical expertise now offered a scientific basis for good cooking and good feeding. Stewing was better than boiling, because you customarily consumed the stewing juices. Skilled and conscientious cooks already knew this in the mode of custom and practical judgment, but now they could know it properly on a sound chemical basis. Subjective taste retained a place in the assessment of goodness but only as a proxy for objective chemical composition.[64]

Accordingly, bouillon, beef soup, or the defatted cold-water extract of finely chopped raw meat known in British usage as "beef tea" were good for you, directly in proportion to the "goodness" taken out of the flesh and contained in the broth. Traditional thinking did not specify what the "goodness" of those nutritional bits might be, but the chemistry of the late eighteenth and early nineteenth centuries sought to identify the nutrients, analyze them, and record their elemental composition. The question remained about whether there might be a way to effectively extract, condense, and concentrate that goodness. The basic procedure that Liebig described—and that had long been known in practical cooking—was to thoroughly mince the meat and then soak it for a while in cold water. (About thirty pounds of lean beef went into making one pound of "true extract," known to medical people in its solid evaporated form as *extractum carnis*, and by 1853 included as a remedy in a new German pharmacopoeia.)[65] This technique effectively extracted all of the albumin contained in the flesh. If you then boil the concoction, the "fibrinous residue" is left behind, and the meat is "perfectly tasteless," since all the odoriferous substances are now in the broth. Roasted meat is tastier than boiled, and the scientific reason is that all the sapid substances, which are in the meat's *juices*, have been sealed in.[66] The truth of this can be practically established, Liebig noted: take a tasteless bit of boiled meat and add to it a portion of the cold-water extract of raw meat, and it again tastes like roasted flesh.[67]

The search for a stable, condensed, and highly nutritious version of animal flesh—beef in a portable package—had been launched at least as early as the French Gelatine Commission of the 1810s.[68] Liebig always had an eye on practical applications of his science: the commercial world was neither foreign nor distasteful to him. It occurred to Liebig—and to his business associates—that something could be done to make the nutritional powers of beef extract available on a global scale, turning it into a profitable product in the marketplace. Through the 1840s and 1850s, and while the extract was being used to some extent in medical contexts, the product was expensive, as the source meat had to be obtained from German butchers. Cost per unit nutrient therefore became a consideration, and in later editions of *Familiar Letters* Liebig noted that there were abundant supplies of cheap meat in Ukraine, in parts of the United States, in Mexico and Australia, and especially in the River Plate area of South America, which might be exploited for producing "immense quantities of the best extract of meat." An extract of beef from such globalized sources "might perhaps acquire a very peculiar importance for the potato-eating population of Europe," even though Liebig turned down a number of approaches in the 1850s and early 1860s from Australia, Mexico, and the United States for launching commercial meat extract ventures.[69] (Cattle in these areas were being slaughtered in huge numbers for their hides alone, while, prior to refrigerated shipping, there had been little use for the fresh meat.) Colonially sourced protein could strengthen the bodies of colonial powers and repair some of the physiological injuries of European poverty.

Soon, a German railway engineer working in Brazil and Uruguay persuaded Liebig to think seriously about making beef extract on a commercial scale using South American meat, proposing that the professor invest in a test factory in a town in Uruguay, which was eventually named Fray Bentos. By 1864, the state-of-the-art plant was up and running. Liebig was made a director of the company—paid £1,000 a year, with a one-off payment of £5,000—on the condition that the company might brand the product with his name, and he received 2 percent of corporate profits for five years. The professor agreed, and Liebig's Extract of Meat Company (LEMCO) was capitalized at the enormous sum of £500,000 on the London Stock Exchange.[70] By the mid-1870s, about five hundred tons a year of the extract were being produced at Fray Bentos. The beef extract was being supplied to armies and hospitals and was also offered on the market for home consumption. Half a teaspoon in half a pint of hot water made a strong beef tea, and it was reckoned to cost about a penny a serving.[71] Liebig worked his contacts to secure medical endorsements, though physicians and public health officials began to push back, and Florence Nightingale evidently had mixed feelings about its benefits.[72] Still, the marketing and advertising were both aggressive and innovative. The company's colorful trading cards and calendars were and remain collectibles (figures 8.1 and 8.2), encouraging cookbook writers to include recipes mentioning the brand.[73] Sometimes LEMCO did more than encourage, commissioning bespoke company recipe books that promoted new ways of using the extract.[74] The company branched out into canning as well, and in 1881 Liebig trademarked the Fray Bentos brand of corned beef. LEMCO was a hugely profitable globalized company: twenty years after Liebig's death in 1873, it had annual revenues of £1.5 million and was paying a yearly dividend of almost 25 percent.[75] Liebig's Meat Extract was among the first branded food products, one of the first based on a specific scientific finding about food constituents and their nutritive power and also one of the first on which a science-based capitalized company was founded. The company owned perhaps the most important brands in the emerging global food industry, arguably the original Big Food corporation.[76] Expert skepticism about the extract's nutritional value inevitably appeared, claiming that what the extract did was to enhance flavor and stimulate appetite without significant nutritional benefits. (Some critical chemists claimed that there wasn't much protein

VERO ESTRATTO DI CARNE LIEBIG.
RAZZE BOVINE: Toro selvatico, epoca degli antichi Germani.
Veggasi a tergo.

Figure 8.1. Italian advertising trading card for Liebig's Extract of Meat (ca. 1900). Here the product is dramatically and mythically linked to masculine strength. Presumably fortified by Liebig's extract, men in the era of the ancient German tribes attack a wild bull with spears. These lithographed cards were produced in thousands of versions in many languages from the early 1870s, and they could be exchanged for coupons to purchase the branded extract.

in the stuff at all.) And branded competition appeared (including Knorr, Maggi, and Bovril), providing a new field of application for organic analysis to distinguish the genuine article from counterfeits.[77] In 1899, LEMCO developed a cheaper version of the meat extract in collaboration with an English chemist that was trademarked as "OXO," first offered as a liquid and later as a stock cube, a product that remains—in the early twenty-first century—a staple of the British domestic larder, having shed almost all of its original scientific and nutritional rationale and now popped into a soup or stew just to save time.[78]

Liebig's Extract of Meat was launched onto the market and into the domestic kitchen from the world of chemical expertise—the chemistry of food components and especially the chemistry

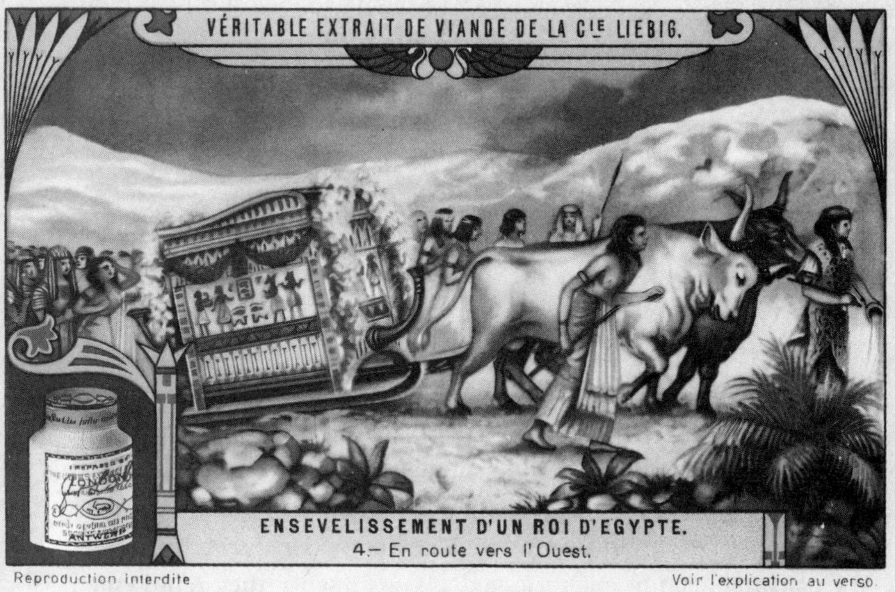

Figure 8.2. French advertising trading card for Liebig's Extract of Meat (ca. 1900). The sarcophagus of an Egyptian pharaoh, on its way to entombment, is pulled by bulls: two sorts of power are here linked by bovine strength. The back of the cards had either an advertisement for another company product or a recipe for a dish using the extract.

of proteins. Liebig and other organic chemists believed that they had identified the nitrogen-containing large molecules that build the body's fabric. Protein was a self-making substance; it made not just bodies but *strong* bodies, stronger than those of the potato-eating classes of Europe and stronger than the rice-eating cultures of Asia. So, part of Liebig-inspired protein worship was traced to power in *making people*: making them sturdy and making them masters of the protein-impoverished. But power was also evident in the networks of institutions and practices in which protein became enfolded. Protein power was manifest in the new bonds forged between the sciences of nutrition and medicine, the world of commerce, and the concerns of the state, efficiently feeding strong soldiers, restoring the ill to health, economically nourishing the poor, and projecting imperial power across the globe. The science of nutritive constituents became part of the new modern constitution, scientifically identified substances that would make healthy bodies and powerful states.

The Constitution of the Calorie

Nineteenth-century chemistry offered a new language for talking about food and a new source of authority for that language. There were foodstuffs that had nitrogen and those that had none; there were entities called proteins that had chemical elements that carbohydrates did not. Knowledge of *nutritiousness* was held courtesy of scientific expertise. Your senses and your digestive tract were not equipped to tell you which things had nitrogen, whether a food contained carbohydrates, and what proportion of carbon, hydrogen, oxygen, and nitrogen an edible thing offered. Foodstuffs had constituents, not qualities, and as constituents, they represented new conditions for knowing about your food and yourself. The powers of food with *qualities* were substantially *apparent* in the world of traditional dietetics, but what were the powers of *constituents*, and what were the conditions of knowing them?

In the chemical schemes developed by Liebig and other nineteenth-century nutrition researchers, proteins built the body, while fats and carbohydrates fueled the body. There was a then-satisfying chemical vocabulary that accounted for bodybuilding—nitrogen in proteinaceous foods was a source of nitrogen-containing proteins in the body's fabric—and other detailed stories about body fueling were emerging in the first part of the nineteenth century. Late eighteenth- and early nineteenth-century work by Lavoisier and others on combustion as oxidation began to supply a framework for understanding the food-fuel-heat-work relationship in terms of chemical combinations.[79] From the point of view of practical nutrition, two salient questions needed to be addressed. First, what were the chemical processes by which the oxidation of foods and the generation of heat occurred in the body? Second, what foods and what constituents of foods were more or less capable of supplying this heat? Both questions occupied physiologists and chemists, while the second was also central to practical advice about what to eat. What was required was an agreed *measure* of the heat-producing capacities of foods.

In both traditional dietetics and the emerging physiological chemistry of the nineteenth century, different types of foods had different capacities to generate body heat. Research agendas were developed to measure the heat-producing powers of foods and their constituents. German researchers—including Liebig students Max-Joseph von Pettenkofer and Carl von Voit at the University of Munich and Max Rubner at Marburg and Berlin—determined the energy values of foods and made findings available for such practical purposes as the rational feeding of animals, soldiers, the institutionalized, and the poor.[80] (Charitably motivated feeding of the impoverished and the imprisoned goes back essentially forever, but nineteenth-century nutrition science could advise the most cost-effective way of feeding, linking efficiency to the constituents of foods.)[81] The researcher who did most to forge twentieth-century relations between nutrition science, lay understandings of food and the body, and the state was the American chemist Wilbur Olin Atwater (1844–1907).[82] Taking his first degree at Wesleyan University in Connecticut and then his doctorate in chemistry at Yale, Atwater eventually returned to Wesleyan as its first professor of chemistry, and there he launched a Liebigian program in the chemical analyses of agricultural materials: fertilizers, maize, meat, fish, and oysters. In 1869–1871 and again in 1882–1883, Atwater traveled to Germany, working with Voit and his associates, and it was there that Atwater became invested in experimental studies of human nutrition and especially in efforts to assess the economic value of foodstuffs in terms of their protein content and potential energy.[83]

The measure of the heat-producing capacity of foods was imported into nineteenth- and early twentieth-century nutrition science from *physics* and particularly from *thermodynamics*, the subdiscipline dealing with the interrelationships of heat and other types of energy. Physicists had established the interconvertibility of different forms of energy: work, for example, can be completely converted into heat, and heat (though only partly) can be converted into work. Research on machine action proceeded alongside investigations of human body action, and the same conceptual resources were called on to explain both. For material machines, the fuel might be, for example, wood or coal or the flow of water; for the body machine, the fuel was food.[84] Nutrition scientists repeatedly told the public that "the body is a machine," and Atwater noted that "like other machines, it requires material to build up its several parts, to repair them as they are worn out, and to serve as fuel."[85] Digestion produced the heat that warmed the body and was transformed into *work*, not only the work of supplying the body's internal processes—the motions of the heart, the circulation of the blood, and even *thinking* about the motions of the heart and blood—but also the work that bodies did in and on the world, such as running, jumping, and moving things about. The goodness of foods was parsed through the language of engine work and engine maintenance: "The best coal for a locomotive is that which will enable it to haul the greatest number of tons over the longest distance. The best food for man is that which will enable him to do the most work in a given time and keep him in perfect condition for further work."[86] In physics, the measure of heat production is the *calorie* (c), defined as the energy needed to raise the temperature of one gram of water through 1°C, or, alternatively, the *kilocalorie* (C), which is the energy needed to raise the temperature of one liter of water through 1°C. The *mechanical equivalent of heat* having been established, the energetic capacity of foods could also be rendered in the physics idiom as *work*, and the settled measure of that capacity in foods became the kilocalorie C—the nutritional calorie—whose measure was either the heat needed to raise water temperature or the force needed to lift a weight a specified height.[87]

It was the kilocalorie—in vernacular usage, just the *calorie*—that by the beginning of the twentieth century became the standard measure of the energy potential of foods.[88] The calorie was the vehicle for translating food powers into numbers, a way of appreciating the direct causal link between human food and the potential for human work.[89] The nutritional calorie was not a *constituent* of foods; it was a *power* that foods possessed by virtue of their constituents and that different foods possessed to different quantifiable degrees. Nor did a specific caloric power qualitatively distinguish one food from another: nineteenth-century nutrition scientists insisted on the interchangeability of foods with respect to their heat-generating capacities. So far as heat production was concerned, you could substitute one food in a certain quantity for another food, perhaps in a different amount, and this was a key principle in emerging nutrition science that became known as the *isodynamic law*.[90] Foods were to be properly understood as energetically fungible. You could determine the quantity of beans that would yield the same heat as a quantity of beef, you could define a scale that would allow you to judge and compare the energetic content of very different American and Italian diets or the diets of hand laborers and those of brain workers, and you could identify the energetic content of different *nutrients*. By the end of the nineteenth century, it was accepted that the ratio of energies produced by the combustion of equal weights of protein, fat, and carbohydrate was 4:9:4—this becoming known as the Atwater factors (or ratios)—and you could measure and compare the energy content of the diets of different nations or occupations or social classes by summing up the quantities of the macronutrients they contained.[91]

Fueling Body and Mind

Atwater launched two related research agendas at Wesleyan. The first was to accurately determine the heat exchanges of the human body under experimental conditions that plausibly resembled those of natural life; the second—and the project that came to occupy most of Atwater's energies—was to establish, tabulate, and disseminate the nutrient makeup of different foodstuffs, their caloric values, and the political economy of obtaining them. There were substantial European precedents for both projects, but Atwater aimed to take them further. He aimed to systematize experimental findings, disseminate and tabulate the results, and bring them to bear on lay dietary decision-making and government policy. Atwater was an entrepreneur as well as an academic scientist, securing

Connecticut state funds for an agricultural experiment station. This was originally attached to Wesleyan and later transferred to Yale before Connecticut opened a second station in connection with the Agricultural College at Storrs. Atwater successfully sought funding from the US Department of Agriculture (USDA), the National Museum, and the Carnegie Institution; he eventually became chief of the Office of Experiment Stations within the USDA, in which capacity he effectively influenced national agricultural policy; and he placed one of his students as his Office of Experiment Stations successor. In 1893, Atwater secured funds for research in human nutrition at the network of state experiment stations over which he presided, and most of his research findings were published as reports to government agencies.[92]

Atwater's first project concerned the principle of energy conservation, established in physics as the first law of thermodynamics. Did conservation obtain in the human body? If it did, then a remaining obstacle would be removed to conceiving of the human body within a wholly naturalistic and mechanical frame. Atwater became familiar with pertinent research during his time in Munich. In the 1860s, Pettenkofer and Voit had worked with a respiration chamber (or room calorimeter) capable of housing a human experimental subject for days, a device further improved by Rubner in 1890. The respiration calorimeter measured gas and heat exchange under varying conditions of human activity and food supply and was intended to answer questions about human metabolism.[93] But now the human body was not just the intellectual *object* of research inquiries, it was a living *part* of the experimental device. This very large-scale apparatus—whose considerable cost was paid by the Bavarian state—allowed Pettenkofer and Voit to make some of the first determinations of calorie-expressed human energy requirements.[94] Atwater described the Munich respiration calorimeter as "one of the most interesting devices of modern experimental science," and, returning to Wesleyan, he pushed to have one constructed, aiming to refine its accuracy and to use it to securely establish whether energy conservation obtained in the human body.[95] Atwater, working with his Wesleyan colleague the physicist Edward Bennett Rosa, constructed a respiration calorimeter that became operational in the mid-1890s (figure 8.3).[96] Its advertised purpose was to study the "fundamental laws of animal nutrition" and "the transformation of energy" in the body. It was now scientifically settled that the body could neither create nor destroy *matter*, so the conservation of matter could be presumed, but *energy* conservation in the body was still a live scientific issue, and establishing this conservation was a state-of-the-art measuring problem that the respiration calorimeter was meant to address. You needed to measure the potential energy—sometimes called chemical energy—that the body received in food, the potential energy it expelled in excrement, the heat the body gave off, the kinetic energy it expended, and any changes in the body's temperature and stored energy (for example, whether it came to contain more or less fat or protein or other potential energy-containing stuff) during the duration of the experiment. Energy was conserved if the budget *balanced*, where balance here was not a moral quality, not an equipoise of *qualities*, but rather a matching of energetic input and energetic output. "The proper adjusting of food to the wants of the body," Atwater wrote, "is in reality a balancing of income and outgo." The sums on one side of the vital ledger had to equal those on the other.[97]

These were complex and hugely expensive experiments. The Atwater-Rosa device was funded by a US government grant of $10,000, and more money was required for its continuing operation, much of which came from the Carnegie Institution of Washington. The costs of construction and operation needed strong justifications for funding sources. The Carnegie Institution as well as the USDA were interested in problems of animal and human feeding so that theoretical interest was supported by promised practical outcomes. Acknowledging German predecessors, Atwater nevertheless advertised the novelty of what he and his associates were doing: "Previous to [these experiments], no attempt had been made, so far as the writers are aware, to measure the total income and outgo of energy in the body of an animal or a man of the conservation of energy, or to show that it does not obtain." There were

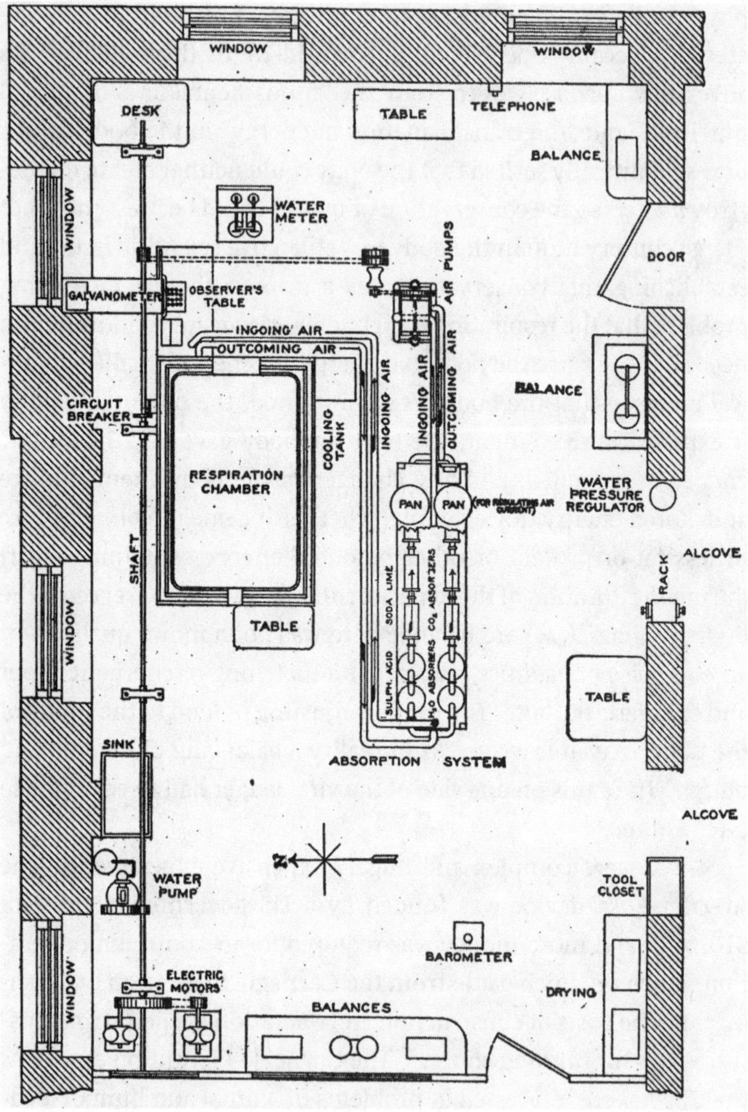

Figure 8.3. Respiration calorimeter. A schematic representation of a respiration calorimeter—the so-called Atwater-Rosa design—used around 1900 at Wesleyan University in Middletown, Connecticut. The experimental subjects, most of whom were laboratory assistants or "mature" students, sat for periods of up to ten days in the respiration chamber and were directed to engage in various types of physical or mental work. Food supplied to the subject was weighed, as were excreta. Heat produced by the subject was measured, and respiratory products were captured and weighed. Source: Atwater and Benedict, *The Respiration Calorimeter* (1905), 208.

Figure 8.4. Male experimental subject at lunch in a respiration calorimeter (ca. 1915).

many measurement difficulties to be overcome in establishing conservation, but one that was particular to the human body was a species of energy that, suitably qualified, was not considered to *exist* in either inanimate systems or nonhuman animate entities, and that was *mental energy*. This was the work done in feeling, concentrating, reflecting, deliberating, and, more generally, *thinking*. Atwater and his collaborators had devised protocols to assess energy expenditure while subjects were, for example, riding a stationary bicycle or weightlifting or (for women) washing, ironing, and knitting, and experimental subjects in the respiration calorimeter were directed to perform or refrain from performing various types of physical activity over defined periods (figures 8.4, 8.5, and 8.6).[98] But muscular activity did not exhaust all the types of work done by human subjects: "It is not impossible that there may be other forms of energy in the outgo besides the heat radiated or otherwise given off from the body and the mechanical energy of external muscu-

Figure 8.5. Women's work in a respiration calorimeter. Respiration experiments were meant, among other things, to measure the energy expenditures of different sorts of human activity. Here, in a Washington, D.C., calorimeter (ca. 1907), a young female subject is knitting, doing what was understood as typical women's work. The energy expenditures of other sorts of female activity, including washing and ironing, were also investigated.

lar work. It may be that the mental and nervous activities involve the expenditure of some form of physical energy which escapes our measurement."[99] And indeed, nineteenth-century materialist conceptions of mind and body presumed that if there was such a thing as mental energy, it would have physical manifestations that

Figure 8.6. Measuring muscular work in a respiration calorimeter. Atwater used a bicycle ergometer to study energy expenditure during muscular activity (ca. 1907). The bicycle was placed inside the respiration chamber and ridden by the experimental subject. The rear wheel was an electromagnetic device allowing pedal resistance to be varied and the work involved to be precisely measured.

ought to be measurable by, for example, the pulse, the body heat produced, and so on.

Measurements of both the outputs of mental work and the required dietary inputs were delicate, and before systematically setting out to address these questions, Atwater acknowledged that determining "what substances are consumed to produce brain and nerve force, and how much of each is required for a given quantity of intellectual labor, are questions which the chemist's balance and calorimeter do not answer."[100] Nevertheless, from the mid-1890s he designed a series of experiments in the respiration calorimeter meant to assess the energy expended in mental as well as physical labor. In one experiment, a student subject was engaged in hard mental work for eight hours a day over several days "calculating results of previous experiments"—an ingenious Matryoshka doll of reflexive design—by taking the college French examination or "studying a German treatise on physics," with the "mental application as intense as it could be made."[101] Measuring the outward physical manifestations of inner mental work proved to be extremely difficult, though related research projects persisted for decades, ultimately settling on a surprisingly small value. As colleagues of Atwater later put it, the extra energy expenditure for an hour of intense thought was about 4 percent over the metabolism of a resting subject. That sort of energy could be supplied, they noted, by four grams of a banana or half a peanut, findings that allowed early twentieth-century nutrition scientists to dismiss traditional ideas that thinking drew significant energies away from digesting or other metabolic modes.[102] The possibility of "other forms of energy" was here bracketed, if not wholly dismissed, and Atwater reckoned that the experiments in the respiration calorimeter powerfully supported energy conservation: the nutrients in food and the consumption of body materials "are the sole sources of heat in the animal body." There are inevitable variations, approximations, experimental errors, and uncertainties, but there is no doubt that "the law of conservation of energy applies in the living organism."[103] Experiments in the respiration calorimeter displayed the human body as a wholly natural object, its energetic economy subject to the same laws as any other natural system: it "actually obeys, as we should expect it to obey, this great law which dominates the physical universe."[104] So far as the human body was concerned—in all its circumstances and taking account of its mental as well as physical activities—the output was heat and work and the input was food.

There could be no doubt that "the labor of the brain is just as dependent upon food and the substances formed from it in the body as the labor of hands," and while there was intense scientific interest in the question of *which* foods and *which* nutrients specially supplied the mind's activity, this was, Atwater wrote, "a problem yet unsolved."[105] (When Atwater instructed popular readers to consider the human body as a machine, he acknowledged that mechanical explanation had not *yet* given adequate accounts of "the higher intellectual and spiritual faculties" but was nevertheless confident that, in general terms, "the right exercise of these depends upon the right nutrition of the body.")[106] Some maintained that fish was good for thinking due to its high phosphorus content and the high phosphorus content of the brain, suggesting a substantive role. A then-fashionable German tag from the materialist Jacob Moleschott asserted "Ohne Phosphor, kein Gedanke" (without phosphorus, no thought). Atwater was skeptical, but at the time there were branded foods and medicines pitching phosphorus-containing "brain food," and the causal association between fish and thought that circulated widely in the culture marked one notable nineteenth-century vernacularization of chemical language.[107] The respiration calorimeter experiments allowed Atwater and his colleagues to establish the energy requirements for different sorts of human activity. While proving conservation was, so to speak, at the theoretical end of Atwater's enterprise, compiling lists of the caloric value of foods, distributing those findings, and inserting them into statecraft were its major practical end products.

To assess the calorie content of foods, Atwater used a *bomb calorimeter*, a modified version of a tabletop-sized instrument that had been developed by the French chemist Pierre Eugène Marcellin Berthelot and was already widely used in European physical and nutritional research. You put a defined weight of some foodstuff in a sealed, pressurized, oxygen-enriched metal vessel surrounded by a water bath; the food was then electrically ignited and burned, and the resulting heat output was taken from the rise in the temperature of the surrounding water.[108] The bomb calorimeter was used in physics to measure heat production in the combustion of such things as wood and coal, and the same device was used to quantify the capacity of human foods to produce heat when digested. The credibility of the bomb calorimeter's findings depended on an acceptance that burning in the experimental apparatus of the device was energetically equivalent to digestion in the stomach.[109] Tabulated calorimetric results could then be used to assess the energetic input of subjects in the respiration calorimeter and, knowing the nature and quantity of foods, the caloric intake of ordinary eaters.

On the Tables: Types of Food and Types of People

The products of Atwater's research on food energy content took the form of graphic *tables*, all sorts of tables, reproduced again and again and commonly embedded either in reports to the federal government and state governments or as illustrations of points of principle in popular writings. There were tables that provided the quantity of nutrients delivered for a specific amount of money (figure 8.7) and documented the proportions of nutritive aliment and waste in commonly purchased foodstuffs.[110] Atwater and his colleagues aimed to encyclopedically cover as many American foods as was practical and to produce tables showing the macronutrient makeup of foods (protein, fats, carbohydrates), their caloric content, and the amounts of nonnutritious mineral ash and water they contained. Other tables compiled by nutrition scientists represented *dietaries*, the actual foods consumed over periods of time by different types of people and foods considered requisite for those types. Categories of interest in these dietaries included people of different races, nationalities, sexes, ages, and especially occupations.[111] The resulting tables allowed many sorts of comparisons to be made, comparisons with scientifically established standards and among social groups. How did the actual consumption patterns of a given group compare with the intake of energy and of nutrients that experts deemed necessary? How were calories and nutrient groups actually consumed distributed among different sorts of foodstuffs? How was energy consumption related to body types and to modes of work? Were energy and protein consumed in the most cost-effective way? How much food purchased was actually eaten,

TABLE B.—*Amounts of nutrients furnished for 25 cents in food materials at ordinary prices.*

Food materials as purchased.	Prices per pound.	Twenty-five cents will pay for—					
		Total food materials.	Nutrients.				Fuel value.
			Total.	Protein.	Fats.	Carbohydrates.	
	Cents.	*Pounds.*	*Pounds.*	*Pounds.*	*Pounds.*	*Pounds.*	*Calories.*
Beef, sirloin	10	2.50	.79	.38	.41	....	2,425
Do	15	1.67	.52	.25	.27	....	1,620
Do	20	1.25	.39	.19	.20	....	1,215
Do	25	1	.31	.15	.16	....	970
Beef, round	8	3.13	.95	.56	.39	....	2,675
Do	12	2.08	.63	.37	.26	....	1,780
Do	16	1.56	.47	.28	.19	....	1,335
Beef, neck	4	6.25	1.85	.98	.87	....	5,500
Do	6	4.17	1.23	.65	.58	....	3,670
Do	8	3.13	.93	.49	.44	....	2,755
Mutton, leg	8	3.13	.96	.47	.49	....	2,925
Do	14	1.79	.55	.27	.28	....	1,675
Do	20	1.25	.38	.19	.19	....	1,170
Ham, smoked	10	2.50	1.23	.37	.86	....	4,340
Do	16	1.56	.77	.23	.54	....	2,705
Salt pork	10	2.50	2.09	.02	2.07	....	8,775
Do	14	1.79	1.50	.02	1.48	....	6,285
Do	18	1.39	1.16	.01	1 15	....	4,880
Codfish, fresh	6	4.17	.45	.44	.01	....	855
Do	10	2.50	.27	.27	....	....	510
Codfish, dried salt	6	4.17	.68	.67	.01	....	1,315
Do	8	3.13	.51	.50	.01	....	985
Mackerel, salt	10	2.50	.74	.37	.37	....	2,275
Do	15	1.67	.49	.24	.25	....	1,520
Oysters, 25 cents per quart	12.5	2	.24	.13	.03	.08	520
Oysters, 35 cents per quart	17.5	1.43	.17	.09	.02	.06	370
Oysters, 50 cents per quart	25	1	.12	.06	.02	.04	260
Eggs, 15 cents per dozen	8.8	2.84	.63	.34	.29	....	1,860
Eggs, 25 cents per dozen	14.7	1.70	.38	.21	.17	....	1,115
Eggs, 35 cents per dozen	20.6	1.21	.27	.15	.12	....	790
Milk, 3 cents per quart	1.5	16.67	2.05	.60	.67	.78	5,420
Milk, 6 cents per quart	3	8.33	1.02	.30	.23	.39	2,705
Milk, 8 cents per quart	4	6.25	.77	.23	.25	.29	2,030
Cheese, whole milk	12	2.08	1.36	.59	.74	.03	4,305
Do	15	1.67	1.09	.47	.59	.03	3,455
Do	18	1.39	.91	.39	.49	.03	2,875
Cheese, skim milk	6	4.17	2.25	1.60	.28	.37	4,860
Do	8	3.13	1.69	1.20	.21	.28	3,645
Do	10	2.50	1.35	.96	.17	.22	2,910
Butter	15	1.67	1.45	.02	1.42	.01	6,035
Do	25	1	.86	.01	.85	....	3,615
Do	35	.71	.61	.01	.60	....	2,565
Sugar	5	5	4.89	....	....	4.89	9,100
Do	7	3.57	3.50	....	....	3.50	6,495
Wheat flour	2	12.50	10.87	1.37	.14	9.36	20,565
Do	2.5	10	8.70	1.10	.11	7.49	16,450
Do	3	8.33	7.24	.91	.09	6.24	13,705
Wheat bread	3	8.33	5.56	.73	.14	4.69	10,660
Do	5	5	3.34	.44	.08	2.82	6,400
Do	8	3.13	2.09	.28	.05	1.76	4,005
Corn meal	2	12.50	10.45	1.15	.47	8.83	20,565
Do	3	8.33	6.97	.77	.32	5.88	13,705
Oatmeal	3	8.23	7.51	1.22	.59	5.70	15,370
Do	5	5	4.52	.74	.36	3.42	9,225
Rice	6	4.17	3.64	.31	.02	3.31	6,795
Do	8	3.13	2.73	.23	.01	2.49	5,100
Beans	5	5	4.22	1.16	.10	2.96	8,075
Potatoes, 45 cents per bushel	0.75	33.33	5.70	.60	.03	5.07	10,665
Potatoes, 60 cents per bushel	1	25	4.27	.45	.02	3.80	8,000
Potatoes, 90 cents per bushel	1.5	16.67	2.85	.30	.02	2.53	5,335

Figure 8.7. The price of nutrients. This table assembles much of the information that Atwater reckoned consumers needed in order to make prudent eating decisions: the quantity of different foods you could buy for twenty-five cents and their macronutrient and caloric content. This was what Atwater called "the pecuniary economy of foods." Source: W. O. Atwater, *Foods: Nutritive Value and Cost* (Washington, DC: US Government Printing Office, 1894), 28.

and how much was wasted? Dietaries sometimes also engaged with questions about knowledge, belief, preference, and tastes. *Why* did different sorts of people eat as they did? What were their beliefs about food and nutrition? Was there evidence of dietary *change* over time, and if so what were the causes of change? And if there was no positive change in the dietaries of a group, why not, and what might be done to bring about the right sort of change?

In nutrition science, pertinent differences between types of people were invariably seen quantitative, not qualitative. All human bodies were heat engines. All bodies functioned on the same physical principles, their caloric requirements depending on body mass and surface area, the energy required to keep the body in working order, and the external work bodies actually did: cutting stone, rowing a boat, knitting, cooking, thinking, and so on. Among the most fundamental and presumably natural categories of human subjects were the young and the old, the male and the female. The dietary standards for children up to one and a half years old was 765 calories and 28 grams of protein per day; for ages two to six, it was 1,420 calories and 55 grams of protein; and between six and fifteen years, it was 2,040 calories and 75 grams of protein. Energy and protein requirements were taken to vary according to patterns of work in adulthood and generally declined as the body aged.[112] Following the long tradition of dietary writing, Atwater took the adult male body as the norm. Some of the tables of dietaries that he compiled made no mention of women, but others did.[113] Yet, neither Atwater nor his fellow nutrition scientists ignored the female body. Very much of what was encompassed by the linguistic category "men" was presumed to include both sexes. The average female body was smaller than the average male body, so adjustments to energetic requirements had to be made on that basis. The size of those adjustments was debated, but expert opinion settled on the idea that adult women required from 80–90 percent of the energy needed by adult men. Atwater's group used the 80 percent figure in its dietaries, taking that as an estimate of what women actually consumed when direct observational evidence was unavailable.[114] This figure was partly ascribed to differences in mass: "On an average, women are only four-fifths as large as men, and, consequently, dietaries for groups of women will require about four-fifths of the amount of food." Yet, some evidence was accumulating that the basal metabolism of men and women of a similar weight were the same or even that women ran hotter, so to speak.[115] However, another part of these differences in energetic requirements was attributed to typical patterns of activity, which were established by a male-dominant culture and were therefore potentially revisable matters of custom, convention, and social power. In the 1890s, a New York physician and dietitian reckoned that there was little that needed to be said about sex differences in consumption, which were considered as a straightforward combination of physical facts and social arrangements: "Women eat less food than men relatively because their average size is smaller, and also absolutely because they do less work and lead a more indoor life." That difference was in his assessment "slight": "when other conditions are equalised, the question of sex has very little influence upon the quantity of food consumed," with no references made to any qualitative differences.[116] Some nutrition scientists treated the 80 percent figure as if it were entirely natural; others judged that the lesser caloric requirements of women were "largely due to the indoor and sedentary life led by so many women. Under equal conditions a woman of the same size requires the same amount of food as a man."[117] Through the early part of the twentieth century, the question of whether differences in male and female patterns of activity were matters of custom or of nature was increasingly contested. By the 1910s, female physicians disputed findings that purportedly showed men to possess "a greater fund of energy" than women of the same age and weight. If that was the case, was it because of inherent physiological differences or because men were, in practice, better fed? And here a political and economic argument was made that women *should* be better nourished, and this mainly meant in practice that they should consume more calories.[118] In Atwater's circle, dietaries for women (and for children) tended to be subsumed within overall treatments of *families*, and associates conceded that little specific attention had been given to female diets, though there were a few dietaries specifically compiled for female college students in different part of the United States. Investigators acknowledged a common assumption "that women prefer different foods, being more fond of pastry and sweets and less fond of meat," but such studies as were devoted to women did not find convincing evidence that this was so.[119]

In traditional dietetics, medical engagement with women focused on those physiological functions that distinguished female from male bodies: menstruation, pregnancy, lactation and nursing, and, to a lesser extent, menopause. But here too considerations of dietary differences were almost wholly addressed in quantitative terms. Apart from the mass and surface areas of women's bodies and the work they usually did, the relative silence of late nineteenth- and early twentieth-century nutrition scientists about women proceeded from the basic assumption that all human bodies were heat engines, all working on the same principles. Explicit accounts of women's nutritional requirements mainly surfaced in connection with pregnancy, where it was understood as a matter of course that the nourishment of pregnant women bore on the nourishment and subsequent health of their offspring. (In the index of an influential early twentieth-century nutrition science textbook, the entry "women" just referred the reader to "lactation and parturition.")[120] The pregnant woman was an object requiring nutrition, but she was also—through her placenta and later through her milk—food for the child she was bearing, and much expert attention was paid to the relationship between the woman's diet and the nutrients contained in her milk.[121] Physicians, obstetricians, and nutrition scientists were aware of the folkish presumption that the pregnant woman should "eat for two," but for the most part that notion found little favor. She should follow the principles of moderation, eat "that which pleases her most," and avoid foods that caused indigestion or nausea. Experts disagreed about the nature of foods suitable for pregnancy and nursing: some had no special advice; some cautioned against alcohol and foods that were fatty, spicy, or strongly flavored; and some considered that adding a moderate intake of beer, claret, or port to the diet of the nursing woman improved the quality of her milk and gave her strength for "her extra duties."[122] Pregnancy, it was repeatedly insisted, was "a normal condition," but it was also "one in which diet is of special importance." In the idiom of nutrition science, the diet of the pregnant and nursing woman needed to contain the right nutrient constituents in the right quantities; deficiencies at these stages of a woman's life would determine whether her child would become strong and healthy.[123]

Race and ethnicity also figured in dietary compilations especially in the United States, though these were not central concerns. Male and female bodies were heat engines, with pertinent differences treated quantitatively, and so too were the bodies of whites and African Americans, Irish and Italians, Russian Jews, and Native Americans in New Mexico. Interest in the diversity of American race and ethnicity was in part an extension of a program of dietary surveillance over ever-widening social domains, but it was also informed by the Progressive political agenda. Since diet figured in both productivity and physical and mental well-being, knowledge of the dietaries of different racial and ethnic groups could be recruited in efforts to further integrate them into the modern social and economic order. Atwater professed interest in "the elevation of the poor whites and the negroes of the South," and the government-sponsored dietary surveys he initiated aimed to reveal what role diet might play in the continuing reconstruction of the Confederate states.[124] Were the former slaves getting the right amounts of protein and calories? Dietaries compiled for African American field laborers in Alabama documented a diet heavy in corn meal, flour, molasses, bacon, and salt pork. This provided adequate supplies of dietary fuel but inadequate levels of protein, and it was a diet much poorer than those of neighboring whites and of white laborers in the North. Black people were "both underfed and overfed," with too little flesh-forming nutriment and too much fat even for those doing hard labor. "It is evident," Atwater judged, that "the diet of the negro in the South is a very important factor of his character and condition." It was one of "the evils under which the colored people live," with its evident effects on their "physical, mental, and moral character." Poor diet was a substantial obstacle to the "upward progress" of the Negro race, while Atwater conceded that bad food choices and malnutrition could not be separated from the many other handicaps of their circumstances.[125] The mostly poor native populations of New Mexico seemed generally well nourished on their largely corn- and bean-based diet, though initial impressions were of a worrying deficiency in protein and fat, a lower consumption of these nutrients than of any other American group with the possible exception of the "negroes."[126]

Dietaries put together for inner-city Chicago in the 1890s included groups of Italian, French Canadian, and Bohemian (Czech) immigrants. (Three families of what the investigators called "Americans" were also included.) The aim was to find out what various immigrant groups actually ate, how their diet compared to scientifically formulated nutritional standards, whether their food choices were cost-effective, what the immigrant groups understood about the general principles of expert nutritional knowledge, and what their overall health appeared to be. Italian immigrants, for instance, persisted with the olive oil, wine, pasta, and imported cheeses to which they were accustomed, though these were expensive and ought to be rationally replaced by readily available American products. The Italians were so wedded to their customary fare and so "dreaded American food" that they would not go to hospitals or work for employers who could not supply the foods they preferred. The diets of Jews had long been a matter of compelling interest to Christian commentators, but for the Atwater network Jews were just one more ethnic group among the many pouring into the United States in the late nineteenth century, and the Chicago survey included Russian Jews, both orthodox (observing the laws of kashrut) and the nonorthodox (or "liberal") who did not follow religious prohibitions. Despite the peculiarities of religious law, Jewish dietary practices—both orthodox and liberal—broadly met the standards set by nutrition scientists, though the Chicago Jews seemed to consume slightly too much expensive protein and should be counseled to eat less meat and more cereals. Investigators commented on the "intelligence" of the immigrant groups with respect to the "American" norm and to an ideal of nutritional knowledge, absent which food choice would be unsatisfactory, solely determined by custom, taste, and price.[127] There was no question here of qualitative differences between ethnic groups, and there was no suggestion that such deficiencies as were discovered could be addressed by anything other than quantitatively framed interventions: more or less caloric intake; more or less protein, fats, and carbohydrates; and more cost-effective choices.

Some groups treated in dietary surveys were applauded for happening upon admirably cost-effective ways of satisfying energy and protein requirements. The "Scotchman" was typically "shrewd" in making oatmeal, haddock, and herring dietary centerpieces—all economical sources of protein—and "the thrifty inhabitants" of Atwater's own New England used a "frugal and rational" diet, "supplementing the fat of their pork with the protein of beans and the carbohydrates of potatoes," making them "well nourished, physically strong, and distinguished for their intellectual and moral force."[128] The "negroes," whose diet was of such concern to progressively minded nutrition scientists, could nevertheless be congratulated for their prudence: "The weekly ration of the southern negro, which he chooses in preference to any other, to wit: a peck of meal and three and one-half pounds of bacon, is also one which contains the elements of vital force at an exceedingly low cost. There are probably no working people so cheaply supplied as the southern negroes."[129]

For nutrition scientists, food was the fuel for bodily work. Those whose work required great bodily energy needed more fuel than those whose work was less physically demanding. Dietaries recognized this at a fundamental level: a man "at hard work" needed roughly 20 percent more calories than a man "at moderate work." The work of different specific occupations demanded different levels of fueling, and the dietaries compiled by late nineteenth-century nutrition scientists—German, British, and American—detailed the energetic requirements of a great range of economically and politically relevant work (figure 8.8). A Leipzig cabinetmaker was reported to consume 2,757 calories a day; "underfed" manual laborers in northern Italy consumed 2,192 calories; a glass blower in East Cambridge, Massachusetts, consumed 3,590; a "sewing-girl" in London consumed 1,820; a German miner "at very severe work"

CHART II.

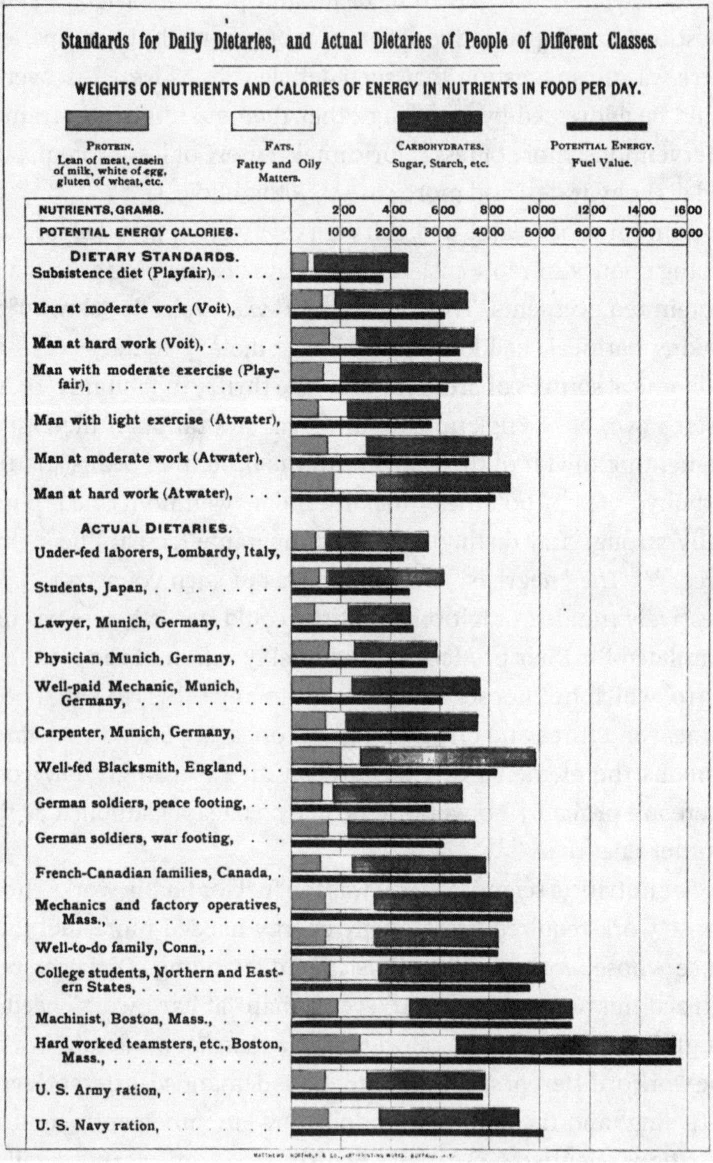

Figure 8.8. Comparative dietaries. The table represents both scientifically established dietary standards and the actual daily food consumption of different sorts of people doing different sorts of work in different countries. The horizontal bars show the distribution of nutrient classes by weight (top bar) and the total calories supplied (lower bar). Source: W. O. Atwater, "Tables of Foods and Dietaries," in *The National Medical Dictionary . . .* , Vol. 1, *A–J*, ed. John S. Billings (Philadelphia: Lea Brothers, 1890), xliii.

consumed 4,195; and a brick maker in Middletown, Connecticut, consumed an astonishing 8,818 calories per day.[130] Fuel had to be accommodated to function, and that was a principle that made as much sense for political and economic well-being as for the health of individual bodies.

How to Cook, How to Eat, How to Be

The scientific concept of work belonged to physics; through research on the energetic content of food and its measurement, work then became a central concern of late nineteenth-century nutrition science, and work formed a powerful link between the sciences of food and political economy. "The role of the political economist is hardly fitting for the chemist," Atwater cautioned,[131] but he and the wealthy Boston industrialist and free trade economist Edward Atkinson (1827–1905) were joined at the hip in an alliance between two disciplines central to the rationalizing impulses of Progressive Era America.[132] Atwater's early research had included topics not directly concerned with human nutrition: the USDA was then more interested in animal feeding and crop growth, and the farmers who were its major constituents encouraged the government to retain those foci. It may have been Atkinson's encouragement that inspired Atwater to shift attention to human nutrition, and it was his wealthy patron who also urged Atwater to write specifically for a popular readership. From 1887, Atwater published a series of pieces for the middle-class *Century Illustrated Monthly Magazine*, to which Atkinson contributed an introduction.[133] It was in these popular essays that Atwater disseminated the scientific views of nutrition that he wanted ordinary eaters to consume, digest, and act upon.

By the 1880s, dietaries tabulating the intakes of different occupational groups in the United States, Europe, and elsewhere were being published by American governmental agencies concerned with the supply of labor and its cost and productive efficiency.[134] Atkinson considered that nutrition science could supply answers to questions central to practical political economy. What fuels human labor? What constituents are the most potent fuels? At what cost could these be supplied? How might a rational diet for human labor be designed? How could workers be informed about that diet and induced to follow its prescriptions? Proper nutrition was the solid basis of productive labor. Food supplied the potential energy that powered the work of body and mind:

> If the measure or quantity of food is not sufficient and if it is not rightly adjusted to the conditions of complete nutrition, both the manual and the mental efficiency of man will be impaired. If the force which is generated by the assimilation of food is not adequate to the complete support of the workman, he will become incapable of making the product out of which his wages are paid in sufficient measure to resupply himself for the subsequent efforts necessary for the maintenance of his productive power. Life is a conversion of force, and if the force or food power supplied to man is inadequate or incomplete, the right conversion of force cannot go on.[135]

Well-nourished workers were productive workers; they produced more profit for employers, and their products were more competitive in the marketplace. Atkinson maintained that "half of the struggle for life is a struggle for food," elsewhere giving the aphorism statistical nuance.[136] The "more thrifty and prosperous" members of the "so-called working-classes" spend half of their income on food, but the percentage for "common laborers" might rise as high as 60 or even 70 percent.[137] The question was how to ensure that their wages were better spent—that they did not purchase food that lacked nutrients, that delivered inferior nutritional value for money spent, or that was inefficiently prepared.

One major way of rationalizing workers' food spending was to identify and eliminate waste. Atwater and Atkinson were convinced that waste was endemic among the laboring classes. Poorly informed consumers bought food much of which did not wind up on the plate, cooked food in wasteful ways, and ate in excess of what the body needed. Their food choices often made little nutritional sense. Beans were a cost-effective way of obtaining protein, while beefsteak was not; too many sweet things were eaten; and too much alcohol was drunk. "A pound of very lean beef and a quart of milk contain about the same quantity of actually nutritious materials," Atwater lectured, but the beef costs much more: drink more milk and eat less beef. It is "important for our purses that the [required] nutrients are obtained at minimum cost."[138] Rational decisions could be read off the tables of the "pecuniary economy" of food prepared by nutritional expertise and circulated through the popular print. It was good to buy food as cheaply as possible, but, uninformed by nutrition science, ordinary eaters' price-value judgments were faulty. Ordinary eaters reckoned that "nutrition is in proportion to price," but it often isn't.[139] Nor did the inverse follow: cheap isn't *always* good (see figure 8.7). You have to know what portion of a purchased food consists of nutriment and what of waste, you have to know the energy content of food per unit cost, and you have to know the quantity of protein you were getting for your money. The waste that followed from ignorance lowered the workers' standard of living and kept wages at an artificially high level.[140] Dietary waste was bad for workers' bodies and bad for employers' profits. The slogan was "consumption limited, production unlimited."[141] And the cure for waste was expertly formulated information about what foods to consume and how foods should be prepared. In traditional dietetics, much of bad eating proceeded from the sin of gluttonous excess; in the new nutritional order, bad eating flowed from a deficit of proper knowledge.

Food not only has to be bought it also has to be cooked, and part of the overall expense involves knowing how to cook efficiently. Atkinson was an innovative industrialist and brought that talent to bear on the workings of the domestic kitchen. He applauded the recent applications of scientific method to so many domains of production and distribution but lamented that there "is one art, perhaps the most important of all in its relations to the material, moral, and intellectual welfare of the community, to which little or no science has as yet been applied," and that was cooking.[142] "Science seems to have stopped outside the door of the kitchen," and it had to be invited in.[143] His associate, Atwater, was wholly on board: poor cooking—through nutrient loss—was bad for the body and also bad for the household budget: "How great are the economic loss and the injury to health from irrational cooking. . . . Certain it is that meats overdone are made less digestible thereby, and it would seem that the methods of baking and frying in vogue . . . must injure the nutritive value of the food." Atwater applauded contemporary initiatives to teach "the daughters of working-people how to do housework and how to select food and cook it," holding out the promise of a "happier home life in the future." Scientifically educated wives and daughters would improve domestic contentment and enhance domestic virtue, Atwater said.[144] But his plans were more concrete than just disseminating nutritional information. Atkinson's publications included recipes for rationally prepared dishes, and he invented, patented, manufactured, and offered for sale the "Aladdin Oven," a fuel-efficient cooking box made of vulcanized wood pulp lined with metal and asbestos. This was offered as a powerful technology for reducing the cost of efficiently getting nutrients onto the workers' plates.[145]

Learning from nutritional expertise brought hard economic benefits, but that expertise could deliver soft benefits too. The proper management of food and drink—one of the nonnaturals in traditional dietetics—had been an important article of virtue as well as a help to health, and while the notion of *balance* in nineteenth-century nutrition science was almost entirely a matter of nutrient and energetic *quantities*, a reformulated connection between the *good* and the *good for you* persisted. The vocabulary of nutrition science had not *yet* supplied a chemical explanation for why a healthy mind followed from a well-fed body, though Atwater did not doubt the fact of the matter—a good state of "the higher intellectual and spiritual faculties . . . depends upon the right nutrition of the body"—but he was sure that between diet and "the intellectual, social, and moral" there was "an important connection, one that reaches down deep into the philosophy of human living."[146] Evidence for that causal link was readily available, transparent to the most casual observer. In a materialist idiom, a diet adequately supplied with protein built up and maintained the fabric of brain and nerves as well as blood, muscles, and other organs, and in a practical economic idiom, everyone knew that workers not obliged to scramble for nutrients had more leisure to develop their mental and moral faculties, with the surplus generated by good diet then available to devote to their "higher needs." So, the condition for self-improvement, for moral and intellectual progress, was reliance on modern expertise. If workers followed "the teachings which the science of nutrition will supply and the teaching of economy will enforce," they will be better off physically, financially, and morally. Cost-effective eating was here identified as an article of modern morality. There was little hope for the development of the "higher Christian graces" in the poorer classes if they failed to feed their bodies rationally, so reinscribing the traditional tag *mens sana in corpore sano* in the idiom of the new nutrition science. The hoped-for future, not just of the nation but also the "race," will "bring that provision for physical wants which is requisite for the best welfare of mind and soul."[147]

Traditional dietetics recognized custom as second nature; it mapped the qualities of foods onto ordinary classifications and considered taste as a generally reliable index of wholesomeness. Nutrition science took a different view. The food decisions of uninstructed people were indeed based on custom, received common sense, and common tastes, and nutrition scientists—especially in a scientific order enshrining Darwinism—did not consider such things as *wholly* useless. To some extent, customary practice had hit upon diets that sustained normal function and could be supplied from available resources. The continued existence of various human groups following various diets testified to that degree of success. "Through a process of natural selection, and by way of experience without scientific knowledge, each race and each nation has found out the kind of food or the combinations that will give the right proportions of nutrients at the least cost." The recognition of this circumstance might limit the scope of dietary interventions proposed by nutrition scientists: to a degree, ordinary eaters were regarded as protoscientists, and there were things they *understood* about nutrition that, so to speak, they did not reflectively *know*. That much conceded, there was still ample room for denying the legitimacy of both custom and taste and for instructing the ordinary eater that the road to health must pass through the scientific laboratory. First, the common categorization of foods by their ordinary names—beef, mutton, cabbage, cow's milk, wheat flour—did not pick out their nutritionally relevant makeup. Right thinking about food and right action in deciding what to eat depended on understanding food *analytically*: you should "consider, not the food as a whole, but the nutriment it actually contains, which is a very different thing."[148] You must understand that your food consists of the macronutrients—proteins, fats, carbohydrates—and these in turn consist of the suite of irreducible chemical elements, as do your organs, tissues, and blood. Since Liebig's time, chemical analyses of food had become familiar in the wider culture, and while, as Atwater noted, "no one has ever made a complete chemical analysis of a human body," scientists have sufficient assurance from the analyses of various organs to tabulate in chemical terms what we really *are*. The chemist finds that the average man of 148 pounds is 92.4 pounds oxygen, 31.3 pounds carbon, 14.6 pounds hydrogen, 4.6 pounds nitrogen, 2.8 pounds calcium, and smaller amounts of phosphorus, potassium, sulfur, and so on. Scripture and science agreed: "dust thou art, and unto dust shalt thou return," whether in the grave or in the laboratory.[149]

Darwinian concessions notwithstanding, nutrition scientists considered that ordinary sensations were a bad basis for good food decisions. Atwater conceded *something* to the role of taste, to the sensory experience of digestion, and to a hollowed-out version of the traditional notion of "agreement" (now rendered in quotation marks): "For people in good health and with good digestion there are two important rules to be observed in the regulation of the diet. The first is to choose things that 'agree' with them, and to avoid those which they can not digest and assimilate without harm. . . . [And] for guidance in this selection, nature provides us with instinct, taste, and experience." But taste and especially taste deformed—as it often was—by custom often seriously misled. It did not pick out wholesomeness and functionality: "In our actual practice of eating, we are apt to be influenced too much by taste, that is, by the dictates of the palate; we are prone to let natural instinct be overruled by acquired appetite."[150] Courtesy of nutrition science expertise, we know that we eat too much, and this alone testifies to the poor guidance given by uncorrected inclinations. "We like the taste of meat," Atwater noted, and indeed, "a moderate amount of meat is, I believe, very desirable." But the "trouble is in our lack of moderation": excess is "injurious to our health."[151] Taste and appetite lead us astray. Neither taste nor uninstructed lay knowledge penetrate to the level of nutritive constituents, which is the level needed to guide good eating. Nutrition science was still acknowledged to be imperfect, but it is to scientific experts that laypeople should turn to repair their knowledge and to improve their practical decisions: "Physiological chemistry adds to [taste and common sense] the knowledge . . . of the composition of food and the laws of nutrition."[152] Early modern commentators had lamented the taste-corrupting effects of fashion, and late nineteenth-century nutrition scientists used new chemical and physiological vocabulary to show how reason and evidence must correct custom: "as man becomes civilized and gains in wealth he loses his instinct and he loses his health at the same time. . . . He must learn to apply his reason to the choice of his food."[153] In the past, taste and immoderation were to be restrained by the moral will; now they were to be corrected by the allied deliverances of chemical and economic expertise. Experts could show people the relevant facts about food, but at the end of the day they had to help themselves, to take acts of will that would in the end build and reinforce an even stronger will. Good food decisions—both nutritional and economic—could be directly read off Atwater's tables, since the tables displayed the constituents that foods contained, what functions they supported, and what those constituents cost. Unexamined custom, uninstructed judgments about goodness, and unmediated experiences of taste and agreement had to give way to the lessons laid out on the tables.

In the Kitchen: Learning the Lessons of Expertise

The tables made the facts manifest; the practical task for politically active experts was to build the institutional means of getting the facts to ordinary eaters and induce them to conform. The kitchen had long provided a metaphor for how digestion worked; now, the kitchen was identified as a site where the findings of nutrition science should be publicly digested and assimilated. The female housekeeper was reckoned to be the person who effectively made family decisions about what was purchased and how it was prepared. She stood at the threshold of the home; she could pay attention to or ignore the advice of nutritional expertise. The housekeeper's ignorance about nutrition was assumed, more so in immigrant families, but nutritional ignorance was also thought to be pervasive among the "Americans."[154] The housekeeper might be worse than ignorant; she could actively oppose expert nutritional knowledge. Speaking to a largely female audience at a Boston cooking school, Atkinson said that his "experiments" in cooking technology "had been undertaken to help overcome the inertia of woman."[155] His colleague Atwater also had moments of despair about the state of women's knowledge and their capacity to absorb new nutritional knowledge: "Of course the good wife and mother does not understand about protein and potential energy and the connection between the nutritive value of food and the price she pays for it, and doubtless she never will." Yet, in a more optimistic moment, Atwater hoped that good information would soon be diffused "among those who have the time and training to get hold of it, [and] the main facts will gradually work their way to the masses."[156] The basic vocabulary of food chemistry, Atwater expected, will "come to be part of that common knowledge, or inherited common sense . . . which actuates the conduct of thoughtful people."[157]

In Boston, the leading local female nutrition scientist was Ellen H. Richards (1842–1911), analytical chemist and, from 1884 until her death, an instructor in "sanitary chemistry" at the Massachusetts Institute of Technology, where she was the first female graduate and the first woman to hold a teaching position. Richards also identified the practical problem posed by long-established dietary taste and custom among the common people, though she did not reckon that these were particular to women. Speaking of the "so-called lower orders or people poorer than ourselves," Richards observed that "*people like best that to which they are accustomed*. Novelty in food does not commend itself to those who have had little variety in their lives." "A higher rule of life than the mere gratification of taste . . . must prevail," Richards wrote. The condition for the efficient working of the human machine was making taste subservient to knowledge of food constituents. "Not until we believe with the positiveness of a creed that health is desirable above all else . . . shall we be willing to restrict ourselves to what is *good for* us, believing also that only that is good for us which enables the human machine to work to its fullest capacity without friction or breakdown." It is not enough to tell the people about food value, though Richards was prominent in doing so. She also recommended a systematic study of obstructing "habits and customs," but little effort was actually given to carrying out these kinds of sociological or psychological inquiries.[158]

Atwater was right about the eventual spread of expert nutritional categories, and indeed, the educational work done by his New England associates had much to do with the domestication of nutrition science categories in American culture. The vocabulary of proteins, fats, and carbohydrates as well as calories and nutritional value per unit cost was rapidly being absorbed into the body of "common knowledge." That vocabulary offered a way of conceiving of food and the body that "agreed with" the rationalizing and modernizing sensibilities of turn-of-the-century culture and was being effectively folded into the practices of government and business in both the United States and Europe. In America, expert nutritional knowledge was increasingly channeled through the state. Professional organizations of medical practitioners and their allies—doctors, dietitians, public health officials—distributed information about the virtues and vices of foods. Expert nutritional knowledge was becoming substantially vernacularized, and modern understandings of the tag "you are what you eat" centered on the food constituents that powered the body and made its substance.

Some of the vehicles for accomplishing this purposeful extension of nutritional knowledge belonged to institutions of civil society, involving women experts addressing women housekeepers. Having been identified as a problem for the dissemination of nutritional knowledge, women could also be seen as a major part of the solution. By the 1880s Atkinson and Atwater joined up with Ellen Richards, collaborating in practical enterprises to feed the Boston working classes and to further disseminate nutritional information.[159] In 1894, Atwater used his position at the USDA to arrange a government pamphlet on practical cooking to be prepared by Atkinson, with sections on food value by Richards, and in 1902 Atwater opened his Wesleyan laboratory to teachers of domestic science for short courses intended to connect them with governmental work on nutrition (figure 8.9). Lessons on how to cook now carried the authority of the state.[160] The home economics movement—some preferred the term "domestic science"—developed from the middle of the nineteenth century and aimed to produce and distribute expertise on rational household management, and, from the 1880s and 1890s, academic nutritional scientific knowledge was a significant resource. Richards was the "engineer" of the Lake Placid Conferences that met from 1899 to 1908, out of which emerged the American Home Economics Association, with a largely but not totally female membership and with Richards as its first president. In 1909, that organization launched the *Journal of Home Economics*, a vehicle for information exchange, an early editor of which was Helen W. Atwater (1876–1947), the daughter of the Wesleyan nutrition scientist.[161] The American home economics movement was academicized in the early years of the twentieth century, with departments founded at Cornell University, Columbia University, the Massachusetts Institute of Technology, the University of Illinois, the University of Wisconsin, and elsewhere.[162]

Home economics was also being politicized, linked up with the bureaucracies and practices of government. The USDA established the Office of Home Economics in 1915, elevated to a bureau in 1923. Helen Atwater worked as a writer in the Office of Home Economics, and by the 1920s she was collaborating on pamphlets instructing consumers about how to buy foods. (These publications were

Figure 8.9. W. O. Atwater's Wesleyan Summer School for Nutritionists (ca. 1902). Atwater is third from left in the front row (man with both hands on his knees). Note that about half the attendees are women.

arguably the first federal government dietary guidelines.) Here, the authority of the state was mobilized to tell consumers how they should think about food values, what language was appropriate to use to refer to food constituents, and what they should eat on a day-to-day basis. *How to Select Foods* disseminated Wilbur Atwater's sentiments, instructing housekeepers not to be guided by taste or habit or hunger and not to believe that cheap foods were necessarily rational purchases. The state was now a major actor in introducing wives and mothers to the vocabulary of macronutrients, calories, and cost-nutrient calculations; women were instructed about the quantities of food that were required for family members of different ages, occupations, and ways of life. Housekeepers were shown pictures of what these foods and amounts looked like, and sample meals satisfying nutrient requirements were laid out. And if doubt remained about the particulars of a proper diet, women were counseled to apply "to [their] State leader in agricultural and home economics and to the home-economics department of her State agricultural college." In these ways, the housekeeper was shown the inadequacies of lay knowledge and how it might be repaired.[163]

By the early years of the twentieth century, there was an American cottage industry producing books intended to familiarize laypeople with the fundamentals of nutrition science and show them the errors of everyday and traditional knowledge. In 1904, Fanny Farmer's cookbook for the feeding of the sick introduced readers to the calorie, cited Atwater as an authority, and reproduced some of his tables of the nutrient content of foods.[164] *Diet and Health: With Key to the Calories* (1918) by the California physician LuLu Hunt Peters sold millions of copies and remained on best-seller lists through the 1920s, becoming one of the most important American vehicles for popularizing the notion of the calorie, of *calorie counting* as a way to manage body weight, and of respecifying the word "diet" from its sense in traditional dietetics as an ordered way of life to the narrow notion of weight loss. Peters explained what calories were, why they were important, how the calorie content of foods was determined, how many calories you needed a day, and, if you wanted, how to reduce calorie intake.[165] By 1927, a home economist observed with satisfaction that "the calorie is a familiar word in the vocabulary of practically every adult" and noted that it was fast becoming standard among children, who even then were counting their calories.[166] By the mid-twentieth century, the vocabulary of nutrition science had become—incompletely but substantially—part of everyday knowledge.[167] It was nothing new for ordinary people to know what their *food* was, what *they* were, how their food made them what they were, and what sorts of people had expertise in such things. Traditional dietetics had told them that for centuries. But now they knew it in a different way.

Nutritional Knowledge on a Global Stage

The domestication of nutrition science—its substantial transition from the expert to the vernacular domain, from the scientific laboratory to the kitchen—makes it a ubiquitous feature of the modern cultural and institutional order. But another association between the new idiom for thinking about food and its role in vital function was global, not local. The outcome of competition between nations and races was understood to depend on the food supply or, strictly speaking, on the food supply mediated by its price, its allocation, and knowledge of its nutrient composition. "The underfed cannot compete with the well-nourished," Atkinson wrote, and at the turn of the century that competition was taking place on the global stage.[168] Global competition was about productivity, but it was also about culture, morals, and mind. Conceiving the context between nations in terms of the food supply and its nutritive content preceded the wide distribution of the notion of the calorie. Writing in the 1880s, the Welsh physician Sir William Roberts causally linked diet to national, racial, and class power, more so in qualities of mind than of body. If we compare "the general characteristics" of the "high-fed and low-fed classes and races," we see a clear distinction: "In regard to bodily strength and longevity the difference is inconsiderable; but in regard to mental qualities the distinction is marked. The high-fed classes display on the whole a richer vitality, more momentum and individuality of character, and a greater brain-power, than their low-fed brethren." It was from the high-fed that genius and "eminent men" emerged. Among the high-fed, the "British races and the other races of Western Europe, together with their descendants in different parts of the globe, are . . . fitted to supply us with a body of dietetic customs which may be regarded as a beneficial model. These races and nations are, on the whole, but especially in intellectual power, far in advance of all others." This was substantially owing to a varied diet—foods summoned from all parts of the globe—and also because high-feeders were said to be moderate consumers of alcoholic beverages, tea, and coffee, all of which, when carefully managed, increase mental acuity and had historically contributed to the "rise and progress of the exact sciences." [169] The scientific army, so to speak, marched forward on its stomach.

One mode of globalization concerned the relative energy, productivity, and—as it was said—"vigor" of nations and races. With what energy did their people work? How much could they produce? What force could they exert? How acutely could they think? Matters belonging to political economy could now be expressed in the language of nutrition science, a science that offered practical answers to questions about efficiency, productivity, and the like. The dietaries compiled by late nineteenth-century nutrition scientists compared the potential energy required by different occupations—masons, cabinet makers, professors—but also compared the potential energy taken in by practitioners of a given occupation in different countries. Globally, the supply of food was the ultimate basis for the wages commanded by workers and for their productivity. This claim was widely articulated, vigorously so by Atkinson: "the rate of wages in different countries vary as the supply of food is abundant or scanty." Those wages are highest, "or, in other words, the effective earning or productive power of man is greatest in the United States, where the supply of food is most abundant." American workers were the most productive, followed by those in Great Britain, Holland, France, Germany, Belgium, Italy, Austria, and Russia. It was an ordering that was taken clearly to map onto differences in the supply of food, with scarcity caused by a limited supply of arable land, by its poor cultivation, or by the diversion of food to unproductive purposes.[170] More precisely, workers' vigor and workers' wages both depended on calories and protein consumed, the fuel for work and the materials to make the bodily fabric that performed work: "the rate of wages depends upon the supply of the nitrogenous element of food."[171] Atkinson offered several conclusions for politicians concerned with maintaining national power and influence. And those lessons drew crucially on the knowledge accumulated by nutrition science: workers must be well fed and efficiently and economically fed, and the food supply of the productive members of society must not be diverted to the feeding of productively useless armies. Atkinson was a member of the American Anti-Imperialist League: he opposed a large standing military; his opposition was informed by appreciations of the food supply and its relations to wages and production.[172]

On the one hand, the categories of calorie and protein provided a universal metric: it allowed observers to statistically order by rank the goodness of the diets of all the world's nations and races. This ranking was concerned with the quantities of interchangeable chemical constituents, and in that way diversity was both recognized and constrained. Discussing the views of C. F. Langworthy, Atwater's successor as the head of the USDA's nutritional department, the historian Nick Cullather writes that "challenging dietary theories of racial difference, Langworthy stressed that the human diet was far less diverse than had formerly been thought": "Broken down into chemicals, the potatoes and cheese that fed the Irish laborer were identical, except in quantity, to the rice and ghee that nourished an Indian coolie." The nutritional constituents of food became an accepted idiom in political economy and governmental policy, both national and international. It was in Europe that the calorie first became a measure of food energy, but it was in America that it was most strongly embedded (as Cullather has written) "in systems of distribution and administration. . . . From the first, the purpose of the calorie was to render food, and the eating habits of populations, politically legible." In the United States more than in European countries, the notion of the calorie allowed food to be seen as "an instrument of power."[173]

The vocabulary of nutritional constituents was also globalized in turn-of-the-century Western views of racial struggle and its connection with the environment. Some of the categories involved still bore some resemblances to those of traditional dietetics. So, for example, the benefits of meat eating and the possibility of vegetarianism traded in ordinary sensibilities about heartiness and strength building that were similar to those embedded in seventeenth- and eighteenth-century views of beef eating and its role in making people brave, stolid, steadfast, and hearty.[174] Roberts, for example, was well aware of the notions of the calorie and of "proteid matter," but in the main, he fleshed out the nature of high feeding in vernacular terms, conforming to price considerations and social value. If a man, Roberts wrote, regularly consumes "meat three times a day, and drinks daily of high class wines and spirits, . . . his diet must be ranked as high. There are distinctions also to be drawn in regard to the several kinds of meat and the several kinds of vegetable articles used. Beef and game should probably be ranked as higher diet than poultry and fish—and oats and wheat as higher than rice and potatoes." In the English-speaking world, the higher classes were the higher feeders; on a global scale, the European races fed higher than the Asiatic races, and these nutritional differences contributed importantly to differences in both bodily and mental power.[175]

Racial dominance depended on an abundant and nutritious food supply. And in this way, considerations of empire were linked to not just nutrient constituents but also the ability of the land to produce the crops that sustained human life, and that ability was expressed, since Liebig's time, in terms of chemical constituents. Constituents in the soil became constituents in the plants growing in the soil, which in turn became constituents in the flesh of livestock and of human beings. Soil, plants, animals, and human beings were linked cyclically through the transportation and transformation of chemical constituents. That cycle ultimately determined the numbers and types of human beings on the planet; it could be influenced by economic policy; and it offered a language for justifying and directing political and military action. In the last years of the nineteenth century, a great panic arose over the future of the global food supply, and this panic implicated the availability of chemical categories whose identities had been established earlier in the century. The historical origins of the panic were contained in the arithmetical foundations of Thomas Malthus's *Essay on Population* (1798). Population, that is, hungry mouths, increased in a geometrical ratio—2, 4, 8, 16, and so on—while agricultural productivity, the food for those mouths, could increase only arithmetically: 1, 2, 3, 4, and so on. The increase in population, if unchecked, would eventually run up against the capacity of the land to feed people (some of Malthus's followers thought it would be very soon), and the result would be catastrophe. Natural law would cruelly and surely reduce the number of people in regard to the ability of Earth to feed them.

Malthus's many nineteenth-century followers addressed practical questions about population growth, its nature, consequences, and limits, with some urging "prudential restraint" (voluntary measures to refrain from sexual relations) and others—delicately or frankly—broaching the subject of birth control.[176] But there was widespread acceptance of Malthus's claim that increases in agricultural productivity could never prevent but at most could delay the inevitable crunch. One arm of the turn-of-the-century alarm concerned the exhaustion of available virgin land. This was essentially a statistical matter: how did you calculate the extent of as yet uncultivated and potentially arable land? There were commentators who reckoned that there was little remaining virgin land in America, Canada, Australasia, and Russia. Many accepted the reality of these looming limits; others—notably Atkinson in Boston—insisted that there was still adequate potentially arable acreage.[177] It was also recognized that the productivity of arable land depended on supplies of nitrogenous fertilizer. And it was quite recently understood that growing leguminous crops such as clover and then plowing them under could increase the available "fixed" nitrogen content of the soil, but there was much skepticism that this would solve the nitrogen problem.[178] There was little more manure from livestock to be obtained, and a series of nasty wars were fought over access to deposits of seabird droppings—guano—and inorganic nitrates from Latin America and the Pacific Islands, the supplies of both of which, it was feared, were running out. In 1856, the US Congress passed the Guano Islands Act, allowing the United States to use military force to take possession of any unclaimed islands containing guano deposits, an act that remains in force. Nitrogen had gone to war.[179]

Toward the end of the nineteenth century technological solutions were envisaged, and their development was aggressively encouraged. If it were possible to artificially fix nitrogen—taken from the atmosphere or separated from coal gas—this would be a powerful solution to the food crisis, and scientists and engineers were urged to work hard to invent such technologies. Some thought that this was unlikely or impossible. Optimism was balanced by skepticism until the German chemist Fritz Haber succeeded in the laboratory fixation of atmospheric nitrogen in 1909 and, together with the industrial chemist Carl Bosch and researchers in the giant chemical firm BASF, achieved industrial scale during World War I.[180] In America, both Atwater and his patron Atkinson were confident that science was the best hope for ensuring the future food supply. Neither rested their optimism on the artificial fixing of nitrogen, but they were not convinced that a Malthusian crisis was inevitable. Malthus's pessimistic followers had ignored or underestimated the effects of science-informed agricultural productivity, waste reduction, more intelligent land use, improved methods of distribution, and a population properly educated in the fundamentals of nutrition science. Atwater did not deny that America was running out of virgin soil.[181] Steps had to be taken to make better use of land that was already cultivated. Too much land was being used for raising food animals, and the carcasses of livestock raised in America had too much fat and too little protein-containing muscle. Americans were also consuming too much meat, and land better fitted to raising crops for human consumption was given over to producing feed for livestock.[182] Informed political decisions, it was hoped, would change all this, but the major resource for solving the problem of the food supply was scientific research. Atkinson wrote that "no one yet knows the productive capacity of a single acre of land anywhere."[183] "We stand at the very beginning of progress in scientific agriculture," he said, "leading to lessened labor and increased product."[184] Atwater had a compelling metaphor to dissuade people from thinking of the land as a limiting resource: "It is right to consider the soil as a laboratory and not as a mine, responding in just proportion to the intelligence and work put upon it," and Atkinson repeatedly elaborated the conceit: "In the mind of Malthus and of Ricardo land appears to have been regarded as a mine—subject like mines of metal to exhaustion. We are but beginning to learn that land is but an instrument or a laboratory responding in its products to the minimum of labor and the maximum of intelligence."[185] Powered by nutritional science and agricultural chemistry, "the possibilities of the food supply in the future are measureless."[186] The same science that promised efficient nutrition was also meant to allay apocalyptic anxieties about the future of the race and the species.

According to Atwater and Atkinson—and others who reckoned that there were scientific solutions to any food crisis—it was wrong to think of the *soil* as an infinitely complex material requirement for plants' nourishment, something needed for plant growth and also required to be in a certain physical condition. The soil was merely a vehicle for the transport of specific chemical compounds into growing plants. The metaphor of the mine seriously misled, and Atkinson explained that "modern science has proved that land is a mere instrument for the conversion of certain elements of nutrition from one form into another." A fertile soil was a container for substances such as nitrates, lime, potash, and phosphoric acid, and the sources of these were "practically unlimited." Plant growth is a matter of chemical elements, not of "soil."[187] The Liebigian formula soon became an institution in the twentieth-century practice of intensive agriculture powered by synthetic fertilizers. Fertility consisted of an adequate supply of nitrogen, phosphorus, and potassium, the elemental ingredients listed on bags of commercially produced inorganic fertilizer. Atwater and his associates wanted no more talk of limits to food supply due to limits of available soil or of its native fertility. NPK fertilizers supply the fertility, and scientifically designed physical arrangements were well in hand to dispense with soil altogether. Agricultural scientists had recently discovered that food plants "can flourish on the most barren soils or even without any soil at all." Only minerals—making up just a tiny proportion of plant substance—came from the soil, with the rest obtained directly or indirectly from the air. In his Wesleyan laboratory, Atwater was growing vegetables in sea sand, carefully washed and heated to ensure sterility and supplemented only with the NPK chemicals and minute amounts of easily obtainable sulfur, iron, chlorine, and magnesium. Here was a plausible future for the global food supply. The future of Earth did not depend on earth.

Atwater and Atkinson were optimists. But at the turn of the century, the voices of the pessimists were louder, and their alarms were more urgently expressed. The crisis that the pessimists envisaged was more specific and more emotionally and politically charged. A general Malthusian panic over the food supply in the late nineteenth century took specific form in anxiety over *wheat*. At the 1898 meeting of the British Association for the Advancement of Science, its incoming president, the chemist William Crookes (1832–1919), warned of an imminent crisis in the supply of wheat, notably to the import-dependent British market but more generally to what he called civilized society.[188] Crookes considered that there was little more virgin land on which to grow wheat and little more guano or South American nitrates to fertilize the soil on which wheat could grow, and he did not even mention the idea of soilless agriculture. While there was hope for new technologies to fix nitrogen and produce artificial fertilizer, those technologies must come very soon if disaster was to be averted. This was an existential crisis for those races whose staff of life was wheaten bread. White people ate wheaten bread: "Wheat is the most sustaining food grain of the great Caucasian race, which includes the peoples of Europe, the United States, British America, the white inhabitants of South Africa, Australasia, parts of South America, and the white population of the European colonies."[189] Britain would face starvation if its imports were blockaded by European rivals, and all the white bread-eating races of the world were vulnerable to the caprices of several bad global harvests. The production of wheat depended on soil nitrogen, so the wheat crisis could be rendered in elemental terms. You needed nitrogen to build strong and civilized bodies and minds, and you needed nitrogen to grow wheat. Like all other people, the civilized races required nitrogen, but it was the civilized races who knew how to obtain it. As a virulently racist American physician wrote in 1909, "the lower the civilization the less able they are to obtain nitrogen. In civilization itself, only the less intelligent classes are unable to obtain nitrogen." Meat was good, of course, but in the choice of bread, it was wheaten white bread that "contains the most available nitrogen," so the progressive tendency of the most intelligent members of civilized society was to turn away from brown bread—whose nitrogen-containing bran was considered indigestible—and toward the white loaf. Fine white bread made fine white men.[190]

Surprisingly for a chemist, Crookes broadly maintained the superior nourishing capacity of wheat, while his "Address of the President" made no mention of either calories or protein. "We are born wheat-eaters," Crookes announced. "If bread fails—not only us, but all the bread-eaters of the world—what are we to do? Other races, vastly superior to us in numbers, but differing widely in material and intellectual progress, are eaters of Indian corn, rice, millet, and other grains; but none of these grains have the food value, the concentrated health and sustaining power of wheat, and it is on this account that the accumulated experience of civilized mankind had set wheat apart as the fit and proper food for the development of muscle and brains."[191] Those remarks apart, Crookes's lecture presented the goodness of wheat eating as a matter of course for the white races, and the celebration of wheat bore a family resemblance to early modern English enthusiasm for beef. Crookes glancingly mentioned the option of consuming alternative grains but clearly did not think that such a thing was practical. Starvation loomed, he said, "with no substitution possible unless Europeans can be induced to eat Indian corn or rye bread," the prospect of which he airily dismissed.[192] Crookes might have appealed in these connections to the durability of long-established custom, but he did not do that. The religiously inclined might also have made something of the sacramental place of bread, but Crookes never made that gesture either. His 1898 lecture was soon fleshed out in book form, reaching its third edition in 1917.[193] The introduction to that edition written by the Welsh Liberal politician Lord Rhondda celebrated wheat in nutritional terms that Crookes himself saw little reason to use. "If we grow food," Rhondda wrote, "we should grow the best food." "Contemporary physiology ratifies the ancient opinion that bread, and above all, wheaten bread, is the staff of life. Recent discoveries of "vitamines," or "accessory factors" of diet, merely confirm this opinion when they demonstrate the richness of wheat not only in all the ingredients hitherto recognised as necessary for our support, but also in these newly identified factors."[194] Crookes had been trained by one of Liebig's students and was wholly familiar with the categories of nutrition science, possibly taking that vocabulary so much as a matter of course that there was no reason for it to intrude in his lecture. Nevertheless, his identification of the wheat crisis elicited much criticism, and some of this focused on the nutritional constituents of wheat.

In the United States, Atkinson summarily rejected Crookes's Malthusianism. Atkinson added nutritional criticisms to his existing views about the great extent of land still available for growing wheat and the likelihood of technological fixes. He saw no convincing reasons to set wheat on a nutritional pedestal. Why, Atkinson asked, were the English farmers unable "to think about agriculture except in terms of wheat?" Now, he said, a distinguished scientist "seems to ignore all other grain and to predict future starvation on an expected deficiency in the supply of wheat."[195] Atkinson did not call out Crookes on his remarks about the superiority of the Caucasian races, but he did not agree with the Englishman's presumption that the fate of the whites depended on the supply of wheat. Atkinson proposed instead to judge the goodness of wheat by its nutritional constituents, and he produced comparative tables of the nutrients actually contained by various food grains. Wheat was indeed nutritious: 11.1 percent protein, 75.6 percent carbohydrates, 1.1 percent fats, and 1,660 calories per pound of flour. But it was not notably more nutritious than maize flour and oatmeal, with the latter containing 15.1 percent protein. Atkinson had long maintained that the Scots knew what they were doing when they treated oats rather than wheat as the staff of life: "oat meal is one of the cheapest foods that we have; that is, it furnishes more nutritive material, in proportion to the cost, than almost any other."[196] And the New Englanders also knew what they were doing with their customary diet of corn bread and baked beans, both cheap and efficient delivery systems for protein, carbohydrates, and calories.[197] (Atkinson had little enthusiasm for rice, falling in with much contemporary Western scientific sentiment that it was a poor source of nutrients and that the Asiatic races were badly served by their rice-based diet, which "may cost less in money, but is deficient in some of the nutrients which are necessary to full vigor.")[198] The American economist was here giving the English chemist a lecture on the chemical constituents of foods, shifting the debate about the fate of the human species and the white races onto the terrain of nutrient tables. Crookes would have none of it. Replying to Atkinson, Crookes rejected the idea that "tables giving the nutrient equivalent of other grains, and the potential energy of given weights thereof" had any relevance, and also dismissed the notion that "other grains and the toothsome 'Boston baked bean' [were] as desirable as wheat." The question was about the goodness and necessity of wheat for white people, not about nutrient-equivalent alternatives. Crookes did not entertain the notion that the feeding of the Caucasian races could or should be resolved by substituting oatmeal, maize, or "toothsome" baked bean protein for wheat protein.[199]

For Atkinson and his scientific associates, answers to questions about what people need to eat and should eat were clear. They were made visible in the tables that represented the nutrient constituents that foods contained, the potential energy foods made available, foods' relative cost per weight of nutrient or yield of energy, and the energetic requirements for different sorts of people doing different sorts of work. It was not a matter of what was suitable for the *nature* of English people, African Americans, Chinese, or observant Russian Jews. Dietary requirements were considered to depend almost entirely on body mass, stage of life, and the type and intensity of work that the human machine was doing. Foods were delivery systems, and the tables compiled by nutrition scientists testified to the nutrients and energy that different foods delivered. The lessons inscribed in the tables set aside considerations that had been central features of traditional dietetics. Uninstructed, lay thinking about foods and their powers was now not only irrelevant; it also constituted an obstacle to the absorption of expert knowledge, with experts assuming the task of effectively instructing people about what their foods were like and offering them a new vocabulary for knowing about foods and their powers. You could *not* be your own physician, and if you thought you could, the physicians would diagnose you as an idiot. What foods people were accustomed to was now beside the point, and for the most part and for the common people, custom led to ill health, ineffective work, and shortened life. Similarly, taste misled. The modern order had produced foods that stimulated appetite and delivered delight but were nutritionally unwise. The sensory experiences that people had in taking their food—sweetness, warmth, bitterness, or even repletion—were wholly unreliable indices of the nature of foods and of their effects on the body. Nutritional experts told you what you should eat if you wanted to be healthy, productive, and long-lived, but they did not—with rare exceptions—tell you how to sleep and excrete, where to situate your house, whether to take exercise by leaping or by walking, and above all how to manage your emotions. Nutrition scientists occasionally spoke about a causal connection between food and mental vigor or acuity, but food no longer had the capacity to make or damage *character*. There were no "choleric" types who needed to be cautioned against taking foods excessively hot and dry and no "phlegmatic" people who should avoid foods that were too moist and cold. Modern nutritional experts tended to be silent about moral makeup and moral action. What experts deemed *good for you* was a different matter than what was *good*, or, in an instrumental idiom, their advice presumed that being a healthy and productive member of society substantially *defined* what it was to be good.

The tables and the institutions that produced and distributed them represented new realities in the culture that talked about food and its fate in the body. One new thing was the energetic conception of the human body and the role of food in both making and fueling the engine. That understanding gave nutrition a key role in the economic and political orders. The health of populations had long been a state concern, but the body-as-engine linked human work to machine work in ways that were new, resonant, and consequential. If you wanted to understand and potentially manage the production of goods and the projection of power, then you had to understand how productive and power-projecting human machines were nourished. This made nutrition pertinent to statecraft. The production of nutritional knowledge was increasingly sponsored by the state and disseminated by the state, and its reliability was warranted by state authority. As the large-scale capitalist food industry developed through the nineteenth and twentieth centuries, the authority of state-sponsored scientific expertise about good eating was increasingly contested by business-backed expertise whose advertised claims about the virtues of foods and about a proper diet were sometimes at odds with governmental policy and sometimes intruded into the making of government policy.[200] But even the resulting instability of food expertise testified to its economic and political significance: there was disagreement among interest groups about what good eating should be, but there was little disagreement about the language in which goodness should be expressed. Finally, inscribed in the new order was a systematic appreciation of the relations between the feeding of people, the work done by people, the numbers of people inhabiting Earth, and the capacity of the land to produce food and support people. This was an appreciation that linked people, food, work, and the land through chemical vocabulary: constituents, not qualities.

CONCLUSION

The Way We Eat Now

Last Course

How do we eat now? What do we think about our food, what it is and what it does? What do we think about the role of our food in making us who we are? What do we believe about the relationships, if any, between the instrumental and the moral aspects of what we eat, between what is good for us and what is good? What is different about late modern notions of eating and being, and what about the past survives? Present-day arrangements are surrounded by the aura of self-evidence, and one of the roles of history is to act as a momentary solvent of the taken-for-granted. So, what does the present look like when it is seen as one end of a strand whose other end disappears over the horizon of the distant past?

Some basic patterns of traditional dietetics should be rehearsed here. Foods were routinely described through their qualities: hot, cold, moist, and dry. Literate people might learn about those qualities from books by physicians and those instructed by physicians, but much knowledge of the qualities and consequences of foods might also be inferred through ordinary sensory experience. You should *know yourself*, and you should *be your own physician*: how foods fared with you and how they nourished and constituted your body belonged to your own sensory experience, and, informed by that experience, they were subject to your own prudential management. External expertise in dietetic matters was widely recognized, but the circumstances in which it was superior to your own competent knowledge were limited. Foods that were wholesome for you—that *agreed with* you—were those that you digested well, did not make you feel unwell, and pleased you.

The management of eating belonged to an overall system of understanding, describing, and prescribing human conduct. The notion of *diet* referred to an ordered way of living, one that included food and drink but incorporated many other aspects of your life that were also regarded as potentially subject to volitional control. In that system, aspects of self-management were causally connected; so, for example, what you ate bore upon your emotional and mental life, patterns of physical activity, sleeping, sexuality, and so on. The sense that you are what you eat proceeded from an appreciation that *qualities of food* ultimately became *qualities of you*. Self-making implicated food-taking. People whose constitution was innately marked by certain qualities should normally consume foods possessing those qualities. You might disturb or even remake your constitution with specific patterns of consumption, for example, consuming foods whose qualities differed from your natural constitution. What was *good for you* tended also to be *good*: moral and medical orders supported each other. The overall counsel for right eating was the same as that for right living: *balance and moderation* made for health and for virtue. The dietary and moral orders were molded to each others' contours.

The emergence of nutrition science in the nineteenth century unsettled much of this. Foods do *not* have qualities; they have—indeed, for nutritional purposes they *are*—an assemblage of chemical *constituents*: the macronutrients (proteins, carbohydrates, and fats), the micronutrients (vitamins, minerals, coenzymes, etc.), and such nonnutritive stuff as water and other indigestible substances. The body's energy is fueled by its food, and that energy—the calorie—is a power of the chemical constituents in food. Your sensory experience of foods is *not* now a reliable index of their composition, and you *cannot* know about such constituents apart from reliance on external expertise. What counts as common sense is now widely taken as no sense at all. The notion of foods that *agree with* you is relegated to the status of folk wisdom and valued accordingly. Expertise on food is now rarely, if ever, connected to advice on sleeping, sexuality, and your mental and emotional life. From the end of the nineteenth century, the notion of *diet* referred to taking food and drink—mainly about *restricting* their intake so as to reduce your weight—and very little, if at all, about an overall pattern of living. The saying "you are what you eat" retains a place among modern cultural tags, but now it refers almost solely to the role of food constituents in the chemical makeup and physiological function of your body. Considerations of what is *good for you* are now disengaged from what is *good*. Medical experts advising what you should eat are no longer in the business of offering moral counsel. The recommendation of dietary *balance and moderation* has lost much of its past authority and now increasingly appears as imprecise, old-fashioned, folkish, and foolish. What is now wanted is knowledge of specific food constituents and their specific physiological role in the body, and if that knowledge points to a diet radically different from customary patterns, then that departure might even be taken as a warrant of its power.

I began this book with a short account of an ordinary late modern meal, a humble bean soup for a light domestic dinner. Its ordinariness was used as a prompt to consider all sorts of things I might be thinking about as I decided on and consumed this meal. What did I find that I believed or took account of about what the food was, its social and cultural meanings and its effects on the human body? On reflection, I acknowledged that on most occasions I did not think very much about such things. Comfortably situated as I am, hunger is rarely an important consideration, choice among many possible foods is normal, and taste preferences are often paramount. But when I reflected on the matter, I found that there were very many considerations that *might* be on my mind. These included what I could recall about nutrients as well as medical benefits and risks, environmental significance, and so on. I accept that although my social and economic circumstances are unexceptional for the modern Western middle classes, my own thinking about food and eating is not typical. After all, I am an academic who has been reflecting on and writing about these sorts of subjects for years. But historical facts and sensibilities were practically absent from my opening remarks. I considered that consciousness of such things was, so to say, special-case awareness specific to my identity as a historian, and I wanted to begin this book by picking out the typical aspects of my thinking about food and eating.

Yet, on still further reflection, I perhaps should not have excluded history from the long list of relevant considerations. The historical sketch I placed at the beginning of this chapter is in some respects *already familiar to readers*. That is because—this book apart—it is a noticeable feature of our present-day culture. Stories about the historical transition from traditional to modern thinking about food and eating have been told and retold in the late twentieth- and early twenty-first centuries. This history is sometimes celebrated as proof of *progress*. Humors and qualities are bracketed with astrological influences and occult forces as instances of unenlightened error; we are now better nourished because we are better informed about what nourishment actually is. But that same history has also been recruited in critical exercises, rendered not as a triumph but rather as an account of *how we went wrong*, about how features of modern nutrition science make for bad eating and associated harms. Sometimes those stories are nostalgically flavored, embracing lost and presumably better ways of dietary thinking and acting. If we didn't actually *eat* better in the past, it is suggested that we *thought* better about food and eating.

Modern Food Thinking Is *Bad for You*

Consider a few of the better-known lay contemporary commentators on food and eating, here taken not as authorities to be commended or criticized but instead as testimony to our own historical moment. In 2013, the Australian sociologist and food policy academic Gyorgy Scrinis made up a name for modern food thinking gone wrong. What he called "nutritionism" designated the notion that food value is and ought to be ascribed to the chemically identified *nutrients* that foods contain. Nutritionism considers foods as collections of constituents, with each constituent having its specific capacities and physiological consequences, and in nutritionist thinking there is nothing more to good eating than ensuring that the right nutrient constituents are taken and in the right quantities.[1] The category "food" (and also the categories "meal" and "diet") is to be replaced by the set of specifiable substances and measurable quantities that scientific expertise discerns *in* food. Scrinis finds this disastrously wrongheaded, and so too do just a few influential medical writers who have called nutritionism "bullshit du jour": it is good science perverted, or simply bad science.[2] It is not good for our thinking, and following its dictates is not good for our bodies.

In the early years of the present century, the tag caught on and resonated through much of the general culture. Nutritionism came to be criticized as pernicious, a mode of modern expertise pathologically infecting the public mind. There were some scientists and many more lay cultural commentators who pushed back against the nurtritionist heresy, none more consequentially than the American journalist Michael Pollan, one of the most listened-to global voices about proper eating whose books have been translated into many of the major languages, both European and Eastern. Pollan's *The Omnivore's Dilemma* (2006) was a best-selling denunciation of the modern food system and was structured by accounts of four imagined meals, each representing a different mode of getting food on the table, ranging from the industrially produced fast-food meal (very bad and easy to get) to a meal whose materials were personally hunted and foraged (very good and harder to get). Pollan quickly followed up with a slimmer book, *In Defense of Food* (2008), supposedly responding to inquiries about what he himself ate, laying out the dietary principles he followed and inviting others to eat as he did.[3]

It was in this popular writing context that Pollan happened on Scrinis's "nutritionism" tag and put it into wider circulation.[4] And this was also the setting in which Pollan offered a potted *history* of ideas about food and eating, a history that in many respects tracks what was summarized at the beginning of this chapter. In the early nineteenth century, Pollan writes, the English chemist William Prout identified proteins, fats, and carbohydrates as "the three principal constituents of food." Prout begat Liebig, and it was the great German chemist who did most to establish nutritionist ways of thinking about food and its capacities, about the fertility of the soil and the nourishment of plants. The first decades of the twentieth century saw the identification of micronutrients, and there were later discriminations between the varieties of fat (saturated, unsaturated, monounsaturated, polyunsaturated), essential minerals, and fatty acids, along with the recognition of the problematic place of cholesterol (and its subspecies). But as Pollan problematically claims, "it really wasn't until late in the twentieth century that nutrients began to push food aside in the popular imagination of what it means to eat."[5] Pollan saw nutritionism as scientific error: explanatory overreach.[6] The repertoire of food constituents known to nutrition scientists is itself, Pollan says, radically incomplete. Food should be thought of as hugely complex, and diets are even more complex. Condemning what he sees as the too-blunt instrument of expert dietary counsel, Pollan insists on the pertinence of individual variation. Nutritionism cannot hope to account for the dense systems of interactions between foods, between foods and the environment, and between the foods that comprise your diet and the living stuff that is *you*. We should stop talking about nutrients and instead revert to talking in a plain sort of way about *food*, the word now increasingly stressed. Nutritionism is faulty science because, Pollan judges, food is more than the sum of its known constituents, and the properly scientific way of thinking about food and the systems that produce it should be holistic, not analytic.[7]

Pollan recognizes that nutritionist frames are deeply ingrained in modern orders of thinking and acting, even if he appears to underestimate how hard it is to give them up. He passionately denounces thinking about food outcomes in terms of specific nutrients, but the grip of his self-denying discipline occasionally loosens. Borrowing, as Pollan concedes, "the nutritionist's reductive vocabulary," he lists a set of micronutrients that cannot be obtained from "a diet of refined seeds," the highly processed cereals that, for example, go into industrial white bread. These include antioxidants, phytochemicals, and the omega-3 and omega-6 fatty acids, "which some researchers believe will turn out to be the most crucial missing nutrients of all." The modern diet is said to have the wrong ratio of omega-3 to omega-6 acids.[8] Then, realizing his lapse, Pollan offers his own deviation as further evidence of the culturally powerful "undertow of nutritionism."[9]

To combat nutritionism, Pollan urges a revaluation of common sense and ordinary experience. Broad maxims of right eating ought to take precedence over the quantified tables of nutrients and requirements produced by nutrition scientists. Pollan's criticism of nutritional expertise has rhetorical style as well as technical substance. In place of nutritionist formulae, Pollan offers three deflationary maxims: "Eat food. Not too much. Mostly plants." That is as much as really needs to be said—or that should be said—about "the supposedly incredibly complicated and confusing question of what we humans should eat in order to be maximally healthy." The rest is detail. In prenutritionist times, people took their food knowledge from "culture, which, at least when it comes to food, is really just a fancy word for your mother."[10] "Mostly plants" is a direct counter to carnivorous ways of eating in rich countries and is indirect dissent from the protein worship that tracks back to Liebig. It is a gesture toward dietary simplicity and restraint associated with both traditionally Mediterranean societies and images of a past less abundantly and pathologically supplied with animal flesh. The apparently bland injunction to "eat food" seems facile, but Pollan aims to give it substance. There is no real need for precision. Do not think of dinner as a quantified collection of nutrients and their capacities: a roast chicken is a roast chicken, and an arugula salad is what it seems to be. Do not reduce, do not analyze, and do not quantify. And if the ingredients listed on modern commercial food packets include things that do not exist "in nature" or unfamiliar substances with arcane names or with names you find hard to pronounce—tricalcium phosphate, high-fructose corn syrup, azodicarbonamide, ethoxylated monoglycerides—don't eat them. You can recognize bad modern food by the modern technical language naming their constituents.

Pollan's subsidiary maxims endorse simplicity. They prohibit eating anything with more than five listed ingredients or anything that "your great-grandmother wouldn't recognize as food."[11] Authenticity is the way to health: eat a chicken; do not eat chicken nuggets. The counsel "not too much" might be a gesture at massive stocks of epidemiological evidence about the risks of obesity, a form of expertise that Pollan matter-of-factly accepts. He finds it sufficiently obvious that modern inhabitants of rich countries overeat and that excess is unhealthy. Pollan is not a signed-up calorie counter; he does not specify how many calories, how many ounces, and how many "units" of alcohol should be taken. "Not too much" signals alliance with ordinary prudence; it is the kind of injunction that continues to circulate in lay society despite the immense authority of the counters and the quantifiers. The history of nutrition science is, Pollan thinks, a history of claiming to know too much. And that is one reason why he laments the decline of maternally transmitted "tradition and habit and common sense," maintaining that laypeople can indeed act as their own dietary doctors: "I speak mainly on the authority of tradition and common sense," he says, where tradition, prudence, and "Mom" should have their proper values reinstated.[12] Much of the history of nutrition science can and should be safely unwound.

Pollan blames Liebig and his nineteenth-century associates for the medicalization of food thinking, the sentiment that the sole purpose of food is the promotion of bodily health, and the view that health is secured by consuming the proper range and quantity of nutrients.[13] Immensely profitable fast-growing industries produce and market "functional foods," dietary supplements (vitamins, protein, fiber, minerals, amino acids, fatty acids, probiotics), and nutraceuticals (commercial products promising specific health benefits through the addition of ingredients or the through the enhancement of naturally occurring ingredients). Pollan condemns such things as extreme modes of nutritionism, their apotheosis being the synthetic food Soylent, a Silicon Valley–designed product engineered to contain all the nutrients required by the human body, its consumption intended to take the place of "the meal" and hopefully to signal "the end of food."[14]

Modern Food Thinking Is *Bad*

Traditional dietetics and traditional practical morality occupied common cultural terrain: what was bad for you about eating was also bad. Certain ways of eating were morally reprehensible; eating certain things at certain times in certain amounts could feed the badness in your own nature. The decline of traditional dietetics disturbed these substantive associations: overeating as a *risk factor* for disease is different from gluttony as a *vice*; moderation is powerful advice when it counts as a virtue, much less so when moderation as a way to health can be trumped by the drive to optimize the intake of nutrients or by the food- or drug-powered quest to be "better than well."[15] The emergence of nutrition science and attendant institutional changes in the food system disturbed traditional dietary links between the instrumental and the moral. A pervasive intellectual sensibility about the modern cultural order points to its supposed amoralism. The post-Darwinian human body is considered to be a natural object like any other, and dominant expert opinion maintains the same for mental function. The consumption of food is an instrumental act—having to do with maintaining health and function—and moral management is a matter wholly different from feeding the body. The *good for you* now is supposed to have nothing to do with the *good*: if you want health, go to your physician or nutritional expert; if you want to know what is moral, go to your priest, imam, rabbi, or professional ethicist, though some now think there is no such thing as moral expertise.

The rise of modern nutrition science disrupted these links between the dietary and moral domains, but there continued to be a path between them. In traditional dietetics, how to *eat* was a substantial answer to questions about how to *be*: it remains so, but the modes of morality have changed. During the past century, three pertinent new moral actors were increasingly propelled onto the cultural scene. One is the *environment*, considered as an entity that human actions can affect. What we should eat figures in urgent discussions about the climate emergency and biodiversity. It used to be said that we have ourselves on our plates, and now the fate of the planet is there too. The second new actor is the *economy*, especially historically recent corporate forms involved in producing and marketing food. The food system is identified as a pillar of modern capitalism. Its globalized forms and the food choices they influence are seen as ideologically inflected ways of sustaining capitalist institutions and modes of acting.[16] And the third actor is the *social order*, the texture of everyday relationships linking agents in the food system. Critics say that our existing social identities are threatened by new ways of producing, marketing, and consuming food. Opponents of the modern food system and of the modern diet rightly represent themselves as speaking from the margins of massive modern institutional and economic realities, but they are the people who historicize, analyze, and probe and also aim to introduce changed moral sensibilities into thinking about food.

Pollan's *Omnivore's Dilemma* portrays bad food as immoral, inscribed within and sustaining a bad society. We should be closer to nature. We should understand what is really at the end of our forks: where the plastic-wrapped supermarket steak really comes from and what price is really paid by the animal, by the environment, and by low-wage workers on its way to our plate. We should understand the enormous environmental damage done by industrially produced food. Apocalypse threatens if we do not change our agricultural ways. The envisaged crisis is not a Malthusian crunch but rather an environmental collapse: "The way we eat now is having a profound effect on climate change, which certainly threatens to bring about the end of the world as we know it."[17] Knowing what is in store for us if we don't radically change our diet, we should aim to eat organically, sustainably, locally, and seasonally. The search for the authentic may make us *feel* virtuous, but authenticity is also an index of a sustainable compact between humanity and the natural environment. We should acknowledge the anonymity of the modern systems linking us as eaters, food animals, the food that the animals eat, and the people who produce food for us and the animals. Reflecting on that anonymity, we should seek to restore and enhance relations of *familiarity* that modernity has erased: we should know the individual animals whose flesh we eat as well as the land on which our food plants grow and the people who grow them. Only then will we realize our full humanity; only then will we be whole. The retail cost of our food does not reflect the real cost of producing it. Modern food is cheap because much of its actual price has been off-loaded as damage to the planet and harm to the poor and the powerless. Bad food is a modern tragedy of the commons. Pollan is an American writer, and his target is largely the peculiarly American way of making and taking food. Yet, he also gestures at something of greater scope, and a section heading of his food manifesto flags a more general way of designating the problem: "The Western Diet and the Disease of Civilization."[18] The American way of eating here serves as shorthand for the global malaise of modernity.

Increasingly vigorous vegetarian movements are now vehicles for many sorts of both instrumental and moral concerns. Some adherents believe that meat eating is bad for bodily health, maintaining, in nutritionist terms, that consuming too much cholesterol and saturated fat clogs the arteries and leads to heart disease and other sorts of ill health. Others shun meat to avoid involvement in cruelty to animals. In the seventeenth century there were writers who worried about a causal *physical* link between eating bloody food and becoming bloody-minded, not because they cared greatly about animal suffering.[19] Present-day physiological science no longer supports that kind of causal story about the dietary roots of human violence—bestial ways of being brought about by eating beasts—but our culture does support vegetarian and vegan diets as a public display of moral sensibility and opposes unfeeling violence against the animals.[20] Meat eating now does not *make* you bloody-minded; rather, it shows that you *are* bloody-minded. In the modern urbanized way of life, we have become disengaged from the routine realities of animal husbandry and slaughter, and one consequence is that many people have acquired heightened sensitivities to what animals *feel*. Some principled meat eaters will consume flesh only if the beasts "have had a good life" and have been humanely slaughtered or if eaters take personal responsibility for killing them and kill only what they intend to eat. But in recent decades, environmental concerns have become increasingly important in meat avoidance, linking food choice and moral choice by way of planetary well-being. It is in these connections that the notion of planetary health has come to sit alongside that of bodily health: eating right is a therapeutic imperative for Earth. Belching cattle and the emissions of other livestock are said to be responsible for about 15 percent of all greenhouse gases produced through human agency.[21] Amazonian rainforests are destroyed to provide pasture for cattle and to grow vast fields of soybeans for feeding animals, manure from industrial pig farms pollutes groundwater, and farmed salmon escape and interbreed with wild populations, endangering genetic diversity. A moral course of action—for those concerned with both the planet and future human generations—demands the rejection of meat and dairy products. The emerging technology of lab-grown (or "cultured") meat is now offered as a solution to the practical-moral problems associated with eating a cut of cow meat, but its future in the marketplace cannot be predicted with any confidence.[22]

It is in the United States, critics say, where foodways have gone most spectacularly wrong: the "Western diet" is the globalization of the worst bits of the "American diet," and McDonald's is the worst bit of that. From the 1980s, the notion of "McDonaldization" came to designate in academic social studies the globalization of not just bad foodways but also a broad pattern of modern *rationalizations* and *standardizations* of production, work, consumption, and ultimately human wants.[23] The Big Mac has spread across the world, and its standardized uniformity is key to that spread, so much so that from 1986 *The Economist* magazine has only semifacetiously charted the "correctness" of the world's currency exchange rates through its "Big Mac Index," the rate implied by the cost of a Big Mac in two countries compared to the actual exchange rate.[24]

American critics were largely preoccupied with the badness of their own country's foodways, but the most consequential pushback against modern bad food emerged in Italy. In 1986, the sociologist and journalist Carlo Petrini was disgusted by the opening of a McDonald's restaurant in the Piazza di Spagna in Rome and helped organize a demonstration that handed protestors plates of proper food: penne pasta. The organization emerging from that demonstration a few years later was Slow Food, one of the most successful global protest movements of recent times. Slowness—the taking of time in preparing and consuming food—was indeed an element in Petrini's thinking. Its graphic symbol is the snail, and the movement's manifesto condemns "enslavement by speed," but what is most opposed to "fast" is not so much a leisurely pace of preparation and consumption but rather the *diversity* of foods and foodways.[25] Fast food is figured as the enemy of not just the slow but also the local and the irreplaceably specific. The existence of unique local foods and local preparations is endangered by the spread of monocultural factory farms, the globally standard Big Mac, and other standardized food preparations, and Slow Food devised a worldwide series of "Arks" and "Presidia" meant to rescue endangered foodstuffs from looming extinction. Big Macs are served up anywhere in the world, but the Slow Food movement aims to secure the survival of foods that are *only* produced in certain places and define the sense of human identity associated with them.[26]

Petrini says that "in taking a stand against McDonald's and Pizza Hut, multinationals that flatten out flavors like steamrollers, we know that we have to fight our battle on their ground, using their weapons: globalization and worldwide reach." The choice of an English-language name for the movement signals its awareness that it is using the tools of the global order to combat globalism: "If globalization has succeeded all over the planet by offering the same meat patty to everyone, then others might achieve the same kind of expansion by emphasizing the culture of diversity."[27] For Petrini and his associates, the preservation of unique foods and foodways is a pushback against globalization and uniformity, but Slow Food has its roots and its special force in Italy, where what counts as local is not so much the "Italian" versus the global-American but rather Piedmontese preparations versus Tuscan, the unique raw *salsiccia* of Petrini's hometown of Bra, the Castlemagno cheese from the mountains above the Valle Grana, and the characteristics of Ligurian olive oil compared to Umbrian. (The French wine world likes to speak of place-specific *terroir*, and the early Slow Food movement, with its strong involvement in Piedmontese wine culture, extended that usage to the role of place in making the uniqueness of all the world's authentic foodstuffs.)[28] "With enough time and enough McDonald's and the like," wrote an early American admirer of Slow Food, "there will be no roots to return to," and the standardizing bureaucracies of the European Union did on that continent what American fast food was doing on a global scale.[29] What would be irretrievably lost if that happened would be not just a fruit, a vegetable, or a pasta but also a way of living, an aspect of social identity and what it means to be a human being of a specific sort. In the seventeenth and eighteenth centuries, the roast beef of Olde England had defined what it was to be English, and in traditional dietetic thinking, it was what helped to *make* English natures. With the decline of traditional dietetics, that substantive link was submerged, but the symbolic and cultural identification of foods with a sense of collective identity persists, defiantly reaffirmed by Slow Food: we are not all Americans, and we don't all eat like Americans. "Food is the product of a region," Petrini writes, and contains the history of the people who live there.[30] Here the tag "you are what you eat" is recruited by the resistance to globalization.

We are not all moderns either, Petrini insists: we don't all eat or live like moderns. The food order, we should understand, is folded into the moral order, and the critics implicitly recover a traditional sense of *diet* as an ordered mode of life: eating as social being. A good society can be restored through good eating, and the Slow Food slogan merges the moral and the descriptive: food should be "good, clean, and fair."[31] Cleanness—in Slow Food speak—refers to the naturalness of food: its freedom from industrial manipulation and its environmental innocuousness. Fairness insists on social justice for all actors in the food system, a pricing regime that allows consumers to purchase what they require and want, producers to have a decent life, and diversity to flourish. And the notion of goodness indexes a lost culture of taste and honest pleasure. The production and consumption of bad food sustains a bad social order. Petrini started out as a communist, Slow Food emerged from the Italian Left, and McDonald's and the globalized food corporate order were well-chosen targets. Modernity has many dimensions, but global capitalism was configured as its evil leading edge. The badness of a fast-food world was attached to political ideology and its analyses of inequality, exploitation, and cultural imperialism, but its badness was also evident in the decline of the dining table as an intimate social scene and in increasing indifference to the sensory experiences of eating and drinking. Against the sensibility that modernity has separated the moral from the instrumental aspects of eating, and against the notion that the relationship between the dietary good for you and the good belong to a world we have lost, it is evident that there are now *new* moral presences at dinner and *new* sensibilities about the virtues of eating what is good for you. The morality of food and eating has not, after all, been lost; it has been culturally relocated and reattached.

The Right (and Left) of Pleasure

Slow Food means to offer "a new culture for eating and living." It is a pattern of thinking and acting that rejects the ways in which modernity has hollowed out, flattened, and homogenized the experience of eating and individualizes its occasions, depersonalizes the relationship between eater and producer, and objectifies thinking about what food is.[32] Alice Waters, chef proprietor of the transformative northern California restaurant Chez Panisse, writes that "eating is a political act, but in the way that the ancient Greeks used the word 'political'—not just to mean having to do with voting in an election, but to mean 'of, or pertaining to, all our interactions with other people'—from the family to the school, to the neighborhood, the nation, and the world. . . . The right choice saves the world."[33] The Kentucky farmer, environmentalist, and poet Wendell Berry famously said that "eating is an agricultural act," and Petrini endorses the sentiment.[34] Slow Food aims to restore the value of food-taking as a mode of subjectivity and to reinstate familiarity relationships between those who work on the land and those who consume what the land produces. The privatization of fast-food eating is opposed by *commensality*: the shared meal. *Conviviality* is celebrated, following the term's derivation from the Latin roots for "living together." Conviviality affirms collective identity, its pleasures including the shared recognition of the tastes of place and the gustatory recognition of *who we are*.[35] "Let us rediscover the flavors and savors of regional cooking," Petrini writes.[36] The Italian celebration of regionality and pleasure fell on fertile soil in northern California. Alice Water's Berkeley restaurant was flourishing long before Carlo Petrini started up Slow Food, but the pair are now allies. Waters's most recent book is branded as a "Slow Food Manifesto" and it is dedicated to Petrini.[37]

Pleasure is not seen as morally bad, and the taking of proper pleasures is affirmed as a virtuous act.[38] Acknowledging the longstanding identification of the political Left with asceticism, Petrini wants to recuperate and revalue the notion of *gastronomy*, which is explicitly modeled on Brillat-Savarin's conception of a practice that encompasses and develops the natural sciences (chemistry, agronomy, ecology), the human sciences (anthropology, sociology, political economy), medicine, technology, practical processes of food preparation, and even epistemology, especially the reflective understanding of sensory experience that draws on Continental philosophical traditions.[39] Petrini conceives of reflective attention to taste as an authentic pleasure belonging to ordinary people, and he announces his intention to "defend the 'right to pleasure.'"[40] Writing about the origins of Slow Food, Petrini says that "the left made up its mind that 'we want to eat well too.'" The traditional link between gluttony and caring about taste was made into a barbed and superficially self-deprecating joke; the Slow Food founders accounted themselves as "*golosi democratici e antifacisti*" (democratic and antifascist gluttons) and "nuovi edonisti" (new hedonists). But the thrust of Slow Food's engagement with taste was to radically *distinguish* legitimate food pleasure from illegitimate gluttony and even from health consequences.[41] Gastronomes who reflectively consider and properly taste their food are *not* prone to excess; reflective taste at table is opposed to stomach stuffing, and there is no antagonism between pleasure and health: "it is all a question of training and moderation."[42] The relevant moral setting is no longer a religiously inflected opposition to vice but is instead a politically inflected opposition to globalized economic systems. Just as eating is an agricultural act, so too, Petrini says, the taking of gastronomic pleasure is a political act. Slow Food proposes "to wed pleasure to awareness and responsibility, study and knowledge, and to offer opportunities for development even to poor and depressed regions."[43] There is nothing trivial about dinner: much depends on it.

Modern Tastes

In traditional dietetics, taste—unless it had been corrupted by habits of bad living or unless dishes had been elaborately compounded or foods had been improperly combined—was considered a robustly reliable index of wholesomeness. Taste told you something about the *nature* of foods and the likely consequences of consuming them. This was the conceptual frame that gave sense and legitimacy to the traditional notion that if it tastes good, it's good for you. That is, there was much of the *objective* contained in this *subjective* mode, and that sort of objective knowledge might be available to you directly through the senses, not requiring external expertise. The submergence of traditional dietetic culture unwound the ties between the subjectivity of taste and its capacity to discern objective realities. In the modern order, the notion of taste as an index of physiological consequence and wholesomeness has been almost wholly lost.

By and large, the nutrients identified from the time of Liebig onward *do not taste*.[44] You cannot use gustatory or olfactory sensation to discern their presence, and insofar as you know things about nutrient substances, you know them by relying on external expertise. The "meaty" osmazome identified by nineteenth-century chemists was said to taste of cooked meat, while sucrose tastes sweet and sodium chloride tastes salty, and the fattiness of many "fatty" substances is so sensed in the mouth, but few other nutrients clearly declare their identities through taste, and many have little or no discernible taste. Vitamins are tasteless, and so are many of the mineral trace elements contained in food. Carbohydrates do not have a "carbohydrate-y" taste, there is no taste you can reliably call proteinaceous, cholesterol cannot be tasted as such, and while there may be expert chemists and sensory scientists who *can* discern the distinctive tastes of polyunsaturated and monounsaturated fats, any such sensory capacities remain a niche accomplishment. Moreover, the modern artificial flavor industry and the scientists of Big Food companies are skilled at spoofing consumers' senses and at manipulating their hedonic responses. These companies' business is to produce artificial colors, textures, smells, and tastes. And all of this has helped generate skepticism that commercially produced foods really are what our senses make us think they are. We now have a range of artificial sweeteners that have no sugar and do not have its physiological effects, we understand that there are many fruit flavors that have never seen a fruit, and some people understand very well that the food industry employs skilled chemists who can engineer consumers' addictive bliss response by finding the sweet spot of the sweet, salty, and fat.[45] In the early modern period, it was considered that natural taste as a guide to wholesomeness might be corrupted by custom; now, it is often said that taste is corrupted by the ingenuity of food science and technology. Both the industrialization of food flavor and the rise of medical testimony about the dangers of the sweet, the fat, and the salty have supported a modern sentiment that is the opposite of *quod sapit nutrit*: there is now a widespread presumption that "if it tastes good, it's bad for you."

For all that, there are scientists who believe that consumers' taste can and should be *trained* to accurately identify food and drink constituents established by laboratory expertise. A number of organoleptic—that is, taste and odor-generating—constituents of wines having been discovered, some wine chemists and sensory scientists have sought to educate drinkers about which constituent chemicals are responsible for the olfactory impressions of, for instance, green peppers or black currants. Drinkers are provided with aroma wheels, standardized vocabularies, and reference samples, all designed to produce a trained community who can talk in an objective style about the substances that are said to cause a range of distinct odors.[46] There are philosophers who insist—against much skepticism—that there are objective "facts of the matter" allowing competent judgment of the *goodness* of foods and drinks, such as what makes one wine *better* than another.[47] And, famously, there are wine writers and marketers who maintain—again against skepticism—that quality can be not only objectified but also precisely *quantified*; for example, one wine is scored 93 points while another gets 81 points.

Attempts to describe and define the sensory characteristics of foods occupy only a small and esoteric part of the vast volume of present-day commentary on food goodness. Marketers and advertisers concentrate on making associations and eliciting emotions, menu writers walk a fine line between telling diners what sensations they may expect from a dish and repelling them through pompous detail, and restaurant critics award stars and points but rarely seek to say what a dish *tastes like*, most commonly comparing its goodness to that of preparations that readers are presumed to already know. And some of the most notable writers on the food system who urge the importance and value of taste have little use for the analytic vocabularies of scientific objectification. Petrini insists that "much knowledge can be gained through the taste buds and the mucous membrane in the nose," but this is not the knowledge of constituents; it is "experience that is closely related to pleasure."[48] The Slow Food world tends to employ only a sparse vocabulary to describe tastes and smells, and its adherents often don't attempt to describe them at all. In this subculture, you may presume that attentive eaters and drinkers can *know* a taste, can *recognize* things they have consumed before, and can detect and value *differences* between otherwise similar things, but it is not thought necessary that this knowledge be parsed through a string of predicates denoting distinct tastes and odors or that discerning the nature and quality of foods should involve objective knowledge of their chemical constituents. Connoisseurship in this mode consists in the sensory *recognition* of the thing, in *distinguishing* one thing from another, not in the identification of its components. As such, taste may be a feature of moral and political judgment. Taste may be treated as a valued index of what is, for example, authentic, natural, regional, local, seasonal, and artisanal; taste judgments and preferences may also serve as evidence of what *people* are like and what sort of society they value.[49]

Terms such as "connoisseur" and "gourmet" are now often felt to be associated with pretention and elitism, and the more demotic notion of the "foodie" emerged in the early 1980s, entering common currency shortly thereafter.[50] The word was evidently intended to designate people with a heightened interest in not just the goodness and taste of food but also the methods of its production and preparation. But both the deflationary resonance of "foodies" and the right-thinking moralism with which they were associated prompted backlash. Virtue was being signaled, and critics were keen on exposing virtue signaling. The tastes and taste judgments of food and wine have become, for many, paradigm instances of undisciplined subjectivity: the tag *De gustibus non est disputandum*—"There's no arguing about taste"—is now deployed in commentary on contests over whether, for instance, New York thin-crust pizza is better than Chicago deep-dish pizza or, indeed, whether there's a solid basis for *any* such goodness judgments. The lost objective aspect of taste in traditional dietetics has left it as, so to speak, an intellectual orphan, with many believing that not only is taste wholly subjective but also that taste judgments are arbitrary: there's no arguing about taste, no accounting for it, no point even in attempting to communicate much about it or persuading others to your preferences. Even as taste vocabulary proliferates and becomes more baroque, skeptical sentiment builds that taste is about nothing at all or that, if it's about anything, it's about social display. The cutting free of taste from its role in discerning the Galenic qualities of *foods* has left taste floating free in the culture to manifest the discernment, sensitivities, refinements, and, in the end, qualities of *people*. In the early nineteenth century, Brillat-Savarin insisted that gastronomy was a science; that recognizing, reflecting upon, and talking about taste belonged to civilized discourse; and that ultimately a scientific basis would be found for both good cooking and good talking about taste. The connoisseurship of the late twentieth and early twenty-first centuries retains only fragments of Brillat-Savarin's sentiments about good food and the civilized condition, while bits and pieces of his scientism survive in the rarefied circles of molecular gastronomy.

The foodies are vocal and well covered by the media, but there has been vigorous resistance. Some recent polemical reactions to connoisseurship focus on the arbitrariness of its judgments; others have targeted the moral status of its attention to food, objecting to the evangelism of Slow Food and refusing to join in the demonization of the Big Mac.[51] In Britain, the novelist and cultural commentator Will Self is disgusted with the "oleaginous ideology" of foodie culture: "gastronomy has replaced social democracy as the prevailing credo of our era." By the 1990s "more and more swallowed the idea that you are, indeed, what you eat" and "jettisoned the troublesome business of acquiring culture by any other means than orally." The whole foodie business is overdone; it is sham politics for those unwilling to do real politics, and it is culture for those too lazy to acquire proper Culture. All that moralism about food is itself immoral: it is self-absorbed and self-indulgent. Stuffing themselves with artisanal opulence, the foodies turn blind eyes to the legions of the world's hungry. If they ever denied themselves, they would find, Self declared, that hunger is, after all, the best sauce.[52] The English novelist John Lanchester loves good food and has spent much of his life writing about it, but he too reckons that matters have gone too far: "There's just too much of everything. Too much hype, too much page space." When Petrini and Waters announce that good food will save the world, Lanchester can't make himself believe it.[53] In the United States, *The Atlantic* published an outburst against the foodies by an international relations academic (and vegan). He too called out their false moralism and "sanctimonious posing": "Even if gourmets' rejection of factory farms and fast food is largely motivated by their traditional elitism, it has left them, for the first time in the history of their community, feeling more moral, spiritual even, than the man on the street." It was ridiculous to talk about a locally sourced meal as—in Pollan's precious turn of phrase—a "secular seder." Foodies are just indulging themselves in a modern version of an old vice. Their insistence to the contrary notwithstanding, they are *gluttons*, and that is just what they should be called. Their planetary and social moralism is nothing but thin cover for "gorging themselves": "Being overstuffed, for the food lover, is not a moral problem."[54]

The Age of Extremes?

In traditional dietetics, the ordering of everyday patterns of living was meant to secure health and long life, and that same ordering constituted and signaled moral worth. The six things nonnatural organized Galenic dietetics and those nonnaturals made up much of the meaningful patterns of everyday life. Moderation was central to both instrumental and moral purposes. Through the early modern period, moderation might be variously interpreted, and the demands of public life meant that occasional departures from strict rule might be excused, but moderation was rarely challenged as a secular civic virtue. That is, observing moderation—the golden mean—was both good for you and good. Balance was a related notion, and like moderation, its sense pertained to a well-tempered body, a well-ordered mind, and prudently regulated behavior. In traditional dietetics, balance was a due equilibrium of *qualities*, and one way to secure and maintain it was through the consumption of foods that had the relevant qualities. While the ability to cite texts in support of moderation belonged to the learned, acknowledging the virtue and prudence of moderation was a common cultural possession. Scholars and peasants, the high and the low, knew how to recognize and criticize excess. As medical and moral counsel, moderation could scarcely be coherently challenged.

How to talk about the historical career of dietary moderation? I have said that moderation has lost its authority and that some of this loss is consequent on the rise of nutrition science and the changing sense of *diet*, from an ordered form of life to an ordered form of eating and, finally, to specific regimes of weight loss. The counsel of balance and moderation is an awkward fit with the sensibilities and vocabulary of nutrition science. The tempering of humoral qualities is different from securing expertly established quantities of nutrient constituents. A balanced temperament of the body is the aim in traditional dietetics, while in nutrition science each of a range of specific body functions is reckoned to require the consumption of the proper quantities of the nutrients supporting those functions: carbohydrates and fats for fuel, protein for bodybuilding, omega-3 fatty acids for nerve health, vitamin D for the immune system, enough vitamin C to avoid scurvy, monounsaturated fats to lower LDL cholesterol, and so on. The recommendations of nutritional expertise are inspectable in government pamphlets and websites, on the labels of food packages, and in the tables of nutrient content made available by some restaurants. Do not drink more than 21 units of alcohol a week; do not exceed 2,500 calories a day; limit daily sodium intake to 2,300 milligrams; and ensure that you get 15 mcg of vitamin D, 1.3 mg of vitamin B_6, and 11 mg of zinc. Modern expertise now also enjoins *variety*, and this presupposes a far wider palette of choice and a far larger community of those who can choose than was the case in the not-so-distant past when dietary exoticism and diversity were commonly frowned upon: they were considered ostentatious and were thought to impose tough digestive demands. Balance and moderation are not now matters of subjective experience, nor are they identified as moral virtues; they ultimately refer to quantified standards laid down by expertise, and trust in that expertise is how they come to individual awareness.

In modern thought, the levels of expertly prescribed nutrients are determined by the amounts considered necessary to avoid specific illnesses or unsatisfactory physiological function. The Nobel Prize–winning chemist Linus Pauling famously campaigned from the 1950s to recommend megadoses of vitamin C to deal with conditions ranging from the common cold to cancer, heart disease, and mental illness. The scheme was called "orthomolecular medicine"—the right molecules in the right concentrations—and the aim was to ensure ideal health by bringing about the most propitious conditions for vital molecular reactions. "The optimum molecular concentrations of substances normally present in the body," Pauling said, "may be different from the concentrations provided by the diet."[55] Lay responses to these sorts of claims included the idea that if a certain amount of a required nutrient was good, then a larger amount ought to be very good. As one critical American nutrition scientist put it, "There has been a transition from focusing on minimum needs to the reality that today our problem is excess—excess calories and, yes, excesses of vitamins and minerals too."[56] Along with the notion that very large intakes of specific nutrients were good for you, there was the claim that commonly present food constituents could harm you. Critics might concede that fast food can, after all, power the body and build its fabric, but because it contains certain constituents—for example, trans fats, high-fructose corn syrup—and excessive amounts of other constituents such as sodium, it may nevertheless be toxic. We are invited to think of the fast-food burger in the same frame as we now think of the cigarette: conferring no benefits and doing significant damage. The currently popular category "superfoods" picks out consumables—goji berries, wheatgrass, chia seeds, quinoa—that are supposed to have specific health benefits because of the specific nutrients they contain or that they contain in special densities. "Functional foods" form a related category: they are advertised to impart additional health benefits because of substances added to them—for example, foods fortified with vitamins or minerals or added fiber—or whose normal content of those substances has been bumped up by breeding programs.[57] The presumed goals of a diet shaped by knowledge of specific nutrients with specific physiological functions are the avoidance of disease, the extension of life, and the optimization of vital function. Food here is not *only* medicine and may even—often combined with radical "calorie restriction"—confer the gift of immortality. Ray Kurzweil, high-tech inventor and one of Silicon Valley's leading futurists, spends about a million dollars a year on nutritional pills supporting "heart health, eye health, sexual health, and brain health," with the aim of living forever.[58] If moderation and balance are guides to normal function and normal longevity, then the diet for immortality *should* be extreme.

The present, it's been said, is an age of political, social, and moral extremes, and moderation has lost its grip on the culture.[59] In the 1950s, Simone Weil wrote that "modern life is given over to immoderation. Immoderation invades everything; actions and thoughts, public and private life. . . . There is no more balance anywhere."[60] "Extremism in the defense of liberty is no vice," declared presidential candidate Barry Goldwater in 1964, and "moderation in pursuit of justice is no virtue," a sentiment that explicitly set itself against long-standing ethical traditions and that is now thoroughly embedded in libertarian thought and in the widening political polarities of Right and Left. The same challenges to the legitimacy and sense of moderation populate the culture of food and eating. Expert nutrition science and radical dietary faddism are increasingly doppelgängers, with the analytic categories of science generating some of the forms of what counts as pseudoscience. The conditions for this apparently paradoxical association of the official and the bizarre are contained in fundamental categories of nutrition science. Accept that foods are delivery systems for specific nutrients, that specific nutrients support specific vital functions, that there are specific food constituents that erode good health, and that health can be secured by optimizing the quantity of good nutrients and minimizing the suspect ones. What may follow from such acceptance are simplistic heresies focusing on specific nutrients and specific physiological consequences, one-size-fits-all regimes for weight loss and good health. These presumptions license and give credibility to any number of modern proprietary diets designed for the major purpose of weight loss: diets prescribing the consumption of grapefruits, cabbage soup, lemon juice, and apple cider vinegar; diets that recommend maximizing or minimizing the intake of specific nutrient classes (gluten, carbohydrates, protein, fiber); diets intended to regulate the body's acidity and alkalinity; diets picking out modes of food preparation (raw, unprocessed); diets based on the supposed place of foods in human evolution; or diets designed to encourage or discourage certain metabolic pathways.

In recent times, no diet more thoroughly penetrated the public culture or was more entangled with the theme of moderation than the Atkins Diet (and its many offshoots, including South Beach and Zone). The Atkins Diet enjoyed enormous global popularity, emerging in the United States in the early 1970s, peaking in the 1990s and early 2000s, and spawning a vast industry publishing Atkins-related books and marketing branded foodstuffs before Atkins Nutritionals Inc. precipitously declined around 2005 and filed for bankruptcy.[61] The Atkins Diet was judiciously positioned in streams of both nutritional expertise and practical ethics. Atkins famously claimed that effective weight loss could be secured from a regime radically restricting carbohydrates and radically increasing protein and fat consumption. It was a diet that flew in the face of then-dominant nutritional expertise that considered fat human bodies to be the result of eating such stuff as fatty meat, regarded carbohydrates as generally benign, and encouraged a carbohydrate-dominant regimen on the lines of the familiar Mediterranean diet. Evidence presented at 1977 US Senate hearings concerned about the poor American diet noted with alarm that the proportion of calories provided by carbohydrates had been declining steadily through the century, while calories derived from fats and animal protein were increasing. This was thought to result in ill health, including an epidemic of heart disease.[62] But analogical reasoning that causally connected fatty foods with the making of fat human bodies was, Atkins said, defective, refuted by modern physiological science: "It's the 'You-are-what-you-eat' argument totally devoid of any reference to known metabolic mechanisms."[63] Atkins's prescriptions for right eating came elaborately wrapped in nutrition science theory. The scientific story concerned a causal connection between "insulin resistance" and obesity. Taking too much in the way of carbohydrates—and especially in their refined forms—triggered the release of large amounts of insulin; cells then became resistant to insulin, inefficiently metabolizing glucose in the normal way and storing the deficit as fat. The carb-induced insulin rush lowered blood glucose too much and made you hungry again. You ate, and you got fat. A damaged metabolism, caused by eating too many carbs and not enough protein and fat, produced, in psychophysiological terms, a form of addiction. Obesity is, then, not the result of failure of will, because the will has already been dietarily damaged; fat bodies are testaments to metabolic failure. The body-mind causal connection central to traditional dietetics was here reconfigured: the routine consumption of carbohydrates enslaves the will, and in that metabolic condition, the will cannot be freed without taking a scientifically evidenced decision to radically change dietary regime. Here as elsewhere, Atkins offered a moral-metabolic pastiche, a mash-up of partially digested modern nutrition science and selected aspects of traditional attitudes toward the goodness of foods.

The formal incoherence of this position seemed not to trouble either Dr. Atkins or his legions of global followers. "Willpower is not required on the Atkins diet," he wrote, "only the wisdom to put yourself in a position where you won't be needing it."[64] You *can* take an act of will to remedy the physiological damage, but it is not through self-denial; quite the opposite. All those foods that had been identified as rich, satiating, and fat making should now be consumed without restraint. Calorie counting was pointless, because bread calories (bad) were different from steak calories (good). You can repair a deranged metabolism, retuning the body to get its energy from stored fat, so offering an unlikely pairing of luxury, indulgence, and weight loss. And as the low-carb diet weaned the body off insulin resistance, the addicted body would be restored to its normal, healthy condition. There is no need to offer an *excuse* for dietary excess; luxury and indulgence are unabashedly celebrated. Atkins offered a reconfigured version of the Hippocratic aphorism about desperate conditions requiring desperate remedies. The idea of balance is inadequate to repair the results of what he saw as an unbalanced modern diet: "*An imbalance cannot be corrected by adding balance; it can only be corrected by counterbalancing with an unbalanced correction.*"[65] Luxury and what seems like excess are good:

you can "eat as much as you like." "Gourmet eating," Atkins promised, "will be yours once you master all the possibilities of offered by a diet allowing butter and cream sauces. That's a beautiful prospect of eating to come."[66] This was *not* a modern version of the traditional Rule of Celsus, the rule of no rule. Although the Atkins regime *was* a form of discipline, the center of emotional gravity here was *misrule*, the carnivalesque. You can indulge your appetites, just on the assumption that your *natural* appetites are for animal flesh and fats and that appetites for carbohydrates are *unnatural* cravings. The intention here was to restore taste to something of its traditional role as a gauge of goodness and to revalue appetite as an index of how much you should consume.[67]

Modernity and Moderation

Is the voice of dietary moderation still audible in late modernity? With what authority—if any—does it speak? What are the meanings of modern moderation, and to what realities does it refer? And does present-day moderation have anything of the instrumental-moral charge with which it was endowed in traditional dietetics? The counsel of moderation was hugely compromised by the decline of traditional dietetics and the rise of nutrition science, but its authority was not that easily rejected, and it has *not* disappeared from contemporary culture. Indeed, you can hear moderation—and its problematically related notion of *balance*—commended in a range of settings, approved by many institutional and lay sources and serving a range of purposes. As in the early modern period, there are still contests about what *counts* as moderation and balance, but there remain substantial obstacles to the straightforward approval of the excessive and the extreme. To that extent, homage—or at the least lip service—is still paid to moderation. Even the founder of the radical Atkins Diet was unwilling to accept critics' charge that his regime was immoderate. The then-current carb-heavy, fat-phobic "normal" Western diet was itself, he said, immoderate, unbalanced, unnatural, and a challenge to robust common sense. Or, Atkins insisted, if there *was* an immoderate aspect to his regime, it applied only to its earliest "induction" phases, after which carb-wary moderation ruled. Atkins did not *oppose* his regime to moderation; rather, he respecified what was genuinely immoderate and therefore in need of urgent repair.[68]

There are other ways in which moderation persists in present-day lay settings. When I was a boy, my mother endorsed a modern folk version of *quod sapit*: "a little of what you fancy does you good." She instructed me to "eat up" but not to "be a pig." I didn't press her for the nutrition-science bases of that advice; its authority was contained within her control of the kitchen, the dining table, and me, though she occasionally used the vocabulary of macro- and micronutrients in instructing me about the powers of foodstuffs. She administered daily doses of cod-liver oil, and she also dispensed then-current folk wisdom about carrots being good for the eyesight, orange juice warding off colds, fish building the brain, and spinach making strong muscles. I recall her saying that carrots contained vitamin A, orange juice had vitamin C, spinach was full of iron, and cod-liver oil had vitamin D, but my memory might not be reliable about this sort of thing, and it was mainly just an idiom for conveying the general sentiment that certain foods had certain powers because they contained certain substances.[69] I understood her directives to be about what was *good for* me, so she was, for me, the major channel for whatever was solid about expert nutritional knowledge. As a child, I did intermittently eat far too much and was reprimanded for it, while my sister ultimately went very far to the other extreme.[70] There was nothing special about my mother's sensibilities, although mid-twentieth-century norms of regular family eating together—now disrupted by increasing individualization, changing patterns of work, and shifting notions of what constitutes a meal—then provided face-to-face contexts in which the moral and the instrumental were more naturally joined up. Michael Pollan explicitly gestures at that sort of bland but substantial maternal authority when he warns against eating "too much." Carlo Petrini has little to say about health and dietary excess. There was, he confesses, a "period of my life when I enjoyed food perhaps to an excessive extent, and in effect proving right those people who regard pleasure as the antagonist of health." But he came to appreciate that there was no such genuine antagonism: "it is all a matter of training and moderation." Conviviality is a virtue; benign excess happens and is meant to happen at celebratory occasions, and Petrini is skeptical that much harm comes from it.[71] The great American prophetess of American fine eating was Julia Child. She found the vocabulary of nutrition science distasteful and was not prone to moralizing, but she was well aware that beliefs about medical risk and physiological benefit were making her readers anxious about the healthiness of bœuf bourguignon. Toward the end of her life Julia was clearly fed up with food faddism, medicalized fastidiousness, and the neglect of pleasure: "A low-fat diet . . . What does that mean? I think it's unhealthy to eliminate things from your diet. Who knows what they have in them that you might need? Of course, I'm addressing myself to normal, healthy people. . . . What I'm trying to do is encourage people to embrace moderation—small helpings, no seconds, no snacking, a little bit of everything, and above all, have a good time."[72]

At the opposite pole to lay common sense, the modern state distributes expert pronouncements on good diet. Beginning in 1980, the US government's *Dietary Guidelines for Americans* condensed nutrition science wisdom into a compact set of recommendations for right eating that spoke with the joint authority of the state and of academic science. From Atwater's time at the end of the nineteenth century, government prescriptions centered on quantified standards of calorie and nutrient intake and reminders of energetic balance; body weight was a matter of "calories-in and calories-out."[73] Yet, perched uneasily alongside twentieth- and twenty-first-century official nutrient-by-nutrient counsel, there are residual but recurrent invocations of balance and moderation. Consider, for instance, a 1996 summary of government dietary guidelines: you should "eat a variety of foods," "balance the food you eat," and "choose a moderate diet." But this was not the moderation of qualities belonging to traditional dietetics: the reference here was to the food constituents whose intake was elsewhere quantified, and dietary *variety* was recommended as the best way to satisfy requirements for all known essential nutrients and to increase the likelihood that any still-unknown nutrients would be supplied.[74] The context of *moderation* was spelled out in terms of a quantified intake of sugars, salt, and alcohol. *Balance* appeared only in connection with the idea that health might demand physical exercise as well as proper aliment, and the function of exercise was understood as calorie expenditure.[75] In the most recent *Guidelines*, "moderation" crops up only with reference to alcohol consumption, "balance" appears in connection with risk-benefit calculations and getting the right number of calories, and "variety," again, is recommended as a way of ensuring the satisfaction of diverse nutrient needs.[76]

From 1992, the US government decided to condense dietary guidelines and graphically represent recommendations in the form of a pyramid, with foods to be consumed in greater quantity at the base and those to be taken in smaller amounts toward the apex (figure C.1). The pyramid convention in dietary matters originated in Sweden, but a range of pictorial schemes—a wheel, a pie, a pagoda, a tribal hut, a beanpot—have been globalized in the nutritional policies of many countries West and East (figure C.2). The counsel they represent has changed over time with pressures from interest groups and as nutrition science itself changes. In 2005, the United States revised its graphic form, offering additional detail about the size of servings and about what sorts of fruits, vegetables, and meats were best, with recommended ways of preparing them and noting the nutrient constituents for which some foods were delivery systems. The oddly placed stair-climbing stick figure at the side was a gesture at the relevance of physical exercise (figure C.3). A further revision followed in 2011: in its current form, the convention represents what a meal might *look* like on the plate. Dietary advice is further individualized, quantification is diluted, and less obligatory-sounding language is preferred: "choose," "look for," "go easy on" (figure C.4).

Possibly in response to Pollan-like sensibilities, official pronouncements now add formal qualifications to earlier analytic and reductionist dietary advice. We are to understand that the relevant object of dietary advice might not be food items in isolation but

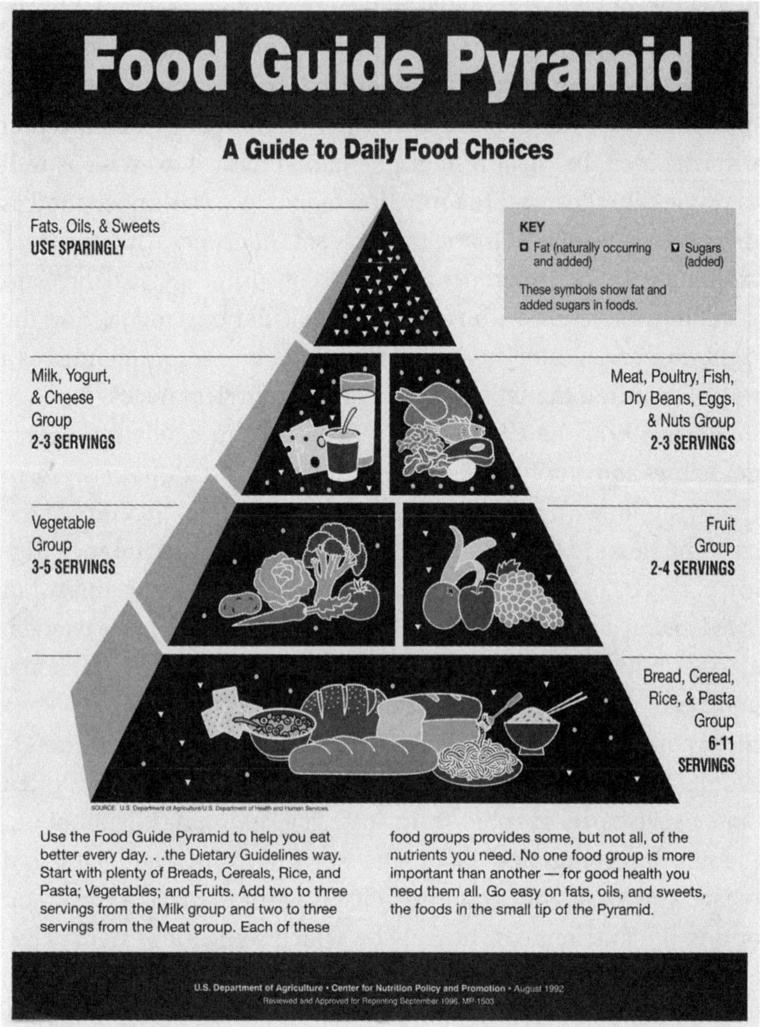

Food Guide Pyramid
A Guide to Daily Food Choices
Fats, Oils, & Sweets
USE SPARINGLY
KEY
Fat (naturally occurring and added)
Sugars (added)
These symbols show fat and added sugars in foods.
Milk, Yogurt, & Cheese Group
2-3 SERVINGS
Meat, Poultry, Fish, Dry Beans, Eggs, & Nuts Group
2-3 SERVINGS
Vegetable Group
3-5 SERVINGS
Fruit Group
2-4 SERVINGS
Bread, Cereal, Rice, & Pasta Group
6-11 SERVINGS
SOURCE: U.S. Department of Agriculture/U.S. Department of Health and Human Services
Use the Food Guide Pyramid to help you eat better every day. . .the Dietary Guidelines way. Start with plenty of Breads, Cereals, Rice, and Pasta; Vegetables; and Fruits. Add two to three servings from the Milk group and two to three servings from the Meat group. Each of these food groups provides some, but not all, of the nutrients you need. No one food group is more important than another — for good health you need them all. Go easy on fats, oils, and sweets, the foods in the small tip of the Pyramid.
U.S. Department of Agriculture • Center for Nutrition Policy and Promotion • August 1992
Reviewed and Approved for Reprinting September 1996, MP-1503

Figure C.1. The USDA Food Guide Pyramid. The graphic representation of government nutrition guidelines as a pyramid was introduced in the United States in 1992. The base of the pyramid is made up of foods that should be taken in greater quantity than those at the top.

rather routine *patterns* of consumption, that complex food-food interactions might weigh against facts about the nutrient constitution of individual foods, that the benefits of eating *foods* might not be satisfied by consuming specific nutrient chemicals as dietary supplements, and that expert advice should take into account

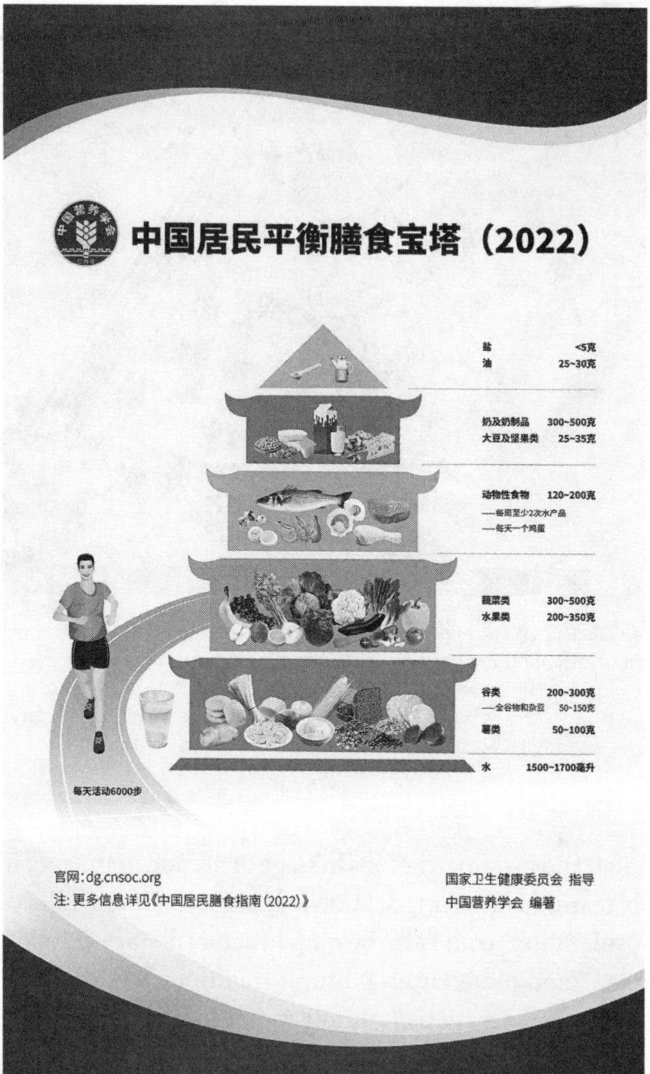

Figure C.2. "Chinese Residents' Balanced Diet Pagoda," a product of the Chinese Nutrition Society. Graphic conventions for communicating government nutritional advice vary widely across the world. Here the form (as of 2022) is a pagoda, but the general idea is, again, that eaters should take more servings of foods at the bottom and fewer of those toward the top. Daily servings are quantified—120–200 grams of meat, 50–100 grams of starches, 25–35 grams of beans and nuts—as is the recommended amount of exercise—6,000 steps of physical movement. Accompanying text links recommendations to required intakes of protein, fats, carbohydrates, and micronutrients. The vocabulary of Traditional Chinese Medicine is absent. (Translation courtesy of Tina Wei.)

Figure C.3. The USDA MyPyramid. In 2005, the USDA changed the graphic form of official nutrition advice. Proportionality was now indicated by the width at the base of the different color-coded food groups, and a concession to critics who wanted food intake to be modified by physical activity was a stick figure climbing stairs at the left.

individual and group variations in stage of life and manner of living. A novel feature is glancing acknowledgment that hard-to-account-for considerations could also bear upon actual dietary decisions, for example, "food preferences, cultural traditions and customs, and budgetary considerations," although nothing is said about what these "preferences" and "customs" might be and what significance they might have. The nonnaturals and the notion of diet as an ordered pattern of living are, of course, absent from present-day state advice, but, again, there has been recent recognition of the importance of exercise—mainly in its effects on the calorie budget—and even rare mentions of the health relevance of sleep.[77] The language of balance, moderation, and variety doesn't take up much space in current official pronouncements; it is rarely fleshed out, and, when it is, it takes its sense from the language of nutrient substances and

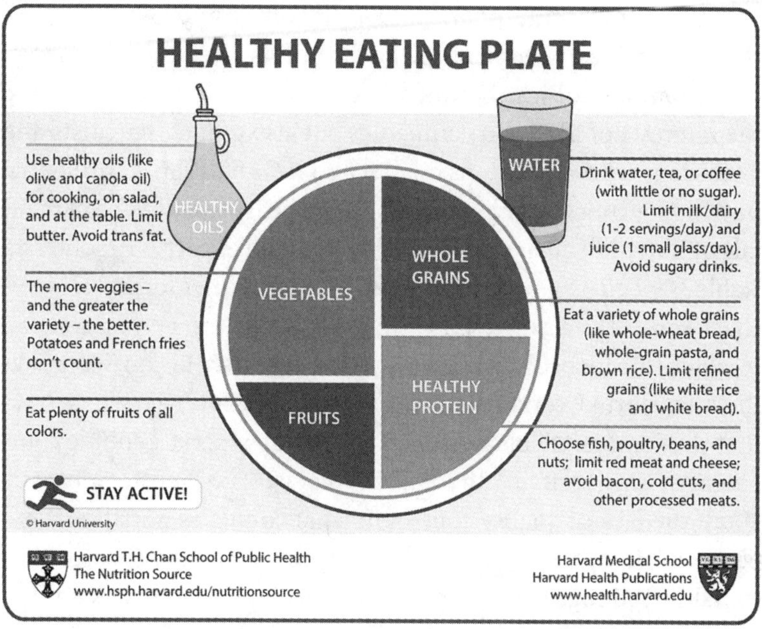

Figure C.4. In 2011, the USDA revised its pyramid convention once more, further individualizing nutritional advice. The conventionalized plate is here divided up among the various food groups. Clicking on each section delivers you to extensive textual resources, refining food categories, and advising proper consumption, such as "Make half your grains whole grains," and offering the vocabulary of constituents to justify choice. It explains, for example, that refining grains removes iron and B vitamins. Additional links allow consumers to personalize their eating plan based on such considerations as age, sex, height, weight, and level of physical activity. The version shown here was developed by the Harvard School of Public Health and differs somewhat from the government plate in the level of textual detail immediately shown.

their quantities and specific functions. That is, moderation seems here mainly a hoped-for connector to what are taken as robust lay sensibilities and values. And while this might be considered a gesture toward joined-up instrumental and moral considerations, that gesture is perfunctory. For the most part, the voice of state dietary expertise speaks the language of nutrients and number, of the instrumental means to achieve what are taken as the universal goals of disease prevention, health, and long life.

The scientific and political work involved in producing these state guidelines is enormous, and contests over their construction continue not just between academic nutrition scientists and the interests of Big Food companies but also among scientists and physicians whose positions on diet and health differ.[78] Influential nutrition writers such as Marion Nestle insist that the fundamentals of nutrition advice dispensed by genuine experts are solid and stable—adequate guides to right eating—while her formal ally Michael Pollan thinks differently: there is, he writes, no reason at all to defer to supposed nutrition science experts. He considers that their purported certainties are unstable and ever-changing and that the complexity of the subject is beyond experts' genuine competences.[79] In Britain, the epidemiologist Tim Spector despairs about the lack of quality control in what counts as nutritional expertise: "No other field of science or medicine sees such professional infighting, lack of consensus and lack of rigorous studies to back up the health claims of the myriad dietary recommendations."[80] The broad findings and recommendations of dietary expertise—whether dispensed by the state or by nutritional and medical professionals—circulate in the culture, but their authority remains limited. One sort of limitation flows from the variety of purposes performed by real-life food-taking. Eating serves body function but also serves social function. If widely accepted beliefs about food, health, and obesity are both stable and correct, why, it is asked, are we not healthier? Why do weight-loss diets almost invariably fail, the "sensible" ones as well as the "fads"? Why are the rates of heart disease and cancer increasing? "Thirty years of nutritional advice," Pollan announces, "have left us fatter, sicker, and more poorly nourished."[81] There is evidence that the effects of state guidance are patchy: government attempts to establish the uptake of its advice point to only partial public awareness that the *Guidelines* even exist and shaky knowledge of what its recommendations are. The government's own measures of effectiveness vary. Some reports indicate that only 10 percent of Americans eat in compliance with those recommendations; others show that despite state efforts, there is little sign that things are getting better.[82] There are, of course, many other types of nutritional advice available to the public, including online sources, the opinions of family and friends, tables of statistics and hearsay about statistics, and the deliverances of diet book writers such as Atkins and his many competitors. It is in that complicated context that eaters make real-world food choices: How many eggs can you safely eat? How much salt? Do megadoses of vitamin E prevent cancer or promote it? Are all fats bad for you or only certain kinds? Does this advice apply to *me* or to some abstract "average person"?[83]

This state of affairs invites skepticism, one influential mode of which is represented by criticisms of nutritionism and endorsements of dietary common sense and moderation. Although *Omnivore's Dilemma* and *In Defense of Food* are partly about healthy eating, they are not medical texts and are more concerned with sustainable food systems and social virtue. But moderation also appears on the margins of the publishing industry that produces medical and quasi-medical texts. There are not many such examples, and compared to the advocacy of fad diets, moderation-commending texts sell in small numbers. There is a physician-written book, *What Happened to Moderation?*, that advertises itself as a "common-sense approach" to health. Fads are condemned, nutrition studies are said to be marked by degrees of uncertainty that ought to be more frankly acknowledged, and holistic approaches to health are advocated. Its author writes that inadequate attention is paid to such life activities as sleep, exercise, and the management of stress and the emotions: "Today we are living in a world of extremes." It is not just fads but also mainstream medicine that have "been guilty of excesses and extremes." No overall advice is better than whatever commonsensically counts as a good diet, regular exercise, and keeping calm, that is, moderation. There *is* nutritional expertise in the modern culture, but it is overrated, and your own sense of how your body works is underrated.[84]

Moderation Incorporated

The greatest density of moderation language emerges not from the cultural fringes but instead from the center of the globalized food system. If you walk into any American fast-food outlet, you can consult a card of Nutrition Facts with the same information available on corporate websites. (In the United States, the Food and Drug Administration made this a requirement in 2014, to be enforced from 2018, but many fast-food chains, including McDonald's, were displaying calorie content and some other nutrient information years before the legal mandate.)[85] If you want to know such things, you can discover how many calories are delivered by your quarter-pound burger with cheese and how many grams of protein, how much total fat (of which how much is saturated and how much is trans fat), and how much cholesterol, sodium, potassium, vitamin D, and so on it contains. You can—if you wish—consult a Nutrition Calculator on the corporate website to select items that have a specific level of various nutrients, and you can construct a meal that satisfies certain officially established requirements.

Alongside this nutritionist idiom, however, the publications and websites of fast-food companies also employ a different idiom for encouraging right thinking about food and health. McDonald's products, the company declares, can be good for you when part of a diet "based on the sound nutrition principles of balance, variety, and moderation."[86] This is now internationally used corporate boilerplate, repeated endlessly in public discussions about the nutritional quality of its offerings. "At McDonald's," the Hong Kong arm of the company says, "we believe in the nutritional principles of balance, variety and moderation."[87] In Ireland and France, it is said that the company "has always advocated the benefits of a balanced diet. The essence of a good diet is based on balance, variety, moderation and physical activity."[88] The standard form gets special stress when McDonald's is charged with damaging customers' health. It is then insisted that the company's burgers, fries, and milkshakes "can be part of a healthy diet based on the sound nutrition principles of balance, variety and moderation," and a company "chef" sums up McDonald's dietary philosophy: "It's all about choice, balance, and moderation."[89]

Many fast-food companies and their influential allies—professional, political, ideological—have subscribed to this formula. KFC announces that its fried chicken can be part of a healthy diet, reminding customers that they "should eat all food in moderation, and balance that with an appropriate amount of exercise." And if customers need textual resources to tell them how to do that, they can pick up a free copy of the brochure *Keep It Balanced* along with their bucket of chicken.[90] Nestlé, having acquired a frozen pizza business from Kraft Foods in 2010, produces glossy publicity assuring consumers that pizza can be "part of a well-balanced diet," and the mozzarella-making American dairy industry echoes the form: "pizza can actually fit into a balanced, healthy diet and provide many nutrients when eaten in moderation."[91] The Papa John's pizza company tries "to satisfy our customers' cravings—from better taste to better for you. And, we believe everyone should eat in moderation, exercise and choose foods from all food groups."[92] A Wendy's public relations executive emphasizes the importance of "a balanced diet coupled with a really good exercise program" and falls in with the motto "everything in moderation."[93] Don't think about foods; think about an overall pattern of eating and, indeed, an overall pattern of living. A website maintained by a group of fast-food enthusiasts goes beyond a mere claim that fast food is innocuous: "Fast Food in Moderation is Good for You"; "The key in life is to keep everything in moderation."[94] In the marketing of alcoholic beverages, pervasive corporate urgings to "enjoy in moderation" and "drink responsibly" may be responses to specific legal requirements, but they are widely viewed with cynicism, not as moral or medical gestures but instead as a tactic used to counter proposed restrictions on advertising.[95]

The industry is not without powerful professional allies, and those allies' defenses of fast food follow the script. The major American professional association of nutritionists receives funding from the fast-food industry—including McDonald's—and seeks to keep a foot in both camps, supporting presidential initiatives to improve children's diet while insisting that there is nothing inherently unwholesome about a Big Mac: "there are no good or bad foods, only good and bad diets."[96] That professional association has been attacked for its Big Food industry links, including a conference for which McDonald's provided the lunch, and an association official responds that "*as dietitians, we are trained to educate our patients/clients on moderation, balance and variety as a means to develop healthy eating habits*," arguing that McDonald's should be applauded for offering more "*better-for-you options*."[97] A California dietitian who had spent seven years managing real estate for McDonald's later self-published a book defending fast food from its critics. Appealing to common sense, she attacked extremism, faddism, one-size-fits-all-ism, and reductionism. The Big Mac isn't toxic; there is much to be said in favor of food that is cheap, quick, convenient, and appealing to the public taste. "Moderation is the key"; a bacon cheeseburger taken occasionally and in moderation does no harm, and you don't need nutritional expertise to know that. True, fast-food restaurants offer food that—routinely consumed or supersized—isn't good for you, but they also sell better stuff. In 2003 and 2005, legislation was introduced in Congress to protect the food industry from "frivolous" lawsuits that claimed—on the model of successful tobacco litigation—that food companies knowingly sold unwholesome and addictive products. The legislation was the Personal Responsibility in Food Consumption Act (colloquially called the "Cheeseburger Bill"), and its presumption was that consumers are possessed of "common sense in a food court," that they have enough nutritional information to know what is good for them and in what quantities, and that "personal responsibility remains a strong American value."[98] The choice is yours; you have free will, can choose to use your common sense, and you can and should choose moderation.[99]

In the past, moderation was central to understandings of religious and civic virtue; in our culture, moderation is enlisted to do political work. Moderation is prominently commended in frankly political texts about healthy eating. According to critics, dietary guidelines laid down by the government, professional bodies, and lackey media emerge from a nanny state. Risk is exaggerated in the cause of restricting individual liberty and devaluing common sense: "Many times, you'll find that most of the risks being hyped by the media are really very small." Some alleged harms from foods and drinks *might* be worth taking account of, "but mostly you'll see that common sense and moderation in all things—and that includes worry—is usually the best course."[100] That cause was embraced by the American Council on Science and Health, a lobbying organization angered by what it took as the alarmism of the anti–fast-food industry activists. There was no justification for requiring warning labels about foods' calorie content; advocacy of organic foods was trendy liberal fetishism; and sugar, salt, and fat were not poisons. The best dietary advice is and always has been moderation in all things and the use of common sense. In principle, that promotion of moderation and common sense could have different textures, but in present-day America it belongs overwhelmingly to the political Right. Given the historically developed links between dietary expertise and the state and between right-wing thought and antigovernment sentiments, common sense, moderation, and suspicion of expertise are now at home in political agendas that see the state as the major threat to individual liberties. One foundation advocating dietary moderation and common sense is funded by the Trump-supporting billionaire Koch brothers and the tobacco industry, while the American Council on Science and Health takes funding from Coca-Cola, Pepsi, and McDonald's and supports their positions.[101]

It is not possible to know what states of belief—about foods, the body, and the effects of foods on the body—inform corporate boilerplate about balance, moderation, and variety. We do know that the message has been received with massive cynicism by some of the most energetic campaigners against fast food, who consider it obvious that moderation was bad-faith counsel. "No amount of eye-rolling can capture how hypocritical it is for food company flacks to talk about 'moderation, balance, and exercise,'" said the director of the most aggressive American nonprofit consumer group advocating for healthier foods: "Anyone who looks at these marketing techniques can see that they encourage excess, not moderation."[102] He detailed the exact content of saturated and trans fats, sodium, cholesterol, and calories in the fried chicken and the burgers, lobbying the federal government to find fast-food companies' TV ads "deceptive" and to judge them illegal.[103] The commendation of moderation was here branded as corporate cynicism. Big Food and fast-food companies stress diet over specific unhealthy items because they want their products on everyone's plates and don't want warning labels attached to a cheeseburger or chicken nuggets. They point to the importance of lifestyle and exercise because you can burn off the burger calories by going for a jog, even offering a free pedometer with a Happy Meal to measure how you're doing, and this at the same time that one of their American outlets was offering "all you can eat french fries."[104]

In the world of traditional dietetics, only the wealthy few could exercise wide choice about what to eat. They could gauge suitability to their constitutions and deliberate on medical grounds whether they should have capon or bustard, whether they should consume beef and if so whether the beef should be roasted or boiled, and whether they could safely eat a peach or should shun it as dangerous. In the just-past and in the global present, hunger and starvation persist, and the choice of foods for many millions of people remains radically limited. But at the same time, choice has been substantially democratized. So too has excess, though the distribution of corpulence has been substantially reversed from what it had been, now more a sign of poverty than of wealth. Gluttony was once a vice; now dietary excess is thought of as the cause of obesity, which in turn is a risk factor for disease. And although overindulgence is sometimes seen as a failure of will, it is often considered a failure to grasp scientific knowledge, an ignorance of proper nutritional principles, or the unavoidable hand dealt you by your genes.

In traditional dietetics, advice about what you should eat was addressed to the individual's rational soul, the entity that considers and chooses. However compromised the soul might be by bad habits, it could still choose to take good advice, could choose to change, and could choose to remake the body that housed it. The same is true in present-day idioms. Food is of course talked about in connection with the body and its functions, and while the notion of an immaterial rational soul has been substantially damaged by materialist science, invitations to food choice are also addressed to the *will*, to a volitional subject that considers and chooses. The notion of the supremacy of the will circulates in commonsense liberal conceptions of *who we are* and how we ought to be treated by others, but it also belongs to views of the marketplace, how the market works and the capacities of those who trade in the market. Those who now most energetically counsel dietary moderation direct attention to consumer choice. Here, food choice continues as an aspect of virtue, but its virtuousness mainly belongs to the secular domain of the market: the good consumer is an informed, choosing consumer, and the well-functioning market supplies the goods that the consumer chooses and at the price the consumer is willing to pay.

Taco Bell reminds consumers that "life is all about choices," that the company is supplying "food for all," and that it is ready to customize its preparations just as the consumer chooses. Like other fast-food outlets, the company assures you that you can "have it your way."[105] You can choose to eat that sort of food or not, and you can choose how much of it to eat and how often. McDonald's notes that "the choice is the customer's" and that customers are in charge of both individual selections and what McDonald's—responding to consumer desires—offers for sale.[106] Apologists for fast food insist that "we must take responsibility for our own choices and actions" and that "our choices are our own," observing that we have choice about what is made available for us to choose: "We can shape what the market offers us. It's called supply and demand."[107] The old dictum is that the consumer can't be wrong and that no one—no governmental or expert institution—should constrain choice. And in the end, consumer choice is informed by not just consumer taste but also consumer well-being. In traditional dietetics, medical expertise was coordinated with individual lay experience, and this dialogue informed the notion that you could and should be your own doctor. The tag no longer circulates as it once did, but the sentiment that validates experience has not disappeared, and this sentiment lends legitimacy to the idea that consumers should choose, that they know what they're doing, that they know that moderation is a good guide to well-being, and that they can effectively choose moderation.

It is not now easy to stand against the idea that the consumer knows best about what to eat. But many of those who condemn "the modern Western diet" do just that. Choice, they say, is compromised; the customer does *not* know best. And that's because individual choice as a guide to goodness has been corrupted. One cause of that corruption is ignorance: bad science and malevolently engineered skepticism about good science have been disseminated by those who profit from bad food. Other causes include the manipulation of taste by the virtuosic technical skills of the flavor industry and the generation of wants through advertising and marketing, especially powerful when directed at the very young. At one time the reliability of the senses as guides to wholesomeness was thought to be corrupted by civilization and bad habits; in present-day critical sensibilities, bad habits have been instilled by a bad political and economic order. The "bad for us" has been ingeniously engineered by the bad. The bad-food industry and its political allies have artfully subjugated human will. The freely choosing human will is celebrated as a foundation of the virtuous and efficient free market, and yet the modern market makes consumers want what is bad for them. The market may be free, but it makes *our* freedom delusory. The consumer, the critics say, has been degraded from a free actor to an addict in urgent need of treatment. Frictions between diverging notions of human volition and of a good society are now played out on the scale of a hamburger.

You Are What You Eat: Revisited

How do we now think about our food and ourselves? We eat modern food, and we think modern thoughts about our eating and our being. Yet, what is specifically modern in our thought about these things is not the whole of what we think. Some ancient, medieval, and early modern categories are rightly said to be dead, erased through cultural change. No one today, except for some historical specialists, knows the six things nonnatural. We say that someone has a "good sense of humor" without any awareness of the four humors and their cosmological framework. We think of melancholy as a slightly arch way of designating sadness, and there are many people now who do not even know the words "phlegmatic," "bilious," and "sanguinary," let alone think of their identity in these terms. If we encounter the notions of "spleen" and "vapors," it is most likely through reading eighteenth- and nineteenth-century novels, and even then many of us might have to consult a dictionary. If we say that peppers are hot, we point not to qualities but instead to chemical constituents that account for their effect on our senses. When T. S. Eliot's Prufrock wondered whether he would "dare to eat a peach," neither the poet nor his readers would have in mind the health dangers posed by the fruit's watery quality. My mother's claim that chili "didn't agree with her" had nothing to do with a matching of qualities, and it is hard to get unqualified assent from many people these days to the claim that "what tastes good is good for you."

But not all the categories and sensibilities of traditional dietetics are lost in just those ways, and we don't have an entirely suitable language to describe the texture of relationships between what we call past and what we call present. William Faulkner famously said that "the past is not dead, it's not even past," and if we want to understand our own beliefs, we should come to terms with not just their genealogy but also some appreciation of *continuity through change*, of the durability of some linguistic categories as their references and meanings alter, and of the persistence of practices as their supposed conceptual bases change. Notions such as "humors," "temperaments," "habits," and "constitutions" continue in currency, while their references mutate and evolve; "moderation" as a Christian virtue and as something pertaining to food qualities means something different from "moderation" as hitting officially prescribed targets of caloric, fat, or sugar intake. We still know what it is to be a glutton, but the secular bits of contemporary culture see gluttony more as disgusting and as medically unwise than as a sin. We *are* moderns with respect to these things, but our modernity is a pastiche, a palimpsest, a potpourri.

One last time: Consider the saying "you are what you eat." The idiom of modern nutrition science informs one widely distributed reading. You are a bag of chemicals. You should think of your food as bags of chemical constituents; taking in those constituents allows you to build your body's fabric, to fuel its functions, and, as a living animal, to *be*. This way of understanding the relations between eating and being is a distinctive mark of the modern cultural order. Its emergence in the nineteenth and twentieth centuries represents substantial historical change in thinking about what we eat, who we are, where knowledge of food and its vital functions is located, and what authority stands behind that knowledge. Knowledge about food has changed, and we have changed with it. Foods that we have eaten in the past become part of our bodies' present fabric; past thinking about foods leaves traces in what we now know and presume. In the matter of eating and being, there are predicaments, cultural attachments and categories that link us to the past. Self-knowledge has always implicated food knowledge, though its modes differ. Food has always been a medical concern, even while food endowed with qualities is different from food made up of constituents. A past culture that promoted the notion that you should be your own doctor is significantly different from one that reposes its trust in the knowledges of external expertise, but eaters have always been aware of those who claim special expertise about food and its functions. In traditional dietetics, it was considered that taste might be a robustly reliable guide to the nature and value of foods, and now it is not. But sensory deliverances are still attached to modern food thinking, testifying now not to objectivity but instead to consequential subjectivity, not to digestion but instead to social distinction. In the modern order, the counsel of moderation has lost much of its authority, and there are morally relevant presences at the table that were absent from past culture, yet the substance and manner of eating remains a matter of right and wrong. In the past, knowledge about what we eat belonged to knowledge about who we are. It still does.

NOTES

Introduction

1. The form is in Benjamin Franklin's *Poor Richard's Almanac*, but it was common in early modern ethical and medical texts (e.g., Thomas Brugis, *The Marrow of Physicke* [London, 1648],63), and in antiquity it was attributed to both Socrates and Cicero. See *Oxford Dictionary of Proverbs*, 6th ed., ed. Jennifer Speake (Oxford: Oxford University Press, 2015), 89.

2. For reflections on the sentiments of present-day food localism, see Fabio Parasecoli, *Gastronativism: Food, Identity, Politics* (New York: Columbia University Press, 2022), esp. chap. 3.

3. Xaq Frohlich, *From Label to Table: Regulating Food in America in the Information Age* (Berkeley: University of California Press, 2023).

4. John Archer, *Every Man His Own Doctor* (London, 1671), 23–24.

5. "Folk Sayings on Eating and Drinking," Italy Revisited, accessed March 22, 2020, http://www.italyrevisited.org/photo/Folk_Sayings_on_Eating_and_Drinking/page14.

6. Jean Anthelme Brillat-Savarin, *The Physiology of Taste, or, Meditations on Transcendental Gastronomy*, trans. M. F. K. Fisher (New York: Counterpoint, 1999), 3; Ludwig Feuerbach, "Das Geheimniss des Opfers, oder Der Mensch ist was er isst," in *Sämmtliche Werke*, Vol. 10, 2nd ed., ed. Wilhelm Bolin and Friedrich Jodl, 41–67 (Stuttgart-Bad Cannstatt: Frommann Verlag, 1959–1960). Feuerbach originally used the tag in an enthusiastic 1850 review of a book on nutrition by Jakob Moleschott and then, pleased with the wordplay, recycled it in the 1862 piece cited here. See also Steven Shapin, "'You Are What You Eat': Historical Changes in Ideas about Food and Identity," *Historical Research* 87 (2014): 377–92.

7. Armand Goubaux, *Etudes sur le cheval considéré comme bête de boucherie . . .* (Paris: Imprimerie et Librairie d'Agriculture et d'Horticulture, 1872), 8. This reference, together with a nice reading of the Brillat-Savarin dictum and its various nineteenth-century interpretations, is in Alan D. Krinsky, "Let Them Eat Horsemeat! Science, Philanthropy, State, and the Search for Complete Nutrition in Nineteenth-Century France," PhD dissertation, University of Wisconsin, 2001, 1–2, a reference I owe to Lisa Haushofer.

8. Victor Hugo Lindlahr, *You Are What You Eat: How to Win and Keep Health through Diet* (New York: National Nutrition Society, 1940). See also Sander L. Gilman, *Diets and Dieting: A Cultural Encyclopedia* (New York: Routledge, 2008), 178; and Hannah Landecker, "Being and Eating: Losing Grip on the Equation," *BioSocieties* 10 (2015): 253–58.

9. Sherman units are an obsolete measure for the minimum dose of a vitamin that protects against deficiency disease.

10. These sorts of things are discussed in Steven Shapin, "Breakfast at Buck's: Intimacy, Informality, and Innovation in Silicon Valley," *Osiris* 35 (2020): 324–47. See also Claude Fischler, "Food, Self, and Identity," *Social Science Information* 27 (1988): 275–92.

11. Max Weber, "Science as a Vocation," in idem, *From Max Weber: Essays in Sociology*, trans. and ed. H. H. Gerth and C. Wright Mills (New York: Oxford University Press, 1958), 139. See also Steven Shapin, "Weber's *Science as a Vocation*: A Moment in the History of *Is* and *Ought*," *Journal of Classical Sociology* 19 (2019): 290–307; idem, "Science and the Modern World," in *The Handbook of Science and Technology Studies*, 3rd ed., ed. Edward Hackett, Olga Amsterdamska, Michael Lynch, and Judy Wajcman, 433–48 (Cambridge, MA: MIT Press, 2007).

12. Steven Shapin, *The Scientific Life: A Moral History of a Late Modern Vocation* (Chicago: University of Chicago Press, 2008), chap. 3; and idem, "The Virtue of Scientific Thinking," *Boston Review* 40, no. 1 (January–February 2015): 32–39.

13. George Rosen, "Nostalgia: A 'Forgotten' Psychological Disorder," *Clio Medica* 10 (1975): 28–51; and Helmut Illbruck, *Nostalgia: Origins and Ends of an Unenlightened Disease* (Evanston, IL: Northwestern University Press, 2012).

14. On the periodization of dietetic thought, see a large body of work by Ken Albala, including his *Eating Right in the Renaissance* (Berkeley: University of California Press, 2002), esp. 7–8, 14–47.

15. See, notably, commendation of this kind of picture building in Allen J. Grieco, *Food, Social Politics, and the Order of Nature in Renaissance Italy* (Florence: Villa I Tatti, 2019), 21.

Chapter 1

1. The most important and probing comparative exercise is Shigehisa Kuriyama, *The Expressiveness of the Body and the Divergence of Greek and Chinese Medicine* (New York: Zone, 2002). See also G. E. R. Lloyd, *Adversaries and Authorities: Investigations into Ancient Greek and Chinese Science* (Cambridge: Cambridge University Press, 1996); idem and Nathan Sivin, *The Way and the Word: Science and Medicine in Early China and Greece* (New Haven, CT: Yale University Press, 2002); Helaine Selin, ed., *Medicine across Cultures: History and Practice of Medicine in Non-Western Cultures* (New York: Kluwer, 2003); Jack Goody, *Food and Love: A Cultural History of East and West* (London: Verso, 1998); idem, *Cooking, Cuisine, and Class: A Study in Comparative Sociology* (New York: Cambridge University Press, 1982); Hormoz Ebrahimnejad, ed., *The Development of Modern Medicine in Non-Western Countries: Historical Perspectives* (London: Routledge, 2009); and Nancy N. Chen, *Food, Medicine, and the Quest for Good Health: Nutrition, Medicine, and Culture* (New York: Columbia University Press, 2009).

2. Jacques Jouanna, "Dietetics in Hippocratic Medicine: Definition, Main Problems, Discussion," in idem, *Greek Medicine from Hippocrates to Galen: Selected Papers*, trans. Neil Allies, ed. Philip van der Eijk (Leiden: Brill, 2012), 137–53.

3. One of Michel Foucault's influential later works centrally recognized the role of regimen in self-making. See Michel Foucault, *The History of Sexuality*, Vol. 2, *The Use of Pleasure*, trans. Robert Hurley (New York: Viking, [1984] 1990), e.g., 101.

4. Ludwig Edelstein, "The Dietetics of Antiquity," in idem, *Ancient Medicine: Selected Papers of Ludwig Edelstein*, ed. Owsei Temkin and C. Lilian Temkin (Baltimore: Johns Hopkins University Press, [1931] 1967), 307–8.

5. Owsei Temkin, "Greek Medicine as Science and Craft," in idem, *The Double Face of Janus and Other Essays in the History of Medicine* (Baltimore: Johns Hopkins University Press, 1977), 147.

6. The English saying—as attributed to Hippocrates—possibly began wide circulation only in the 1970s, and while it does not appear in just this form in the Hippocratic Corpus, it is wholly in accord with the basic principles of Hippocratic dietetics. On aspects of the curious history of this slogan, see Diana Cardenas, "Let Not Thy Food Be Confused with Thy Medicine: The Hippocratic Misquotation," *e-SPEN Journal* 8 (2013): 260–62. See also Juliana Adelman and Lisa Haushofer, "Food as Medicine, Medicine as Food," *Journal of the History of Medicine* 73 (2018): 130. On dietetics and preventive medicine in general, see Heikki Mikkeli, *Hygiene in the Early Modern Medical Tradition* (Helsinki: Academia Scientiarum Fennica, 1999), chap. 2.

7. Francis Fuller, *Medicina Gymnastica: or, A Treatise Concerning the Power of Exercise, with Respect to the Animal Oeconomy* (London: John Matthews, 1705), 56. (Fuller was not a doctor but, suffering from hypochondriasis and dyspepsia, delved into the subject and was much influenced by the physician Thomas Sydenham.)

8. The Hippocratic tract *The Art* addressed and countered widespread skepticism about medical efficacy: some people got better without physicians' attention and some died under medical care. But the author mobilized a series of arguments as to why such skepticism was unwarranted: patients often failed to follow good medical advice, and everyone understood that some conditions were beyond help. Hippocrates, "The Art," in Hippocrates, *Works*, Vol. 2, trans. W. H. S. Jones (Cambridge, MA: Harvard University Press, 1923), 2:185–217.

9. Jacques Jouanna, "Air, Miasma and Contagion in the Time of Hippocrates and the Survival of Miasmas in Post-Hippocratic Medicine," in idem, *Greek Medicine*, 121–36.

10. For example, Hippocrates, "Humours," in Hippocrates, *Works*, 4:88–89, 92–93.

11. W. H. S. Jones, "Introduction," to Hippocrates, *Works*, 4:xlvii.

12. Temkin, "Greek Medicine as Science and Craft," 147. See also Roy Porter, *The Greatest Benefit to Mankind: A Medical History of Humanity* (New York: Norton, 1998), 58–60.

13. Ken Albala, *Eating Right in the Renaissance* (Berkeley: University of California Press, 2002).

14. Steven Shapin, "Why Was 'Custom a Second Nature' in Early Modern Medicine," *Bulletin of the History of Medicine* 93 (2019): 1–26; idem, "How to Eat Like a Gentleman: Dietetics and Ethics in Early Modern England," in *Right Living: An Anglo-American Tradition of Self-Help Medicine and Hygiene*, ed. Charles E. Rosenberg, 21–58 (Baltimore: Johns Hopkins University Press, 2003); idem, "Proverbial Economies: How an Understanding of Some Linguistic and Social Features of Common Sense Can Throw Light on More Prestigious Bodies of Knowledge, Science for Example," *Social Studies of Science* 31 (2001): 731–69, esp. 756–57.

15. Wesley D. Smith, "The Development of Classical Dietetic Theory," in *Hippocratica: Actes du Colloque Hippocratique de Paris (4–9 Septembre 1978)*, ed. M. D. Grmek (Paris: CNRS, 1980), 444. On that stability, see, for example, Sandra Cavallo and Tessa Storey, *Healthy Living in Late Renaissance Italy* (Oxford: Oxford University Press, 2013), 3–7.

16. Temkin, "Greek Medicine as Science and Craft," 147; idem, "Health and Disease," in idem, *Double Face of Janus*, 423.

17. Owsei Temkin, *Galenism: Rise and Decline of a Medical Philosophy* (Ithaca, NY: Cornell University Press, 1973), 39–40; and idem, *Hippocrates*

in a World of Pagans and Christians (Baltimore: Johns Hopkins University Press, 1991), 15, 45–47. See also Henry Sigerist, "Galen's *Hygiene*," in idem, *Landmarks in the History of Hygiene* (London: Oxford University Press, 1956), 1–19, esp. 12. The medical historian Ludwig Edelstein observed that the central dictum of ancient dietetics was that "he who would stay healthy . . . must know how to live rightly." Edelstein, "Dietetics of Antiquity," 303.

18. Lay challenges to professional expertise are specially considered in chapter 5 (below).

19. For recent scholarship specifically about medical care by women, see, e.g., Alisha Rankin, *Panaceia's Daughters: Noblewomen as Healers in Early Modern Germany* (Chicago: University of Chicago Press, 2013); Elaine Leong, *Recipes and Everyday Knowledge: Medicine, Science, and the Household in Early Modern England* (Chicago: University of Chicago Press, 2018); Leigh Ann Whaley, *Women and the Practice of Medical Care in Early Modern Europe, 1400–1800* (London: Palgrave Macmillan, 2011); Mary E. Fissell, "Introduction: Women, Health, and Healing in Early Modern Europe," *Bulletin of the History of Medicine* 82 (2008): 1–17; Edith Snook, "'The Women Know': Children's Diseases, Recipes and Women's Knowledge in Early Modern Medical Publications," *Social History of Medicine* 30 (2017): 1–21; and Sharon Strocchia, *Forgotten Healers: Women and the Pursuit of Health in Late Renaissance Italy* (Cambridge, MA: Harvard University Press, 2019). The problems and opportunities of excavating the knowledge of the uneducated have been recognized by historians meaning to write in the mode of "historical anthropology." See, notably, Carlo Ginzburg, *The Cheese and the Worms: The Cosmos of a Sixteenth-Century Miller*, trans. John and Anne Tedeschi (Baltimore: Johns Hopkins University Press, 1980); and Peter Burke, *Popular Culture in Early Modern Europe* (New York: Harper & Row, 1978).

20. For example, Steven Shapin, "Descartes the Doctor: Rationalism and Its Therapies," *British Journal for the History of Science* 33 (2000): 131–54, esp. 131–32.

21. Roger French, "Where the Philosopher Finishes, the Physician Begins: Medicine and the Arts Course in Thirteenth-Century Oxford," *Dynamis* 20 (2000): 75–106.

22. The distinction between primary and secondary qualities is often associated specifically with the mechanical natural philosophy of the seventeenth century—in versions by Galileo, Boyle, Locke, and others—but the followers of Aristotle also had their version. See, e.g., Robert Pasnau, "Scholastic Qualities, Primary and Secondary," in *Primary and Secondary Qualities: The Historical and Ongoing Debate*, ed. Lawrence Nolan (Oxford: Oxford University Press, 2011), 41–61; and see chapter 6, 217–21 (below).

23. There were also attacks on this account of the four qualities in the Hippocratic Corpus, one of many indications of their heterogeneous nature. However, the cosmological scheme of the qualities persisted and was incorporated in the developing body of dietetic medicine into the early modern period. Hippocrates, "Ancient Medicine," in Hippocrates, *Works*, trans. W. H. S. Jones (Cambridge, MA: Harvard University Press, 1923), 1:12–13.

24. Hippocrates, "The Nature of Man," in Hippocrates, *Works*, 4:11–13. See also Jacques Jouanna, "The Legacy of the Hippocratic Treatise *The Nature of Man*: The Theory of the Four Humours," in idem, *Greek Medicine*, 335–59.

25. Temkin, *Galenism*, 17–18.

26. Galen, "On the Humours," in Mark Grant, ed., *Galen on Food and Diet* (London: Routledge, 2000), 14. A classic historical source for humoral theory is Raymond Klibansky, Erwin Panofsky, and Fritz Saxl, *Saturn and Melancholy: Studies in the History of Natural Philosophy, Religion, and Art* (London: Thomas Nelson, 1964), esp. 3–14, 97–126.

27. For example, Nancy G. Siraisi, *Medieval & Early Renaissance Medicine: An Introduction to Knowledge and Practice* (Chicago: University of Chicago Press, 1990), 104–8. See also Philip Lyndon Reynolds, *Food and the Body: Some Peculiar Questions in High Medieval Theology* (Leiden: Brill, 1999), 107–8.

28. For example, Noga Arikha, *Passions and Tempers: A History of the Humours* (New York: Ecco, 2007), 8–10. Cf. Hans L. Haak, "Blood, Clotting and the Four Humours," in *Blood, Sweat and Tears: The Changing Concepts of Physiology from Antiquity into Early Modern Europe*, ed. Manfred Horstmanshoff, Helen King, and Claus Zittel, 295–305 (Leiden: Brill, 2012).

29. Steven Shapin, "Why Was 'Custom a Second Nature' in Early Modern Medicine?," *Bulletin of the History of Medicine* 93 (2019): 1–26.

30. There are many fairly recent summaries of humoral schemes, their relation to notions of character and disposition, and their use in systems of recognition. See especially Arikha, *Passions and Tempers*, prologue and chap. 1. See also Siraisi, *Medieval & Early Renaissance Medicine*, 101–6; Piers D. Britton, "(Hu)moral Exemplars: Type and Temperament in Cinquecento Painting," in *Visualizing Medieval Medicine and Natural History, 1200–1550*, ed. Jean Ann Givens, Karen Reeds, and Alain Touwaide, 177–203 (Aldershot: Ashgate, 2006); Christopher Allen, "Painting the Passions: The *Passions de l'Âme* as a Basis for Pictorial Expression," in *The Soft Underbelly of Reason: The Passions in the Seventeenth Century*, ed. Stephen Gaukroger, 79–111 (London: Routledge, 1998). For historical treatments of central features of dietetic thinking, see, for example, Klaus Bergdolt,

Wellbeing: A Cultural History of Healthy Living, trans. Jane Dewhurst (Cambridge: Polity, [1999] 2008), 24–26, 87–91; Mikkeli, *Hygiene in the Early Modern Medical Tradition*; and David Gentilcore, *Food and Health in Early Modern Europe: Diet, Medicine, and Society, 1450–1800* (New York: Bloomsbury, 2016).

31. For example, Sujata Iyengar, *Shakespeare's Medical Language: A Dictionary* (London: Continuum, 2011), 73–78. See also Jouanna, "The Legacy of the Hippocratic Treatise *The Nature of Man*," 344–45, 355; and Shapin, "Why Was 'Custom a Second Nature'?," 16–18. For an early modern account of how the body's humors and spirits work on the facial muscles to produce its expressiveness, see J[ohn] B[ulwer], *Athomyotomia or a Dissection of the Significative Muscles of the Affections of the Minde* (London, 1649), 102–4. And see chapter 3, 107–11 (below).

32. William Cavendish, Duke of Newcastle, *A New Method, and Extraordinary Invention, to Dress Horses, and Work Them According to Nature* (London, 1667), 27–28.

33. Steven Shapin, "'You Are What You Eat': Historical Changes in Ideas about Food and Identity," *Historical Research* 87 (2014): 384–90.

34. Following Galen's treatise *Quod animi mores corporis temperamenta sequantur*, the tag became commonplace in the sixteenth and seventeenth centuries. See, e.g., Robert Boyle, "The Aretology," in idem, *The Early Essays and Ethics of Robert Boyle*, ed. John T. Harwood (Carbondale: Southern Illinois University Press, 1991), 33: "The Fitisians, according to that celebrated saying of their master Galen, Mores animi sequuntur Temperamentum Corpis, The Customes of the Mind Follo the Temperament of the Body"; John Selden, *Titles of Honor* (London, 1614), "Preface," sig. b3: "the Minds inclination follows the Bodies Temperature; whereof *Galen* hath a special Treatise"; and many references in Robert Burton, *The Anatomy of Melancholy* (New York: NYRB Books, [1621] 2001), e.g., Part 1, 374. See also Sorana Corneanu, "The Nature and Care of the Whole Man: Francis Bacon and Some Late Renaissance Contexts," *Early Science and Medicine* 22 (2017): 138–41, 145, 148 (and 137 for Western familiarity with this text by the first part of the sixteenth century).

35. Temkin, *Hippocrates in a World of Pagans and Christians*, 15, 45–47. See also Sigerist, "Galen's *Hygiene*," 12.

36. Ben Jonson, *The Comicall Satyre of Every Man Out of His Humor* (London, 1600), sig. Bijv.

37. See Marjorie Garber, *Character: The History of a Cultural Obsession* (New York: Farrar, Straus and Giroux, 2020).

38. For pathologies of religion proceeding from "peccant humors," see, e.g., Richard Allestree, *The Causes of the Decay of Christian Piety* (London, 1667), 406.

39. G. E. R. Lloyd, "The Hot and the Cold, the Dry and the Wet in Greek Philosophy," *Journal of Hellenic Studies* 84 (1964): 92–106.

40. Eric Partridge, *The Routledge Dictionary of Historical Slang*, 6th ed. (London: Routledge/Taylor & Francis e-Library, 2006), 5847.

41. Levinus Lemnius, *The Touchstone of Complexions*, trans. from Latin by T. N. (London, [1561] 1633), 139, 141. See also Klibansky et al., *Saturn and Melancholy*, 13, 103.

42. Siraisi, *Medieval & Early Renaissance Medicine*, 116–17, 137–41.

43. François Rabelais, *Gargantua and Pantagruel*, trans. J. M. Cohen (London: Penguin, [1532–1534] 1955), 376.

44. Lemnius, *Touchstone of Complexions*, 155. See also chapter 3, 111–13, and chapter 7, 265–67 (below).

45. Hippocrates, "Airs Waters Places," in Hippocrates, *Works*, 1:122–23, 136–37. See also Jouanna, "Dietetics in Hippocratic Medicine," 142–45; and John [Juan] Huarte, *Examen de ingenios: The Examination of Mens Wits*, trans. Camillo Camilli (London, [1588] 1594), 57–58, discussing ancient debates about whether people living in hotter climates might be more intelligent and whether those inhabiting colder climates might be more stubborn in their opinions.

46. Aristotle, *On the Generation of Animals*, Book IV, 775a.13–21, in Aristotle, *Complete Works*, 2 vols., ed. Jonathan Barnes (Princeton, NJ: Princeton University Press, 1984), 1:1199. See also Plutarch, *Moralia*, 17 vols., trans. Paul A. Clement, Herbert B. Hoffleit, and Frank Cole Babbitt (Cambridge, MA: Harvard University Press, 1927–2004), 8: 227–33 (for ancient dispute over whether women were colder and moister than men); and Katharine Park, "The Myth of the 'One Sex' Body," *Isis* 114 (2023): 157–58.

47. Hippocrates, "Diseases of Women. I," in Hippocrates, *Works*, trans. Paul Potter (Cambridge, MA: Harvard University Press, 2018), 11:13. See also Jacques Jouanna, "Wine and Medicine in Ancient Greece," in idem, *Greek Medicine*,181 and n. 63.

48. Laz[are] Riverius [Rivière], *The Universal Body of Physick*, trans. William Carr (London, [1640] 1657), 16.

49. See, e.g., Lemnius, *Touchstone of Complexions*, 68; Huarte, *Examen*, 270–82; Thomas Burgis, *The Marrow of Physicke* (London, 1648), 57; Joan Cadden, *The Meanings of Sex Differences in the Middle Ages: Medicine, Science, and Culture* (Cambridge: Cambridge University Press, 1993), 32–33; and Siraisi, *Medieval & Early Renaissance Medicine*, 102–3; Gail Kern Paster, *Humoring the Body: Emotions and the Shakespearian Stage* (Chicago: University of Chicago Press, 2004), chap. 2. Thomas W. Laqueur's influential *Making Sex: Body and Gender from the Greeks to Freud* (Cambridge, MA: Harvard University Press, 1990), esp. 34–35, 40–41,

92, linked females' supposed differences in temperament to views that women were incompletely concocted and less perfect versions of males—the "one sex"—but see the lacerating demolition in Park, "The Myth of the 'One Sex.'"

50. Hippocrates, "Humours," in Hippocrates, *Works*, 4:88–89, 92–93. See also Grant, *Galen on Food*, 17; and Klibansky et al., *Saturn and Melancholy*, 10.

51. For example, Henry Peacham, *Minerva Britanna, or a Garden of Heroical Devises* (London, 1612), 126. Cf. Guillaume de Salluste Du Bartas, *Du Bartas His Divine Weekes and Workes* (London, 1611), 377 (for melancholy and cold, dry winter); and Jean Gailhard, *The Compleat Gentleman: Or Directions for the Education of Youth* (London, 1678), 64–65 (for melancholy and cold, dry autumn). And for the remarkable stability of this scheme, see Klibansky et al., *Saturn and Melancholy*, 10–11. These associations textually migrated to a range of climates from their Mediterranean origins, so a degree of variability is unsurprising.

52. Klibansky et al., *Saturn and Melancholy*, 120.

53. Edelstein, "Dietetics of Antiquity," 304.

54. See, e.g., Temkin, "On Galen's Pneumatology," in idem, *Double Face of Janus*, 154–61. For the persistence of the doctrine of spirits into later periods, see, among many sources, D. P. Walker, "The Astral Body in Renaissance Medicine," *Journal of the Warburg and Courtauld Institute* 21 (1958): 119–33, esp. 120–21; Carol V. Kaske and John R. Clark, "Introduction," in Marsilio Ficino, *Three Books on Life*, ed. and trans. Kaske and Clark (Binghamton, NY: Renaissance Society of America, [1489] 1989), 42–51; Gail Kern Paster, "Nervous Tension: Networks of Blood and Spirit in the Early Modern Body," in *The Body in Parts: Fantasies of Corporeality in Early Modern Europe*, ed. David Hillman and Carla Mazzio, 107–25 (London: Routledge, 1997), 108–15; Paster, *Humoring the Body*, esp. 61–70; and Oriana Walker, "The Breathing Self: Toward a History of Respiration," PhD dissertation, Harvard University, 2016, 65–69. Colloquial invocations of the spirits continue to appear into modern times, as in the notion of "high spirits" and in John Maynard Keynes's famous ascription of ebullient economic action to "animal spirits," albeit without the traditional reference to physical entities and their consequences.

55. A. B., *The Sick-Mans Rare Jewel Wherein Is Discovered a Speedy Way How Every Man May Recover Lost Health, and Prolong Life* (London, 1674), 19–20. See also Daniel Sennert, *The Institutions or Fundamentals of the Whole Art, Both of Physick and Chirurgery* (London, 1656), 12: the spirits are "the purest, thinnest, hottest, most moveable body, proceeding from the most purest and subtilest part of the bloud."

56. The "vapors"—considered as physiologically and mentally conse-

quential fumes produced after consuming hard-to-digest foods—are discussed in detail in chapter 4, 159–64 (below).

57. Loosely following the account in Jean Fernel, *The Physiologia of Jean Fernel (1567)*, trans. John M. Forrester (Philadelphia: American Philosophical Society, 2003), 299–301.

58. See, e.g., Lawrence Babb, *The Elizabethan Malady: A Study of Melancholia in English Literature from 1580 to 1642* (East Lansing: Michigan State University Press, 1951), chap. 1–2.

59. Lemnius, *Touchstone of Complexions*, 7, 14 (see also 11–40); Burgis, *Marrow of Physicke*, 56–57.

60. Rabelais, *Gargantua and Pantagruel*, 375.

61. John Charles Bucknill, *The Medical Knowledge of Shakespeare* (London: Longman, 1860), 82–83; John Moyes, *Medicine & Kindred Arts in the Plays of Shakespeare* (Glasgow: James MacLehose, 1896), 14.

62. Herbert J. C. Grierson, ed., *Metaphysical Lyrics & Poems of the Seventeenth Century: Donne to Butler* (Oxford: Clarendon, 1921), 16–18.

63. Fernel, *Physiologia*, 293–301. See also Antonio Clericuzio, "Chemical and Mechanical Theories of Digestion in Early Modern Medicine," *Studies in History and Philosophy of Biological and Biomedical Sciences* 43 (2012): 329–37, esp. 329.

64. Thomas Burnet[t], "Dr. Burnet's Demonstration of True Religion," Part 2, in *A Defence of Natural and Revealed Religion: Being an Abridgement of Sermons Preached at the Lecture Founded by the Hon. Robert Boyle, Esq.*, Vol. 4 (London: Arthur Bettesworth and Charles Hitch, [1724–1725] 1737), 15.

65. Quoted in Michael Hunter, *Robert Boyle (1627–1691): Scrupulosity and Science* (Woodbridge: Boydell, 2000), 91, 118.

66. Richard Ward, *The Life of the Learned and Pious Dr Henry More*, ed. M. F. Howard (London: Theosophical Publishing Society, [1710] 1911), 82–83, quoted and discussed in Steven Shapin, "The Philosopher and the Chicken: On the Dietetics of Disembodied Knowledge," in *Science Incarnated: Historical Embodiments of Natural Knowledge*, ed. Christopher Lawrence and Shapin (Chicago: University of Chicago Press, 1998), 38.

67. Hippocrates, "The Nature of Man," 262.

68. Grant, *Galen on Food*, 16. See also idem, *Dieting for an Emperor: A Translation of Books 1 and 4 of Oribasius' Medical Compilations with an Introduction and Commentary* (Leiden: Brill, 1997), 6; and Galen, *A Translation of Galen's Hygiene (De Sanitate Tuenda) by Robert Montraville Green* (Springfield, IL: Charles C. Thomas, 1951), 13.

69. Siraisi, *Medieval & Early Renaissance Medicine*, 120–21.

70. For humoral schemes' accommodation of plastic normalcies, see Klibansky et al., *Saturn and Melancholy*, 11–12.

71. Grant, *Galen on Food*, 15.

72. Hippocrates, "Regimen in Health," 46–49.

73. Hippocrates, "Breaths," in *Works*, 2:228–29. See also Temkin, "Byzantine Medicine: Tradition and Empiricism," in idem, *The Double Face of Janus*, 204; and idem, *Hippocrates in a World of Pagans and Christians*, 12–13.

74. Hippocrates, "Regimen in Health," 44–45. See also Siraisi, *Medieval & Early Renaissance Medicine*, 121.

75. Grant, *Galen on Food*, 124, 161, 163.

76. Quoting Edelstein, "Dietetics of Antiquity," 304–5, 311, 313–14. See also Temkin, "Medicine and Moral Responsibility," in idem, *The Double Face of Janus*, 53.

77. Hippocrates, "Regimen, I," in Hippocrates, *Works*, 4:226–27.

78. Hippocrates, "Regimen in Health," in Hippocrates, *Works*, 4:58–59.

79. Montaigne liked the twenty-years version: "Tiberius was wont to say, that *whosoever had lived twenty yeares, should be able to answer himselfe of all such things as were either wholesome or hurtfull for him, and know how to live and order his body without Phisicke.*" Michel Eyquem de Montaigne, "Of Experience," in idem, *The Essayes of Michael Lord of Montaigne*, 3 vols., trans. John Florio (London: J. M. Dent, [1580–1588] 1910), 3:339. The tag appeared widely and persistently and in all sorts of texts—medical, ethical, prudential—sometimes attributed to the Tiberius, sometimes treated as an unauthored, but authoritative ancient saying. The cited version is in a representative collection of proverbs: W. Gurney Burnham, *A Book of Quotations, Proverbs and Household Words* (London: Cassell, 1907), 775. In the early modern period, the story was prominently invoked by Descartes. See Steven Shapin, "Descartes the Doctor: Rationalism and Its Therapies," *British Journal for the History of Science* 33 (2000): 139; and Francis Bacon, *Historie Naturall and Experimentall of Life and Death* (London, 1638), 92–93.

80. Aulus Cornelius Celsus, *De Medicina*, trans. W. G. Spencer, 3 vols. (Cambridge, MA: Harvard University Press, 1960), 1:57. (This is not the same Celsus as the second-century CE Greek philosopher and opponent of Christianity.)

81. Plutarch, "Rules for the Preservation of Health," in *Plutarch's Lives and Miscellanies*, 5 vols., ed. A. H. Clough and William W. Goodwin (New York: Colonial Company, 1905), 1:277–78. See also Bergdolt, *Wellbeing*, 82–84.

82. Hippocrates, "Aphorisms," in *Works*, 4:101–2, 121; and idem, "Regimen, III," in *Works*, 4:369, 381–83.

83. Celsus, *De Medicina*, 1:43.

84. Appeals to the Rule of Celsus in the early modern are discussed in detail in chapter 5, 192–95 (below). See also Mikkeli, *Hygiene*, 92–96.

85. Plato, *Republic*, Book III, 406b-c, in Plato, *The Collected Dialogues*, ed. Edith Hamilton and Huntington Cairns (Princeton, NJ: Princeton University Press, 1963), 650–51.

86. [Michael Scot], *The Philosophers Banquet Newly Furnished and Decked Forth with Much Variety of Many Severall Dishes . . . by W. B. Esquire*, 3rd ed. (London, [1609] 1633), 89–90. See also Mikkeli, *Hygiene*, 48–54; and further treatment of the quest for health in relation to the call of civic duty in chapter 5, 187–90 (below).

Chapter 2

1. See Mary Douglas's anthropological discussion of the four great cultural storehouses of legitimation—God, Money, Time, and Nature—in "Environments at Risk," in Douglas, *Implicit Meanings: Essays in Anthropology* (London: Routledge & Kegan Paul, 1975), 209. See also Lorraine Daston, *Against Nature* (Cambridge, MA: MIT Press, 2019).

2. The historical origins and medical role of the nonnaturals in dietetic medicine have been widely and well treated. Among very many examples are L. J. Rather, "The 'Six Things Non-Natural': A Note on the Origins and Fate of a Doctrine and a Phrase," *Clio Medica* 3 (1968): 337–47; Luis García Ballester, "On the Origin of the 'Six Things Non-Natural' in Galen," in *Galen und das hellenistische Erbe*, ed. Jutta Kollesch and Diethard Nickel, 105–15 (Stuttgart: Steiner, 1993); Saul Jarcho, "Galen's Six Non-Naturals: A Bibliographic Note and Translation," *Bulletin of the History of Medicine* 44 (1970): 372–76; Peter H. Niebyl, "The Non-Naturals," *Bulletin of the History of Medicine* 45 (1971): 486–92; Heikki Mikkeli, *Hygiene in the Early Modern Medical Tradition*, Annals of the Finnish Academy of Sciences and Letters, Humaniora, no. 305 (Helsinki: Academia Scientiarum Fennica, 1999), chap. 1–2; Guenter B. Risse, "In the Name of Hygieia and Hippocrates: A Quest for the Preservation of Health and Virtue," in idem, *New Medical Challenges during the Scottish Enlightenment* (Amsterdam: Rodopi, 2005), 135–69; David Gentilcore, *Food and Health in Early Modern Europe: Diet, Medicine, and Society, 1450–1800* (New York: Bloomsbury, 2016), 11–15; and Sandra Cavallo and Tessa Storey, "Conserving Health: The Non-Naturals in Early Modern Culture and Society," in *Conserving Health in Early Modern Culture: Bodies and Environments in Italy and England*, ed. Cavallo and Storey, 1–19 (Manchester: Manchester University Press, 2017).

3. Jerome J. Bylebyl, "The Medical Side of Harvey's Discovery: The Normal and the Abnormal," in *William Harvey and His Age: The Medical and Social Context of the Discovery of the Circulation*, ed. Jerome J. Bylebyl, 28–102, Supplement to *Bulletin of the History of Medicine*, new series, 2 (Baltimore: Johns Hopkins University Press, 1979), esp. 29–30.

4. Humphrey Brooke, *Ugieine: Or A Conservatory of Health* (London, 1650), 22.

5. Among very many eighteenth-century British medical presentations of the nonnaturals, see, e.g., John Burton, *A Treatise on the Non-Naturals; in Which the Great Influence They Have on Human Bodies Is Set Forth* (York: A. Staples, 1738); William Forster, *A Treatise on the Causes of Most Diseases Incident to Human Bodies, and the Cure of Them, First, by a Right Use of the Non-Naturals: Chiefly by Diet* (Leeds: James Lister, 1745); Richard Brookes, *The General Practice of Physic . . . to Which Is Prefixed . . . the Use of the Non-Naturals*, 2 vols., 2nd ed. (London: J. Newbery, [1751] 1754); Alexander Sutherland, *Attempts to Revive Antient Medical Doctrines . . . V. Of the Non-Naturals*, 2 vols. (London: A. Millar, 1763); Francis de Valangin, *A Treatise on Diet, or the Management of Human Life; by Physicians Called the Non-Naturals* (London: J. and W. Oliver, 1768); William Smith, *A Sure Guide in Sickness and Health . . . Directions How to Use the Non-Naturals for the Preservation of Health* (London: J. Bew, 1776); William Buchan, *Domestic Medicine* (various editions from 1769 into the nineteenth century); and A. G. Sinclair, *Artis Medicinæ Vera Explanatio . . . to Which Are Added Many . . . Remarks and Observations, on . . . the Non-Naturals* (London: J. Johnson, 1791). These are texts—Buchan's excepted—that belonged to academic medicine: the persistence of the nonnaturals in lay thought about the body and its functions is another matter. The reformulation of the nonnaturals in response to changes in scientific thinking during the mid to late-seventeenth century is treated in chapter 7, 254–57 (below).

6. Steven Shapin, "How to Eat Like a Gentleman: Dietetics and Ethics in Early Modern England," in *Right Living: An Anglo-American Tradition of Self-Help Medicine and Hygiene*, ed. Charles E. Rosenberg (Baltimore: Johns Hopkins University Press, 2003), 31–33.

7. Hippocrates, "Aphorisms," in Hippocrates, *Works*, trans. W. H. S. Jones (Cambridge, MA: Harvard University Press, 1931), 4:101. See also chapter 1, 54–57 (above), and chapter 5, 190–92 (below).

8. For excess as "an enemy to Nature," see Daniel Sennert, *The Institutions or Fundamentals of the Whole Art, Both of Physick and Chirurgery* (London, 1656), 227. For moderation in practical ethics, see Shapin, "How to Eat Like a Gentleman." See also Noga Arikha, "'Just Life in a Nutshell': Humours as Commmon Sense," *Philosophical Forum* 39 (2008): 303–14; Michel Jeanneret, *A Feast of Words: Banquets and Table Talk in the Renaissance*, trans. Jermey Whiteley and Emma Hughes (Chicago: University of Chicago Press, [1987] 1991), 73–77; and Sandra Cavallo and Tessa Storey, *Healthy Living in Late Renaissance Italy* (Oxford: Oxford University Press, 2013), chap. 1–2.

9. William Perkins, *The Whole Treatise of the Cases of Conscience* (Lon-

don, [1608] 1642), Book III, 323. See also David N. Harley, "Medical Metaphors in English Moral Theology, 1560–1660," *Journal of the History of Medicine and Allied Sciences* 48 (1993): 396–435, esp. 409–11.

10. Concentration here on airs and relative inattention to waters reflects both their representation in traditional dietetic writings and their place in recent scholarship. See, however, Joyce E. Chaplin, "Why Drink Water? Diet, Materialisms, and British Imperialism," *Osiris* 35 (2020): 99–122, esp. 102–10.

11. Robert Burton, *The Anatomy of Melancholy* (New York: NYRB Books, [1621] 2001), Part 1, 237.

12. Andrew Boorde, "A Compendyous Regyment of Helth," in *Andrew Boorde's Introduction and Dyetary*, ed. F. J. Furnivall, Early English Text Society, Extra Series, no. X (London: N. Trübner, [1542] 1870), 235. See also William Vaughan, *Approved Directions for Health, Both Naturall and Artificiall* (London, 1612), 2–8; and André Du Laurens, *A Discourse of the Preservation of Sight: Of Melancholicke Diseases; of Rheumes, and of Old Age*, 2nd ed. (London, [1594] 1599), 59–61.

13. John Archer, *Every Man His Own Doctor* (London, 1671), 15.

14. Everard Maynwaringe, *Tutela Sanitatis, Sive Vita Protracta: The Protection of Long Life, and Detection of Its Brevity* (London, 1663), 18; and idem, *The Method and Means of Enjoying Health, Vigour, and Long Life* (London, 1683), 41.

15. A. B., *The Sick-Mans Rare Jewel Wherein Is Discovered a Speedy Way How Every Man May Recover Lost Health, and Prolong Life* (London, 1674), 23. See also Thomas Cogan, *The Haven of Health* (London, [1584] 1630), "To the Reader," sig. B5; William Bullein, *The Government of Health* (London, [1562] 1595), 29v–30r; and Jean Goeurot, *The Regiment of Life, Whereunto Is Added a Treatise of the Pestilence* (London, [1544] 1550), sig. Diiii.

16. Maynwaringe, *Tutela Sanitatis*, 18; and idem, *Method and Means*, 38.

17. Thomas Cock, *Hygieine, or, A Plain and Practical Discourse upon the First of the Six Non-Naturals, viz. Air* (London, 1665), 2.

18. For example, Annick Le Guérer, *Scent: The Mysterious and Essential Powers of Smell*, trans. Richard Miller (New York: Turtle Bay, 1992), 63–72; Carla Mazzio, "The History of Air: Hamlet and the Trouble with Instruments," *South Central Review* 26, nos. 1–2 (2009): 153–96, esp. 175–76; Sujata Iyengar, *Shakespeare's Medical Language: A Dictionary* (London: Continuum, 2011), 142, 249–52, 272; and Mark S. R. Jenner, "Follow Your Nose? Smell, Smelling, and Their Histories," *American Historical Review* 116 (2011): 335–52.

19. Archer, *Every Man His Own Doctor*, 19.

20. In eighteenth-century Paris, the artificial exhalations of innovative distilleries were pointed to as causes of what commentators thought

was increasing mental and moral disorders in urban populations. See E. C. Spary, *Eating the Enlightenment: Food and the Sciences in Paris, 1670–1760* (Chicago: University of Chicago Press, 2012), 158, 161, 172–73. See also the classic work of Alain Corbin, *The Foul and the Fragrant: French Social Imagination*, trans. Aubier Montaigne (Cambridge, MA: Harvard University Press, [1982] 1986).

21. William Buchan, *Domestic Medicine: or, A Treatise on the Prevention and Cure of Diseases*, 11th ed. (London: A. Strahan, T. Cadell, J. Balfour, W. Creech, 1790), 76–79.

22. Archer, *Every Man His Own Doctor*, 15, 21–22.

23. John Armstrong, *The Art of Preserving Health: A Poem* (London: A. Millar, 1744), 12.

24. Vaughan, *Approved Directions*, 2. See also [Michael Scot], *The Philosophers Banquet*, 3rd ed. (London, 1633), 8–11; and Joshua Scodel, *Excess and the Mean in Early Modern English Literature* (Princeton, NJ: Princeton University Press, 2002), chap. 1 and 3. For northerly and easterly winds, see, for example, S. H., *The Preservation of Health, or a Dyet for the Healthfull Man* (London, 1624), 16, bound together with Sir John Harington, trans., *The English Mans Doctor, or, The Schoole of Salerne* (London, [trans. 1607] 1624; and Thomas Elyot, *The Castell of Health* (London, [1539] 1587), 12r.

25. Maynwaringe, *Method and Means*, 37, 44; and idem, *Tutela Sanitatis*, 18.

26. Archer, *Every Man His Own Doctor*, 15, 21–22. See also Bullein, *Government of Health*, 29v–30r.

27. John Evelyn, *Fumifugium, or, The Inconveniencie of the Aer and Smoak of London Dissipated Together with Some Remedies Humbly Proposed* (London, 1661), 1, 25.

28. Maynwaringe, *Method and Means*, 145.

29. Boorde, "Compendyous Regyment of Helth," 247–48. For six or seven hours, see Buchan, *Domestic Medicine* (1790), 87. For a survey of early modern English dietary views on sleep, see Sasha Handley, *Sleep in Early Modern England* (New Haven, CT: Yale University Press, 2016), 18–38. For Renaissance medical counsel, see Karl H. Dannenfeldt, "Sleep: Theory and Practice in the Late Renaissance," *Journal of the History of Medicine* 41 (1986), 415–41.

30. Archer, *Every Man His Own Doctor*, 99.

31. James Cleland, *The Instruction of a Young Noble-Man* (Oxford, 1612), 213–14. See also Platina, *On Right Pleasure and Good Health*, trans. Mary Ella Milham (Tempe, AZ: Medieval & Renaissance Texts & Studies, [1470] 1998), 111–13; and Buchan, *Domestic Medicine* (1790), 87–88.

32. Boorde, "Compendyous Regyment of Helth," 247 (left before right);

A. B., *Sick-Mans Rare Jewel*, 26 (right before left); and Handley, *Sleep in Early Modern England*, 25–28.

33. Maynwarynge, *Method and Means*, 148–49. See also Thomas Tryon, *A Treatise of Dreams & Visions* (London, 1689), 49, 219–20, 292.

34. Boorde, "Compendyous Regyment of Helth," 248–49. For the dietetic virtues of devotions before bed and on rising, see Vaughan, *Approved Directions*, 147–50.

35. Maynwaringe, *Method and Means*, 146–47, 149; Boorde, "Compendyous Regyment of Helth," 248–49; and Sasha Handley, "Sleep-Piety and Healthy Sleep in Early Modern English Households," in *Conserving Health in Early Modern Culture: Bodies and Environments in Italy and England*, ed. Sandra Cavallo and Tessa Storey (Manchester: Manchester University Press, 2017), 195–99.

36. James I, King of England, *Basilikon Doron: Or His Maiesties Instructions to His Dearest Sonne, Henrie the Prince*, 2nd ed. (London, [1599] 1603), 108.

37. Obadiah Walker, *Of Education, Especially of Young Gentlemen* (London, 1673), 58. See also Elyot, *Castell of Health*, 83–88; Thomas Brugis, *The Marrow of Physicke* (London, 1648), 63; and William Drage, *A Physical Nosonomy, or, A New and True Description of the Law of God (Called Nature) in the Body of Man* (London, 1664), 332 (for strong exercise and the relief of constipation).

38. Walker, *Of Education*, 58–59, 68. For glandular obstructions and weak nerves in particular, see Buchan, *Domestic Medicine* (1790), 83.

39. Brooke, *Ugieine*, 143–46.

40. Anon., *A Treatise Concerning the Plague and the Pox* (London, 1652), 17; and A. M., *Queen Elizabeths Closset of Physical Secrets* (London, 1656), 17.

41. Cogan, *Haven of Health*, 3–4 (pro tennis); and Francis Bacon, *Historie Naturall and Experimetall, of Life and Death* (London, 1638), 221. See also [William Ramesey], *The Gentlemans Companion, or, A Character of True Nobility and Gentility* (London, 1672), 126.

42. Nicolaas Fonteyn, *The Womans Doctour, or, An Exact and Distinct Explanation of All Such Diseases as Are Peculiar to That Sex* (London, 1652), 196; and Anon., *Every Woman Her Own Midwife* (London, 1675), 17. (The relevant passages and their caution about opening up pores to infected air were evidently copied out of earlier texts, where they were not specifically addressed to women.) The link between "leaping" and miscarriage was considered Aristotelian: Aristotle, *The Problemes of Aristotle* (London, 1595), sig. F3.

43. Vaughan, *Approved Directions*, 65–66.

44. Cogan, *Haven of Health*, 7–8.

45. For recruitment of the proverb and for recognition of contrary senti-

ments, see, for example, Cogan, *Haven of Health*, 215 (also 9). For approval of walking after dinner, see Harington, *English Mans Doctor*, 1.

46. Vaughan, *Approved Directions*, 66.

47. Roger Ascham, *Toxophilus, the Schole of Shootinge* (London, 1545), 14. For approval of tennis, see Cogan, *Haven of Health*, 4. "Coyting" (or "quoiting") was probably a form of what is now called horseshoes, throwing rings of iron, rope, or rubber at pegs in the ground.

48. Santorio Santorio, *Medicina Statica: Being the Aphorisms of Sanctorius*, 2nd ed. (London: W. and J. Newton, 1720), 270 (for jumping); and Walker, *Of Education*, 68–69 (for a law of nature and elaborate steps).

49. James I, *Basilikon Doron*, 120–21. See also Cleland, *Instruction of a Young Noble-Man*, 217–18; Brooke, *Ugieine*, 168. For "leaping" in the specific context of the formation of gentle youth and in military training, see Thomas Elyot, *The Boke Named the Governour* (London, 1537), 59; Jean Gailhard, *The Compleat Gentleman: Or Directions for the Education of Youth* (London, 1678), 51; and Edward Cooke, *The Character of Warre, or the Image of Martial Discipline* (London, 1626), chap. 6–8 (on running, leaping, and vaulting). Even future soldiers might, however, possess constitutions that enforced moderation in such exercises. See, e.g., Santorio, *Medicina Statica* (1720), 269. (It is not clear exactly what sort of exercise constituted leaping, possibly some activity related to mounting a horse and possibly a version of what is now called leapfrog. The "Ten Lords a-leaping" in the Christmas carol were presumably engaging in just this type of exercise.) More puritanically inclined practical ethical writers condemned a genteel life given over to "Hawking, Hunting, Drinking, Ranting, &c. which are the sole exercises almost of many of our Gentry, in which they are too immoderate." Ramesey, *Gentlemans Companion*, 122.

50. Baldesar Castiglione, *The Book of the Courtier*, ed. Daniel Javitch, trans. Charles Singleton (New York: Norton, [1528] 2002), 73–74.

51. James I, *Basilikon Doron*, 120–21.

52. Brooke, *Ugieine*, 182–83. See also Archer, *Every Man His Own Doctor*, 101.

53. Cogan, *Haven of Health*, 279. For concoctions and their appropriate "excrements," see Karine van 't Land, "Sperm and Blood, Form and Food: Late Medieval Medical Notions of Male and Female in the Embryology of *Membra*," 374–75, 382–85; Barbara Orland, "White Blood and Red Milk: Analogical Reasoning in Medical Practice and Experimental Physiology (1560–1730)," 448–49, 461–65; and Michael Stolberg, "Sweat, Learned Concepts and Popular Perceptions, 1500–1800," 504–7, all in *Blood, Sweat and Tears: The Changing Concepts of Physiology from Antiquity into Early Modern Europe*, ed. Manfred Horstmanshoff, Helen King, and Claus Zittel (Leiden: Brill, 2012).

54. Buchan, *Domestic Medicine* (1790), 121–22. See also Brooke, *Ugieine*, 189–90.

55. Maynwaringe, *Tutela Sanitatis*, 38. See also Cavallo and Storey, *Healthy Living*, chap. 8.

56. Archer, *Every Man His Own Doctor*, 102. See also Brooke, *Ugieine*, 193.

57. Archer, *Every Man His Own Doctor*, 101.

58. Buchan, *Domestic Medicine* (1790), 259, 411, 420, 426.

59. John Locke, *Some Thoughts Concerning Education* (London, 1693), 25–30.

60. Buchan, *Domestic Medicine* (1790), 312.

61. [Jean Feyens], *A New and Needful Treatise of Wind Offending Mans Body*, trans. William Rowland (London, 1676), 18–20, 33–51.

62. For example, Buchan, *Domestic Medicine* (1790), 443. For general comments on wind and diet, see, e.g., [Feyens], *New and Needful Treatise*, 5–6, 21–29.

63. Harington, *English Mans Doctor*, 2.

64. Benjamin Franklin, "From Benjamin Franklin to the Royal Academy of Brussels, [after 19 May 1780]," Founders Online, 1780, https://founders.archives.gov/documents/Franklin/01-32-02-0281.

65. Buchan, *Domestic Medicine* (1790), 443–46.

66. By the early eighteenth century, notably in France but also in Britain, a relatively new disease, now called "the vapors," emerged and eventually reached epidemic proportions. Its dietary causes and mental and moral effects were matters of major medical attention and are treated in chapter 4, 159–67 (below). But see Michael Stolberg, *Experiencing Illness and the Sick Body in Early Modern Europe* (New York: Palgrave Macmillan, 2011), 164–70. See also Richard Blackmore, *A Treatise of the Spleen and Vapours* (London: J. Pemberton, 1725); Nicholas Robinson, *A New System of the Spleen, Vapors, and Hypochondriack Melancholy* (London: A. Bettesworth, 1729); and John Purcell, *A Treatise of Vapours; or, Hysterick Fits* (London: Nicholas Cox, 1702).

67. Brooke, *Ugieine*, 210–15. See also Stolberg, "Sweat, Learned Concepts and Popular Perceptions," 511–12, 516–18.

68. Brooke, *Ugieine*, 217–19.

69. For example, Sara Read, *Menstruation and the Female Body in Early Modern England* (London: Palgrave Macmillan, 2013); and Cathy McClive, *Menstruation and Procreation in Early Modern France* (New York: Routledge, 2016).

70. Peter H. Niebyl, "The English Bloodletting Revolution, or Modern Medicine before 1850," *Bulletin of the History of Medicine* 51 (1977): 464–83. For Galenic sources, see Peter Brain, *Galen on Bloodletting: A Study of the*

Origins, Development and Validity of His Opinions (Cambridge: Cambridge University Press, 1986). For cross-cultural comparison, see Shigehisa Kuriyama, "Interpreting the History of Bloodletting," *Journal of the History of Medicine* 50 (1995): 11–46.

71. Brugis, *Marrow of Physicke*, 68.

72. Brooke, *Ugieine*, 184.

73. For example, Francis Bacon, *Historie of Life and Death*, 156–57 (for whether or not abstinence from venery did conduce to long life).

74. Vaughan, *Approved Directions*, 69. See also Maynwaringe, *Tutela Sanitatis*, 39.

75. Brooke, *Ugieine*, 186–87.

76. For example, Robert Bayfield, *Enchiridion Medicum: Containing the Causes, Signs, and Cures of All Those Diseases, That Do Chiefly Affect the Body of Man* (London, 1655), 47, 250, 284–85.

77. Archer, *Every Man His Own Doctor*, 102–3.

78. On expert medical treatment of female "seed" and fluid release during sex, see, e.g., Levinus Lemnius, *The Secret Miracles of Nature* (London, [1559] 1658), 17–18; Joanna B. Korda, Sue W. Goldstein, and Frank Sommer, "The History of Female Ejaculation," *Journal of Sexual Medicine* 7 (2010): 1965–1975; and Thomas W. Laqueur, *Making Sex: Body and Gender from the Greeks to Freud* (Cambridge, MA: Harvard University Press, 1990), 46–51, 99–100. Any statement about what is *absent from* a body of writings has to be qualified, and similar qualifications must apply to a distinction between the dietetic literature and expert medical texts aimed mainly at establishing anatomical and physiological facts. I have not seen dietetic texts treating female fluid release during sex along the same lines as male ejaculation, but it is possible that there may be some such mention in texts I have not seen.

79. For example, Brooke, *Ugieine*, 185–87; Bayfield, *Enchiridion Medicum*, 291; and Cogan, *Haven of Health*, 278.

80. Archer, *Every Man His Own Doctor*, 103. See also Brooke, *Ugieine*, 188.

81. Boorde, "Compendyous Regyment of Helth," 300–301.

82. Quoted in James Lees-Milne, *The Age of Inigo Jones* (London: Batsford, 1953), 103.

83. Archer, *Every Man His Own Doctor*, 103–4.

84. Théophile Bonet, *A Guide to the Practical Physician* (London, 1686), 694. See also Bradford Bouley, "Digesting Faith: Eating God, Man, and Meat in Seventeenth-Century Rome," *Osiris* 35 (2020): 55–56.

85. There are very many such indications in the dietary literature. See, for example, Boorde, "Compendyous Regyment of Helth," 279, 282–82; Archer, *Every Man His Own Doctor*, 47, 60–62; and Cogan, *Haven of Health*, 101, 123, 155. See also Allen J. Grieco, "Vegetable Diets, Hermits and

Melancholy in Late Medieval and Renaissance Italy," in idem, *Food, Social Politics and the Order of Nature in Renaissance Italy*, Villa I Tati Series, 34 (Milan: Officina Libraria, for Villa I Tati, 2019), 243–62.

86. Henricus Ronsovius, *De Valetudine Conservanda, or The Preservation of Health, or A Dyet for the Healthfull Man* (London, 1624), 33–35 (bound together with Harington, *English Mans Doctor*).

87. On the history of the emotions and aspects of their dietetic management, see Susan James, *Passion and Action: The Emotions in Seventeenth-Century Philosophy* (Oxford: Clarendon, 1997); Stephen Gaukroger, ed., *The Soft Underbelly of Reason: The Passions in the Seventeenth Century* (London: Routledge, 1998); Fay Bound Alberti, "Emotions in the Early Modern Medical Tradition," in idem, *Medicine, Emotion and Disease, 1700–1950* (London: Palgrave Macmillan, 2006), 1–21; and Naama Cohen-Hanegbi, "A Moving Soul: Emotions in Late Medieval Medicine," *Osiris* 31 (2016): 46–66.

88. See, e.g., Edward Reynolds, *A Treatise of the Passions and Faculties of the Soule of Man* (London, 1640), 48. See also Paster, *Humoring the Body*, 1–6; and James, *Passion and Action*, chap. 1.

89. Vaughan, *Approved Directions*, 110.

90. Thomas Tryon, *A New Art of Brewing Beer, Ale, and Other Sorts of Liquors* (London, 1690), 11.

91. For an exploration of the idea of comfort eating and the complex relations between humoral and cultural views of foods, see Rachel Winchcombe, "Comfort Eating: Food, Drink and Emotional Health in Early Modern England," *English Historical Review* 138, nos. 590–591 (February–April 2023): 61–91.

92. Temkin, *Hippocrates in a World of Pagans and Christians*, 13–14.

93. Burton, *Anatomy of Melancholy*, "Democritus Junior to the Reader," 37.

94. Maynwaringe, *Tutela Sanitatis*, 51.

95. Elyot, *Castell of Health*, 64r.

96. Archer, *Every Man His Own Doctor*, 97.

97. Maynwaringe, *Tutela Sanitatis*, 59–61.

98. Henry Peacham, *The Compleat Gentleman, Fashioning Him Absolute in the Most Necessary & Commendable Qualities* (London, 1622), 185.

99. Cleland, *Instruction of a Young Noble-Man*, 206.

100. Peacham, *Compleat Gentleman*, 185.

101. Richard Brathwait, *The English Gentleman, Containing Sundry Excellent Rules or Exquisite Observations, Tending to Direction of Every Gentleman, of Selecter Ranke and Qualitie* (London, 1630), 62–63, 305.

102. Richard Brathwait, *The English Gentlewoman* (London, 1631), 139.

103. Vaughan, *Approved Directions*, 89.

104. Desiderius Erasmus, *The Education of a Christian Prince*, trans. Lester K. Born (New York: Octagon, [1516] 1965), 160.

105. Thomas Wright, *The Passions of the Minde* (London, [1601] 1630), 82.

106. Burton, *Anatomy of Melancholy*, Part 2, 103–7; see also 119–21.

107. Burton, *Anatomy of Melancholy*, "Democritus Junior to the Reader," 20–21. See also Angus Gowland, *The Worlds of Renaissance Melancholy: Robert Burton in Context* (Cambridge: Cambridge University Press, 2006), esp. chap. 1; and Mary Ann Lund, *Melancholy, Medicine and Religion in Early Modern England* (Cambridge: Cambridge University Press, 2010), esp. chap. 3–4.

108. Boorde, "Compendyous Regyment of Helth," 235.

109. Maynwaringe, *Tutela Sanitatis*, 51–52.

110. Buchan, *Domestic Medicine* (1790), 117.

111. Wright, *Passions of the Minde*, 159–72 (for music). See also Penelope Gouk, "Music and Spirit in Early Modern Thought," in *Emotions and Health, 1200–1700*, ed. Elena Carrera, 221–39 (Leiden: Brill, 2013).

112. Wright, *Passions of the Minde*, 88–103, 183–93.

113. Vaughan, *Approved Directions*, 90–91, 99.

114. The best survey of these dietaries is in Ken Albala, *Eating Right in the Renaissance* (Berkeley: University of California Press, 2002), esp. chap. 1 and 3. See also Sandra Cavallo and Tessa Storey, "Regimens, Authors and Readers: Italy and England Compared," in *Conserving Health*, ed. Cavallo and Storey, 23–52; Heikki Mikkeli, *Hygiene in the Early Modern Medical Tradition*, Annals of the Finnish Academy of Sciences and Letters, Humaniora, no. 305 (Helsinki: Academia Scientiarum Fennica, 1999), chap. 3; Klaus Bergdolt, *Wellbeing: A Cultural History of Healthy Living*, trans. Jane Dewhurst (Cambridge: Polity, [1999] 2008), chap. 5; and Paul Slack, "Mirrors of Health and Treasures of Poor Men: The Uses of the Vernacular Medical Literature of Tudor England," in *Health, Medicine and Mortality in the Sixteenth Century*, ed. Charles Webster, 237–73 (Cambridge: Cambridge University Press, 1979).

115. Boorde, "Compendyous Regyment of Helth," 250–52.

116. Anon., *A Discourse Translated Out of Italian, That a Spare Diet Is Better Than a Splendid and Sumptuous: A Paradox*, bound together with Leonardus Lessius, *Hygiasticon: or, The Right Course of Preserving Life and Health into Extream Old Age*, 2nd ed., trans. T[imothy] S[mith][?] (Cambridge, [1614] 1634), 58.

117. For example, Ken Albala, "The Ideology of Fasting in the Reformation Era," in *Food and Faith in Christian Culture*, ed. Albala and Trudy Eden, 41–58 (New York: Columbia University Press, 2012).

118. Brathwait, *The English Gentleman*, 327.

119. Matthew Griffith, *Bethel: or, a Forme for Families* (London, 1634), 180–82.

120. For example, Steven Shapin, "Was Luigi Cornaro a Dietary Expert?," *Journal of the History of Medicine* 73 (2018): 135–49.

121. Burton, *Anatomy of Melancholy*, Part 2, 26. For gluttony, restraint, and civility in the early modern, see the fine treatment in Viktoria von Hoffmann, *From Gluttony to Enlightenment: The World of Taste in Early Modern Europe* (Urbana: University of Illinois Press, 2016), chap. 2.

122. Cogan, *Haven of Health*, 193.

123. Harington, *English Mans Doctor*, 6, 24.

124. Maynwaringe, *Tutela Sanitatis*, 22. On moderation and early modern gentlemanly culture, see Shapin, "How to Eat Like a Gentleman," 27–33.

125. Archer, *Every Man His Own Doctor*, 12.

126. [Thomas Gainsford], *The Rich Cabinet furnished with varieties of Excellent Discriptions, Exquisite Characters, Witty Discourses, and Delightfull Histories, Devine and Morrall* . . . (London, 1616), 143v.

127. For example, Thomas Tryon, *The Way to Health, Long Life and Happiness* (London, 1683), 46–48, 228–29; and idem, *Monthly Observations for the Preserving of Health* (London, 1688), 20–21.

128. Brooke, *Ugieine*, 111–12.

129. [Richard Allestree], *The Gentleman's Calling* (London, 1660), 85.

130. For "correction," cuisine, and taste, see Albala, *Eating Right in the Renaissance*, 6–7, 88–90, 98–99; Brian Cowan, "New Worlds, New Tastes: Food Fashions after the Renaissance," in *Food: The History of Taste*, ed. Paul Freedman, 196–231 (Berkeley: University of California Press, 2007), 230–32; and chapter 3, 125–26 (below).

131. Ambroise Paré, *The Workes of That Famous Chirurgion Ambrose Parey*, trans. Thomas Johnson (London, 1649), 25.

132. Jean Anthelme Brillat-Savarin, *The Physiology of Taste*, trans. Anne Drayton (London: Penguin, [1825] 1994), 50–56, 132–40 (quoted passages 132–33), though there were broadly similar approvals of variety by Adam Smith on grounds of political economy and the benefits of trade. For the categories of traditional medical dietetics and cuisine, see Albala, *Eating Right in the Renaissance*, esp. chap. 8. See also Massimo Montanari, *Medieval Tastes: Food, Cooking, and the Table*, trans. Beth Archer Brombert (New York: Columbia University Press, [2012] 2015), esp. chap. 3; Jean-Louis Flandrin, "Dietary Choices and Culinary Technique, 1500–1800," in *Food: A Culinary History*, ed. Flandrin and Massimo Montanari, 403–17 (New York: Penguin, 2000); and idem, "From Dietetics to Gastronomy: The Liberation of the Gourmet," in *Food: A Culinary History*, 418–34.

133. Diogenes Laërtius, *The Lives, Opinions, and Remarkable Sayings of the Most Famous Ancient Philosophers*, 2 vols. (London, 1688), 1:115, 121. See also Steven Shapin, "The Philosopher and the Chicken: On the Dietetics of Disembodied Knowledge," in *Science Incarnated: Historical Embodiments of Natural Knowledge*, ed. Christopher Lawrence and Shapin, 21–50 (Chicago: University of Chicago Press, 1998).

134. This sensibility is associated with the work of the historical sociologist Norbert Elias, *The Civilizing Process: Sociogenetic and Psychogenetic Investigations*, Vol. 1, *The History of Manners*, trans. Edmund Jephcott (Oxford: Blackwell, [1939] 1978). See also Stephen Mennell, *All Manners of Food: Eating and Taste in England and France from the Middle Ages to the Present* (Oxford: Blackwell, 1985); and Montanari, *Medieval Tastes*, chap. 17.

135. Castiglione, *Book of the Courtier*, 99; see also 102.

136. James I, *Basilikon Doron*, 107.

137. Burton, *Anatomy of Melancholy*, Part 1, 226–27. Burton was here quoting Livy and Agrippa while criticizing the gluttonous tendencies of his own time. For a concise survey of Elizabethan attitudes about gluttony, see Joan Fitzpatrick, *Food in Shakespeare: Early Modern Dietaries and the Plays* (Aldershot: Ashgate, 2007), 12–23.

138. Brathwait, *English Gentleman*, 305, 311. A brilliant study of the early modern career of moderation in political and religious settings briefly treats dietary moderation: Ethan H. Shagan, *The Rule of Moderation: Violence, Religion and the Politics of Restraint in Early Modern England* (Cambridge: Cambridge University Press, 2011), 53–56.

139. Lessius, *Hygiasticon*, sig. A3v; see also 11–12; for equation of sobriety and temperance, see 15.

140. Burton, *Anatomy of Melancholy*, Part 2, 27.

141. Richard Baxter, *A Christian Directory, or, a Summ of Practical Theologie and Cases of Conscience . . .* (London, 1673), 374. See also Joan P. Alcock, "Gluttony: The Fifth Deadly Sin," in *The Fat of the Land: Proceedings of the Oxford Symposium on Food and Cookery 2002* (Bristol: Footwork, 2003), 11–21.

142. Philippians 3:19. See also Jeanneret, *Feast of Words*, 79–82; and Susanne Scholz, *Body Narratives: Writing the Nation and Fashioning the Subject in Early Modern England* (New York: St. Martin's, 2000), 26–28.

143. For example, Gregorius Magnus (Pope Gregory I, ca. 540–604 CE), quoted by St. Thomas Aquinas: "With eating, pleasure is mixed with necessity; we are not able to discern what is required by necessity and what is claimed by pleasure." This passage is quoted in Luca Vercelloni, *The Invention of Taste: A Cultural Account of Desire, Delight and Disgust in Fashion, Food and Art*, trans. Kate Singleton (London: Bloomsbury, [2005] 2016), 28.

144. For example, Jack Hartnell, *Medieval Bodies: Life, Death and Art in the Middle Ages* (London: Profile, 2018), 204–5.

145. C. M. Woolgar, *The Senses in Late Medieval England* (New Haven, CT: Yale University Press, 2006), 108, 111–12. See also James N. Davidson, *Courtesans and Fishcakes: The Consuming Passions of Classical Athens* (Chicago: University of Chicago Press, 1997), chap. 1.

146. Ramesey, *Gentlemans Companion*, 118: "Eat not till thou hast an appetite; and then, eat not till thou hast none."

147. Locke, *Some Thoughts Concerning Education*, 12–14.

148. See, for example, Fitzpatrick, *Food in Shakespeare*, 15.

149. Thomas Fuller, *Collected Sermons, D.D. 1631–1659*, Vol. 1, ed. John Eglington Bailey and William E. A. Axon (London: Gresham Press, 1891), 200; Louis Lémery, *A Treatise of All Sorts of Food*, trans. D. Hay (London: T. Osborne, [1702] 1745), 10–11; John Harris, *The Divine Physician: Prescribing Rules for the Prevention, and Cure of Most Diseases, as Well of the Body, as the Soul* (London, 1676), 22; and Richard Stock, *The Doctrine and Use of Repentance* (London, 1610), 317: "fewer die of the sword, then of surfetting by a plentie, and full diet."

150. Archer, *Every Man His Own Doctor*, sig. A3r.

151. Harley, "Medical Metaphors in English Moral Theology," 411.

152. For example, Sylvester Graham, *Lectures on the Science of Human Life*, 2 vols. (Boston: Marsh, Capen, Lyon, and Webb, 1839). The exciting nature of certain foods and condiments has been a continuing theme in dietary writing. See, for example, the sentiments of the Seventh-day Adventists and the nutritional principles of the American cereal manufacturer John Harvey Kellogg (1852–1943) in J. H. Kellogg, *The New Dietetics: What to Eat and How; A Guide to Scientific Feeding in Health and Disease* (Battle Creek, MI: Modern Medicine Publishing, 1921).

153. The original source was the Roman comedian Terence in the 2nd century CE, and the saying became proverbial in the early modern period. A mid-sixteenth-century translation of Erasmus's *Proverbs* gave the Latin tag as "Without meate and drinke the lust of the body is colde"; alternatively, "The beste way to tame carnall lust, is to kepe abstinence of meates and drinkes"; and "A licorouse [licentious] mouth a licourouse taile." Desiderius Erasmus, *Proverbs or Adages*, ed. and trans. Richard Taverner (London, 1569), 34v. See also Juliann Vitullo, "Taste and Temptation in Early Modern Italy," *Senses & Society* 5 (2010): 106–18, esp. 107–9, 113.

154. Geoffrey Chaucer, "The Wife of Bath's Prologue," in *The Works of Our Ancient, Learned, & Excellent Poet, Jeffrey Chaucer* . . . (London, 1687), 66, quoted in Burton, *Anatomy of Melancholy*, Part 3, 63.

155. François Rabelais, *Gargantua and Pantagruel*, trans. J. M. Cohen (London: Penguin, 1955), 373. See also Baxter, *Christian Directory*, 374.

156. Cleland, *Instruction of a Young Noble-Man*, 209.

157. Wright, *Passions of the Minde*, 128–29.

158. Nathaniel Wanley, *The Wonder of the Little World: Or, a General History of Man* (London, 1678), 178–79.

159. Some attribute the saying to Aristotle, while others attribute it

to Ovid, Plutarch, or Augustine. Certainly, by the Renaissance it was a commonplace, though mainly invoked in moral commentary, politics, and law. Donald R. Kelley, "'Second Nature': The Idea of Custom in European Law, Society, and Culture," in *The Transmission of Culture in Early Modern Europe*, ed. Anthony Grafton and Ann Blair, 131–72 (Philadelphia: University of Pennsylvania Press, 1990). The medical career of the saying and its uses and meanings is treated in Steven Shapin, "Why Was 'Custom a Second Nature' in Early Modern Medicine?," *Bulletin of the History of Medicine* 93 (2019): 1–26.

160. Brooke, *Ugieine: or A Conservatory of Health*, 34–35.

161. Michael Schoenfeldt, *Bodies and Selves in Early Modern England: Physiology and Inwardness in Spenser, Shakespeare, Herbert, and Milton* (Cambridge: Cambridge University Press, 1999), 11–13. See also Sorana Corneanu, *Regimens of the Mind: Boyle, Locke, and the Early Modern Cultura Animi Tradition* (Chicago: University of Chicago Press, 2012), 75–78; and Phil Withington, "Addiction, Intoxicants, and the Humoral Body," *Historical Journal* 65 (2022): 68–90, esp. 72–74, 82–83.

162. Paré, *Workes*, 24.

163. Thomas Muffett, *Healths Improvement: or, Rules Comprizing and Discovering the Nature, Method, and Manner of Preparing All Sorts of Foods* (London, [comp. ca. 1600] 1655), 285. See also Tryon, *A New Art of Brewing*, 11.

164. John Arbuthnot, *An Essay Concerning the Nature of Aliments, and the Choice of Them, according to the Different Constitutions of Human Bodies . . .*, 4th ed. (London: J. and R. Tonson, [1731] 1756), 196.

165. Buchan, *Domestic Medicine* (1790), 62.

166. Jacques Jouanna, "Politics and Medicine: The Problem of Change in *Regimen in Acute Diseases* and Thucydides (Book 6)," in idem, *Greek Medicine from Hippocrates to Galen: Selected Papers*, trans. Neil Allies, ed. Philip van der Eijk, 21–38 (Leiden: Brill, 2012).

167. Harington, *English Mans Doctor*, 16.

168. Rabelais, *Gargantua and Pantagruel*, 86–87.

169. Paré, *Workes*, 24. See also Lessius, *Hygiasticon*, 42–44; and Francis Bacon, "Of Regiment of Health," in idem, *The Essayes or Counsels, Civill and Morall*, ed. Michael Kiernan (Cambridge, MA: Harvard University Press, [1597] 1985), 100.

170. Michel Eyquem de Montaigne, "Of Experience," in idem, *The Essayes of Michael Lord of Montaigne*, trans. John Florio, 3 vols. (London: J. M. Dent, [1580–1588] 1910), 3:340.

171. Montaigne, "How One Ought to Governe His Will," in idem, *Essayes*, 3:261.

172. Lessius, *Hygiasticon*, 43–44.

173. Herman Boerhaave, *Dr. Boerhaave's Academical Lectures on the Theory of Physic*, 6 vols. (London: W. Innys, 1742–1746), 6:240–41.

174. Laurent Joubert, *Popular Errors*, trans. Gregory David de Rocher (Tuscaloosa: University of Alabama Press, [1579] 1989), 247–68; idem, *The Second Part of the Popular Errors*, trans. Gregory David de Rocher (Tuscaloosa: University of Alabama Press, [1587] 1995), 127–72; and Alice Leonard and Sarah E. Parker, "'Put a Mark on the Errors': Seventeenth-Century Medicine and Science," *History of Science* 61, no. 3 (2023): 287–307. See also chapter 5, 198–200 (below).

175. For example, Hippocrates, "Aphorisms," 98–99, 140–41, 108–9, 120–21, 102–3.

176. Paul Oskar Kristeller, "The School of Salerno," *Bulletin of the History of Medicine* 17 (1945): 138–94; Owsei Temkin, "Nutrition from Classical Antiquity to the Baroque," in idem, *'On Second Thought' and Other Essays in the History of Medicine and Science* (Baltimore: Johns Hopkins University Press, 2002),182–83; and Melitta Weiss Adamson, *Medieval Dietetics: Food and Drink in Regimen Sanitatis Literature from 800 to 1400* (Frankfurt-am-Main: Peter Lang, 1995), 97–102.

177. D. H. Craig, *Sir John Harington* (Boston: Twayne, 1985).

178. J. C. Drummond and Anne Wilbraham, *The Englishman's Food: Five Centuries of English Diet* (London: Pimlico, [1939] 1991), 67–69.

179. For example, "Early to bed and early to rise makes a man healthy, wealthy, and wise" and "Hunger is the best sauce."

180. For orality and forms of knowledge, see, e.g., Walter J. Ong, *Orality and Literacy: The Technologizing of the Word* (New York: Methuen, 1982); Adam Fox, *Oral and Literate Culture in England, 1500–1700* (Oxford: Clarendon, 2000); and Steven Shapin, "Proverbial Economies: How and Understanding of Some Linguistic and Social Features of Common Sense Can Throw Light on More Prestigious Bodies of Knowledge, Science for Example," *Social Studies of Science* 31 (2001): 731–69.

181. Harington, *English Mans Doctor*, 1–2, 6, 16, 33.

Chapter 3

1. Levinus Lemnius, *The Touchstone of Complexions*, trans. Thomas Newton (London, [1561] 1633), 9.

2. Sir John Harington, *The English Mans Doctor, or, The Schoole of Salerne* (London, [trans. 1607] 1624), 35–39. Listing and characterizing the temperaments was a feature of practically every dietetic text and many ethical works about human natures and dispositions. See, e.g., Thomas Tryon, *The Way to Health, Long Life and Happiness* (London, 1683), 14–38; and John Archer, *Every Man His Own Doctor* (London, 1671), 4–11.

3. Thomas Elyot, *The Castell of Health* (London, [1539] 1587), 2–4. For dreams as diagnostic of temperament and therefore of medical interest, see Thomas Wright, *The Passions of the Minde* (London, [1601] 1630), 65; and Thomas Tryon, *A Treatise of Dreams & Visions* (London, 1689), esp. 6.

4. Lemnius, *Touchstone of Complexions*, 145.

5. Erwin Panofsky, *The Life and Art of Albrecht Dürer* (Princeton, NJ: Princeton University Press, 1955), 168.

6. The classic study of Dürer's *Melencolia I* and its cultural setting and historical career is Raymond Klibansky, Erwin Panofsky, and Fritz Saxl, *Saturn and Melancholy: Studies in the History of Natural Philosophy, Religion, and Art* (London: Thomas Nelson, 1964).

7. Henry Peacham, *Minerva Britanna, or a Garden of Heroical Devises* (London, 1612), 126–29. The "Pestane rose" is an allusion to the extraordinarily fragrant damask roses of ancient Paestum, near Salerno. The geometric cube under Melancholy's foot may be a nod to the earlier Dürer image, or it may be a more generally recognized emblem of the scholarly life. See also Elena Carrera, "Anger and the Mind-Body Connection in Late Medieval and Early Modern Medicine," 95–146 (esp. 102–5), and Angus Gowland, "Medicine, Psychology, and the Melancholic Subject in the Renaissance," (185–220), both in *Emotions and Health, 1200–1700*, ed. Carrera (Leiden: Brill, 2013).

8. Aristotle, *Prior Analytics*, Book II, 70b.7–10, in Aristotle, *Complete Works*, 2 vols., ed. Jonathan Barnes (Princeton, NJ: Princeton University Press, 1984), 1:113.

9. [Pseudo-]Aristotle, *Physiognomonica*, ed. and trans. Sabine Vogt (Berlin: Akademie Verlag, 1999).

10. Giambattista della Porta, *De humana physiognomonia libri IIII* (Vici Equense, 1586).

11. James Pilkington, *The Works of James Pilkington, B.D., Lord Bishop of Durham*, ed. James Schofield (Cambridge: University Press, [1575] 1842), 292.

12. Sir Thomas Browne, *Christian Morals* (Cambridge: University Press, [comp. ca. 1650] 1716), 59–60.

13. George Orwell, "Extracts from a Manuscript Notebook," in *The Collected Essays, Journalism and Letters of George Orwell*, ed. Sonia Orwell and Ian Angus, 4 vols. (London: Secker & Warburg, 1968), 4:515 (entry for April 17, 1949).

14. For comparative attentiveness to surface signs and inner states, see Shigehisa Kuriyama, *The Expressiveness of the Body and the Divergence of Greek and Chinese Medicine* (New York: Zone, 2002), chap. 3–4.

15. Steven Shapin, "Why Was 'Custom a Second Nature' in Early Modern Medicine?," *Bulletin of the History of Medicine* 93 (2019): 17–18.

16. John Donne, “To a Lady of a Dark Complexion,” in *Poems of John Donne*, 2 vols. ed. E. K. Chambers (London: Scribner, 1896), 2:267–68.

17. John Donne, “To a Painted Lady,” in Donne, *Poems*, 2:261–62.

18. Charles Le Brun, *A Method to Learn to Design the Passions, Proposed in a Conference on Their General and Particular Expression*, trans. John Williams (London: J. Huggonson, [1698] 1734), 12, and see 20–22 for the eyebrows and the eyes themselves as the privileged facial sites for presenting the passions.

19. Charles Gildon, *The Life of Mr. Thomas Betterton, the Late Eminent Tragedian* (London: Robert Jeeb, 1710), 37. See also Joseph R. Roach, *The Player's Passion: Studies in the Science of Acting* (Ann Arbor: University of Michigan Press, 1993), 66–68.

20. Johann Caspar Lavater, *Essays on Physiognomy*, 8th ed., trans. Thomas Holcroft (London: William Tegg, [1789] 1853), 13–16. The categories of humors and temperaments were not, however, notable features of Lavater's system.

21. George Combe, *A System of Phrenology* (Edinburgh: John Anderson, 1830), 32–33. See also Nicholas Morgan, *Phrenology, and How to Use It in Analyzing Character* (London: Longmans, Green, 1871), 73–74 (black bile “had only a fancied existence, and a belief in it has been given up”).

22. There are very many sources for the food qualities recounted in this and the next several paragraphs. See, for example, Andrew Boorde, “A Compendyous Regyment of Helth,” in *Andrew Boorde's Introduction and Dyetary*, ed. F. J. Furnivall, Early English Text Society, Extra Series, no. X (London: N. Trübner, [1542] 1870); Guglielmo Gratarolo, *A Direction for the Health of Magistrates and Studentes*, trans. T. N. (London, 1574); Thomas Cogan, *The Haven of Health* (London, [1584] 1630); Thomas Muffett, *Healths Improvement: or, Rules Comprizing and Discovering the Nature, Method, and Manner of Preparing All Sorts of Foods* (London, [ca. comp. ca. 1600] 1655); and Tobias Venner, *Via recta ad vitam longam* (London, 1620). On the notion of gradus, or qualitative degrees, see, e.g., Melitta Weiss Adamson, *Food in Medieval Times* (Westport, CT: Greenwood Press, 2004), 207–11.

23. Thomas Tryon, *Friendly Advice to the Gentlemen-Planters of the East and West Indies* (London, 1684), 24–25; and Venner, *Via recta ad vitam longam*, 141–42. On the potato, see Rebecca Earle, *Feeding the People: The Politics of the Potato* (New York: Cambridge University Press, 2020); idem, “Potatoes and the Hispanic Enlightenment,” *The Americas* 75 (2017): 639–60, esp. 639–41; and Sara Pennell, “Recipes and Reception: Tracking ‘New World’ Foodstuffs in Early Modern British Culinary Texts, c. 1650–1750,” *Food & History* 7 (2009): 11–34. The tomato—belonging to the nightshade family—had a more bumpy path to acceptance. See David Gentilcore, *Pomodoro! A History of the Tomato in Italy* (New York: Columbia University

Press, 2010); and Alfred W. Crosby Jr., *The Columbian Exchange: Biological and Cultural Consequences of 1492* (Westport, CT: Greenwood, 1972), chap. 3 and 5. For treatment of maize, see chapter 4, 143–45 (below).

24. Cogan, *Haven of Health*, 118; and Henry Butts, *Dyets Dry Dinner Consisting of Eight Severall Courses* (London, 1599), sig. C5.

25. Cogan, *Haven of Health*, 65, 67; and Butts, *Dyets Dry Dinner*, sig. B5, H4. See also Venner, *Via recta vitam longam*, 142. Adamson (*Food in Medieval Times*, 209–10) notes some disagreement about food qualities among medieval dietary writers—as there continued to be in the early modern period—but says that differences usually extended no further than one degree. The fourth degree and many of those in the third degree "almost all belong to a group that can be described as seasonings, herbs, spices, with the occasional fruit or vegetable thrown in."

26. Robert Burton, *The Anatomy of Melancholy* (New York: NYRB Books, [1621] 2001), Part 1, 217. For drying, see also William Bullein, *The Government of Health* (London, [1562] 1595), 60v–61r; and Archer, *Every Man His Own Doctor*, 27.

27. For example, William Vaughan, *Approved Directions for Health, Both Naturall and Artificiall* (London, 1612), 33–34; Bullein, *Government of Health*, 61v; and Venner, *Via recta vitam longam*, 49–50.

28. Archer, *Every Man His Own Doctor*, 40.

29. For example, William Turner, *A New Boke of the Natures and Properties of All Wines That Are Commonly Used Here in England* (London, 1568), esp. sig. Bii–Biii. For the history of describing wines in terms of their sensory qualities, see Steven Shapin, "The Tastes of Wine: Towards a Cultural History," *Rivista di Estetica*, new series, 51 (2012): 49–94.

30. James Howell, *Epistolæ Ho-Elianæ: The Familiar Letters of James Howell*, 4 vols. (Boston: Houghton, Mifflin, [1645–1655] 1907), 3:119–20 (letter written to Henry, Lord Clifford, October 17, 1634).

31. Archer, *Every Man His Own Doctor*, 85–86; and Cogan, *Haven of Health*, 210.

32. Cogan, *Haven of Health*, 238. See also Boorde, "Compendyous Regyment of Helth," 254–55.

33. See, e.g., Allen J. Grieco, "Medieval and Renaissance Wines: Taste, Dietary Theory, and How to Choose the 'Right' Wine (14th–16th Centuries)," *Mediaevalia* 30 (2009): 15–42, esp. 28–29.

34. From the Middle Ages through the eighteenth century, there was much scholarly attention to what were called "the internal senses," and these sometimes were the object of medical management. The internal senses, in these usages, referred to such capacities as imagination and memory, not to the sensations of internal bodily states. See Andreas Rydberg, "Michael Alberti and the Medical Therapy of the Internal Senses," *Journal*

of the History of Medicine 74 (2019): 245–66; and Katharine Park, “The Organic Soul,” in *The Cambridge of Renaissance Philosophy*, ed. Charles B. Schmitt et al., 464–84 (Cambridge: Cambridge University Press, 1988).

35. Etymologically, concoction is boiling + together. For a sound source for thought about digestion and digestibility in traditional dietetics, see Ken Albala, *Eating Right in the Renaissance* (Berkeley: University of California Press, 2002), chap. 2, esp. 56–58 (for distinctions between the stages of concoction). See also idem, “Food for Healing: Convalescent Cookery in the Early Modern Era,” *Studies in History and Philosophy of Biological and Biomedical Sciences* 43 (2012): 323–28, esp. 323–24.

36. Thomas Stanley, *The History of Philosophy, the Second Volume* (London, 1656), 66, quoting Aristotle, “Meteorology,” 379b11–381b21, in Aristotle, *Complete Works*, 2 vols., ed. Jonathan Barnes (Princeton, NJ: Princeton University Press, 1984), 1:609–12.

37. Francis Quarles, “On the Body of Man,” in idem, *Divine Fancies Digested into Epigrammes, Meditations, and Observations* (London, 1633), 22–23. See also Burton, *Anatomy of Melancholy*, Part 1, 151.

38. Timothie Bright, *A Treatise of Melancholy* (London, [1586] 1613), 4–5.

39. See Burton, *Anatomy of Melancholy*, Part 1, 156; Jean Fernel, *The Physiologia of Jean Fernel (1567)*, trans. John M. Forrester (Philadelphia: American Philosophical Society, 2003), 403–11; and Walter Charleton, *Natural History of Nutrition, Life, and Voluntary Motion* (London, 1659), 11–32.

40. See, for example, concise accounts in Lawrence Babb, *The Elizabethan Malady: A Study of Melancholia in English Literature from 1580 to 1642* (East Lansing: Michigan State University Press, 1951), 5–6, 8–9; and Jack Hartnell, *Medieval Bodies: Life, Death and Art in the Middle Ages* (London: Profile, 2018), 205–6. The term “chyme” seems to have entered the English language as a medical category in the 1590s, with “chyle” appearing in the 1650s, the latter penetrating the literate vernacular in the eighteenth century. See, notably, William Buchan, *Domestic Medicine: or, A Treatise on the Prevention and Cure of Diseases*, 11th ed. (London: A. Strahan, T. Cadell, J. Balfour, W. Creech, 1790), 73, 117, 182, 455, 709; George Cheyne, *The English Malady: or, a Treatise of Nervous Diseases of All Kinds, as Spleen, Vapours, Lowness of Spirits, Hypochondriacal, and Hysterical Complaints* (London: G. Strahan, 1733), 135, 159, 188, 301–3, 337, 342; and idem, *Essay of Health and Long Life*, 3rd ed. (London: George Strahan, [1724] 1725), 27, 35, 67–69, 79–80, 110–41. Seventeenth-century “mechanical” frameworks for understanding the transformation of food into the stuff of the human body are nicely treated in Justin E. H. Smith, “Diet, Embodiment, and Virtue in the Mechanical Philosophy,” *Studies in History and Philosophy of*

Biological and Biomedical Sciences 43 (2012): 338–48, esp. 340–41. See also chapter 6, 232–36 (below).

41. For example, Archer, *Every Man His Own Doctor*, 29, 33–34, 37, 44, 52, 58–59, 64–65; and Elyot, *Castell of Health*, 29r, 30v.

42. Elyot, *Castell of Health*, 76. See also Schoenfeldt, "Fables of the Belly," 245.

43. Venner, *Via recta vitam longam*, 61, 63, 72, 75–76, 80–81, 92, 138–41, 155; Archer, *Every Man His Own Doctor*, 27, 31, 33, 35, 44, 46, 52, 60–61, 64; and Harington, *English Mans Doctor*, 13. See also Anita Guerrini, "A Natural History of the Kitchen," *Osiris* 35 (2020): 37.

44. Archer, *Every Man His Own Doctor*, 25.

45. Venner, *Via recta vitam longam*, 56, 61, 76, 82; and Archer, *Every Man His Own Doctor*, 25, 27, 29, 31, 42, 46, 52.

46. Venner, *Via recta vitam longam*, 138; Butts, *Dyets Dry Dinner*, sig. D4. See also Albala, *Eating Right in the Renaissance*, 88–91; and Steven Shapin, *Changing Tastes: How Foods Tasted in the Early Modern Period and How They Taste Now*, The Hans Rausing Lecture 2011, Salvia Småskrifter, No. 14 (Uppsala: Tryck Wikströms, for the University of Uppsala, 2011), 18–19.

47. Archer, *Every Man His Own Doctor*, 25.

48. Cogan, *Haven of Health*, 150; and Venner, *Via recta vitam longam*, 61. See also Archer, *Every Man His Own Doctor*, 35, 43; and Sophie Gee, *Making Waste: Leftovers and the Eighteenth-Century Imagination* (Princeton, NJ: Princeton University Press, 2010), 64–65.

49. Among very many examples of these descriptions, see, e.g., Archer, *Every Man His Own Doctor*, 30, 36, 60, 63, 73–74; Venner, *Via recta vitam longam*, 27; and Thomas Brugis, *The Marrow of Physicke* (London, 1648), 62.

50. Cogan, *Haven of Health*, "Epistle Dedicatorie," sig. A4 and 24. Here "blood" usually designated the fluid coursing through your veins and arteries. This was generally understood to include the humor called "blood," while the actual fluid in your vessels contained all the four humors. See Philip Lyndon Reynolds, *Food and the Body: Some Peculiar Questions in High Medieval Theology* (Leiden: Brill, 1999), esp. 109.

51. Guglielmo Gratarolo, *A Direction for the Health of Magistrates and Studentes*, trans. Thomas Newton (London, 1574), sig. I ii, sig. H ii. See also Venner, *Via recta vitam longam*, 54–61; [Scot], *The Philosophers Banquet*, 33; and Elyot, *Castell of Health*, 12–13.

52. There are many sources for the attributions in this paragraph. See, e.g., Archer, *Every Man His Own Doctor*, 28, 33, 38, 41, 67–69, 72; Cogan, *Haven of Health*, 53, 63, 85, 135, 159, 182; and Venner, *Via recta vitam longam*, 137. For the digestibility of cheese, see Paolo Savoia, "Cheesemaking in the Scientific Revolution: A Seventeenth-Century Royal Society Report on Dairy Products and the History of European Knowledge," *Nuncius* 34

(2019): 427–55, esp. 437–38. Beef was of special interest to early modern English commentators, and its juice-making powers are treated in chapter 4, 145–54 (below).

53. Elyot, *Castell of Health*, 19r.

54. Archer, *Every Man His Own Doctor*, 67–71. See also Platina, *On Right Pleasure and Good Health*, trans. Mary Ella Milham (Tempe, AZ: Medieval & Renaissance Texts & Studies, [1470] 1998), 127–29; and Daniel Margócsy, "The Pineapple and the Worms," *KNOW: A Journal on the Formation of Knowledge* 5 (2021): 53–81, esp. 60–62 (for contemporary medical worries that fresh fruit might breed worms in the body).

55. Harington, *The English Mans Doctor*, 12.

56. Venner, *Via recta vitam longam*, 92, 112, 117; and Archer, *Every Man His Own Doctor*, 76.

57. Elyot, *Castell of Health*, 13.

58. Burton, *Anatomy of Melancholy*, Part 1, 217. Burton was here quoting and summarizing Jean Fernel, *The Physiologia of Jean Fernel (1567)*

59. Bright, *Treatise of Melancholy*, 31–32; Cogan, *Haven of Health*, 53; Archer, *Every Man His Own Doctor*, 27, 32–34 (32, for spleen); and Vaughan, *Approved Directions*, 35–36.

60. Burton, *Anatomy of Melancholy*, Part 1, 217–23. See also Venner, *Via recta vitam longam*, 44–45. There were many sorts of strong or fortified wines popular in Renaissance and early modern England that are now obscure or unknown but that were objects of medical concern at the time. *Alicante* was a dark red Spanish wine, *metheglin* was a spiced mead, *hippocras* was a sweetened and spiced wine, *rumney* was a sweet Greek wine, and *brown bastard* was also a sweet wine, often Spanish.

61. Burton, *Anatomy of Melancholy*, 217–18.

62. For example, [Thomas Cock], *Kitchin-Physick: or, Advice to the Poor* (London, 1675), 31–32, 38, 41. See also idem, *Miscelanea Medica: or, a Supplement to Kitchin-Physick* (London, 1675), 1–3 (for injunction "to *cure* with *contraries*"). Cock, a Cambridge-educated physician, was here writing polemically against those "chymical" practitioners who supposed that their specific remedies could supplant Galenic dietetic regimes.

63. Venner, *Via recta vitam longam*, 71, 75; Archer, *Every Man His Own Doctor*, 33, 40; and Butts, *Dyets Dry Dinner*, sig. K3.

64. Venner, *Via recta vitam longam*, 83, 120.

65. Harington, *English Mans Doctor*, 19; and Vaughan, *Approved Directions*, 55.

66. Vaughan, *Approved Directions*, 57.

67. Harington, *English Mans Doctor*, 22; and Venner, *Via recta vitam longam*, 107.

68. Venner, *Via recta vitam longam*, 118. See also Butts, *Dyets Dry Dinner*, sig. C5.

69. Venner, *Via recta vitam longam*, 176.

70. Boorde, "Compendyous Regyment of Helth," 255; Humphrey Brooke, *Ugieine, or A Conservatory of Health* (London, 1650), 25; A. B., *The Sick-Mans Rare Jewel Wherein is Discovered a Speedy Way How Every Man May Recover Lost Health, and Prolong Life* (London, 1674), 101. See also Charles Ludington, "'Claret Is the Liquor for Boys; Port for Men': How Port Became the 'Englishman's Wine,' 1750s–1800," *Journal of British Studies* 48 (2009): 364–90.

71. Albala, *Eating Right in the Renaissance*, 139, 241–43.

72. For leeks, see Harington, *English Mans Doctor*, 22–23. For lettuce, see Everard Maynwaringe, *The Method and Means of Enjoying Health, Vigour, and Long Life* (London, 1683), 87; and Francis Bacon, *Historie Naturall and Experimentall of Life and Death* (London, 1638), 218. For cheese, see Harington, *English Mans Doctor*, 10; and Lémery, *Treatise of Food*, 221–24. For walnuts, see Archer, *Every Man His Own Doctor*, 77. For oysters, see Lémery, *Treatise of Food*, 310.

73. Venner, *Via recta vitam longam*, 142–43.

74. John Evelyn, *Acetaria, a Discourse of Sallets* (London, 1699), 10, 16, 27, 30, 47.

75. Cf. chapter 1, 31–33, 40–42 (above), and chapter 2, 95–102 (above).

76. See chapter 2, 99–101 (above).

77. François Rabelais, *Gargantua and Pantagruel*, trans. J. M. Cohen (London: Penguin, [1532–1534] 1955), 88. See also Michel Jeanneret, *A Feast of Words: Banquets and Table Talk in the Renaissance*, trans. Jermey Whiteley and Emma Hughes (Chicago: University of Chicago Press, [1987] 1991), 74–77.

78. See, e.g., Joan Fitzpatrick, *Food in Shakespeare: Early Modern Dietaries and the Plays* (Aldershot: Ashgate, 2007).

79. See John Russell Brown, *The Shakespeare Handbooks. Hamlet* (London: Palgrave Macmillan, 2006), 18–19; J. Dover Wilson, *What Happens in Hamlet* (Cambridge: Cambridge University Press, [1935] 2003), 311; W. F. Bynum and Michael Neve, "Hamlet on the Couch," in *The Anatomy of Madness: Essays in the History of Psychiatry*, ed. William F. Bynum, Roy Porter, and Michael Shepherd, 3 vols. (London: Tavistock, 1985), 1:289–304; and Bright, *Treatise of Melancholy* (London, [1586] 1613), 312–13: "The ayre, meete for melancholicke folke, ought to bee thinne, pure and subtile, open, and patent to all windes: in respect of their temper, especially to the South, and South-east."

80. Steven Shapin, "'You Are What You Eat': Historical Changes in Ideas About Food and Identity," *Historical Research* 87 (2014), 377–92, esp. 381–82. A historian of London public dining has pointed out that Petruchio's offerings map onto standard choices offered to guests in an inn or tavern. See Martha Carlin, "'What Say You to a Piece of Beef and Mustard?': The

Evolution of Public Dining in Medieval and Tudor London," *Huntington Library Quarterly* 71 (2008): 199–217.

81. Venner, *Via recta vitam longam*, 27.

82. India Mandelkern, "The Politics of the Palate: Taste and Knowledge in Early Modern England," PhD dissertation, University of California, Berkeley, 2015, 41–42; idem, "Taste-Based Medicine," *Gastronomica* 15, no. 1 (2015): 8–21; Steven Shapin, *Changing Tastes*; and chapter 5, 200–207 (below).

83. For brief but perceptive remarks on the humoral basis of agreement and taste in traditional dietetics, see David Howes, "Preface: Accounting for Taste," in Luca Vercelloni, *The Invention of Taste: A Cultural Account of Desire, Delight and Disgust in Fashion, Food and Art*, trans. Kate Singleton (London: Bloomsbury, [2005] 2016), ix.

84. The notion of "gratefulness to the taste" appeared in early modern medical texts as well as recipe books, often in connection with signs of digestibility and wholesomeness. See, e.g., Venner, *Via recta vitam longam*, 69; and Thomas Willis, *A Medical-Philosophical Discourse of Fermentation*, trans. S. P. (London, 1681), 19.

85. For example, Francis Bacon, *Sylva Sylvarum: or A Naturall Historie in Ten Centuries* (London, 1627), 125, 141, 144, 158, 172, 175, 233; Nehemiah Grew, *The Anatomy of Plants* (London, 1682); Thomas Tryon, *The Good House-Wife Made a Doctor* (London, 1692), 157–62 (for qualified enthusiasm about bitterness); and John Floyer, *Pharmako-Basanos; or the Touch-Stone of Medicines* (London, 1687), e.g. 4, 21–33, 81–82 (for the cleansing effect of bitters). See also Mandelkern, "Politics of the Palate," 13–14, 32–36; and idem, "Taste-Based Medicine," 12–16. The tag *Quod sapit* is further discussed in chapter 5, 201–7 (below).

86. Magninus of Milan, quoted in Jean-Louis Flandrin, "Seasoning, Cooking, and Dietetics in the Late Middle Ages," in *Food: A Culinary History*, ed. Flandrin and Massimo Montanari (New York: Columbia University Press, 1999), 320.

87. Brooke, *Ugieine*, 82.

88. Michael Schoenfeldt, "Fables of the Belly in Early Modern England," in *The Body in Parts: Fantasies of Corporeality in Early Modern* Europe, ed. David Hillman and Carla Mazzio, 242–61 (New York: Routledge, 1997), esp. 244–47.

89. Walter Pagel, *Paracelsus: An Introduction to Philosophical Medicine in the Era of the Renaissance*, 2nd rev. ed. (New York: S. Karger, [1958] 1982); and Michel Foucault, *The Order of Things: An Archaeology of the Human Sciences* (New York: Vintage, [1966] 1994), 25–45.

90. Tryon, *Way to Health*, 16.

91. Venner, *Via recta vitam longam*, 86, 90. See also Butts, *Dyets Dry Dinner*, sig. L4.

92. Cătălin Avramescu, *An Intellectual History of Cannibalism*, trans. Alistair Ian Blyth (Princeton, NJ: Princeton University Press, 2009).

93. Boorde, "Compendyous Regyment of Helth," 275.

94. William Vaughan, *The Newlanders Cure: Aswell of Those Violent Sicknesses Which Distemper Most Minds* (London, 1630), 3.

95. Johann Amos Comenius, *Naturall Philosophie Reformed by Divine Light, or, A Synopsis of Physicks* (London, 1651), 166.

96. Lémery, *Treatise of Food*, 179.

97. Specific reasoning from the characteristics of cattle to those of human eaters is discussed in chapter 4, 149–54 (below).

98. Archer, *Every Man His Own Doctor*, 23–24.}

Chapter 4

1. For a brief account of seventeenth-century dietetic "tuning" involving the Cambridge Platonist Henry More and his friend, the philosopher Lady Ann Conway, see Steven Shapin, "The Philosopher and the Chicken: On the Dietetics of Disembodied Knowledge," in *Science Incarnate: Historical Embodiments of Natural Knowledge*, ed. Christopher Lawrence and Steven Shapin (Chicago: University of Chicago Press, 1998), 39–40.

2. Nicholas Culpeper, Adbiah Cole, and William Rowland, *The Practice of Physick in Seventeen Several Books, . . . Being Chiefly a Translation of That Learned and Reverend Doctor, Lazarus Riverius [Lazare Rivière]* (London, 1655), 400–403.

3. William Buchan, *Domestic Medicine; or, the Family Physician* (Edinburgh: Balfour, Auld, and Smellie, 1769), 558, 565, 576–78.

4. William Buchan, *Advice to Mothers, on the Subject of Their Own Health; and on the Means of Promoting the Health . . . of Their Offspring*, 2nd ed. (London: T. Cadell and W. Davies, [1803] 1811), 2–3. In the southern United States, the most popular medical text similarly advised light exercise and cheerfulness in the case of menstrual obstruction as well as bleeding and drugging. See John C. Gunn, *Gunn's Domestic Medicine, or Poor Man's Friend*, 4th ed. (Springfield, OH: John M. Gallagher, [1830] 1835), 391–400.

5. Charles-Louis de Secondat, Baron de Montesquieu, *The Complete Works*, Vol. 1, *The Spirit of Laws* (London: T. Evans and W. Davis, 1777), 302–3.

6. Robert Burton, *The Anatomy of Melancholy* (New York: NYRB Books, [1621] 2001), Part 1, 230–31.

7. Burton, *Anatomy of Melancholy*, Part 1, 231; see also ibid., "Democritus Junior to the Reader," 92.

8. Henry Butts, *Dyets Dry Dinner consisting of Eight Severall Courses* (London, 1599), sig. A2.

9. Michel Eyquem de Montaigne, "Of Experience," in idem, *The Essayes of Michael Lord of Montaigne*, 3 vols., trans. John Florio (London: J. M. Dent, [1580–1588] 1910), 3:340. For the dietetic sense of custom as "second nature," see chapter 2, 95–99 (above).

10. Montaigne, "Of Custom," in *Essayes*, 1:106.

11. John Bostock, *An Elementary System of Physiology*, new ed. (London: Henry G. Bohn, [1826] 1836), 554–55.

12. Thomas Tryon, *The Good House-wife Made a Doctor* (London, 1692), 170–71, 174.

13. William Smith, *A Sure Guide in Sickness and Health, in the Choice of Food, and Use of Medicine* (London: J. Bew and J. Walter, 1776), 81–82. For treatments of spirituous drinks as well as tobacco and coffee and the historical understanding of addiction, see Phil Withington, "Addiction, Intoxicants, and the Humoral Body," *Historical Journal* 65 (2022): 68–90; and idem, "Remaking the Drunkard in Early Stuart England," *English Language Notes* 60 (2022): 16–38.

14. Thomas Morton (1632) and William Wood (1634), both quoted in Joyce E. Chaplin, "Natural Philosophy and an Early Racial Idiom in North America: Comparing English and Indian Bodies," *William and Mary Quarterly*, 3rd series, 54 (1997): 236–37; and idem, *Subject Matter: Technology, Science, and the Body on the Anglo-American Frontier, 1500–1676* (Cambridge, MA: Harvard University Press, 2001), 152.

15. Francis Higginson, *New-Englands Plantation: Or, a Short and True Description of the Commodities and Discommodities of that Countrey* (London, 1630), 9–10.

16. Christopher Columbus, *The Four Voyages*, ed. and trans. J. M. Cohen (Harmondsworth: Penguin, 1969), 81 (for "vomiting"), 166 (for "disagreeing with them"). However, Columbus did note that Europeans could, over time, adapt to new foods. See also Nicolás Wey-Gómez, *The Tropics of Empire: Why Columbus Sailed South to the Indies* (Cambridge, MA: MIT Press, 2008), 411–16, 420–23.

17. Chaplin, *Subject Matter*, 149–51. See also Susan Scott Parrish, *American Curiosity: Cultures of Natural History in the Colonial British Atlantic World* (Chapel Hill: University of North Carolina Press, 2012), chap. 2.

18. John Gerard, *The Herball or Generall Historie of Plantes* (London, 1597), 71. Gerard's account, largely copied out of an earlier Flemish herbal, was reprinted through the seventeenth century and was well known among the American colonists.

19. For French patterns, see Robert Launay, "Maize Avoidance? Colonial French Attitudes towards Native American Foods in the Pays des Illinois (17th and 18th Century)," *Food and Foodways* 26 (2018): 92–104. For Spanish attitudes, see Rebecca Earle, *The Body of the Conquistador: Food, Race and*

the Colonial Experience in Spanish America, 1492–1700 (Cambridge: Cambridge University Press, 2012), esp. chap. 1, 4; and idem, "'If You Eat Their Food . . .': Diets and Bodies in Early Colonial Spanish America," *American Historical Review* 115 (2010): 688–713.

20. John Winthrop Jr. to Robert Boyle, July 27, 1662, reproduced in Fulmer Mood, "John Winthrop, Jr., on Indian Corn," *New England Quarterly* 10 (1937): 125.

21. [Benjamin Franklin], "Second Reply to 'Vindex Patriae,' 2 January 1766," Founders Online, https://founders.archives.gov/documents/Franklin/01-13-02-0003.

22. [Benjamin Franklin], "Further Defense of Indian Corn, 15 January 1766," Founders Online, https://founders.archives.gov/documents/Franklin/01-13-02-0014.

23. Gianamar Giovannetti-Singh, "Galenizing the New World: Joseph-François Lafitau's 'Galenization' of Canadian Ginseng, *ca*. 1716–1724," *Notes and Records of the Royal Society* 75 (2021): 59–72. For the pineapple, see Sean B. Silver, "Locke's Pineapple and the History of Taste," *Eighteenth Century: Theory and Interpretation* 49 (2008): 43–65. For the potato, see chapter 3, 115 (above).

24. Jacobus Bontius, *An Account of the Diseases, Natural History, and Medicines of the East Indies*, trans. anon. (London: T. Noteman, [comp. 1642] 1769), 120, 125.

25. For example, Thomas Tryon, *Friendly Advice to the Gentlemen-Planters of the East and West Indies* (London, 1684), 36–37.

26. Earle, *The Body of the Conquistador*; idem, "'If You Eat Their Food . . .'"; and Alfred Crosby, *The Columbian Exchange: Biological and Cultural Consequences of 1492* (Westport, CT: Greenwood, 1972). For India, see E. M. Collingham, *Imperial Bodies: The Physical Experience of the Raj, c. 1800–1947* (Oxford: Blackwell, 2001); Mark Harrison, *Colonizing the Body: State Medicine and Epidemic Disease in Nineteenth-Century India* (Berkeley: University of California Press, 1993), 36–43; and idem, *Climate and Constitutions: Health, Race, Environment and British Imperialism in India 1600–1850* (New York: Oxford University Press, 1999), esp. 80–83.

27. For historical treatments, see Ben Rogers, *Beef and Liberty: Roast Beef, John Bull and the English Nation* (New York: Vintage, 2004); Roy Porter, "Consumption: Disease of the Consumer Society?," in *Consumption and the World of Goods*, ed. John Brewer and Roy Porter (London: Routledge, 1993), 58–84; Menno Spiering, "Food, Phagophobia and English National Identity," *European Studies* 22 (2006): 31–48; Anita Guerrini, "Health, National Character and the English Diet in 1700," *Studies in History and Philosophy of Biological and Biomedical Sciences* 43 (2012): 349–56, esp. 350–52, 355–56; and Steven Shapin, "'You Are What You Eat': Historical

Changes in Ideas about Food and Identity," *Historical Research* 87 (2014): 377–92, esp. 384–90. "Beef-eaters" as a nickname for the Yeomen Warders of the Tower of London seems to date from the 1660s, and while its precise origin is unclear, the link between beef and brave stolidity is evident.

28. There were indeed some dietary writers who found beef "hot," though much here seems to have depended on the mode of preparation.

29. See chapter 3, 133–136 (above).

30. Edward Chamberlayne, *Angliae notitia; or, The Present State of England* (London, 1669), 21, 23–24.

31. Giovanni Botero, *Relations of the Most Famous Kingdomes and Common-wealths thorowout the World . . .*, 2nd ed. (London, [1597] 1630), 87.

32. Robert Burns, "Address to a Haggis" (1786). English translation: "Is there that over his French ragout, / Or olio that would sicken a sow, / Or fricassee would make her vomit / With perfect disgust, / Looks down with sneering, scornful view / On such a dinner." Robert Burns, *Poems and Songs* (Edinburgh: William P. Nimmo, 1868), 13–14. For eighteenth-century French attitudes to beef eating, see E. C. Spary, *Feeding France: New Sciences of Food, 1760–1815* (Chicago: University of Chicago Press, 2014), 112–13.

33. Franklin, "Further Defense of Indian Corn."

34. J. A. Paris, *A Treatise on Diet: with A View to Establish, on Practical Grounds a System of Rules for the Prevention and Cure of the Diseases Incident to a Disordered State of the Digestive Functions* (New York: E. Duyckinck, Collins & Co., [1826] 1828), 5. Complaints about the depraving consequences of innovative French cooking on both health and natural taste appeared in France as well. See, for example, E. C. Spary, *Eating the Enlightenment: Food and the Sciences in Paris, 1670–1760* (Chicago: University of Chicago Press, 2012), esp. chap. 5.

35. Jeremiah Wainewright, *A Mechanical Account of the Non-Naturals: Being a Brief Explication of the Changes Made in Humane Bodies, by Air, Diet, &c.* (London: Ralph Smith, 1707), 160.

36. Thomas Dekker, *The Cold Yeare 1614 . . . Written Dialogue-wise, in a Plaine Familiar Talke between a London Shop-keeper and a North-Country-man* (London, 1615), sig. A3v. The capon-beef contrast was common in early modern England, and it bore a sexual as well as an analogical load, capons being castrated roosters (or cocks). See also Massimo Montanari, *The Culture of Food*, trans. Carl Ipsen (Oxford: Basil Blackwell, 1994), 76. In *As You Like It* (Act 2, scene 7), the "fifth age" of man is exemplified not by the hot-tempered "fourth age" soldier but instead by a sedentary judge "In fair round belly with good capon lined."

37. For example, Mark Dawson, *Plenti and Grase: Food and Drink in a Sixteenth-Century Household* (Totnes, Devon: Prospect Books, 2009), chap.

4; David Gentilcore, *Food and Health in Early Modern Europe: Diet, Medicine and Society, 1450–1800* (London: Bloomsbury, 2016), 85–86; Joan Thirsk, *Food in Early Modern England: Phases, Fads, Fashions, 1500–1760* (London: Continuum, 2007), 45–46; Stephen Mennell, *All Manners of Food: Eating and Taste in England and France from the Middle Ages to the Present* (Oxford: Basil Blackwell, 1985), 40–61, 102–3; Andrew B. Appleby, "Diet in Sixteenth-Century England: Sources, Problems, Possibilities," in *Health, Medicine and Mortality in the Sixteenth Century*, ed. Charles Webster (Cambridge: Cambridge University Press, 1979), 97–116, esp. 97–99, 108–9; and Spiering, "Food, Phagophobia and English National Identity," 32–35.

38. Thirsk, *Food in Early Modern England*, 237–40. See also C. M. Woolgar, *The Culture of Food in England, 1200–1500* (New Haven, CT: Yale University Press, 2016), 65–76; Norbert Elias, "On the Eating of Meat," *Food & History* 2, no. 2 (2004): 11–16, esp. 12; and Allen J. Grieco, "Food and Social Classes in Late Medieval and Renaissance Italy," in idem, *Food, Social Politics and the Order of Nature in Renaissance Italy*, Villa I Tatti, Series 34 (Milan: Officina Libraria, for Villa I Tatti, 2019), 104–5. "Butcher's meat" was and remains a common designation for uncured fresh meat as opposed to salted meat or sausages and usually excluding poultry and game.

39. Alessandro Magno, "The London Journals of Alessandro Magno 1562," trans. and ed. Caroline Barron, Christopher Coleman, and Claire Gobbi, *London Journal* 9, no. 2 (Winter 1983): 143. See also Martha Carlin, "'What Say You to a Piece of Beef and Mustard?': The Evolution of Public Dining in Medieval and Tudor London," *Huntington Library Quarterly* 71 (2008): 214; and Spiering, "Food, Phagophobia and English National Identity," 32–34.

40. Henri Misson, *Memoirs and Observations in His Travels over England*, trans.[?] Ozell (London: Dr. Browne, A. Bell, et al., [1698] 1719), 311.

41. Peter [Pehr] Kalm, *Kalm's Account of His Visit to England on His Way to America in 1748*, trans. Joseph Lucas (London: Macmillan, [1752, in Swedish] 1892), 14–15. Kalm's qualification about "masters" apparently recognizes that the better off ate more meat than the poor.

42. Thomas Cogan, *The Haven of Health* (London, [1584] 1636), 129.

43. Samuel Pepys, *The Diary of Samuel Pepys: Companion*, ed. William Matthews and Robert Latham (London: HarperCollins, 2000), 143–49.

44. A[lexander] Hunter, *Culina Famulatrix Medicinæ: or Receipts in Modern Cookery; with a Medical Commentary*, 3rd ed. (York: T. Wilson and R. Spence, 1806), 6.

45. Ralph Holinshed, *The First and Second Volumes of Chronicles Comprising 1. The Description and Historie of England, 2. The Description and Historie of Ireland, 3. The Description and Historie of Scotland* (London, 1587), 110.

46. Cogan, *Haven of Health*, 128–29.

47. Tryon, *Friendly Advice*, 54.

48. The Frenchman Lémery endorsed English enthusiasm about the quality of their beef: "Every body knows that [cattle] are bigger in *England* than in *France*; and that their Flesh is better." Louis Lémery, *A Treatise of All Sorts of Food*, trans. D. Hay (London: T. Osborne, [1702] 1745), 185–86.

49. Eleazar Duncon, *The Copy of a Letter Written by E. D. Doctour of Physicke to a Gentleman, by Whom It Was Published* (London, 1606), 6. See also Thomas Muffett, *Healths Improvement: or, Rules Comprizing and Discovering the Nature, Method, and Manner of Preparing All Sorts of Foods* (London, [comp. ca. 1600] 1655), 61. For the proverb, see John Clarke, *Parœmiologia Anglo-Latina* (London, 1639), 33.

50. Tobias Venner, *Via recta ad vitam longam* (London, 1620), 55.

51. Lémery, *Treatise of Food*, 184. See also Massimo Montanari, *Medieval Tastes: Food, Cooking, and the Table*, trans. Beth Archer Brombert (New York: Columbia University Press, 2015), chap. 6.

52. Manningtree is a town in Essex on the road from London to Harwich, and it is thought that the reference here was to the tradition of roasting a whole ox at the annual town fair. The local cattle were also reputed for their enormous size. Robert Nares, *A Glossary; or, Collection of Words, Phrases, Names, and Allusions to Customs and Proverbs. Etc.*, 2 vols., new ed. (London: John Russell Smith, 1859), 2:546.

53. See J. B. Fisher, "Digesting Falstaff: Food and Nation in Shakespeare's *Henry IV* Plays," *Early English Studies* 2 (2009): 1–23.

54. See chapter 2, 93–95, and chapter 3, 125–26 (above). Erasmus's *Adages* included "Wythout meate and drynke the lust of the body is colde" and "The beste way to tame carnal lust, is to kepe abstinence of meates and drynkes." [Desiderius Erasmus], *Proverbes or Adagies*, ed. and trans. Richard Taverner (London, 1539), fol. xxxvi. For seventeenth-century Roman clerical appreciations of meat eating, sexuality, and vigor, see Bradford Bouley, "Digesting Faith: Eating God, Man, and Meat in Seventeenth-Century Rome," *Osiris* 35 (2020): 55–57. For meat eating, carnality, and spirituality, see Allen J. Grieco, "Vegetable Diets, Hermits and Melancholy in Late Medieval and Renaissance Italy," in idem, *Food, Social Politics and the Order of Nature in Renaissance Italy*, Villa I Tati Series, 34 (Milan: Officina Libraria, for Villa I Tati, 2019), 243–62.

55. Henry Peacham, *The Truth of Our Times* (London, 1638), 92–93.

56. William Vaughan, *The Newlanders Cure: Aswell of Those Violent Sicknesses Which Distemper Most Minds* (London, 1630), 3.

57. Into the 1670s, those praising English military valor felt it necessary to say that this did *not* depend on adequate supplies of beef: "'Tis a base and malicious scandal to say, That his Valour ebbs and flows with the condition

of his Snapsack, or that he can never fight well unless Beef and Bag-pudding be his Seconds." Anon., *The Character of a True English Souldier. Written by a Gentleman of the New-rais'd Troops* (London, 1678). See also Joan Fitzpatrick, "Diet and Identity in Early Modern Dietaries and Shakespeare: The Inflections of Nationality, Gender, Social Rank, and Age," in *Shakespeare Studies*, Vol. 42, ed. James R. Siemon and Diana E. Henderson (Madison, NJ: Fairleigh Dickinson University Press, 2014), 75–90, esp. 77. On meat, the military virtues, and social standing, see Montanari, *Culture of Food*, 14–15, 73–74.

58. [Daniel Defoe], *The True-Born Englishman: A Satyr* (London, 1700), 26.

59. Thomas Tryon, *Healths Grand Preservative: or The Womens Best Doctor* . . . (London, 1682), 15–19. See also idem, *Miscellania: or, A Collection of Necessary, Useful, and Profitable Tracts on Variety of Subjects* (London, 1696), 35–36. The gesture was biblical and understood to be so: "Only be sure that thou eat not the blood: for the blood is the life; and thou mayest not eat the life with the flesh" (Deuteronomy 12:23) and "And whatsoever man there be of the house of Israel, or of the strangers that sojourn among you, that eateth any manner of blood; I will even set my face against that soul that eateth blood, and will cut him off from among his people. For the life of the flesh is in the blood" (Leviticus 17:10–11). The notion of "life" here included "soul," that which animated a living thing, carrying its unique vital attributes.

60. Thomas Tryon, *Tryon's Letters, Domestick and Foreign, to Several Persons of Quality* (London: Geo. Conyers, 1700), 87.

61. Tryon, *Healths Grand Preservative*, 15–19.

62. Charles Blount, *Great is Diana of the Ephesians, or, The Original of Idolatry together with the Politick Institution of the Gentiles Sacrifices* (London, 1680), 44, quoted in Tristram Stuart, *The Bloodless Revolution: A Cultural History of Vegetarianism from 1600 to Modern Times* (New York: Norton, 2006), 125.

63. See Keith Thomas, *Man and the Natural World: Changing Attitudes in England, 1500–1800* (London: Allen Lane, 1983), 294; and Anita Guerrini, "The Ghastly Kitchen," *History of Science* 54 (2016): 89–90.

64. Frederick Accum, *Culinary Chemistry, Exhibiting the Scientific Principles of Cookery* (London: R. Ackermann, 1821), 54–56.

65. Edwin Lankester, *On Food. Being Lectures Delivered at the South Kensington Museum* (London: Robert Hardwicke, 1861), 173.

66. Justus von Liebig, "On the Nutritive Value of Different Sorts of Food," *The Lancet* 93 (January 2, 9, 23, 1869; February 6, 1869; March 13, 1869), 4–5, 36–38, 113–15, 186–87, 357–58 (quoting 186). For treatment of Liebig on the chemistry of nutrition, see chapter 8, 286–304 (below).

67. J. Milner Fothergill, *A Manual of Dietetics* (New York: William Wood, 1886), 53.

68. Friedrich Nietzsche to Franziska Nietzsche, May 1885, in *Selected Letters of Friedrich Nietzsche*, ed. and trans. Christopher Middleton (Indianapolis: Hackett, 1969), 242.

69. Friedrich Nietzsche, "Why I Am So Clever," in *Ecce Homo*, trans. Anthony M. Ludovici (Mineola, NY: Dover, [1908] 2004), 30. For late nineteenth- and early twentieth-century scientific claims about diet and racial dominance, see chapter 8, 337–49 (below).

70. Daniel Defoe, *Due Preparations for the Plague, As Well for Soul as Body* (London: E. Matthews, 1722), 45.

71. Jean-Jacques Rousseau, *Émile*, trans. Barbara Foxley (London: J. M. Dent, [1762] 1950), 118.

72. Julien Offray de la Mettrie, *Man a Machine* (Dublin: W. Brien, [1747] 1749), 13. "One nation, we may observe, is generally heavy and stupid; and another is sprightly, gay, and sagacious. What is this owing to, unless it be in part to the food they live on . . . ?" (16).

73. William Buchan, *Observations Concerning the Diet of the Common People* (London: A. Strahan, 1797), 12. See also idem, *Domestic Medicine* (1769), 67. Sadly, it seems that the notion of "having a beef" with someone is an American usage, dating from the late nineteenth century.

74. For example, William Bullein, *The Government of Health* (London, [1562] 1595), 60v–61r; and Butts, *Dyets Dry Dinner*, sig. Ir.

75. Andrew Boorde, "A Compendyous Regyment of Helth," in *Andrew Boorde's Introduction and Dyetary*, ed. F. J. Furnivall, Early English Text Society, Extra Series, no. X (London: N. Trübner, [1542] 1870), 271–72; Sir John Harington, trans., *The English Mans Doctor, or, The Schoole of Salerne* (London, [trans. 1607] 1624), 8; and Muffett, *Healths Improvement*, 58–61.

76. Harington, *English Mans Doctor*, 3.

77. George Cheyne, *The English Malady: or, a Treatise of Nervous Diseases of All Kinds, as Spleen, Vapours, Lowness of Spirits, Hypochondriacal, and Hysterical Complaints* (London: G. Strahan, 1733), esp. 168–70.

78. Bullein, *Government of Health*, 60v; and Antoine Le Grand, *An Entire Body of Philosophy* (London, 1694), 236.

79. Robert Appelbaum, *Aguecheek's Beef, Belch's Hiccup, and Other Gastronomic Interjections: Literature, Culture, and Food among the Early Moderns* (Chicago: University of Chicago Press, 2006), chap. 1.

80. Thomas Wright, *The Passions of the Minde* (London, [1601] 1630), 129–30. Early eighteenth-century dietary writers continued to cite Galenic authority for the link between flesh eating in general and stupidity: "those who live upon the Flesh of Asses or Camels, are usually heavy, and dull of Understanding." Lémery, *Treatise of Food*, 179.

81. Buchan, *Observations Concerning the Diet of the Common People*, 5, 10–11. A "diathesis" was understood as a tendency toward a certain pathological condition.

82. Diogenes Laërtius, *The Lives, Opinions, and Remarkable Sayings of the Most Famous Ancient Philosophers*, 2 vols. (London, 1688), 1:420.

83. Plutarch, "Of Eating of Flesh," in *Plutarch's Lives and Miscellanies*, ed. A. H. Clough and William W. Goodwin, 5 vols. (New York: Colonial Company, 1905), 5: 9.

84. John Dryden, *Troilus and Cressida* (London, 1679), 22.

85. Washington Irving, "John Bull," in *The Works of Washington Irving*, Vol. 2, *The Sketch Book* (New York: Putnam, 1890), 433. The character of John Bull was created in 1712 by Dr. John Arbuthnot, the Scottish physician who also wrote a treatise on the nature of foods. See John Arbuthnot, *The History of John Bull*, new ed. (Oxford: Clarendon, [1712, 1727] 1976).

86. The next several paragraphs follow Steven Shapin, "'The Mind Is Its Own Place': Science and Solitude in Seventeenth-Century England," *Science in Context*, 4 (1991): 191–218; and especially, idem, "The Philosopher and the Chicken."

87. Burton, *Anatomy of Melancholy*, "Democritus Junior to the Reader," 75; Part 1, 169–70, 385, 411. For "more terrible than Death," see John Arbuthnot, *An Essay Concerning the Nature of Aliments, and the Choice of Them, according to the Different Constitutions of Human Bodies . . .*, 4th ed. (London: J. and R. Tonson, [1731] 1756), 319–20. See also Levinus Lemnius, *The Touchstone of Complexions* (London, [1576] 1633), 215–48. A distinction was often made between natural and supernatural forms of melancholy, that is, melancholy born of natural constitution and abuse of the nonnaturals versus the sort that was visited by God or Satan. It was largely the natural sort that was addressed by dietetic thought and practices.

88. Aristotle, "Problems," Book 30, 953a.10–15, in Aristotle, *Complete Works*, 2 vols., ed. Jonathan Barnes (Princeton, NJ: Princeton University Press, 1984), 2:1498–1499; and Seneca, *Moral Essays*, 3 vols., trans. John W. Basore (Cambridge, MA: Harvard University Press, 1958), 1:29. That question echoed through the Renaissance and beyond. See Marsilio Ficino, *Three Books on Life*, trans. Carol V. Kaske and John R. Clark (Binghamton, NY: Renaissance Society of America, [1489] 1989), 117; Burton, *Anatomy of Melancholy*, Part 1, 422; and Darrin M. McMahon, *Divine Fury: A History of Genius* (New York: Basic Books, 2013), 54–56.

89. For Stoic dietetics, see Epictetus, *The Discourses of Epictetus, with the Encheiridion and Fragments*, trans. George Long (New York: A. L. Burt, [1888?]), 434, 439, 443. See also Seneca, *Moral Essays*, 1:128, 151.

90. Thomas Stanley, *The History of Philosophy, the Third and Last Volume* (London, 1660), 45–46. See also Mirko D. Grmek, *Diseases in the Ancient*

Greek World, trans. Mireille Muellner (Baltimore: Johns Hopkins University Press, 1989), chap. 9.

91. For austere dietetics and religiosity, see notably Caroline Walker Bynum, *Holy Feast and Holy Fast: The Religious Significance of Food to Medieval Women* (Berkeley: University of California Press, 1987), 3.

92. Peter Brown, *The Body and Society: Men, Women and Sexual Renunciation in Early Christianity* (New York: Columbia University Press, 1988), 218–21. See also Michel Foucault, "On the Genealogy of Ethics: An Overview of a Work in Progress," in *Michel Foucault: Beyond Structuralism and Hermeneutics*, ed. Hubert L. Dreyfus and Paul Rabinow, 2nd ed. (Chicago: University of Chicago Press, 1983), 229.

93. Desert Fathers, *The Sayings of the Desert Fathers (Apophthegmata Patrum)*, trans. Benedicta Ward (London: A. R. Mowbray, 1975), 43–44.

94. Desert Fathers, *The Lives of the Desert Fathers (The Historia Monachorum in Aegypto)*, trans. Norman Russell (Oxford: Mowbray, 1981), 109; and Piero Camporesi, *The Anatomy of the Senses: Natural Symbols in Medieval and Early Modern Italy*, trans. Allan Cameron (Cambridge: Polity, 1994), esp. chap. 4.

95. Brown, *Body and Society*, 223.

96. For example, *Arbeit, Musse, Meditation: Betrachtungen zur Vita Activa und Vita Contemplativa*, ed. Brian Vickers (Zurich: Verlag der Fachvereine Zürich, 1985); idem, ed., *Public and Private Life in the Seventeenth Century: The Mackenzie-Evelyn Debate* (Delmar, NY: Scholars' Facsimiles & Reprints, 1986); and Steven Shapin, "'A Scholar and a Gentleman': The Problematic Identity of the Scientific Practitioner in Early Modern England," *History of Science* 29 (1991): 279–327.

97. André Du Laurens, *A Discourse of the Preservation of Sight: Of Melancholicke Diseases; of Rheumes, and of Old Age*, 2nd ed. (London, [1594] 1599), 84. It has recently been asserted that it is a "misconception" that early modern humoral pathology "was based primarily on the idea of a balance between the natural humours or the corresponding qualities (hot, cold, wet and dry)." Rather, disease was then referred principally to "impure, morbid matter." There is no doubt that notions of matter corrupting the humors were historically significant, but much of the material referred to here and in subsequent chapters establishes the continuing early modern significance of qualitative balance. Michael Stolberg, "Medical Popularization and the Patient in the Eighteenth Century," in *Cultural Approaches to the History of Medicine*, ed. Willem de Blécourt and Cornelie Usborne (London: Palgrave Macmillan, 2004), 91.

98. Burton, *Anatomy of Melancholy*, Part 1, 245.

99. Burton, *Anatomy of Melancholy*, Part 1, 234.

100. Anon., *The Compleat Doctoress: or, A Choice Treatise of All Diseases Insident to Women* (London, 1656), 14, 91.

101. [Thomas Gainsford], *The Rich Cabinet Furnished with Varietie of Excellent Discriptions* (London, 1616), 134–35; cf. chapter 2, 93–95 (above). Helicon was a Greek mountain where there were springs sacred to the Muses.

102. For a representative dietary regime to be observed by melancholics, see Laurens, *Discourse of Sight*, 104–7.

103. For example, William Bullein, *Gouernment of Health*, 60v; Lémery, *Treatise of Food*, 184–85, 192, 206; Cogan, *Haven of Health*, 113–14; Burton, *Anatomy of Melancholy*, Part 1, 217–19; Laurens, *Discourse of Sight*, 105–7; and Thomas Brugis, *The Marrow of Physicke* (London, 1648), 62.

104. Muffett, *Healths Improvement*, 137–38.

105. Lemnius, *Touchstone of Complexions*, 218.

106. Burton, *Anatomy of Melancholy*, Part 1, 219.

107. John Purcell, *A Treatise of Vapours, or, Hysterick Fits* (London: Nicholas Cox, 1702), 22.

108. Thomas Elyot, *The Castell of Health* (London, 1587), 76.

109. Michael Heyd, *'Be Sober and Reasonable': The Critique of Enthusiasm in the Seventeenth and Early Eighteenth Centuries* (Leiden: E. J. Brill, 1995), 99–100 (on the views of Henry More). See also Koen Vermeir, "The 'Physical Prophet' and the Powers of the Imagination: Part I. A Case-Study on Prophecy, Vapours and the Imagination (1685–1710)," *Studies in History and Philosophy of Biological and Biomedical Sciences* 35 (2004): 561–91; and Oriana Walker, "The Breathing Self: Toward a History of Respiration," PhD dissertation, Harvard University, 2016, 70–72.

110. William Basse, *A Helpe to Memory and Discourse with Table-Talke as Musicke to a Banquet of Wine* (London, 1630), 10.

111. For example, Cogan, *Haven of Health*, 185; Archer, *Every Man His Own Doctor*, 62–63; Burton, *Anatomy of Melancholy*, Part 1, 220–21, and Part 2, 101; and Basse, *Helpe to Memory*, 10–11. On attitudes toward the digestibility of cheese, see Paolo Savoia, "Cheesemaking in the Scientific Revolution: A Seventeenth-Century Royal Society Report on Dairy Products and the History of European Knowledge," *Nuncius* 34 (2019): 427–55, esp. 437–38.

112. Thomas Reid, *Essays on the Powers of the Human Mind* (London: Thomas Tegg, [1785] 1827), 222. See also Humphrey Brooke, *Ugieine: or A Conservatory of Health* (London, 1650), 135 (for vapors interfering with sleep); and Brugis, *Marrow of Physick*, 65.

113. For detailed treatment of the mental and emotional consequences of digestion and indigestion, see Michael Walkden, "Digestion and Emotion in Early Modern Medicine and Culture, c. 1580–c. 1740," PhD dissertation, University of York, 2018. See also Ian Miller, *A Modern History of the Stomach: Gastric Illness, Medicine, and British Society, 1800–1950* (London: Pickering & Chatto, 2011); and James Kennaway and Jonathan Andrews, "'The Grand Organ of Sympathy': 'Fashionable' Stomach Complaints and

the Mind in Britain, 1700–1850," *Social History of Medicine* 32 (2019): 57–79.

114. Thomas Adams, *Diseases of the Soul: A Discourse Divine, Morall, and Physicall* (London, 1616), 8–9. Adams was an Anglican minister, author of a study of the mad and the sinful *Mysticall Bedlam, or the World of Mad-Men* (London, 1615). See also Eleanor Barnett, "Reforming Food and Eating in Protestant England, c. 1560– c. 1640," *Historical Journal* 63 (2020): 507–27, esp. 519–20.

115. For *Flatus Hypochondriacus*, see John Archer, *Secrets Disclosed of Consumption* (London, 1684), 20. For the gendering of spleen and vapors, see Thomas Tryon, *Monthly Observations for the Preserving of Health* (London, 1688), 18–19; and Joseph Addison, "Spectator, no. 115" (July 12, 1711), in *The Works of the Right Honourable Joseph Addison*, 6 vols. (London: George Bell, 1888), 2:449: it is to "a neglect" of the rules regulating the interaction between body and soul "that we must ascribe the spleen, which is so frequent in men of studious and sedentary tempers, as well as the vapours to which those of the other sex are so often subject." See also Richard Blackmore, *A Treatise of the Spleen and Vapours: Hypochondriacal and Hysterical Affections* (London: J. Pemberton, 1725), iii–iv, 16, 97; Nicholas Robinson, *A New Theory of Physick and Diseases Founded on the Principles of the Newtonian Philosophy* (London: C. Rivington, 1725), 56 ("the Vapours in Women are term'd the *Hypocondria* in Men"); Sabine Arnaud, *On Hysteria: The Invention of a Medical Category between 1670 & 1820* (Chicago: University of Chicago Press, 2015), 17–19; and Alexander Wragge-Morley, "Medicine, Connoisseurship, and the Animal Body," *History of Science* 60 (2022): 481–99. Note, however, that Robinson Crusoe said that he was beset with anxiety and that "my Head was full of Vapours." Daniel Defoe, *Robinson Crusoe*, ed. Henry Kingsley (London: Macmillan [1719] 1868), 61; Mark S. Micale, *Hysterical Men: The Hidden History of Male Nervous Illness* (Cambridge, MA: Harvard University Press, 2008), esp. chap. 1; and chapter 2, 71–74 (above).

116. Purcell, *Treatise of Vapours*, 1–2. See also Fredrik Albritton Jonsson, "The Physiology of Hypochondria in Eighteenth-Century Britain," in *Cultures of the Abdomen: Diet, Digestion, and Fat in the Modern World*, ed. Christopher Forth and Ana Carden-Coyne, 15–30 (Basingstoke: Palgrave, 2005).

117. For example, Christopher Lawrence, "The Nervous System and Society in the Scottish Enlightenment," in *Natural Order: Historical Studies of Scientific Culture*, ed. Barry Barnes and Steven Shapin, 19–40 (Beverly Hills, CA: Sage, 1979); and Kenneway and Andrews, "'The Grand Organ of Sympathy,'" esp. 71–72.

118. See, e.g., Andrew Scull, *The Insanity of Place/The Place of Insanity: Essays on the History of Psychiatry* (London: Routledge, 2006), 44; and Roy

Porter, "Introduction," in George Cheyne, *The English Malady*, ed. Roy Porter (London: Tavistock/Routledge, 1991), xi, xxii, xxv–xxxii.

119. Cheyne, *The English Malady*, 194–95. See also Anon., *An Account of the Causes of Some Particular Rebellious Distempers, viz. the Scurvey, Cancers in Women's Breasts, &c. Vapours, and Melancholy, &c.* (London, 1670), 33; and Guenter B. Risse, "Mind-Body Enigma: Hysteria and Hypochrondriasis at the Edinburgh Infirmary," in idem, *New Medical Challenges during the Scottish Enlightenment*, 311–49 (Amsterdam: Rodopi, 2005). For gout, see George Cheyne, *An Essay on the Gout*, 3rd ed. (London: G. Strahan, 1721), 78–79; and Roy Porter and G. S. Rousseau, *Gout: The Patrician Malady* (New Haven, CT: Yale University Press, 1998), 55–57.

120. Harington, *The Englishmans Doctor*, 2; Sir Thomas More, "In Efflatum Ventris" (1518), in idem, *Memoirs of Sir Thomas More*, Vol. 2, ed. Arthur Cayley (London: Cadell and Davis, 1808), 284; and Montaigne, "Of the Force of Imagination," in idem, *Essayes*, 1:99. (The emperor is said to have been Claudius.)

121. G. A. Lindeboom, *Descartes and Medicine* (Amsterdam: Rodopi, 1979), 40. See also Brooke, *Ugieine*, 174.

122. For example, D. Border, *Polypharmakos kai chymistes, or, The English Unparalell'd Physitian and Chyrurgian* (London, 1651), 57–58; Samuel Boulton, *Medicina magica tamen physica: Magical, but Natural Physick* (London, 1656), 152, 175–76; Robert Boyle, *Occasional Reflections upon Several Subjects* (London, 1665), 235–36; and idem, *Some Considerations touching the Usefulnesse of Experimentall Naturall Philosophy* (London, 1663), 35, 60.

123. Harington, *The English Mans Doctor*, 39.

124. Anon., *The Fathers Legacy: or Counsels to His Children* (London, 1678), 30. For recent treatments, see, e.g., Andrew Scull, *Hysteria: The Biography* (Oxford: Oxford University Press, 2009), 9, 25, 35, 51.

125. Leonardus Lessius, *Hygiasticon: or, The Right Course of Preserving Life and Health into Extream Old Age*, 2nd ed., trans. T[imothy] S[mith][?] (Cambridge, [1614] 1634), 32–33. For biographical remarks and the origin of this text, see Cristiano Casalini and Laura Madella, "The Jesuit Cultivation of Vegetative Souls: Leonard Lessius (1554–1623) on a Sober Diet," in *Vegetative Powers: The Roots of Life in Ancient, Medieval and Early Modern Natural Philosophy*, ed. Fabrizio Baldassarri and Andreas Blank (Zurich: Springer, 2021), 181–84. Michael Stolberg documents how the vapors resulting from poor digestion figured in face-to-face communication between physician and patient in sixteenth-century Germany. Michael Stolberg, "'You Have No Good Blood in Your Body': Oral Communication in Sixteenth-Century Physicians' Medical Practice," *Medical History* 59 (2015): 63–82.

126. For coffee, see Pierre Pomet, *A Compleat History of Druggs*, 2 vols.

(London: R. Bonwick, 1712), 1:129. Tobacco was in worse repute in the seventeenth century, sometimes condemned for "throwing such poysonous vapours into the Brain, . . . that it sends many untimely unto their Graves." Anon., *The Women's Complaint against Tobacco* (London, 1675), 3.

127. Georg Andreas Agricola, "The Virtues and Uses of the Cordial Spirit of Saffron," broadside (London, 1680).

128. Butts, *Dyets Dry Dinner*, sig. F11.

129. For cinchona, herbs, seeds, and gruel, see Anon., *An Account of the Causes of Some Particular Rebellious Distempers, viz. the Scurvey, Cancers in Women's Breasts, &c. Vapours, and Melancholy, &c.* (London, 1670), 37–38, 41; Robert Talbor, *The English Remedy: or, Talbor's Wonderful Secret for Cureing of Agues and Feavers* (London, 1682), 25–26; and Burton, *Anatomy of Melancholy*, Part 2, 259–61. For waters, see Thomas Tryon, *A Discourse of Waters* (London, 1696), 23; and Benjamin Allen, *The Natural History of the Chalybeat and Purging Waters of England* (London, 1699), 59. Chalybeat waters were those containing significant amounts of iron.

130. Patrick Anderson, "Grana Angelica: or, The Rare and Singular Vertues and Uses of Those Angelical Pills . . . ," broadside (London, 1677); and Blackmore, *Treatise of the Spleen and Vapours*, esp. 60–89.

131. Samuel Clarke, *Medulla theologiæ, or, The Marrow of Divinity* (London, 1659), 189.

132. Cogan, *Haven of Health*, 246–47, 250–51.

133. Bernardino Ramazzini, "Of the Diseases of Learned Men," in idem, *A Treatise of the Diseases of Tradesmen* (London: Andrew Bell, 1705), 246–74. See also Burton, *Anatomy of Melancholy*, Part 1, 302, and Part 2, 233; Kennaway and Andrews, "'The Grand Organ of Sympathy'"; and Anne C. Vila, *Suffering Scholars: Pathologies of the Intellectual in Enlightenment France* (Philadelphia: University of Pennsylvania Press, 2018), esp. 22–23.

134. Ramazzini, "Diseases of Learned Men," 247–48. On learned hypochondriasis ("and what in *High-Dutch* is call'd *Der Gelahrten Kranckheydt*," see Bernard Mandeville, *A Treatise of the Hypochondriack and Hysterick Diseases*, 3rd ed. (London: J. Tonson, [1711] 1730), 106–12.

135. Marsilio Ficino, *Three Books on Life*, trans. Carol V. Kaske and John R. Clark (Binghamton, NY: Renaissance Society of America, [1489] 1989). See also chapter 1, 46–49 (above).

136. Cogan, *Haven of Health*, 18.

137. Among many sources for this sentiment, see Lessius, *Hygiasticon*, 35.

138. Lessius, *Hygiasticon*, 18–19.

139. Talbor, *The English Remedy*, 106–7.

140. Ficino, *Three Books on Life*, 113.

141. Burton, *Anatomy of Melancholy*, Part 1, 302. See also C. J. Sprengell, *The Aphorisms of Hippocrates and the Sentences of Celsus* (London: R. Bonwick et al., 1708), 243–44.

142. Ficino, *Three Books on Life*, 125.

143. Mandeville, *Hypochondriack and Hysterick Diseases*, 162–63.

144. Ramazzini, "Diseases of Learned Men," 249.

145. Ramazzini, "Diseases of Learned Men," 262. See also Ficino, *Three Books on Life*, 139.

146. Samuel-Auguste Tissot, *An Essay on Diseases Incident to Literary and Sedentary Persons*, 2nd ed., trans. J. Kirkpatrick (London: J. Nourse, [1768] 1769), 21, 23, 30, 53. See also Antoinette Suzanne Emch-Dériaz, "The Non-Naturals Made Easy," in *The Popularization of Medicine, 1650–1850*, ed. Roy Porter (London: Routledge, 1992), 145–48; and Vila, *Suffering Scholars*, 32–45, 127–28, 152–53.

147. Lémery, *Treatise of Food*, 15.

148. James I, King of England, *Basilikon Doron: Or His Maiesties Instructions to his Dearest Sonne, Henrie the Prince*, 2nd ed. (London, 1603), 108, and essentially repeated in James Cleland, *The Instruction of a Young Noble-Man* (Oxford, 1612), 211.

149. Montaigne, "Of Friendship," in idem, *Essayes*, 1:206. See also Lémery, *Treatise of Food*, 15, where avoiding "too serious Applications of the Mind" was also recommended during the whole process of digestion.

150. Michael MacDonald, *Mystical Bedlam: Madness, Anxiety, and Healing in Seventeenth-Century England* (Cambridge: Cambridge University Press, 1981), 15–153. For the common people disqualified for melancholia, see John Earle, *Micro-cosmographie, or, A Peece of the World Discovered in Essayes and Characters* (London, 1628), sig. F3v. "A plaine Country Fellow" has "reason enough to do his businesse, and not enough to be idle or melancholy."

Chapter 5

1. Andrew Boorde, "A Compendyous Regyment of Helth," in *Andrew Boorde's Introduction and Dyetary*, ed. F. J. Furnivall, Early English Text Society, Extra Series, no. X (London: N. Trübner, [1542] 1870), 300. The tag was a commonplace and was the title of or a substantial topic in books by physicians, such as John Archer, *Every Man His Own Doctor* (London, 1671).

2. For example, C. J. Sprengell, *The Aphorisms of Hippocrates and the Sentences of Celsus* (London: R. Bonwick et al., 1708), 236–37.

3. For example, Laurent Joubert, *Popular Errors*, trans. Gregory David de Rocher (Tuscaloosa: University of Alabama Press, [1579] 1989); and idem, *The Second Part of the Popular Errors*, trans. Gregory David de Rocher (Tuscaloosa: University of Alabama Press, [1587] 1995). See also chapter 2, 97–102 (above), and this chapter, 196–98.

4. There were many alternative English titles of the same text, including *A Treatise on Temperance and Sobriety*, *On a Sober Life*, *Sure Methods of*

Attaining a Long and Healthful Life, *The Art of Living Long*, *The Immortal Mentor*, and *How to Live a Hundred Years*. A modern English translation, together with much useful bibliographic information and historical commentary, is Alvise Cornaro, *Writings on the Sober Life: The Art and Grace of Living Long*, ed. and trans. Hiroko Fudemoto, 75–101 (Toronto: University of Toronto Press, 2014). There are many modern summaries of Cornaro's tract and historical reactions to it: for example, Gerald J. Gruman, "A History of Ideas about the Prolongation of Life," *Transactions of the American Philosophical Society*, new series, 56, no. 9 (1966): esp. 66–74; Hillel Schwartz, *Never Satisfied: A Cultural History of Diets, Fantasies and Fat* (New York: Free Press, 1986), 9–16; Federico Bellini, "Diet and Hygiene between Ethics and Medicine: Evidence and the Reception of Alvise Cornaro's *La Vita Sobria* in Early Seventeenth-Century England," in *Evidence in the Age of the New Sciences*, ed. James A. T. Lancaster and Richard Raiswell, 251–68 (Cham, Switzerland: Springer, 2018). The specific issue of medical expertise and its authority is treated in Steven Shapin, "Was Luigi Cornaro a Dietary Expert?," *Journal of the History of Medicine* 73 (2018): 135–49, and that material is drawn on in this account.

5. Friedrich Wilhelm Nietzsche, *Twilight of the Idols*, trans. Anthony M. Ludovici (Ware: Wordsworth, [1889] 2007), 29. Nietzsche thought that Cornaro's long life had nothing to do with his low diet and much to do with "the extraordinary slowness of his metabolism." "A scholar in our own time," Nietzsche wrote, "would simply destroy himself on Cornaro's diet. *Crede experto* [believe one who has tried it]."

6. The publishing history as well as much about the Anglophone reception of Cornaro's book has been well covered in Marisa Milani, "How to Attain Immortality Living One Hundred Years, or, The Fortune of the *Vita Sobria* in the Anglo-Saxon World," in Cornaro, *Writings on the Sober Life*, 183–213 (orig. publ. in Italian in 1980 and trans. by Hiroko Fudemoto).

7. Sir John Sinclair, *The Code of Health and Longevity*, 4 vols. (Edinburgh: Arch. Constable, 1807), 3:27–49.

8. Ernest Van Someren, "Was Luigi Cornaro Right?," in Horace Fletcher, *The A.B.-Z. of Our Own Nutrition* (New York: Frederick A. Stokes, 1903), 27–46. See also Horace Fletcher, *The New Glutton or Epicure* (New York: Frederick A. Stokes Co., 1903), 17, 56, 61, 87, 294, 303–4. The edition of Cornaro sponsored by Fletcher and Kellogg apparently appeared as Louis Cornaro, *The Art of Living Long* (Milwaukee: William F. Butler, 1903). See also [William Dean Howells], "Editor's Easy Chair," *Harper's Monthly Magazine* 108, no. 647 (April 1904), 802–6; and L. Margaret Barnett, "Fletcherism: The Chew-Fad of the Edwardian Era," in *Nutrition in Britain: Science, Scientists, and Politics in the Twentieth Century*, ed. David F. Smith, 6–28 (London: Routledge, 1997), esp. 9–10.

9. Luigi Cornaro, *A Treatise of Temperance and Sobrietie*, trans. George Herbert, bound together with and separately paginated in Leonard Lessius, *Hygiasticon*, 2nd ed. (Cambridge, 1634), 24.

10. Cornaro, *Temperance*, 5–7. For Cornaro and the Italian cultural context of thought about taste and gluttony, see Laura Giannetti, "Of Eels and Pears: A Sixteenth-Century Debate on Taste, Temperance, and the Pleasures of the Senses," in *Religion and the Senses in Early Modern Europe*, ed. Wietse de Boer and Christine Göttler (Leiden: Brill, 2013), 298–301.

11. Cornaro, *Temperance*, 23.

12. Cornaro, *Temperance*, 7.

13. Sander L. Gilman, *Diets and Dieting: A Cultural Encyclopedia* (New York: Routledge, 2008), 62.

14. See chapter 3, 131–33 (above).

15. Cornaro, *Temperance*, 8–9.

16. Cornaro, *Temperance*, 33. See also Marisa Milani, "Introduction to Cornaro," in Cornaro, *Writings on the Sober Life*, 3–69, on 22–23.

17. Cornaro, *Temperance*, 19.

18. Cornaro, *Temperance*, 17. For the force of custom conceived as second nature, see Steven Shapin, "Why Was 'Custom a Second Nature' in Early Modern Medicine?," *Bulletin of the History of Medicine* 93 (2019): 1–26.

19. Cornaro, *Temperance*, 16–20.

20. Cornaro, *Temperance*, 35.

21. Cornaro, *Temperance*, 22.

22. Cornaro, *Temperance*, 25–26.

23. For treatment of one of the earliest dietary writers impressed by Cornaro, see an account of his near contemporary Girolamo Cardano in Nancy G. Siraisi, *The Clock and the Mirror: Girolamo Cardano and Renaissance Medicine* (Princeton, NJ: Princeton University Press, 1997), 79–85. See also Bellini, "Diet and Hygiene"; and Shapin, "Was Luigi Cornaro a Dietary Expert?," 142–44.

24. Among many examples, see Robert Crofts, *The Way to Happinesse on Earth Concerning Riches, Honour, Conjugall Love, Eating, Drinking* (London, 1641), 250–51; and Richard Steele, *A Discourse Concerning Old Age* (London, 1688), 119.

25. Leonardus Lessius, *Hygiasticon: or, The Right Course of Preserving Life and Health unto Extream Old Age*, 2nd ed., trans. T[imothy] S[mith] [?] (Cambridge: University of Cambridge, [1614] 1634), 11. (I cannot find a reference to a first English edition.) For bibliographic discussion of this and subsequent editions, see Milani, "How to Attain Immortality," 189–98. See also Cristiano Casalini and Laura Madella, "The Jesuit Cultivation of Vegetative Souls: Leonard Lessius (1554–1623) on a Sober Diet," in *Vegeta-*

tive Powers: The Roots of Life in Ancient, Medieval and Early Modern Natural Philosophy, ed. Fabrizio Baldassarri and Andreas Blank (Zurich: Springer, 2021), 177–98.

26. Lessius, *Hygiasticon*, 55–56; Cornaro, *Writings on the Sober Life*, 114 and n62; and Milani, "How to Attain Immortality," 192.

27. Lessius, *Hygiasticon*, esp. 47–55, 81, 88–89.

28. Joseph Addison, *The Spectator*, no. 195 (October 13, 1711), in *The Spectator*, 4 vols. (London: Thomas Bosworth, 1853), 1:135.

29. George Cheyne, *An Essay of Health and Long Life* (London: George Strahan, 1724), 206, 226; idem, *The Natural Method of Cureing the Diseases of the Body and the Disorders of the Mind*, 4th ed. (London: George Strahan, 1742), 74; and letter from Cheyne to Samuel Richardson, December 23, 1741, in *The Letters of Doctor George Cheyne to Samuel Richardson (1733–1743)*, ed. Charles F. Mullett, University of Missouri Studies, Volume XVIII, no. 1 (Columbia: University of Missouri Press, 1943), 77.

30. These mathematizing movements are treated in chapter 6 (below).

31. Samuel-Auguste Tissot, *An Essay on Diseases Incident to Literary and Sedentary Persons*, 2nd ed. (London: J. Nourse, [1768] 1769), 131–32.

32. Christoph Wilhelm Hufeland, *The Art of Prolonging Human Life*, trans. anon. (London: Simpkin and Marshall, [1798] 1829), 11, 280.

33. William Buchan, *The New Domestic Medicine: or, Universal Family Physician*, new ed. (London: Alex. Hogg, 1802), 95–96. (This passage does not appear in earlier editions of Buchan and may have been added by the editor, Dr. George Wallis.)

34. Pierre Flourens, *On Human Longevity and the Account of Life upon the Globe*, trans. from 2nd French edition by Charles Martel (London: H. Bailliere, [1854] 1855), 1–6.

35. Francis Bacon, *Historie Naturall and Experimentall of Life and Death* (London, 1638), 156. Other skeptics recruited Hippocratic authority to support the view that "great abstinence [is] hurtful." Benito Jerónimo Feijóo y Montenegro, *Rules for Preserving Health, Particularly with Regard to Studious Persons*, trans. anon. from Spanish (London: R. Faulder, [1727] 1800[?]), 83.

36. Thomas Muffett, *Healths Improvement: or, Rules Comprizing and Discovering the Nature, Method, and Manner of Preparing All Sorts of Foods* (London, [comp. ca. 1600] 1655), 278–79.

37. Robert Burton, *The Anatomy of Melancholy* (New York: NYRB Books, [1621] 2001), Part 1, 230; see also Part 3, 342–43. The seventeenth-century sense of "cockney" (a hen's egg and, by analogy, a coddled child) did indeed pick out town dwellers in general and Londoners in particular but more directly pointed to people who were effete and squeamish: milk-sops. The "merry-thought" of a fowl was what is now called the wishbone,

carrying the idea that the person who wound up with the longer bone in the customary pulling contest would marry first. On the restorative powers of chicken soup and other highly extracted meat broths, see Ken Albala, "Food for Healing: Convalescent Cookery in the Early Modern Era," *Studies in History and Philosophy of Biological and Biomedical Sciences* 43 (2012): 326–28.

38. Muffett, *Healths Improvement*, 278–79.

39. Andrew Combe, *The Physiology of Digestion, Considered with Relation to the Principles of Dietetics*, 2nd ed. (Edinburgh: Maclachlan & Stewart, [1834] 1836), 226.

40. Thomas Bersford, *Theories and Facts for Students of Longevity and Health* (San Francisco: Thomas Bersford, 1908), 46–47 (on "the Cornaro Theory").

41. Woods Hutchinson, "The Dangers of Undereating," *Cosmopolitan Magazine* 47, no. 1 (June 1909), 390.

42. For example, Arthur V. Everitt et al., eds., *Calorie Restriction, Aging and Longevity* (New York: Springer, 2010), 15–16.

43. See, e.g., Elaine Leong, *Recipes and Everyday Knowledge: Medicine, Science, and the Household in Early Modern England* (Chicago: University of Chicago Press, 2018); and Michelle DiMeo and Sara Pennell, eds., *Reading and Writing Recipe Books, 1550–1800* (Manchester: Manchester University Press, 2013).

44. George Cheyne, *The English Malady: or, a Treatise of Nervous Diseases of All Kinds, as Spleen, Vapours, Lowness of Spirits, Hypochondriacal, and Hysterical Complaints* (London: G. Strahan, 1733), 342 and 354 (for weight loss). On the history of ideas about obesity, see, for example, Ken Albala, "The Apparition of Fat in Western Nutritional Theory," in *The Fat of the Land: Proceedings of the Oxford Symposium on Food and Cookery 2002*, ed. Harlan Walker, 1–10 (Bristol: Footwork, 2003), esp. 4–5.

45. Cheyne, *English Malady*, 342; and letter from Cheyne to Samuel Richardson, December 23, 1741, in *Letters of Cheyne to Richardson*, 76–77.

46. As early as the 1720s, the Irish physician Bryan Robinson also knew his weight in pounds and its variation over time. See Bryan Robinson, *A Dissertation on the Food and Discharges of Human Bodies* (Dublin: S. Powell, 1747), 34–35, 68, 83–85. But the personal possession of this sort of knowledge was exceptional. See Lucia Dacome, "Living with the Chair: Private Excreta, Collective Health and Medical Authority in the Eighteenth Century," *History of Science* 39 (2001): 467–500, esp. 483. Kula's classic work on the history of measuring practices contained much about weighing commodities but nothing about weighing human beings. See Witold Kula, *Measures and Men*, trans. R. Szreter (Princeton, NJ: Princeton University Press, 1986). On the early history of weighing instruments, see Bruno

Kisch, *Scales and Weights: A Historical Outline* (New Haven, CT: Yale University Press, 1965).

47. Dacome, "Living with the Chair," 488–89; and Leopold Wagner, *A New Book about London* (New York: E. P. Dutton, 1921), 90.

48. See Schwartz, *Never Satisfied*, 164–73. See also Deborah I. Levine, "Managing American Bodies: Diet, Nutrition, and Obesity in America, 1840–1920," PhD dissertation, Harvard University, 2008, esp. chap. 2; and idem, "Measure, Record, Share: Weight Loss, Biometrics, and Self-Tracking in the U.S.," *American Journal of Preventive Medicine* 55, no. 5 (2018): e147–e151.

49. J[ohn] B[ulwer], *Athomyotomia or a Dissection of the Significative Muscles of the Affections of the Minde* (London, 1649), 102–4.

50. Jonathan Barry and Fabrizio Bigotti, eds., *Santorio Santori and the Emergence of Quantified Medicine, 1614–1790* (Cham, Switzerland: Springer, 2022). Albala, "Apparition of Fat," 4, calls Santorio "the first dietary writer to have systematically weighed himself."

51. Here the outstanding historical work is by Lucia Dacome, especially her "Living with the Chair"; idem, "Balancing Acts: Picturing Perspiration in the Long Eighteenth Century," *Studies in History and Philosophy of Biological and Biomedical Sciences* 43 (2012): 379–91; and idem, "Resurrecting by Numbers in Eighteenth-Century England," *Past and Present*, no. 193 (November 2006): 73–110, esp. 88–92. For details on the chair's construction and operation, see Teresa Hollerbach, "The Weighing Chair of Sanctorius Sanctorius: A Replica," *NTM* 26 (2018): 121–49.

52. For treatment of insensible transpiration and "skin breathing," see Oriana Walker, "The Breathing Self: Toward a History of Respiration," PhD dissertation, Harvard University, 2016, 56–60. See also Michael Stolberg, "Sweat, Learned Concepts and Popular Perceptions, 1500–1800," in *Blood, Sweat and Tears: The Changing Concepts of Physiology from Antiquity into Early Modern Europe*, ed. Manfred Horstmanshoff, Helen King, and Claus Zittel, 503–22 (Leiden: Brill, 2012); and Salvatore Ricciardo, "'An Inquisitive Man, Considering When and Where He Liv'd': Robert Boyle on Santorio Santori and Insensible Perspiration," in Barry and Bigotti, eds., *Santorio Santori*, 239–72.

53. Santorio Santorio, *Medicina Statica: or, Rules of Health*, trans. J. D. (London, 1676), sig. A5.

54. Dr. John Quincy, "Preface," in Santorio Santorio, *Medicina Statica: Being the Aphorisms of Sanctorious, Translated into English by Dr. John Quincy*, 5th ed. (London: T. Longman and J. Newton, [1614; trans. 1712] 1737), v–vi.

55. James Mackenzie, *The History of Health, and the Art of Preserving It*, 3rd ed. (Edinburgh: William Gordon, [1758] 1760), 269.

56. James Keill, *Medicina Statica Britannica*, appended to Santorio, *Medicina Statica* (1737), 321–44, quoting 338. And see here Dacome, "Living with the Chair," 479, 488; and Fabiola Zurlini, "The Uncertainty of Medicine: Readings and Reactions to Santorio between Tradition and Reformation (1615–1721)," in Barry and Bigotti, ed., *Santorio Santori*, 103–17, esp. 111–13.

57. Addison, *The Spectator*, no. 25 (March 29, 1711), 77–80. See also Dacome, "Living with the Chair," 478–79.

58. Henry Peacham, *The Compleat Gentleman, Fashioning Him Absolute in the Most Necessary & Commendable Qualities* (London, 1622), 195.

59. William Buchan, *Domestic Medicine; or, the Family Physician* (Edinburgh: Balfour, Auld, and Smellie, 1769), 64.

60. Richard Baxter, *A Christian Directory, or, A Summ of Practical Theologie and Cases of Conscience* (London, 1673), 374.

61. François, duc de La Rochefoucauld, *The Maxims of La Rochefoucauld*, trans. Louis Kronenberger (New York: Random House, [1655] 1959), 142.

62. William Rawley, "The Life of the Honourable Author," in Francis Bacon, *Resuscitatio, or, Bringing Into Publick Light Severall Pieces of the Works . . . of the Right Honourable Francis Bacon* (London, 1657), sig. b2–c2.

63. Michel Eyquem de Montaigne, "Of Experience," in idem, *The Essayes of Michael Lord of Montaigne*, trans. John Florio, 3 vols. (London: J. M. Dent, [1580–1588] 1910), 3:340, 344.

64. Guglielmo Gratarolo, *A Direction for the Health of Magistrates and Studentes*, trans. Thomas Newton (London, 1574), sig Biii.

65. Feijóo y Montenegro, *Rules for Preserving Health*, 79–80. Feijóo lived from 1676 to 1764, and the Spanish original was probably published in 1727.

66. James Cleland, *The Instruction of a Young Noble-Man* (Oxford, 1612), 210. See also James I, King of England, *Basilikon Doron. Or His Maiesties Instructions to his Dearest Sonne, Henrie the Prince*, 2nd ed. (London, 1603), 108.

67. Pierre Charron, *Of Wisdome*, trans. Samson Lennard (London, [1601] 1608), 541. See also Montaigne, "Of Experience," 344. For Celsus-citing into the nineteenth century on the importance of flexibility, see, e.g., Anon. ["The Physician"], "On the Power of Habit," *New Monthly Magazine* 8 (1823), 326–34, esp. 328–29.

68. Montaigne, "Of Experience," 348.

69. Plutarch, "Rules for the Preservation of Health," in *Plutarch's Lives and Miscellanies*, ed. A. H. Clough and William W. Goodwin, 5 vols. (New York: Colonial Company, 1905), 1:263.

70. Montaigne, "Of Experience," 344, 346.

71. Lessius, *Hygiasticon*, 1–3.

72. Burton, *Anatomy of Melancholy*, Part 2, 29.

73. Humphrey Brooke, *Ugieine: or A Conservatory of Health* (London, 1650), 101. See also Walter Charleton, *Enquiries into Human Nature* (London, 1680), 13.

74. George Cheyne, *An Essay of Health and Long Life* (London: G. Strahan, 1724), 4. See also idem, *An Essay on Regimen* (London: C. Rivington, 1740), xxxvi–xxxvii, xlvi.

75. Francis Bacon, "Of Regiment of Health," in idem, *The Essayes or Counsels, Civill and Morall*, ed. Michael Kiernan (Cambridge, MA: Harvard University Press, [1597] 1985), 101.

76. Bacon, *Life and Death*, 157–58. See also Cogan, *Haven of Health*, 202; and Feijóo y Montenegro, *Rules for Preserving Health*, 83. There was also medical opposition to these conceptions of moderation. See, among other examples, Santorio, *Medicina Statica* (1737), 188: "The Opinion of *Celsus*, that the Non-naturals ought sometimes to be used sparingly, and at others very liberally, is not safe for all Persons."

77. Burton, *Anatomy of Melancholy*, Part 1, 230 (citing *Aphorisms* I, no. 5). See also Louis Lémery, *A Treatise of All Sorts of Food*, trans. D. Hay (London: T. Osborne, [1702] 1745), 10.

78. Bacon, "Regiment of Health," 101.

79. Rawley, "Life of the Honourable Author," sig. c2. See also Sprengell, *Aphorisms of Hippocrates and Sentences of Celsus*, 239–40.

80. Feijóo y Montenegro, *Rules for Preserving Health*, 85.

81. John Aubrey, "Thomas Hobbes," in *Aubrey's Brief Lives*, ed. Oliver Lawson Dick (Ann Arbor: University of Michigan Press, 1975), 155.

82. A source in Avicenna is O. Cameron Gruner, *A Treatise on the Canon of Medicine by Avicenna Incorporating a Translation of the First Book* (London: Luzac, 1930), 412, where Avicenna reported what had been said on the matter: "Some persons claim that it is an advantage to become intoxicated once or twice a month, for, they say, it allays the animal passions, inclines to repose, provokes the urine and sweat, and gets rid of effete matters." In this connection, "effete" meant digested stuff from which the nutritive virtues had been extracted, hence excremental. There was no specific reference here to vomiting, though later commentators understood Avicenna to be taking about just that. Published Latin versions of Avicenna's *Canon medicinae* were available to Europeans from 1484.

83. Montaigne, "Of Experience," 344. See also idem, "Of Drunkennesse," in idem, *Essayes*, 2:18 (quoting the sixteenth-century Parisian physician Jacobus Sylvius to the effect that getting drunk once a month helped "to rowze up . . . the vigor of our stomake.")

84. For drinking to excess as good fellowship, see, for example, Keith Thomas, *In Pursuit of Civility: Manners and Civilization in Early Modern England* (Waltham, MA: Brandeis University Press, 2018), 81, 229.

85. Cogan, *Haven of Health*, 245 (also 207–8, where Galen is the supposed authority). On the virtue of vomiting after unavoidable excess, see Obadiah Walker, *Of Education, Especially of Young Gentlemen* (London, 1673), 68.

86. Santorio, *Medicina Statica* (1737), 215.

87. Among many critical comments, see James Primrose, *Popular Errours: Or the Errours of the People in Physick* (London, 1651), 189.

88. Gratarolo, *Direction for the Health*, sig. Kii.

89. Tobias Venner, *Via recta ad vitam longam* (London, 1620), 36.

90. Thomas Browne, *Pseudodoxia Epidemica: Or, Enquiries into Very Many Received Tenents, and Commonly Presumed Truths* (London, [1646] 1650), 229. See also chapter 2, 90–95 (above).

91. Robert Boyle, *Some Considerations Touching the Usefulnesse of Experimental Naturall Philosophy* (London, 1663), 408. See also John Harris, *The Divine Physician: Prescribing Rules for the Prevention, and Cure of Most Diseases, as Well of the Body, as the Soul* (London, 1676), 30–32.

92. Mackenzie, *History of Health*, 125–26. See also Sprengell, *Aphorisms of Hippocrates, and the Sentences of Celsus*, 244–45.

93. Boniface Oinophilos, de Monte Fiascone, *Ebrietatis Enconium: or, the Praise of Drunkenness . . .*, 2nd ed., trans. anon. (London: E. Curll, [French, 1714; trans. 1723] 1743), 27, 29, 32. The author was Albert-Henri de Sallengre (1694–1723), a French lawyer and man of letters living in the Netherlands. Versions of this text remain in print.

94. Joubert, *Second Part of the Popular Errors*, 117–18. Other texts in the "popular medical errors" tradition include Gaspard Bachot, *Erreurs populaires touchant la médecine et régime de santé* (Lyons, 1626); James Primrose, *Popular Errours. Or the Errours of the People in Physick* (London, [1639] 1651); and Jean-Luc d'Iharce, *Erreurs populaires sur la médecine* (Paris, 1783).

95. Joubert, *Second Part of Popular Errors*, 262.

96. Joubert, *Popular Errors*, 41–42.

97. Joubert, *Popular Errors*, 247–48. See also idem, *Second Part of Popular Errors*, 127–72.

98. Samuel-Auguste David Tissot, *Advice to the People in General, with Regard to Their Health*, trans. J. Kirkpatrick (London: T. Becket and P. A. De Hondt, [1761] 1765), 30–35, 61, 63–64, 576 (quotation). See also Antoinette Suzanne Emch-Dériaz, *Tissot: Physician of the Enlightenment* (New York: Peter Lang, 1992), esp. 66–71.

99. For suggestions about how, in general, to think of medical efficacy in relation to medical authority, see Charles E. Rosenberg, "The Therapeutic Revolution: Medicine, Meaning, and Social Change in Nineteenth-Century America," *Perspectives in Biology and Medicine* 20 (1977): 485–506.

100. Feijóo y Montenegro, *Rules for Preserving Health*, 88; see also 105.

101. For introduction to both *agreement* and the tag *Quod sapit*, see chapter 3, 131–33 (above).

102. For Hippocrates, see Bernard Mandeville, *A Treatise of the Hypochondriack and Hysterick Diseases*, 3rd ed. (London: J. Tonson, [1711] 1730), 154. The attribution must be a free one; the closest version found in Hippocrates's *Aphorisms* (II, no. 38) is "An article of food or drink which is slightly worse, but more palatable, is to be preferred to such as are better but less palatable." See also Aristotle, *Sense and Sensibilia*, 436b15, in *Complete Works*, 2 vols., ed. Jonathan Barnes (Princeton, NJ: Princeton University Press, 1984), 1:694.

103. This is further treated in chapter 6, 217–20 (below).

104. Thomas Tryon, *Tryon's Letters, Domestick and Foreign, to Several Persons of Quality* (London: Geo. Conyers, 1700), 10–11.

105. For both philosophical disdain and qualified medical approval of the reliability of taste, see Charles Burnett, "The Superiority of Taste," *Journal of the Warburg and Courtauld Institutes* 54 (1991): 230–38.

106. Aristotle said that sweetness indexed the capacity to nourish: "nourishment is effected by the sweet." Aristotle, *Sense and Sensibilia*, 422a8, 701. For medieval sensibilities toward the concept of sweetness, see Mary Carruthers, "Sweetness," *Speculum* 81 (2006): 999–1013. For sugar as medicine, see Sidney W. Mintz, *Sweetness and Power: The Place of Sugar in Modern History* (New York: Penguin, 1986), 30, 45, 79, 87.

107. Venner, *Via recta vitam longam*, 104–5. See also Mintz, *Sweetness and Power*, 104.

108. William Cullen, *Lectures on the Materia Medica . . .* (Philadelphia: Robert Bell, [1760] 1775), 45–47, 55, 93–94, 166 (quoting 45–47, 93). This text was based on Edinburgh University lecture notes from the early 1760s. In the late seventeenth century, the English physician John Floyer was one of several writers who not only offered a typology of plants' medicinal powers in terms of their tastes but also allowed that taste might be inferred from other sensory aspects, including sight and touch. John Floyer, *Pharmako-Basanos; or the Touch-Stone of Medicines* (London, 1687), e.g. 74–81. See also chapter 7, 275–77 (below).

109. The *Oxford English Dictionary* notes that the term "disgust"—from the French *desgoust* or the Italian *disgusto*—does not appear in English before about 1600 and is not found in Shakespeare. Much the same applies to the term "distaste."

110. Nicolas Malebranche, *The Search after Truth*: With *Elucidations of the Search after Truth*, trans. and ed. Paul S. Olscamp and Thomas M. Lennon (Cambridge: Cambridge University Press, [1674–1678] 1997), 645–46. See also Kenelm Digby, *Two Treatises, in One of Which, the Nature of Bodies . . . Is Looked Into . . .* (Paris, 1644), 246–47: smell is "nothing but

a passage of . . . exhalations and little bodies unto the braine, fittly accommodated to discerne, what is good, or hurtfull for it, and accordingly to move the body to admit or reject" the foods. Foods that smell bad are, in general, bad for you. People would find the sense of smell more reliable in assessing quality "were they not continually stuffed and clogged with grosse vapours of steamy meates, which are dayly reeking from the table and their stomackes."

111. Tryon, *Tryon's Letters*, 14–15.

112. Jean-Jacques Rousseau, *Émile*, trans. Barbara Foxley (London: J. M. Dent, [1762] 1961), 115–17. See also Rousseau, *Julie, or the New Heloise: Letters of Two Lovers Who Live in a Small Town at the Foot of the Alps*, ed. and trans. Philip Stewart and Jean Vaché (Hanover, NH: University Press of New England, [1760] 2010), esp. 372–73; and S. K. Wertz, "Taste and Food in Rousseau's *Julie, or the New Heloise*," *Journal of Aesthetic Education* 47 (2013): 24–35, esp. 28–29. For dietary adaptability as a practical necessity for a range of public actors, see also chapter 4, 142–45 (above).

113. Carolyn Korsmeyer, *Making Sense of Taste: Food and Philosophy* (Ithaca, NY: Cornell University Press, 1999), esp. chap. 2; George Dickie *The Century of Taste: The Philosophical Odyssey of Taste in the Eighteenth Century* (Oxford: Oxford University Press, 1996); and Steven Shapin, "The Sciences of Subjectivity," *Social Studies of Science* 42 (2012): 170–84, esp. 172–76.

114. Thomas Reid, *The Works of Thomas Reid, D.D.*, ed. Sir William Hamilton, 7th ed., 3 vols. (Edinburgh: Maclachlan and Stewart, [1785] 1872), 1:491.

115. Montaigne, "Of Experience," 340, 348.

116. Montaigne, "Of Experience," 348.

117. Malebranche, *Search After Truth*, 647.

118. Cheyne, *English Malady*, 157.

119. *Dr. Boerhaave's Academical Lectures on the Theory of Physic*, 6 vols. (London: W. Innys, 1742–1746), 4:21. This text was assembled from lecture notes taken by his students at the University of Leiden.

120. Feijóo y Montenegro, *Rules for Preserving Health*, 76–78. See also Shapin, "Why Was 'Custom a Second Nature?,'" 14–16.

121. For example, Combe, *Physiology of Digestion*, 325–26.

122. Cornaro, *Temperance*, 8.

123. John Armstrong, *The Art of Preserving Health: A Poem* (London: A. Millar, 1744), 35, 41.

124. David Hartley, *Observations on Man, His Frame, His Duty, and His Expectations*, 6th ed. (London: Thomas Tegg, [1749] 1834), 459. Hartley generally warned of the unwholesomeness of any aliment having a "high relish."

125. See, in these connections, Massimo Montanari, *Medieval Tastes: Food, Cooking, and the Table*, trans. Beth Archer Brombert (New York: Columbia University Press, 2015), chap. 18.

Chapter 6

1. Nicholas Robinson, *A New Theory of Physick and Diseases Founded on the Principles of the Newtonian Philosophy* (London: C. Rivington, 1725), 192.

2. For immortality in relation to the capacities and aims of early modern medical care, see, e.g., Gerald J. Gruman, "A History of Ideas about the Prolongation of Life," *Transactions of the American Philosophical Society*, new series, 51, no. 9: 1–102; Lucian Boia, *Forever Young: A Cultural History of Longevity from Antiquity to the Present*, trans. Trista Selous (London: Reaktion, [1998] 2004); and David Boyd Haycock, *Mortal Coil: A Short History of Living Longer* (New Haven, CT: Yale University Press, 2008.

3. [Thomas Baker], *Reflections upon Learning, Wherein Is Shewn the Insufficiency Thereof, in Its Several Particulars*, 2nd ed. (London: A. Bosvile, [1699] 1700), 185.

4. Michel Eyquem de Montaigne, "Of Experience," in idem, *The Essayes of Michael Lord of Montaigne*, 3 vols., trans. John Florio (London: J. M. Dent, [1580–1588] 1910), 3:349, 351. See also Max Neuburger, *The Doctrine of the Healing Power of Nature throughout the Course of Time*, trans. Linn J. Boyd (New York: privately printed, 1932); and Hannah Newton, "'Nature Concocts & Expels': The Agents and Processes of Recovery from Disease in Early Modern England," *Social History of Medicine* 28 (2015): 465–86.

5. Montaigne, "Of Experience," 340.

6. Robert Burton, *The Anatomy of Melancholy* (New York: NYRB Books, [1621] 2001), e.g., "Democritus Junior to the Reader," 36. Severinus (1542–1602) was a follower of the polemical Swiss German chemist Paracelsus, who commended mineral remedies over the herbals favored by the Galenists and, in general, expressed contempt for Galenic authority. See Jole Shackelford, *A Philosophical Path for Paracelsian Medicine: The Ideas, Intellectual Context, and Influence of Petrus Severinus (1540–1602)* (Copenhagen: Museum Tusculanum Press, University of Copenhagen, 2004).

7. Benito Jerónimo Feijóo y Montenegro, *Rules for Preserving Health, Particularly with Regard to Studious Persons*, trans. anon. from Spanish (London: R. Faulder, [1727?] 1800?), 1–2, 6–9.

8. Francis Bacon, "The Advancement of Learning [Books I–II]," in *The Philosophical Works of Francis Bacon*, ed. James Spedding, Robert Leslie Ellis, and Douglas Denon Heath, 5 vols. (London: Longman and Co., 1857–1858), 3:367, 373.

9. Francis Bacon, *Historie Naturall and Experimetall, of Life and Death* (London, 1638); and idem, "Of the Dignity and Advancement of Learning, Book IV," in *Philosophical Works*, 4:390. See also Graham Rees's introduction to his edition of Bacon's *De vijs mortis* in Francis Bacon, *Philosophical Studies c. 1611–c. 1619*, ed. Rees (Oxford: Clarendon, 1996), xvii–cx. For documentation of seventeenth-century English medical concern with the prolongation of life, see Charles Webster, *The Great Instauration: Science, Medicine and Reform 1626–1660* (London: Duckworth, 1975), 246–323.

10. Roger K. French, *William Harvey's Natural Philosophy* (Cambridge: Cambridge University Press, 1994); and Robert G. Frank, *Harvey and the Oxford Physiologists: Scientific Ideas and Social Interaction* (Berkeley: University of California Press, 1980).

11. René Descartes, "Discourse on the Method," in *The Philosophical Writings of Descartes*, trans. John Cottingham, Robert Stoothoff, and Dugald Murdoch, 3 vols. (Cambridge: Cambridge University Press, 1984–1991), 1:142–43, 151. This material is treated at length in Steven Shapin, "Descartes the Doctor: Rationalism and Its Therapies," *British Journal for the History of Science* 33 (2000): 131–54.

12. Descartes, "Principles of Philosophy (extracts)," in idem, *Philosophical Writings*, 1:186.

13. René Descartes, "Preface," in *The Passions of the Soul*, ed. and trans. Stephen Voss (Indianapolis: Hackett, [1649] 1989), 7.

14. For example, Jole Shackelford, *William Harvey and the Mechanics of the Heart* (New York: Oxford University Press, 2003), 107–9.

15. René Descartes, "The Description of the Human Body," in idem, *Philosophical Writings*, 1:319. See also idem, "Discourse on Method," 136–39.

16. Descartes, "Description of the Human Body," 316, 319.

17. René Descartes, *Treatise of Man*, trans. Thomas Steele Hall (Cambridge, MA: Harvard University Press, [ca. 1629–1633] 1972), 34–35. See also Jessica Riskin, *The Restless Clock: A History of the Centuries-Long Argument over What Makes Living Things Tick* (Chicago: University of Chicago Press, 2016), 47–57.

18. Descartes, *Treatise of Man*, 5–9.

19. Robert Boyle, *Some Considerations Touching the Usefulnesse of Experimentall Naturall Philosophy* (London, 1663), 6–7, 270, 299. See also idem, *Of the Reconcileableness of Specifick Medicines to the Corpuscular Philosophy* (London, 1685).

20. Boyle, *Usefulnesse of Experimentall Naturall Philosophy*, 95, 113.

21. Archibald Pitcairne, *The Works of Dr. Archibald Pitcairn* (London: E. Curll, 1715), 7, 10, 17.

22. Pitcairne, *Works*, 13.

23. See Owsei Temkin, *Hippocrates in the World of Pagans and Christians* (Baltimore: Johns Hopkins University Press, 1991, 8–9.

24. Galileo Galilei, "The Assayer [excerpts]," in *Discoveries and Opinions of Galileo*, trans. Stillman Drake (Garden City, NY: Doubleday, [1623] 1957), 274.

25. John Locke, *An Essay Concerning Humane Understanding* (London, 1690), 57.

26. Riskin, *Restless Clock*, 53–56.

27. Locke, *Humane Understanding*, 271, 278.

28. Robert Hooke, *Micrographia, or, Some Physiological Descriptions of Minute Bodies Made by Magnifying Glasses* (London, 1665), "Preface," sig. b3. See also Walter Charleton, *Physiologia Epicuro-Gassendo-Charltoniana, or, A Fabrick of Science Natural, upon the Hypothesis of Atoms founded by Epicurus repaired [by] Petrus Gassendus* (London, 1654), 117–19. Micromechanism supported by inference from the then microscopically visible to the not yet microscopically visible is suggested by the work of the philosopher Catherine Wilson, *The Invisible World: Early Modern Philosophy and the Invention of the Microscope* (Princeton, NJ: Princeton University Press, 1995).

29. Jacques Rohault, *Rohault's System of Natural Philosophy*, vol. 1, trans. John Clarke, (London: James Knapton, [1671] 1723), "Preface," sig. b2; see also 113–14.

30. On this, see Alan Gabbey, "The Mechanical Philosophy and Its Problems: Mechanical Explanations, Impenetrability, and Perpetual Motion," in *Change and Progress in Modern Science*, ed. Joseph C. Pitt, 9–84 (Dordrecht: D. Reidel, 1985).

31. Specific chemical sources for changing thought about food and its characteristics and functions are treated in chapter 7 (below).

32. Descartes, *Treatise of Man*, 99, 102, 108.

33. Robert Boyle, *A Free Enquiry into the Vulgarly Receiv'd Notion of Nature* (London, 1686), 304–5, 330–31. See also Walter Charleton, *Enquiries into Human Nature* (London, 1680), "Preface," sig. B1r.

34. Locke, *Humane Understanding*, 278.

35. [George Cheyne], *A New Theory of Continu'd Fevers* (Edinburgh: John Vallange, 1701), "Preface," [2]. See also Anita Guerrini, "Isaac Newton, George Cheyne and the 'Principia Medicinae,'" in *Medical Revolution of the Seventeenth Century*, ed. French and Wear, 227.

36. Julian Offrey de la Mettrie, *Man a Machine*, trans. anon. (Dublin: W. Brien, [1747] 1749), 12–13, 61, 63–64. See also Riskin, *Restless Clock*, 152–60.

37. Shapin, "Descartes the Doctor," esp. 145–48 (for the passions). See also Peter Dear, "A Mechanical Microcosm: Bodily Passions, Good Manners, and Cartesian Mechanism," in *Science Incarnate: Historical Embodiments of*

Natural Knowledge, ed. Christopher Lawrence and Steven Shapin, 51–82 (Chicago: University of Chicago Press, 1998).

38. Isaac Newton, "Some Thoughts about the Nature of Acids" [a translation by John Harris of the paper *De natura acidorum* (1692)]," in *Isaac Newton's Papers & Letters on Natural Philosophy*, 2nd ed., ed. I. Bernard Cohen and Robert E. Schofield (Cambridge, MA: Harvard University Press, 1978), 257.

39. Isaac Newton, *Opticks: or, A Treatise of the Reflections, Refractions, Inflections and Colours of Light*, 4th ed. (London: William Innys, 1730), 350–53, 355, 362, 369.

40. See, for example, Theodore M. Brown, "The College of Physicians and the Acceptance of Iatro-mechanism in England, 1665–1695," *Bulletin of the History of Medicine* 44 (1970): 12–30; idem, "Physiology and the Mechanical Philosophy in Mid-Seventeenth-Century England," *Bulletin of the History of Medicine* 51 (1977): 25–54; idem, "Medicine in the Shadow of the *Principia*," *Journal of the History of Ideas* 48 (1987): 629–48; and Harold J. Cook, "The New Philosophy and Medicine in Seventeenth-Century England," in *Reappraisals of the Scientific Revolution*, ed. David C. Lindberg and Robert S. Westman (Cambridge: Cambridge University Press, 1990), 397–436.

41. The term "iatromathematics" was introduced in English in the seventeenth century (though initially linking medicine to astrology); "iatromechanism" was a later coining. Historical treatments include Anita Guerrini, "James Keill, George Cheyne, and Newtonian Physiology, 1690–1740," *Journal of the History of Biology* 18 (1985): 247–66; idem, "The Tory Newtonians: Gregory, Pitcairne, and Their Circle," *Journal of British Studies* 25 (1986): 288–311; idem, "Archibald Pitcairne and Newtonian Medicine," *Medical History* 31 (1987), 70–83; idem, "Newtonianism, Medicine, and Religion," in *Religio Medici: Medicine and Religion in Seventeenth-Century England*, ed. Ole Peter Grell and Andrew Cunningham (Aldershot: Scolar Press, 1996), 293–312; and Simon Schaffer, "The Glorious Revolution and Medicine in Britain and the Netherlands," *Notes and Records of the Royal Society of London* 43 (1989): 167–90.

42. French, *Medicine before Science*, 197–99; and Anna Marie Roos, *The Salt of the Earth: Natural Philosophy, Medicine, and Chymistry in England, 1650–1750* (Leiden: Brill, 2007), esp. 132–33.

43. Richard Mead, "Preface," in *A Mechanical Account of Poisons*, in idem, *The Medical Works of Richard Mead, M.D.* (London: C. Hitch and L. Lawes, [1702] 1762), x.

44. Cheyne, *A New Theory of Fever*, 11–13. In contemporary medical usage, a "gland" was taken to be an organ that either extracted usable constituents from the blood or secreted them into the blood. See also Andrew

Cunningham, "Sydenham versus Newton: The Edinburgh Fever Dispute of the 1690s between Andrew Brown and Archibald Pitcairne," *Medical History*, suppl. 1 (1981): 71–98.

45. Jeremiah Wainewright, *A Mechanical Account of the Non-Naturals: Being a Brief Explication of the Changes Made in Humane Bodies, by Air, Diet, &c.* (London: Ralph Smith, 1707), "Preface," sig. A2, 169.

46. Wainewright, *Mechanical Account*, 69–70.

47. Robinson, *New Theory of Physick*, 6.

48. Robinson, *New Theory of Physick*, ix; see also 7–8, 18.

49. Robinson, *New Theory of Physick*, viii, 9–10, 14–15, 20, 49. On the emerging eighteenth-century focus on the body's "solid" fibers, as opposed to traditional stress on the fluid humors, see, e.g., Hisao Ishizuka, *Fiber, Medicine, and Culture in the British Enlightenment* (London: Palgrave Macmillan, 2016).

50. [William Burton], *An Account of the Life and Writings of Herman Boerhaave* (London: Henry Lintot, 1743), 28–29.

51. See, for example, Andrew Cunningham, "Medicine to Calm the Mind: Boerhaave's Medical System and Why It Was Adopted in Edinburgh," in *The Medical Enlightenment of the Eighteenth Century*, ed. Andrew Cunningham and Roger French, 40–66 (Cambridge: Cambridge University Press, 1990), 49–54.

52. Bernard Mandeville, *A Treatise of the Hypochondriack and Hysterick Diseases*, 3rd ed. (London: J. Tonson, [1711] 1730), 115, 168–74 (quoting 115). See also Harold J. Cook, "Bernard Mandeville and the Therapy of 'The Clever Politician,'" *Journal of the History of Ideas* 60 (1999): 119–20.

53. For example, Thomas Sydenham, "The Author's Preface," in *The Entire Works of Dr Thomas Sydenham, Newly Made English from the Originals*, 5th ed., trans. John Swan (London: F. Newbery, 1769), xxvii.

54. Rohault, *Rohault's System*, 1:171–74. In the 1650s, the English atomist Walter Charleton maintained that the only intelligible way of explaining how aromatic substances brought about olfactory sensations was by supposing that there must be some sort of geometrical (lock-and-key type) fit between "the Particles" that caused smell and the olfactory nerves, the same applying to the connection between sapid particles and the sensations of taste. Charleton, *Physiologia Epicuro-Gassendo-Charltonia*, 236, 245. See also Kenelm Digby, *Two Treatises, in One of Which, the Nature of Bodies . . . Is Looked Into . . .* (Paris, 1644), 245–49 (on the causes of taste and smell by way of "reall and solide parts" of substances, which work their effects on the sensing organ, "and not by imaginary qualities," though Digby did not here specify particulate shapes, sizes, textures, and the like [246]).

55. George Cheyne, *Essay on Regimen*, 2nd ed. (London: C. Rivington, 1740), ii–iii. See also Anita Guerrini, *Obesity and Depression in the*

Enlightenment: The Life and Times of George Cheyne (Norman: University of Oklahoma Press, 2000), 57. The material here and in the next several paragraphs is drawn from Steven Shapin, "Trusting George Cheyne: Scientific Expertise, Common Sense, and Moral Authority in Early Eighteenth-Century Dietetic Medicine," *Bulletin of the History of Medicine* 77 (2003): 263–97.

56. Cheyne, *Essay on Regimen*, iv.

57. George Cheyne, *An Essay of Health and Long Life* (London: George Strahan, 1724), 220, 222. See also John Arbuthnot, *An Essay Concerning the Nature of Aliments, and the Choice of Them, according to the Different Constitutions of Human Bodies*, 4th ed. (London: J. and R. Tonson, [1731] 1756), 167: "The great Secret of Health is keeping the Fluids in due Proportion to the Capacity and Strength of the Channels, through which they pass."

58. Cheyne, *Health and Long Life*, 172–75; and idem, *English Malady*, 5.

59. Cheyne, *Health and Long Life*, 224.

60. Cheyne, *Essay on Regimen*, x.

61. Cheyne, *Health and Long Life*, 224–25.

62. Cheyne, *Essay on Regimen*, xlix.

63. George Cheyne, *The English Malady: or, a Treatise of Nervous Diseases of All Kinds* (London: G. Strahan, 1733), 26–27. "Aqua fortis" designated what is now called nitric acid; and "aqua regia" was a mixture of nitric and hydrochloric acids.

64. On this pervasive but little noticed feature of seventeenth- and eighteenth-century micromechanism, see Gabbey, "Mechanical Philosophy and Its Problems."

65. George Cheyne, *The Natural Method of Cureing the Diseases of the Body*, 4th ed. (London: George Strahan, 1742), 16; and idem, *Philosophical Principles of Religion, Natural and Revealed*, 5th ed. (London: George Strahan, [1725] 1736), 62.

66. Cheyne, *English Malady*, 44–45; see also 124.

67. For example, Cheyne, *Essay on Regimen*, xiii–xiv; cf. 116.

68. Cheyne, *Health and Long Life*, 224.

69. Cheyne, *Essay on Regimen*, xxiii–xxiv, lviii.

70. Cheyne, *Health and Long Life*, 36, 43–44, 224; and idem, *Essay on Regimen*, 56, 67.

71. Cheyne, *Natural Method*, 57–58.

72. Cheyne, *Natural Method*, lxiv, 51–53, 117; idem, *Essay on Regimen*, 102; and idem, *Philosophical Principles*, 60–66. As a fashionable spa physician, Cheyne advertised special expertise in knowing the constituents and effects of mineral waters from different sources, especially those of Bath, Bristol, Cheltenham, Clifton, Islington, Tunbridge, Pyrmont (in Westphalia), and Spa (in the Ardennes). See George Cheyne, *Observations*

Concerning the Nature and Due Method of Treating the Gout, . . . Together With an Account of the Nature and Qualities of the Bath Waters (London: George Strahan, 1720).

73. Cheyne, *Essay on Regimen*, 266–68 (see also 109–11); idem, *Natural Method*, 119–26, 150; and idem, *English Malady*, 125–37.

Chapter 7

1. Harold J. Cook, *The Decline of the Old Medical Regime in Stuart England* (Ithaca, NY: Cornell University Press, 1986), esp. 28–69; and Mark S. R. Jenner and Patrick Wallis, ed., *Medicine and the Market in England and Its Colonies, c. 1450–c. 1850* (New York: Palgrave Macmillan, 2007). See also N. D. Jewson, "Medical Knowledge and the Patronage System in Eighteenth-Century England," *Sociology* 8 (1974): 369–85 (for the power relations between physicians and their elite patients).

2. Theodore M. Brown, "From Mechanism to Vitalism in Eighteenth-Century English Physiology," *Journal of the History of Biology* 7 (1974): 179–216, esp. 180; Christopher Lawrence, "Ornate Physicians and Learned Artisans: Edinburgh Medical Men, 1726–1776," in *William Hunter and the Eighteenth-Century Medical World*, ed. W. F. Bynum and Roy Porter, 153–76 (Cambridge: Cambridge University Press, 1985), esp. 155–59.

3. For example, Francis Clifton, *The State of Physick, Ancient and Modern* (London: W. Bowyer, 1732), 130–33, 152–54, 157–58, 165. See also David Cantor, ed., *Reinventing Hippocrates* (New York: Routledge, [2001] 2016).

4. Brown, "From Mechanism to Vitalism," esp. 181–85; and Robert E. Schofield, *Mechanism and Materialism: British Natural Philosophy in an Age of Reason* (Princeton, NJ: Princeton University Press, 1970), esp. 191–209.

5. John Arbuthnot, *An Essay Concerning the Nature of Aliments, and the Choice of Them, according to the Different Constitutions of Human Bodies*, 4th ed. (London: J. and R. Tonson, [1731] 1756), 143–44.

6. Brown, "From Mechanism to Vitalism," 187–89.

7. See, e.g., Arnold Thackray, *Atoms and Powers: An Essay on Newtonian Matter-Theory and the Development of Chemistry* (Cambridge, MA: Harvard University Press, 1970).

8. Julian Offrey de la Mettrie, *Man a Machine*, trans. anon. (Dublin: W. Brien, [1747] 1749), 61. See also Sergio Moravia, "From *Homme Machine* to *Homme Sensible*: Changing Eighteenth-Century Models of Man's Image," *Journal of the History of Ideas* 39 (1978): 45–60; and Jessica Riskin, *The Restless Clock: A History of the Centuries-Long Argument over What Makes Living Things Tick* (Chicago: University of Chicago Press, 2016), 152–61, 171–86.

9. Much of the material here and in the following several paragraphs is adapted from Steven Shapin, "Descartes the Doctor: Rationalism and Its Therapies," *British Journal for the History of Science* 33 (2000): 131–54.

10. Letter from Descartes to Princess Elizabeth, December 1646, in René Descartes, *The Philosophical Writings of Descartes*, trans. John Cottingham, Robert Stoothoff, and Dugald Murdoch, 3 vols. (Cambridge: Cambridge University Press, 1984–1991), 3:304–5.

11. René Descartes, *Descartes' Conversations with Burman*, ed. and trans. John Cottingham (Oxford: Oxford University Press, 1976), 51.

12. Adrien Baillet, *The Life of Monsieur Des Cartes Translated from the French by* S. R. (London, [1691] 1693), 259.

13. Baillet, *Life of Des Cartes*, 259–60.

14. Descartes, *Conversations with Burman*, 50.

15. Baillet, *Life of Des Cartes*, 260.

16. Letter from Descartes to Princess Elizabeth, May or June 1645, in Descartes, *Philosophical Writings*, 3:249–51.

17. Steven Shapin, "Why Was 'Custom a Second Nature' in Early Modern Medicine," *Bulletin of the History of Medicine* 93 (2019): 1–26.

18. René Descartes, *Treatise of Man*, trans. Thomas Steele Hall (Cambridge, MA: Harvard University Press, [1629–1633] 1972), 5–10, 17–19.

19. Descartes, *Treatise of Man*, 70 ("when the blood that goes into the heart is more pure and subtle and is kindled more easily than usual, this arranges the little nerve that is there in the manner that is required to cause the sensation of *joy*"), and 111 (for dry air); letter Descartes to Elizabeth, May or June 1645, in Descartes, *Philosophical Writings*, 3:250; and René Descartes, "Principles of Philosophy (extracts)," in idem, *Philosophical Writings*, 1:280–81.

20. Descartes, *Treatise of Man*, 108–12.

21. For example, George Cheyne, "A Philosophical and Practical Essay in the General Method and Medicins . . . ," in idem, *Essay on Regimen*, 2nd ed. (London: C. Rivington, 1740), separately paginated, xliii; and idem, *An Essay of Health and Long Life* (London: George Strahan, 1724), 45, 75.

22. This section draws on Steven Shapin, "Trusting George Cheyne: Scientific Expertise, Common Sense, and Moral Authority in Early Eighteenth-Century Dietetic Medicine," *Bulletin of the History of Medicine* 77 (2003): 263–97.

23. Here and elsewhere for Cheyne's theory and practice, see Anita Guerrini, *Obesity and Depression in the Enlightenment: The Life and Times of George Cheyne* (Norman: University of Oklahoma Press, 2000), esp. 99–113.

24. Cheyne, *Health and Long Life*, 36, 90–92; and idem, "Philosophical and Practical Essay," lix. "Seeds" in Cheyne's usage designated grains and

grain-derived foods in general, from sago and rice puddings to porridge and bread.

25. Cheyne, *Health and Long Life*, 30–32, 226; and idem, *The Natural Method of Cureing the Diseases of the Body*, 4th ed. (London: George Strahan, 1742), 42, 67, 74, 86, 297.

26. Edward Strother, *An Essay on Sickness and Health; . . . in Which Dr. Cheyne's Mistaken Opinions . . . Are Occasionally Taken Note Of*, 2nd ed. (London: C. Rivington, 1725), esp. 28–32, 85–86, 209–23. See also Anon., *Remarks on Dr. Cheyne's Essay of Health and Long Life*, 3rd ed. (Dublin: J. Watts, [1724] 1725), 19.

27. Cheyne, *Health and Long Life*, 38, 231.

28. Cheyne, *Natural Method*, 207–9.

29. Cheyne, "Philosophical and Practical Essay," xiii–xiv.

30. Cheyne, *Natural Method*, 13–14.

31. René Descartes, "Discourse on the Method," in *The Philosophical Writings of Descartes*, trans. John Cottingham, Robert Stoothoff, and Dugald Murdoch, 3 vols. (Cambridge: Cambridge University Press, 1984–1991), 1:118, 122–23.

32. R. H. Meade, *In the Sunshine of Life: A Biography of Dr. Richard Mead, 1673–1754* (Philadelphia: Dorrance, 1974).

33. Richard Mead, *The Medical Works of Richard Mead, M.D.* (London: C. Hitch and L. Lawes, 1762), vii, ix–x.

34. Mead, *Medical Works*, 570–71.

35. See chapter 5, 183–85 (above).

36. Santorio Santorio, *Medicina Statica: Being the Aphorisms of Sanctorious, Translated into English by Dr. John Quincy*, 5th ed. (London: T. Longman and J. Newton, [1614] 1737), 176, 182, 184, 187.

37. See Lucia Dacome, "Resurrecting by Numbers in Eighteenth-Century England," *Past and Present*, no. 193 (November 2006): 73–110.

38. Jeremiah Wainewright, *A Mechanical Account of the Non-Naturals: Being a Brief Explication of the Changes Made in Humane Bodies, by Air, Diet, &c.* (London: Ralph Smith, 1707), "Preface," sig. A2.

39. Wainewright, *Mechanical Account of the Non-Naturals*, 78–79, 155–56, 159, 174. On these sensibilities, see also Francis Fuller, *Medicina Gymnastica: or, a Treatise Concerning the Power of Exercise* (London: John Matthews, 1705), 153–55. Wainewright disputed Fuller's commendation of vigorous exercise, pointing to the harmful consequences of immoderation.

40. John Burton, *A Treatise on the Non-Naturals, in Which the Great Influence They Have on Human Bodies is Set Forth and Mechanically Accounted For* (York: A. Staples, 1738), xiii–xiv, 2, 10–12 (a "merry-andrew" is a clown or fool).

41. Burton, *Treatise on the Non-Naturals*, 213. For the superior nourishment afforded by animal flesh because of its structure and greater ease of digestion and assimilation, see Wainewright, *Mechanical Account of the Non-Naturals*, 159–61. And for eighteenth-century French thinking about the transformation and nutritiousness of foodstuffs, see E. C. Spary, *Feeding France: New Sciences of Food, 1760–1815* (Chicago: University of Chicago Press, 2014), 95–101.

42. Burton, *Treatise on the Non-Naturals*, 81–82, 190–202, 210, 235, 242, 255, 331, 334, 337.

43. Nicholas Robinson, *A New Theory of Physick and Diseases Founded on the Principles of the Newtonian Philosophy* (London: C. Rivington, 1725), viii, 51, 70, 74, 90, 191.

44. Robinson, *New Theory of Physick*, 51, 64.

45. Robinson, *New Theory of Physick*, 65, 197—209.

46. Robinson, *New Theory of Physick*, 63.

47. Robinson, *New Theory of Physick*, 64, 67, 198, 204 (quoted passage), 205.

48. James Mackenzie, *The History of Health, and the Art of Preserving It*, 3rd ed. (Edinburgh: William Gordon, [1758] 1760), 3–5, 90–92, 94–95, 115–16, 158–59, 161–62, 181–86, 235–40, 269, 274.

49. John Fothergill, *Rules for the Preservation of Health* (London: John Pridden, 1762), 45.

50. Ephraim Chambers, *Cyclopædia: or, an Universal Dictionary of Arts and Sciences*, 2 vols. (London: James and John Knapton 1728), 1:211, 2:635, 985. There was little change in these entries in editions appearing through the 1780s.

51. See "Diet," The Encyclopedia of Diderot and d'Alembert, http://quod.lib.umich.edu/d/did/did2222.0003.038/—diet?rgn=main;view=fulltext;q1=non-naturals#idno_did2222.0003.038; "Health," The Encyclopedia of Diderot and d'Alembert, http://quod.lib.umich.edu/d/did/did2222.0002.721/—health?rgn=main;view=fulltext;q1=non-naturals. See also William Coleman, "Health and Hygiene in the *Encyclopédie*: A Medical Doctrine for the Bourgeoisie," *Journal of the History of Medicine* 29 (1974): 399–421.

52. *Encyclopædia Britannica: or, A Dictionary of Arts and Sciences*, 3 vols. (Edinburgh: A. Bell and C. Macfarquhar, 1771), 2:440, 3:122, 143, 147, 149, 159, 170, 403, 891. For remarks on the continuing place of the nonnaturals in eighteenth-century German medicine, see Mary Lindemann, *Health & Healing in Eighteenth-Century Germany* (Baltimore: Johns Hopkins University Press, 1996), esp. 249, 264–66.

53. John Bostock, *An Elementary System of Physiology*, new ed. (London: Henry G. Bohn, [1824–1834] 1836), 806–8.

54. James Makittrick Adair, *An Essay on Diet and Regimen* (London: James Ridgway, 1812), chap. 1–3.

55. For example, J. A. Paris, *A Treatise on Diet: With A View to Establish, on Practical Grounds a System of Rules for the Prevention and Cure of the Diseases Incident to a Disordered State of the Digestive Functions* (New York: E. Duyckinck, Collins & Co., [1826] 1828), 64.

56. Andrew Combe, *The Physiology of Digestion Considered with Relation to the Principles of Dietetics* (Edinburgh: Maclachlan & Stewart, [1834] 1836), 238–41.

57. For the much written about Renaissance withdrawal of high from low culture, see, notably, Peter Burke, *Popular Culture in Early Modern Europe* (London: Maurice Temple Smith, 1978); and Carlo Ginzburg, *The Cheese and the Worms: The Cosmos of a Sixteenth-Century Miller*, trans. John and Anne Tedeschi (Baltimore: Johns Hopkins University Press, [1976] 1980).

58. Mary E. Fissell, "Making a Masterpiece: The Aristotle Texts in Vernacular Medical Culture," in *Right Living: An Anglo-American Tradition of Self-Help Medicine and Hygiene*, ed. Charles E. Rosenberg, 59–87 (Baltimore: Johns Hopkins University Press, 2003); idem, "Readers, Texts, and Contexts: Vernacular Medical Works in Early Modern England," in *The Popularization of Medicine, 1650–1850*, ed. Roy Porter, 72–96 (London: Routledge, 1992); idem, "The Marketplace of Print," in Mark S. R. Jenner and Patrick Wallis, ed., *Medicine and the Market in England and Its Colonies c. 1450-c. 1850*, 108–32 (New York: Palgrave Macmillan, 2007), esp. 110; and William Coleman, "The People's Health: Medical Themes in 18th-Century French Popular Literature," *Bulletin of the History of Medicine* 51 (1977): 55–74.

59. John Wesley, *Primitive Physick: or, An Easy and Natural Method of Curing Most Diseases* (London: Thomas Trye, 1747), viii–xi.

60. Wesley, *Primitive Physick*, xx–xxiv.

61. Wesley, *Primitive Physick*, 31, 51, 72, 84–85.

62. Wesley, *Primitive Physick*, xix. See also Deborah Madden, *'A Cheap, Safe and Natural Medicine': Religion, Medicine and Culture in John Wesley's Primitive Physic*, Clio Medica 83 (Amsterdam: Rodopi, 2007), 35–36, 155–60, 169–76.

63. Madden, *'A Cheap, Safe and Natural Medicine,'* 11.

64. G. S. Rousseau, "John Wesley's *Primitive Physic* (1747)," *Harvard Library Bulletin* 16, no. 3 (July 1968): 242.

65. Heikki Mikkeli, *Hygiene in the Early Modern Medical Tradition* (Helsinki: Academia Scientiarum Fennica, 1999), 168. See also Fissell, "Readers, Texts, and Contexts," 81: "we cannot know who actually read these works."

66. William Buchan, *Domestic Medicine; or, the Family Physician* (Edinburgh: Balfour, Auld, and Smellie, 1769), x.

67. Accounts of *Domestic Medicine* include Christopher J. Lawrence, "William Buchan: Medicine Laid Open," *Medical History* 19 (1975): 20–36; and Charles E. Rosenberg, "Medical Text and Social Context: Explaining Buchan's *Domestic Medicine*," *Bulletin of the History of Medicine* 57 (1983): 22–42.

68. For European versions, see Elisabetta Lonati, "The Dissemination of Medical Practice in Late Modern Europe: The Case of Buchan's *Domestic Medicine*," *Status Quaestionis: A Journal of European and American Studies* 17 (2019), https://doi.org/10.13133/2239-1983/16393. For the publishing history of Buchan's book, see Richard B. Sher, "William Buchan's *Domestic Medicine*: Laying Book History Open," in *The Human Face of the Book Trade: Print Culture and Its Creators*, ed. Peter Isaac and Barry McKay, 45–64 (Newcastle, DE: Oak Knoll, 1999). For Buchan in Brazil, see Kalle Kananoja, "Doctors, Healers and Charlatans in Brazil: A Short History of Ideas, c. 1650–1950," in *Healers and Empires in Global History*, ed. Markku Hokkanen and Kananoja (Cham, Switzerland: Springer International, 2019), 192–93; and Marcos Cueto and Steven Palmer, *Medicine and Public Health in Latin America: A History* (Cambridge: Cambridge University Press, 2019), 51. For Spain, see Enrique Perdiguero, "The Popularization of Medicine in the Spanish Enlightenment," in *The Popularization of Medicine, 1650–1850*, ed. Roy Porter, 160–93 (London: Routledge, 1992). For Russia, see Robert L. Nichols, "Orthodoxy and Russia's Enlightenment, 1762–1825," in *Russian Orthodoxy under the Old Regime*, ed. Nichols and Theofanis George Stavrou (Minneapolis: University of Minnesota Press, 1978), 70. For Japan, see Grant Kohn Goodman, *Japan and the Dutch, 1600–1853* (London: Routledge, 2000), 142.

69. Samuel-Auguste David Tissot, *Advice to the People in General, with Regard to Their Health*, trans. J. Kirkpatrick (London: T. Becket and P. A. De Hondt, [1761] 1765), 41, 62, 80, 88, 99, 102, 117, 136, 162, 164, 221, 266 (all for regimen orientated to aliment and occasionally to bleeding), 566 (for "way of living"). See also Antoinette Suzanne Emch-Dériaz, *Tissot: Physician of the Enlightenment* (New York: Peter Lang, 1992); Patrick Singy, "The Popularization of Medicine in the Eighteenth Century: Writing, Reading, and Rewriting Samuel Auguste Tissot's *Avi au peuple sur sa santé*," *Journal of Modern History* 82 (2010): 769–800; and Michael Stolberg, "Medical Popularization and the Patient in the Eighteenth Century," in *Cultural Approaches to the History of Medicine*, ed. Willem de Blécourt and Cornelie Usborne, 89–107 (London: Palgrave Macmillan, 2004).

70. Tissot, *Advice to the People*, 75–76, 167, 551 (quoted passage).

71. Buchan, *Domestic Medicine* (1769), 63, 64, 74, 78, 96, 120.

72. For the nineteenth-century place of temperaments in physiognomy and phrenology, see chapter 3, 111–13 (above).

73. Buchan, *Domestic Medicine* (1769), 47–48, 83, 96, 101, 168, 183, 189, 200, 256–57, 259, 398–99, 442, 506, 529, 539, 605.

74. Buchan, *Domestic Medicine* (1769), 416. Scurvy at the time was, like gout, a more protean condition than the nutritional-deficit disease afflicting mariners on long voyages and could also encompass such symptoms as debility, bad breath, and aching limbs. In the early nineteenth century, a physician described scorbutic humors as those "which give origin to cutaneous eruptions of various sorts, erysipelatous affections of different sorts, chronic rheumatisms, the gout, sciatica, ill-conditioned ulcers in the legs, of long duration, phagedænic or cancerous ulcers." Letter Dr. [Charles] Bissett to Dr. John Lettsom, October 16, 1785, in *Selections from the Medical Papers and Correspondence of the Late John Coakley Lettsom*, ed. Thomas Joseph Pettigrew (London: Nichols, Son, and Bentley, 1817), 305.

75. William Buchan, *Buchan's Domestic Medicine, Enlarged and Improved: The New Domestic Medicine: or, Universal Family Physician*, new ed., by George Wallis (London: Alex. Hogg, 1802), 60–67.

76. Among many instances, see, for example, Johann Gaspar Spurzheim, *Phrenology in Connexion with the Study of Physiognomy*, 3rd American ed. (Boston: Marsh, Capen & Lyon, [1826] 1836), 191–93; and A. Schoepf Merei, *On the Disorders of Infant Development, and Rickets, Preceded by Observations on the Nature, Peculiar Influence, and Modifying Agency of Temperaments* (London: John Churchill, 1855), 3–8.

77. John C. Gunn, *Domestic Medicine, or Poor Man's Friend* (Knoxville, TN: Printed for the author, 1830; facsimile edition edited by Charles E. Rosenberg (Knoxville: University of Tennessee Press, 1986), esp. 74–94 (for extensive treatment of "intemperance"), 125 (for "fat, plethoric, or sanguine" constitutions), 125–27 (for the nutritiousness of various sorts of animal flesh). See also Ben H. McClary, "Introducing a Classic: 'Gunn's Domestic Medicine,'" *Tennessee Historical Quarterly* 45 (1986): 210–16.

78. William Cullen, *Lectures on the Materia Medica* (Philadelphia: Robert Bell [1760] 1775), 18–20.

79. On the persistence of traditional dietetic categories into the eighteenth century and beyond, see, e.g., Maria Pia Donato, "Galen in an Age of Change (1650–1820)," in *Brill's Companion to the Reception of Galen*, ed. Petros Bouras-Vallianatos and Barbara Zipser, 487–507 (Leiden: Brill, 2019), esp. 498–99.

80. Paracelsus was born as Philippus Aureolus Theophrastus Bombastus von Hohenheim, and the adopted name of Paracelsus was meant to signify his goal of a reformed medicine that went beyond that of Celsus.

81. Georgiana D. Hedesan, "Theory Choice in the Seventeenth Century:

Robert Boyle against the Paracelsian *Tria Prima*," in *Theory Choice in the History of Chemical Practices*, ed. Emma Tobin and Chiara Ambrosio (Zurich: Springer, 2016), 19–20.

82. Walter Pagel, *Paracelsus: An Introduction to Philosophical Medicine in the Era of the Renaissance*, 2nd rev. ed. (Basel: Karger, 1982), esp. 129.

83. See John C. Powers, "Chemistry without Principles: Herman Boerhaave on Instruments and Elements," in *New Narratives in Eighteenth-Century Chemistry*, ed. Lawrence M. Principe (Dordrecht: Springer, 2007), 51–52. See also E. C. Spary, "Liqueurs and the Marketplace in Eighteenth-Century Paris," in *Materials and Expertise in Early Modern Europe: Between Market and Laboratory*, ed. Ursula Klein and Spary (Chicago: University of Chicago Press, 2010), 237–42, 245–46.

84. Lawrence M. Principe, *The Aspiring Adept: Robert Boyle and His Alchemical Quest* (Princeton, NJ: Princeton University Press, 1996), esp. 36–37, 43.

85. Thackray, *Atoms and Powers*, chap. 2, 4, 6.

86. William R. Newman, *Atoms and Alchemy: Chymistry and the Experimental Origins of the Scientific Revolution* (Chicago: University of Chicago Press, 2006), 157–58; and William R. Newman and Lawrence M. Principe, *Alchemy Tried in the Fire: Starkey, Boyle, and the Fate of Helmontian Chymistry* (Chicago: University of Chicago Press, 2002), 21–22.

87. See, notably, Isaac Newton, "Some Thoughts about the Nature of Acids" [a translation by John Harris of the paper *De natura acidorum* (1692)]," in *Isaac Newton's Papers & Letters on Natural Philosophy*, 2nd ed., ed. I. Bernard Cohen and Robert E. Schofield (Cambridge, MA: Harvard University Press, 1978), 255–58. Here Newton talked about attractive forces while also supposing that alkaline substances consisted of "*Earthy* and *Acid* united together."

88. The best treatment of eighteenth-century chemical categories in connection with naturally occurring substances is Ursula Klein and Wolfgang Lefèvre, *Materials in Eighteenth-Century Science: A Historical Ontology* (Cambridge, MA: MIT Press, 2007), esp. 15–19, 33–37.

89. Louis Lémery, *A Treatise of All Sorts of Food*, trans. D. Hay (London: T. Osborne, [1702] 1745), vi–viii.

90. Lémery, *Treatise of All Sorts of Food*, 6–7. "Balsam" was also a Paracelsian category, referring to substances that were oily and penetrating.

91. Lémery, *Treatise of All Sorts of Food*, 185, 51. See also Marie Boas Hall, "Acid and Alkali in Seventeenth Century Chemistry," *Archives internationales d'histoire des sciences* 9 (1956): 13–28; Antonio Clericuzio, "Chemical and Mechanical Theories of Digestion in Early Modern Medicine," *Studies in History and Philosophy of Biological and Biomedical Sciences* 43 (2012): 329–37, esp. 334–36.

92. Principe agrees that "essentially discordant" chemical concepts coexisted in early eighteenth-century chemistry, with Paracelsian categories persisting along with (limited) gestures at micromechanical explanations: Lawrence M. Principe, "A Revolution Nobody Noticed? Changes in Early Eighteenth-Century Chymistry," in *New Narratives in Eighteenth-Century Chemistry*, ed. Principe (Dordrecht: Springer, 2007), 4–5.

93. Cheyne, *Essay on Regimen*, 56–57, 60.

94. Arbuthnot, *Essay of Aliments*, xxiii–xxiv.

95. Arbuthnot, *Essay of Aliments*, 65, 67, 69, 143–53. See also Richard Blackmore, *A Treatise of the Spleen and Vapours: Hypochondriacal and Hysterical Affections* (London: J. Pemberton, 1725), 32, 41–45, 50, 60, 110, 147, 236 (where acid and acrimonious humors are implicated in "the spleen").

96. Arbuthnot, *Essay of Aliments*, 188, 220, 222.

97. Friedrich Hoffman[n], *A Treatise on the Nature of Aliments, or Foods, in General* (London: L. Davis and C. Reymers, [1730] 1761), esp. 4–5, 8–10, 14, 20. See also William Forster, *A Treatise on the Various Kinds and Qualities of Foods* (Newcastle-upon-Tyne: John White, 1738), 13–21, 27, 33, 52, 67–68, 75.

98. Burton, *Treatise on the Non-Naturals*, esp. 215–18, 234.

99. Cullen, *Lectures on Materia Medica*, e.g., 54–55, 103, 150.

100. George Fordyce, *A Treatise on the Digestion of Food* (London: J. Johnson, 1791), esp. 36–52, 104–11.

101. See, for instance, Thomas Willis, *Five Treatises viz. 1. Of Urines, 2. Of the Accension of the Blood, 3. Of Musculary Motion, 4. The Anatomy of the Brain, 5. The Description and Use of the Nerves* (London, 1681), 21–23 (on the diagnostic uses of the smell of urine).

102. Herman Boerhaave, *Dr. Boerhaave's Academical Lectures on the Theory of Physic*, 6 vols. (London: W. Innys, 1742–1746), 4:21. See also Klein and Lefèvre, *Materials in Eighteenth-Century Science*, 215, 239–41.

103. Arbuthnot, *Essay of Aliments*, 47–52.

104. Cullen, *Lectures on Materia Medica*, 54–55.

105. Cullen, *Lectures on Materia Medica*, 45–46, 93–94. Cullen's generalizations about the relationship between taste and nutritiousness reached a lay audience through some editions of Buchan's *Domestic Medicine*. See, for example, William Buchan, *Domestic Medicine* (Exeter: J. & B. Williams, 1828), 39.

106. But see Klein and Lefèvre, *Materials in Eighteenth-Century Science*, 58–59, for the changing pertinence among eighteenth-century chemists of such sensible properties of substances as smell and taste.

Chapter 8

1. J. A. Paris, *A Treatise on Diet: With a View to Establish, on Practical Grounds a System of Rules for the Prevention and Cure of the Diseases Incident to a Disordered State of the Digestive Functions* (New York: E. Duyckinck, Collins & Co., [1826] 1828), 70.

2. Harmke Kamminga and Andrew Cunningham, "Introduction: The Science and Culture of Nutrition, 1840–1940," in *The Science and Culture of Nutrition, 1840–1940*, ed. Kamminga and Cunningham, 1–14 (Amsterdam: Rodopi, 1995).

3. See, among many Renaissance and early modern examples, Lazarus Ercker, *Fleta Minor: The Laws of Art and Nature in . . . Metals* (London, 1683), sig. N2^r. On early chemical characterizations of albumin, see William Thomas Brande, *A Manual of Chemistry*, 2 vols., 3rd ed. (London: John Murray, [1819] 1830), 2:618–21; and John Bostock, *An Elementary System of Physiology*, 3rd ed. (London: Baldwin and Craddock, [1824–1836] 1836), 288–91. A standard modern distinction is between "albumen," designating the whites of an egg, and "albumin" as the specific protein in egg whites. The state of nineteenth-century science makes that clear spelling distinction hard to consistently sustain.

4. E. C. Spary, *Feeding France: New Sciences of Food, 1760–1815* (Cambridge: Cambridge University Press, 2014), 216–31.

5. Brief surveys of discoveries in this area include Louis Rosenfeld, *Origins of Clinical Chemistry: The Evolution of Protein Analysis* (New York: Academic Press, 1982), 2–17; Kenneth J. Carpenter, *Protein and Energy: A Study of Changing Ideas in Nutrition* (Cambridge: Cambridge University Press, 1994), 22–39; and Elmer Verner McCollum, *A History of Nutrition: The Sequence of Ideas in Nutrition Investigation* (Boston: Houghton Mifflin, 1957), 41–62. For food and pertinent chemistry in France, see detailed accounts in Spary, *Feeding France*, esp. chap. 6.

6. Paris, *Treatise on Diet*, 105.

7. For example, Brande, *Manual of Chemistry*, 2:448–49, 564, 566–67, 626, 649. On meat chemistry before the development of the protein concept, see Spary, *Feeding France*, 212–34.

8. See, among many examples, Jonathan Pereira, *A Treatise on Food and Diet*, ed. Charles A Lee (New York: Fowler and Wells, 1843), 218–19.

9. Paris, *Treatise on Diet*, 21, 95–96.

10. For example, Brande, *Manual of Chemistry*, 2:407–8, 447–50, 660.

11. William Prout, "On the Ultimate Composition of Simple Alimentary Substances; with Some Preliminary Remarks on the Analysis of Organized Bodies in General," *Philosophical Transactions of the Royal Society of London* 117 (1827):356–57. (There was also a fourth category, the aqueous principle,

and while it—like salt—was regarded as a necessity for life, it did not figure in nutrition, understood as aliment that fueled or constituted the human frame.) Other writers on food and nutrition deviated from Prout's categorizations. It was said that there must be more than these few alimentary types, and critics argued against the rightness of assigning "new meanings to common and familiar terms": Pereira, *Treatise on Food and Diet*, iii–iv; see also 38–41. (Pereira was a London physician and lecturer on materia medica.)

12. Prout, "On the Ultimate Constitution," 33.

13. Prout, *Chemistry, Meteorology*, esp. 473–83. The notion of "staminal principles" appears here (479), identified as the fundamental parts of alimentary substances. Sales of the Bridgewater series totaled sixty thousand by 1850, and Prout's text was among several extracted in more popular publications such as the *Penny Cyclopedia*. Jonathan R. Topham, "Science and Popular Education in the 1830s: The Role of the *Bridgewater Treatises*," *British Journal for the History of Science* 25 (1992): 397–430, esp. 397n2.

14. Prout, *Chemistry, Meteorology*, 118–19.

15. Hubert Bradford Vickery, "The Origin of the Word Protein," *Yale Journal of Biology and Medicine* 22 (1950): 387–93, esp. 389. See also Harold Hartley, "Origin of the Word 'Protein,'" *Nature* 168 (1951): 244; and Justus von Liebig, *Animal Chemistry or Organic Chemistry in Its Application to Physiology and Pathology*, facsimile of the Cambridge edition of 1842, ed. William Gregory (New York: Johnson Reprint, 1964), 101.

16. G. J. Mulder, *The Chemistry of Vegetable & Animal Physiology*, trans. P. F. H. Fromberg (Edinburgh: William Blackwood, 1849), 291–94, 301, 304–5. This text collected, revised, and summarized work that was conducted from 1835 and published from 1838.

17. The chemical constitution of proteins from amino acids was soon realized—indeed, some of these compounds were identified by Liebig himself—but while chemists understood that many if not all proteins were isomers, with the same elemental constitution, and speculated that their different physiological functions followed from their different spatial arrangements, knowledge of proteins' three-dimensional shape and the sequence of amino acids was not secured until well into the twentieth century. See, e.g., Justus von Liebig, *Researches on the Chemistry of Food, and the Motions of the Juices in the Animal Body*, ed. William Gregory and Eben N. Horsford (Lowell, MA: Daniel Boxby, 1848), 18, 22.

18. Edwin Lankester, *On Food: Being Lectures Delivered at the South Kensington Museum* (London: Robert Hardwicke, 1861), 132.

19. For the factory-like program of organic analysis at Giessen, see J. B. Morrell, "The Chemist Breeders: The Research Schools of Liebig and Thomas Thomson," *Ambix* 19 (1972): 1–46. Morrell showed how the ability

of Liebig's school to churn out huge quantities of organic analyses not only responded to social and economic demands for such knowledge but also created further demand.

20. This concerned Liebig's eventual failure, even after following Mulder's experimental protocols, to isolate the "protein radical" apart from the added sulfur and phosphorus atoms that Mulder postulated, and the dispute soon descended into accusations of incompetence and misrepresentation. The morality of traditional dietetics was about how people ought to live; the origins of nineteenth-century nutrition science were attended by questions about the morality and competence of technical expertise. See G. J. Mulder, *Liebig's Question to Mulder Tested by Morality and Science*, trans. P. F. H. Fromberg (Edinburgh: William Blackwood, 1846). See also E. Glas, "The Liebig-Mulder Controversy: On the Methodology of Physiological Chemistry," *Janus* 63 (1976): 27–46; and William H. Brock, *Justus von Liebig: The Chemical Gatekeeper* (Cambridge: Cambridge University Press, 1997), 195–97.

21. See Elizabeth Neswald, "Nutritional Knowledge between the Lab and the Field: The Search for Dietary Norms in the Late Nineteenth and Early Twentieth Centuries," in *Nutritional Standards: Theory, Policies, Practices*, ed. Elizabeth Neswald, David F. Smith, and Ulrike Thoms, 29–51 (Rochester, NY: University of Rochester Press, 2017), esp. 30–33.

22. Justus von Liebig, *Familiar Letters on Chemistry, and Its Relation to Commerce, Physiology, and Agriculture*, trans. John Gardner (New York: D. Appleton, 1843), 72–75, 99–100.

23. Liebig, *Animal Chemistry*, 92; and idem, *Familiar Letters* (1843), 109. (These texts from the early 1840s overlap considerably, and the latter in particular went through a series of modified editions.) Liebig himself did not then use the word "carbohydrate" even though he understood the elemental composition of the set of alimentary substances—starches and sugars—that were later so categorized. "Carbohydrate" emerged as a chemical term of art in the 1850s, becoming standard by the end of the century. The term certainly was reaching a wide reading public by the end of the 1850s. See, e.g., George Henry Lewes, *The Physiology of Common Life*, 2 vols. (Edinburgh: William Blackwood, 1859–1860), 1:70–71. Expounding Liebig's work, Lewes wrote that "the chemists have named this group," that is, the starchy or saccharine substances, which all contain hydrogen and oxygen in the same proportion as water, "the *hydrates of carbon*, or *carbohydrates*."

24. For example, Liebig, *Familiar Letters* (1843), 84; and idem, *Animal Chemistry*, 38–42.

25. Liebig, *Familiar Letters* (1843), 87–89, 94, 108–9; and idem, *Animal Chemistry*, 43–44 (see also 38, 46, 75, 124).

26. Liebig, *Animal Chemistry*, 92. See also idem, *Familiar Letters* (1843), 64. The German terms are *plastische Nahrungsmittel* and *Respirationsmittel*.

27. Liebig, *Familiar Letters* (1843), 89. It is true, Liebig acknowledged, that at least in certain circumstances, the animal body may use protein to support respiration (oxidizing it), but it normally does not. Liebig, *Animal Chemistry*, 158; and idem, *Familiar Letters* (1843), 107–8. This qualified acknowledgment was more clear in later editions of *Familiar Letters*.

28. Justus von Liebig, "On the Nutritive Value of Different Sorts of Food," *The Lancet* 93 (January 2, 9, 23, 1869; February 6, 1869; March 13, 1869): 5.

29. Liebig, *Animal Chemistry*, 209–10, 231–33. See also Brock, *Justus von Liebig*, 192–94, 199–201.

30. On protein policies in nineteenth-century Germany, see Corinna Treitel, "How Vegetarians, Naturopaths, Scientists, and Physicians Unmade the Protein Standard in Modern Germany," in *Setting Nutritional Standards*, ed. Elizabeth Neswald, David F. Smith, and Ulrike Thoms, (Rochester, NY: University of Rochester Press), 52–73; and Ulrike Thoms, "Setting Standards: The Soldier's Food in Germany, 1850–1960," in *Setting Nutritional Standards*, 97–118.

31. Brock, *Justus von Liebig*, 139–40.

32. See, among many examples, James F. W. Johnston, *The Chemistry of Common Life*, 2 vols., new ed. (Edinburgh: William Blackwood, [1854] 1859), Vol. 1, chap. 5–6 (serially analyzing foodstuffs according to their content of fibrin, albumin, gluten, casein, fat, sugar, and starch). The text is heavily Liebigian without mentioning him by name. Revised and expanded, this text was in print to the end of the nineteenth century. The Google Ngram Viewer shows a sharp uptick in usages of "protein," peaking first in the 1850s and again during World War I and then a generally smooth increase to the present.

33. Anon., "Food," *Penny Cyclopædia*, Vol. 9 (London: Charles Knight, 1837), 342–46; and idem, "Food," *Penny Cyclopædia: The Second Supplement* (London: Knight & Co., 1858), 235–44.

34. Herbert Spencer, *The Principles of Biology*, 2 vols. (London: Williams and Norgate, 1864), e.g., 1:44–45, 182, 483–87.

35. Lewes, *Physiology of Common Life*, 1:72–73. A few years later, Liebig's work was enthusiastically disseminated in Edwin Lankester's lectures on food at the South Kensington Science Museum in London. Lankester, *On Food*, esp. 47–49, 107–9, 131–32, 189–90.

36. In this matter, Lewes was evidently echoing the criticisms of Liebig offered some years earlier by the Dutch materialist physiologist Moleschott. See Jacob Moleschott, *Der Kreislauf des Lebens: Physiologische Antworten auf Liebigs chemische Briefe* (Mainz: von Zabern, 1852). See also Harmke Kam-

minga, "Nutrition for the People, or the Fate of Jacob Moleschott's Contest for a Humanist Science," in *Science and Culture of Nutrition*, 15–47, esp. 32–34; and Frederick Gregory, *Scientific Materialism in Nineteenth Century Germany* (Dordrecht: D. Reidel, 1977), 80–99.

37. Lewes, *Physiology of Common Life*, 1:56, 58, 111.

38. Lewes, *Physiology of Common Life*, 1:115.

39. Lewes, *Physiology of Common Life*, 1:61–63.

40. Lewes, *Physiology of Common Life*, 1:65. The remark about "probabilities" was quoted from Jöns Jacob Berzelius, *Traité de Chimie*, Vol. 5 (1831), quoted in William Prout, *Chemistry, Meteorology and the Function of Digestion Considered with Reference to Natural Theology* (London: William Pickering, 1834).

41. Lewes, *Physiology of Common Life*, 1:80.

42. Especially Sir Albert Howard, *An Agricultural Testament* (London: Oxford University Press, 1940), 37, 168, 181–85. Liebig's views shifted over time about how plants got their nitrogen and whether nitrogen needed to be supplied to the soil through chemical fertilizers.

43. Brock, *Justus von Liebig*, 120–29, 159–67.

44. See, for example, Blaxter and Paterson's sociological study of attitudes to food and cooking in the northeast of Scotland. Mildred Blaxter and Elizabeth Paterson, *Mothers and Daughters: A Three-Generational Study of Health Attitudes and Behaviour* (London: Heinemann, 1982), esp. 138; and idem, "The Goodness Is Out of It: The Meaning of Food to Two Generations," in *The Sociology of Food and Eating: Essays on the Sociological Significance of Food*, ed. Anne Murcott, 95–105 (Aldershot: Gower, 1983).

45. Rebecca L. Spang, *The Invention of the Restaurant: Paris and Modern Gastronomic Culture* (Cambridge, MA: Harvard University Press, 2000). See also Amy B. Trubek, *Haute Cuisine: How the French Invented the Culinary Profession* (Philadelphia: University of Pennsylvania Press, 2000), chap. 2.

46. Denis Papin, *A New Digester or Engine for Softning Bones* (London, 1681).

47. Justus von Liebig, *Familiar Letters on Chemistry*, 4th ed., ed. John Blyth (London: Walton and Maberly, 1859), 433–48.

48. Brock, *Justus von Liebig*, 220–21. See also McCollum, *History of Nutrition*, 75–83.

49. For these sorts of usages, see, for example, Frederick Accum, *Culinary Chemistry, Exhibiting the Scientific Principles of Cookery* (London: R. Ackermann, 1821), 91; and M. H. Braconnot, "On the Conversion of Animal Matter into New Substances," *Philosophical Magazine* 56 (1820), 136. There are several historical accounts of the disputes over osmazome and gelatine, of which the best is Spary, *Feeding France*, 226–32, to which this brief account is indebted. See also Lisa Haushofer, *Wonder Foods: The Science*

and Commerce of Nutrition (Oakland: University of California Press, 2023), 99–100.

50. J. S. Forsyth, *A Dictionary of Diet*, 2nd ed. (London: Henry Cremer, 1834), 227 (art. "Osmazome"). See also "Food," *Penny Cyclopædia*, 9:343: osmazome is "the principle to whom meat owes its sapid taste and odour when dressed."

51. David Howes, "Preface: Accounting for Taste," in Luca Vercelloni, *The Invention of Taste: A Cultural Account of Desire, Delight and Disgust in Fashion, Food and Art*, trans. Kate Singleton (London: Bloomsbury, [2005] 2016), ix.

52. For example, Thomas Thomson, *A System of Chemistry*, 6th ed., 4 vols. (London: Baldwin, Craddock, and Joy, [1802] 1820), 4:416–17.

53. Brock, *Justus von Liebig*, 216–18; and Justus von Liebig, "On the Nutritive Value of Different Sorts of Food," *The Lancet* 93, no. 2371 (February 6, 1869): 186–87.

54. Midcentury chemical texts described osmazome as "an ill-defined compound," and while they routinely detailed how to extract it and included it in treating the chemical makeup of foods and organic tissues, it was not further analyzed. Edward L. Youmans, *A Class-Book of Chemistry* (New York: D. Appleton, 1858), 283. A Google Ngram Viewer chart for osmazome shows peaks of usages in the 1820s and 1840s, declining by the end of nineteenth century, and by the mid-twentieth century it is likely that few but historians knew what osmazome was.

55. Jean Anthelme Brillat-Savarin, *The Physiology of Taste*, trans. Anne Drayton (London: Penguin, [1825] 1994), 63–64 and (for aphorism) 13.

56. For present-day gastronomical treatment, see, for example, Hervé This, *Molecular Gastronomy: Exploring the Science of Flavor*, trans. M. B. Debevoise (New York: Columbia University Press, 2006), 23–24; and Harold McGee, "Osmazome, the Maillard Reaction & the Triumph of the Cooked," in *Taste: Proceedings of the Oxford Symposium on Food & Cookery 1987*, ed. Tom Jaine (London: Prospect Books, 1988), 133–35.

57. Brillat-Savarin, *Physiology of Taste*, 64. He then offered a concise summary of other food substances identified or analyzed by the chemists, including gelatine, albumin, fibrin, gluten, mucilage, gum, sugars, and starches (65–68).

58. Thomas Love Peacock, "Gastronomy and Civilization," in *The Works of Thomas Love Peacock*, Vol. 9, *Critical & Other Essays*, ed. H. F. B. Brett-Smith and C. E. Jones (London: Constable [essay, 1851] 1926), 394. The essay was published, under the initials M. M. in *Fraser's Magazine* for 1851, and contributions were apparently made by his daughter Mary Ellen Nicolls.

59. Eliza Acton, *Modern Cookery, for Private Families*, new and revised ed. (London: Longman, Green, Longman, and Robert, [1845] 1860), 8.

60. Isabella Beeton, *Beeton's Book of Household Management* (London: S. O. Beeton, 1861), 261.

61. Beeton, *Beeton's Book of Household Management*, 51.

62. Anon., "Osmazome," in *Everyday Housekeeping: A Magazine for Practical Housekeepers and Mothers* 23, no. 2 (September 1906): 121–23. See also Anon., "What Is Osmazome?," *Table Talk* 21, no. 7 (July 1906): vi.

63. Note here an extended discussion of osmazome and Brillat-Savarin in Roland Barthes, *The Rustle of Language*, trans. Richard Howard (Berkeley: University of California Press, [1984] 1989), 259–62.

64. Liebig, *Researches on the Chemistry of Food*, 101–3 (William Gregory's remarks are at 103n). See also Acton, *Modern Cookery*, 167–68 (following Liebig and Gregory on "scientific boiling").

65. Liebig, *Researches on the Chemistry of Food*, 110. See also Brock, *Justus von Liebig*, 223, 227; and Mark R. Finlay, "Quackery and Cookery: Justus von Liebig's Extract of Meat and the Theory of Nutrition in the Victorian Age," *Bulletin of the History of Medicine* 66 (1992): 406.

66. Swinburne, "Von Liebig Condensed," 250; and Harold McGee, *On Food and Cooking: The Science and Lore of the Kitchen* (New York: Simon & Schuster, 2003), 113–14, 434–35.

67. Liebig, *Researches on the Chemistry of Food*, 102–3.

68. For wider-ranging treatment of the nineteenth-century search for concentrated meat and its commercial setting, see Haushofer, *Wonder Foods*, chap. 1.

69. Liebig, *Familiar Letters* (1859), 442. See also Brock, *Justus von Liebig*, 222–27; and Mark R. Finlay, "Early Marketing of the Theory of Nutrition: The Science and Culture of Liebig's Extract of Meat," in *Science and Culture of Nutrition*, 48–74.

70. Brock, *Justus von Liebig*, 226–27. See also David Edgerton, *The Shock of the Old: Technology and Global History since 1900* (Oxford: Oxford University Press, 2007), 170–72.

71. Lance E. Davis and Robert A. Huttenback, *Mammon and the Pursuit of Empire: The Political Economy of British Imperialism, 1860–1912* (Cambridge: Cambridge University Press, 1986), 94.

72. Florence Nightingale, *Notes on Nursing: What It Is, and What It Is Not*, new ed. (London: Harrison, [1859] 1860), 98.

73. Finlay, "Early Marketing of the Theory of Nutrition," 63–64.

74. For example, Hannah M. Young, comp., *Liebig Company's Practical Cookery Book: A Collection of New and Useful Recipes in Every Branch of Cookery* (London: Liebig's Extract of Meat Company, 1893); and Maria Parloa, *One Hundred Ways to Use Liebig Company's Extract of Beef: A Guide for American Housewives* (London: Liebig's Extract of Meat Company, 1897). See also Finlay, "Early Marketing of the Theory of Nutrition," 61–63.

75. Brock, *Justus von Liebig*, 229–30.

76. For treatment of the emerging nineteenth-century business of branded nutritional products, see Haushofer, *Wonder Foods*; and idem, “Between Food and Medicine: Artificial Digestion, Sickness, and the Case of Benger’s Food,” *Journal of the History of Medicine and Allied Sciences* 73 (2018): 168–87. See also Spary, *Feeding France*, 125–66.

77. For example, Stephen Darby, *On Fluid Meat* (London: John Churchill, 1870), esp. 12–16; and Whitman H. Jordan, *Principles of Human Nutrition: A Study in Practical Dietetics* (New York: Macmillan, 1913), 301–3.

78. The product is still made by the company Liebig Benelux.

79. McCollum, *History of Nutrition*, 115–33; and Everett Mendelsohn, *Heat and Life: The Development of the Theory of Animal Heat* (Cambridge, MA: Harvard University Press, 1964), 108–39, esp. 134–39.

80. See, e.g., Frederic L. Holmes, “The Formation of the Munich School of Metabolism,” in *The Investigative Enterprise: Experimental Physiology in Nineteenth-Century Medicine*, ed. William Coleman and Holmes (Berkeley: University of California Press, 1988), 179–210; and Dietrich Milles, “Working Capacity and Calorie Consumption: The History of Rational Physical Economy,” in *Science and Culture of Nutrition*, 75–96, esp. 77–78.

81. Tom Scott-Smith, *On an Empty Stomach: Two Hundred Years of Hunger Relief* (Ithaca, NY: Cornell University Press, 2020), 46–48. See also James Simpson, *A Letter to the Right Honourable Henry Labouchere . . . on the More Effective Application of the System of Relief by Means of Soup Kitchens* (London: Whittaker & Co., 1847), esp. 7–8; Ian Miller, “The Chemistry of Famine: Nutritional Controversies and the Irish Famine, c. 1845–7,” *Medical History* 56 (2012): 444–62; and Kenneth J. Carpenter, “Nutritional Studies in Victorian Prisons,” *Journal of Nutrition* 136 (2006): 1–8.

82. Aspects of Atwater’s work have been well treated but in bits and pieces, still awaiting a systematic account. See, e.g., Kenneth J. Carpenter, “The Life and Times of W. O. Atwater (1844–1907),” *Journal of Nutrition* 124, supplement 9 (1994): 1707S–1714S. See also Edward C. Kirkland, “‘Scientific Eating’: New Englanders Prepare and Promote a Reform, 1873–1907,” *Proceedings of the Massachusetts Historical Society*, 3rd series, 86 (1974): 28–52; Jessica J. Mudry, *Measured Meals: Nutrition in America* (Albany, NY: SUNY Press, 2009), 29–46; Harvey Levenstein, “The New England Kitchen and the Origins of Modern American Eating Habits,” *American Quarterly* 32 (1980): 369–86, esp. 370–72, 385–86; idem, *Revolution at the Table: The Transformation of the American Diet* (Berkeley: University of California Press, 2003), 44–59, 74–76; Naomi Aronson, “Nutrition as a Social Problem: A Case Study of Entrepreneurial Strategy in Science,” *Social Problems* 29 (1982): 474–87; Hamilton Cravens, “Establishing the Science of Nutrition at the USDA: Ellen Swallow Richards and Her Allies,” *Agricultural*

History 64 (1990): 123–26; Charlotte Biltekoff, *Eating Right in America: The Cultural Politics of Food and Health* (Durham, NC: Duke University Press, 2013), 16–20; and Charles E. Rosenberg, "Atwater, Wilbur Olin," in *Dictionary of Scientific Biography*, ed. Charles Coulston Gillispie, 325–26 (New York: Scribner, 1970), 1:325–26.

83. A. C. True, "Wilbur Olin Atwater, 1844–1907," *Proceedings of the Washington Academy of Sciences* 10 (1908): 194–98; Carpenter, "Life and Times of Atwater," 1709S–1710S.

84. W. O. Atwater, "What the Coming Man Will Eat," *The Forum* 13 (June 1892): 492.

85. W. O. Atwater, "What We Should Eat," *Century Illustrated Monthly Magazine* 36, no. 2 (June 1888): 257–64. See also Rachel Louise Moran, *Governing Bodies: American Politics and the Shaping of the Modern Physique* (Philadelphia: University of Pennsylvania Press, 2018), 13.

86. Edward Atkinson, with introductory statements regarding the nutritive value f common food materials by Mrs. Ellen H. Richards, *Suggestions Regarding the Cooking of Food* (Washington, DC: US Government Printing Office, 1894), 3 (here quoting Ellen Richards).

87. W. O. Atwater, "The Potential Energy of Food: The Chemistry and Economy of Food. III," *Century Illustrated Monthly Magazine* 34, no. 3 (July 1887): 401. It was during his time in Germany that Atwater encountered the category of the nutritional calorie; Liebig had not mentioned the term in his 1842 book on animal chemistry or in his 1845 paper on food and animal warmth. James R. Hargrove, "History of the Calorie in Nutrition," *Journal of Nutrition* 136 (2006): 2959; and Justus Liebig, "Ueber der thierische Wärme," *Annalen der Chemie und Pharmacie* 53 (1845): 63–77.

88. Thomas S. Hall, *History of General Physiology*, 2 vols. (Chicago: University of Chicago Press, 1969), 2:267–73.

89. Nick Cullather, "The Foreign Policy of the Calorie," *American Historical Review* 112 (2007): 340 (for calories "rendering food into hard figures"). See also Scott-Smith, *On an Empty Stomach*, 48–49.

90. Atwater, "Potential Energy of Food," 401. This principle was established by Max Rubner in the 1880s and over time has been (problematically) rendered by the tag "calories in, calories out," without necessary reference to the specific source of the calories. See, e.g., Graham Lusk, *The Elements of the Science of Nutrition*, 3rd ed. (Philadelphia: W. B. Saunders, [1906] 1917), 36–37. See also Corinna Treitel, "Max Rubner and the Biopolitics of Rational Nutrition," *Central European History* 41 (2008): 1–25, esp. 3–4; and Anson Rabinbach, *The Human Motor: Energy, Fatigue, and the Origins of Modernity* (New York: Basic Books, 1990), 126–27.

91. W. O. Atwater, *Foods: Nutritive Value and Cost* (Washington, DC: US Government Printing Office, 1894), 8, 11; idem, "Potential Energy of

Food," 400–401; and idem, *Principles of Nutrition and Nutritive Value of Food* (Washington, DC: US Government Printing Office, 1910), 11. See also Mudry, *Measured Meals*, 37–38. Atwater's work here extended that of Max Rubner, and the heat values of nutrients were sometimes known as Rubner's factors. See Lusk, *Elements of the Science of Nutrition*, 37–43; and W. O. Atwater and H. C. Sherman, *The Effect of Severe and Prolonged Muscular Work on Food Consumption, Digestion, and Metabolism* (Washington, DC: US Governmental Printing Office, 1901), 39–41.

92. True, "Wilbur Olin Atwater," 195–96; and Aronson, "Nutrition as a Social Problem," esp. 475–76, 479–81.

93. Anon., "Editorial Notes: The Development of the Respiration Calorimeter," US Department of Agriculture, Office of Experiment Stations, *Experiment Station Record* 11, no. 6 (1899–1900), 501–4.

94. Lusk, *Elements of the Science of Nutrition*, 24–30; and Holmes, "Formation of the Munich School."

95. Atwater, "How Food Nourishes the Body," 241–42.

96. W. O. Atwater, C. D. Woods, and F. G. Benedict, *Report of Preliminary Investigations on the Metabolism of Nitrogen and Carbon in the Human Organism, with a Respiration Calorimeter of Special Construction* (Washington, DC: US Government Printing Office, 1897), 11–22; W. O. Atwater and E. B. Rosa, *Description of a New Respiration Calorimeter and Experiments on the Conservation of Energy in the Human Body* (Washington, DC: US Government Printing Office, 1899); W. O. Atwater and F. G. Benedict, *A Respiration Calorimeter with Appliances for the Direct Determination of Oxygen* (Washington, DC: Carnegie Institution of Washington, 1905); and Anon., "The People's Food—A Great National Inquiry: Professor W. O. Atwater and His Work," *Review of Reviews* 13 (1896): 683. Respiration calorimeters were objects of public as well as expert scientific fascination. Visitors showed up unannounced to volunteer as experimental subjects, and two other devices, at the Carnegie Institution in Washington and the Boston Nutrition Laboratory, were soon constructed. See Francis G. Benedict and Thorne M. Carpenter, *Respiration Calorimeters for Studying the Respiratory Exchange and Energy Transformations in Man*, Carnegie Institution of Washington Publication No. 123 (Washington, DC: Carnegie Institution of Washington, 1910). See also Nick Cullather, *The Hungry World: America's Cold War Battle against Poverty in Asia* (Cambridge, MA: Harvard University Press, 2010), 11–12.

97. Atwater, "What We Should Eat," 257. See also Atwater and Rosa, *Description of a Respiration Calorimeter*, 7–10.

98. W. O. Atwater and F. G. Benedict, "The Respiration Calorimeter," in *Yearbook of the United States Department of Agriculture 1904* (Washington, DC: US Government Printing Office, 1905), 205–20.

99. Atwater and Rosa, *Description of a Respiration Calorimeter*, 11.

100. Atwater, *Foods*, 9.

101. Atwater, Woods, and Benedict, *Report of Preliminary Investigations on the Metabolism of Nitrogen and Carbon*, 51.

102. Francis G. Benedict and Cornelia Golay Benedict, *Mental Effort in Relation to Gaseous Exchange, Heart Rate, and Mechanics of Respiration* (Washington, DC: Carnegie Institution of Washington, 1933), 83 (for the peanut); and Francis G. Benedict and Thorne M. Carpenter, *The Influence of Muscular and Mental Work on Metabolism and the Efficiency of the Human Body as a Machine* (Washington, DC: US Government Printing Office, 1909), 45–50, 100. See also idem, "The Energy Requirements of Intense Mental Effort," *Proceedings of the National Academy of Sciences* 16 (1930): 438–43, esp. 433.

103. Atwater and Rosa, *Description of a Respiration Calorimeter*, 90.

104. Atwater, *Principles of Nutrition*, 11.

105. Atwater, "How Food Nourishes the Body," 249–51.

106. Atwater, "What We Should Eat," 257.

107. Atwater, "How Food Nourishes the Body," 251. See also Kamminga, "Nutrition for the People," 29–30.

108. W. O. Atwater and O. S. Blakeslee, *Improved Forms of Bomb Calorimeter and Accessory Apparatus* (Middletown, CT: Pelton & King, 1897); W. O. Atwater and J. F. Snell, "Description of a Bomb-Calorimeter and Method of Its Use," *Journal of the American Chemical Society* 25 (1903): 659–99; for Atwater's account of European developments of the bomb calorimeter, see 659–60.

109. Atwater, "Potential Energy of Food," 399.

110. W. O. Atwater, *Methods and Results of Investigations of the Chemistry and Economy of Food* (Washington, DC: US Government Printing Office, 1895), 22 (Chart I).

111. "It is by means of dietaries that the most reliable data concerning the food consumption of people of different nationality, sex, and occupation, and under different financial and hygienic conditions, can be obtained." W. O. Atwater and A. P. Bryant, *Dietary Studies in Chicago in 1895 and 1896* (Washington, DC: US Government Printing Office, 1898), 7.

112. W. O. Atwater, "Tables of Foods and Dietaries," in John S. Billings, ed., *The National Medical Dictionary*, Vol. 1, *A–J* (Philadelphia: Lea Brothers, 1890), xxxix.

113. Cf. Moran, *Governing Bodies*, 14: "Atwater's food plans mostly ignored women's own nutritional needs. Atwater used 'a man at moderate muscular work' as a physiological norm, the basis of his nutrition math." W. O. Atwater, "Tables of Foods and Dietaries," in Billings, *The National Medical Dictionary*, 1:xxxv–xliii. Compare Chart II (xliii), having no female

categories, with Tables V and VI (xl), where there are a few, albeit drawn from European studies.

114. A[rmand] Gautier, *Diet and Dietetics*, ed. and trans. A. J. Rice-Oxley (Philadelphia: J. B. Lippincott, [1904] 1906), 394, 514. For a representative use of the 80 percent standard, see Atwater and Bryant, *Dietary Studies in Chicago*, 8: "These factors are based in part upon actual investigations and in part upon arbitrary assumption."

115. For example, Gertrude I. Thomas, *The Dietary of Health and Disease* (Philadelphia: Lea & Febiger, 1923),57–58; and Whitman H. Jordan, *Principles of Human Nutrition: A Study in Practical Dietetics* (New York: Macmillan, 1913), 217–18.

116. W. Gilman Thompson, *Practical Dietetics, with Special Reference to Diet in Disease* (New York: D. Appleton, 1895), 290.

117. Julius Friedenwald and John Ruhräh, *Diet in Health and Disease*, 4th ed. (Philadelphia: W. B. Saunders, [1904] 1913), 57.

118. Martha Tracy and Caroline Croasdale, "Do Women Eat Enough?," *Forty-First Annual Meeting of the Alumnæ Association of the Woman's Medical College of Pennsylvania, June 1 and 2, 1916* (Philadelphia: Published by the Association, 1916), 81.

119. H. S. Grindley, J. L. Sammis, et al., *Nutrition Investigations at the University of Illinois, North Dakota Agricultural College, and Lake Erie College, Ohio, 1896–1900* (Washington, DC: US Government Printing Office, 1900), 21–26 (for North Dakota women students).

120. Lusk, *Elements of the Science of Nutrition*, 640.

121. Lusk, *Elements of the Science of Nutrition*, 391–402. See also G. A. Sutherland, ed., *A System of Diet and Dietetics*, 2nd ed. (New York: Physicians and Surgeons Book Company, [1908] 1925), 744–47.

122. Gautier, *Diet and Dietetics*, 394–96. For the possible beneficial effects of alcohol, see Sutherland, *System of Diet*, 747.

123. There are many sources for this sensibility, but see, for example, L. Jean Bogert, *Dietetics Simplified: The Use of Foods in Health and Disease*, 2nd ed. (New York: Macmillan, [1937] 1940), 147.

124. Atwater, "What We Should Eat," 261.

125. W. O. Atwater and Charles D. Woods, *Dietary Studies with Reference to the Food of the Negro in Alabama in 1895 and 1896*, US Department of Agriculture, Office of Experiment Stations, Bulletin No. 38 (Washington, DC: US Government Printing Office, 1897), 7, 64, 68–69. However, studies of African Americans in eastern Virginia documented satisfactory intake of protein—mainly from locally available salted and fresh fish—comparable in this setting to that of local whites. H. B. Frissell and Isabel Bevier, *Dietary Studies of Negroes in Eastern Virginia in 1897 and 1898* (Washington, DC: US Government Printing Office, 1899), 4–41.

126. Arthur Goss, *Dietary Studies in New Mexico in 1895* (Washington, DC: US Government Printing Office, 1897), 22.

127. Atwater and Bryant, *Dietary Studies in Chicago*, 7, 15–16, 42, 71–72.

128. Atkinson, *Science of Nutrition*, 143. See also Paul Freedman, *American Cuisine, and How It Got This Way* (New York: Liveright, 2019), 188–89.

129. Edward Atkinson, *The Application of Science to the Production and Consumption of Food* (Salem, MA: Salem Press, 1885), 26.

130. Atwater, "Tables of Foods and Dietaries," xl, xliii.

131. W. O. Atwater, "The Food-Supply of the Future," *Century Illustrated Monthly Magazine* 43, no. 1 (November 1891): 111.

132. For Atkinson, see Harold Francis Williamson, *Edward Atkinson: The Biography of an American Liberal, 1827–1905* (Boston: Old Corner Book Store, 1934).

133. Edward Atkinson, "The Food Question in America and Europe; or the Public Victualing Department," *Century Illustrated Monthly Magazine* 33, no. 2 (December 1886): 238–47, esp. 241–42; and idem, "Low Prices, High Wages, Small Profits: What Makes Them?," *Century Illustrated Monthly Magazine* 34, no. 4 (August 1887): 568–84. In addition to the popular pieces already referenced, see Wilbur Olin Atwater, "The Digestibility of Food: The Chemistry of Foods and Nutrition. IV," *Century Illustrated Monthly Magazine* 34, no. 5 (September 1887): 733–40; idem, "Pecuniary Economy of Food: The Chemistry of Foods and Nutrition. V," *Century Illustrated Monthly Magazine* 35, no. 3 (January 1888): 437–46; and idem, "Foods and Beverages: The Chemistry of Foods and Nutrition. VI," *Century Illustrated Monthly Magazine* 36, no. 1 (May 1888): 135–40.

134. For example, [W. O. Atwater], "Food Consumption: Quantities, Costs, and Nutrients of Food-Materials," in *Seventeenth Annual Report of [Massachusetts] Bureau of Statistics of Labor, March 1886*, Public Document No. 15 (Boston: Wright & Potter, 1886), 237–328. This was the first systematic dietary survey in the United States.

135. Atkinson, *Science of Nutrition*, 121.

136. What Atkinson actually said was that "one-half the struggle for life measured in money . . . is the price paid for food." Atkinson, *The Application of Science*, 7. But Atwater preferred the catchier version: e.g., W. O. Atwater, "The Chemistry of Foods and Nutrition. I. The Composition of Our Bodies and Our Food," *Century Illustrated Monthly Magazine* 34, no. 1 (May 1887): 59.

137. For example, Edward Atkinson, *The Industrial Progress of the Nation: Consumption Limited, Production Unlimited* (New York: Putnam, 1890), 39.

138. Atwater, "Chemistry of Foods and Nutrition. I," 59–60. See also Freedman, *American Cuisine*, 180–81.

139. Atkinson, *Science of Nutrition*, 4–5.

140. Atkinson, *Science of Nutrition*, 23.

141. The slogan was the subtitle of Atkinson's *Industrial Progress of the Nation*.

142. Atkinson, *Industrial Progress of the Nation*, 340–41.

143. Edward Atkinson, *Suggestions Regarding the Cooking of Food (with Introductory Statements Regarding the Nutritive Value of Common Food Materials by Mrs. Ellen H. Richards)* (Washington, DC: US Government Printing Office, 1894), 9.

144. Atwater, "What the Coming Man Will Eat," 489–90.

145. For the Aladdin Oven and recipes suited to its use, see Atkinson, *Science of Nutrition*, 58–108; and idem, *Industrial Progress of the Nation*, 339–42. For Atwater's measured endorsement, see Atwater, "What the Coming Man Will Eat," 490. See also Levenstein, "The New England Kitchen." The Aladdin Oven worked better in theory than in practice. It took a very long time to reach cooking temperature, and Atkinson was disappointed that only a few hundred were sold, at about $30 each. Atkinson, *Suggestions Regarding the Cooking of Food*, 30.

146. Atwater, "What We Should Eat," 261. See Charlotte Biltekoff, "What Does It Mean to Eat Right? Nutrition, Science, and Society," in *Food Fights: How History Matters to Contemporary Food Debates*, ed. Charles C. Ludington and Matthew Morse Booker (Chapel Hill: University of North Carolina Press, 2019), 128–32; and John Coveney, *Food, Morals and Meaning: The Pleasure and Anxiety of Eating*, 2nd ed. (London: Routledge, 2006), esp. 60–62.

147. Atwater, "What the Coming Man Will Eat," 498. See also Atkinson, "The Food Question," 242.

148. Atwater, "What the Coming Man Will Eat," 491.

149. Atwater, "Chemistry of Foods and Nutrition. I," 60, 64–65.

150. W. O. Atwater, *Food and Diet* (Washington, DC: US Government Printing Office, 1895), 367–68 (reprinted from *Yearbook of the United States Department of Agriculture, 1894*, 357–88, 547–58). See also W. O. Atwater and Charles D. Woods, *Dietary Studies in New York City in 1895 and 1896* (Washington, DC: US Government Printing Office, 1898), 64. For Atwater's occasional invocations of the notion of "agreement," see idem, "What We Should Eat," 262; and idem, "The Digestibility of Food," 733.

151. Atwater, "Food-Supply of the Future," 102.

152. Atwater, *Principles of Nutrition*, 38.

153. Ellen H. Richards, in Atkinson, *Suggestions Regarding the Cooking of Food*, 3.

154. For example, Atwater and Bryant, *Dietary Studies in Chicago*, 72.

155. Quoted in Anon., "Somewhat Personal," *The New England Kitchen*

Magazine 1 (1894): 109. Atkinson liked the idea of the "inertial woman" as dietary problem and he used the form repeatedly. See, e.g., Atkinson, *Suggestions Regarding the Cooking of Food*, 31.

156. Atwater, "Pecuniary Economy of Food," 438.

157. Atwater, "What the Coming Man Will Eat," 489.

158. Ellen H. Richards, "The Food of Institutions," in Anon., ed., *Plain Words about Food: The Rumford Kitchen Leaflet 1899* (Boston: Rockwell and Churchill Press, 1899), 167. On the obstructions presented by custom and uninstructed taste, see idem, "Good Food for Little Money," in *Plain Words about Food*, 123–38, esp. 123–24.

159. For practical feeding arrangements, see Levenstein, "New England Kitchen"; and Kirkland, "'Scientific Eating,'" 41–43. For Richards, see Caroline L. Hunt, *The Life of Ellen H. Richards* (Boston: Whitcomb & Barrows, 1912); Margaret W. Rossiter, "'Women's Work' in Science, 1880–1910," *Isis* 71 (1980): 391–95; and Sarah Stage, "Ellen Richards and the Social Significance of the Home Economics Movement," in *Rethinking Home Economics: Women and the History of a Profession*, ed. Stage and Virginia B. Vincenti (Ithaca, NY: Cornell University Press, 2018), 17–33.

160. Atkinson, *Suggestions Regarding the Cooking of Food*. For the joint involvement of Richards, Atkinson, and Atwater in producing the pamphlet and for the Wesleyan laboratory, see Hunt, *Life of Ellen Richards*, 229–30, 284.

161. See, e.g., Anon., "Lake Placid Conferences on Home Economics, 1899–1908," *Journal of Home Economics* 1, no. 1 (February 1909): 3–6. See also Freedman, *American Cuisine*, 184–87. Some biographical notes are in Melissa J. Wilmarth and Sharon Y. Nickols, "Helen Woodward Atwater: A Leader of Leaders," *Family and Consumer Sciences Research Journal* 41 (2013): 314–24.

162. Rossiter, "'Women's Work' in Science," 395–96. For the history of American home economics, see Carolyn M. Goldstein, *Creating Consumers: Home Economists in Twentieth-Century America* (Chapel Hill: University of North Carolina Press, 2012); and Rima D. Appel, "Home Economics in the Twentieth Century: A Case of Lost Identity," in *Remaking Home Economics*, ed. Sharon Y. Nickols and Gwen Kay, 54–69 (Athens: University of Georgia Press, 2015).

163. Caroline L. Hunt and Helen W. Atwater, *How to Select Foods I: What the Body Needs*, Farmers' Bulletin 808, United States Department of Agriculture (Washington, DC: US Government Printing Office, 1917), 12 (for quoted passage). See also idem, *How to Select Foods II: Cereal Foods*, Farmers' Bulletin 817 (Washington, DC: US Government Printing Office, 1917); and idem, *How to Select Foods III: Foods Rich in Protein*, Farmers' Bulletin 824 (Washington, DC: US Government Printing Office, 1917).

Hunt was the first professor of home economics at the University of Wisconsin, a participant in the Lake Placid Conferences, and the biographer of Ellen Richards. By 1912, Atwater and colleagues had prepared recipe books distributed under the aegis of a New York newspaper, labeled as government policy and liberally salted with advertisements for branded food products. The recipes were framed by short lessons about macronutrients, the calorie, and the heat engine conception of the body. W. O. Atwater, C. F. Langworthy, and Mari[a] Parloa, *Brooklyn Eagle Government Cook Book* (Brooklyn, NY: Brooklyn Daily Eagle, 1912).

164. Fannie Merritt Farmer, *Food and Cookery for the Sick and Convalescent* (Boston: Little, Brown, 1904), e.g. 8–10, 126, 134, 151.

165. LuLu Hunt Peters, *Diet and Health with Key to the Calories* (Chicago: Reilly & Lee Co., 1918), 23–29.

166. Chin Jou, "The Progressive Era Body Project: Calorie-Counting and 'Disciplining the Stomach' in 1920s America," *Journal of the Gilded Age and Progressive Era* 18 (2019): 423, quoting Lydia J. Roberts, *Nutrition Work with Children* (Chicago: University of Chicago Press, 1927), 55.

167. See Victor Hugo Lindlahr, *You Are What You Eat: How to Win and Keep Health through Diet* (New York: National Nutrition Society, 1940); and also the introduction to this book, 13–15 (above). There were very many similar popular tracts in early twentieth-century America that set out what you ought to think about the nature of foods, among which see especially Alida Frances Pattee, *Practical Dietetics with Reference to Diet in Disease* (New York: A. F. Pattee, 1903). More than twenty editions of Pattee's book appeared into the 1940s, progressively introducing the findings of nutrition science.

168. Atkinson, *The Application of Science*, 44.

169. Sir William Roberts, *Lectures on Dietetics and Dyspepsia* (London: Smith, Elder, & Co., 1886), 6–7, 10, 17–18.

170. Atkinson, *Science of Nutrition*, 121. See also Atwater, "Chemistry of Foods and Nutrition. I," 59.

171. Edward Atkinson, *Taxation and Work* (New York: Putnam, 1892), 189. The calorie—as Cullather nicely put it—"gained official recognition as a measure of national vitality." Cullather, "The Foreign Policy of the Calorie," 354. Quantified protein intake also featured in these connections, though less centrally than calories.

172. Atkinson, *Taxation and Work*, 185–91; and idem, *The Application of Science*, 74.

173. Cullather, "The Foreign Policy of the Calorie," 338–39, 345.

174. See chapter 4, 145–54 (above).

175. Roberts, *Lectures on Dietetics and Dyspepsia*, 17.

176. See, e.g., Robert J. Mayhew, *Malthus: The Life and Legacies of an Unlikely Prophet* (Cambridge, MA: Harvard University Press, 2013).

177. William Crookes, "Address of the President before the British Association for the Advancement of Science, Bristol, 1898," *Science*, new series, 8, no. 200 (October 28, 1898): 561–75 (on serious arable land scarcity); and Edward Atkinson, "Wheat-Growing Capacity of the United States," *Appleton's Popular Science Monthly* 54, no. 2 (December 1898), 145–62 (a reply to Crookes on the considerable extent of still-available land).

178. Abundant atmospheric nitrogen was of no use to plant growth; it needed to be "fixed" by transformation into nitrates or ammonia, and it had recently been discovered that bacteria in nodules on the roots of legumes could accomplish fixation.

179. Gregory T. Cushman, *Guano and the Opening of the Pacific World: A Global Ecological History* (Cambridge: Cambridge University Press, 2013); and G. J. Leigh, *The World's Greatest Fix: A History of Nitrogen and Agriculture* (Oxford: Oxford University Press, 2004), esp. 80–86.

180. Among the many accounts of the Haber-Bosch process, see, e.g., Vaclav Smil, *Enriching the Earth: Fritz Haber, Carl Bosch, and the Transformation of the World's Food Production* (Cambridge, MA: MIT Press, 2001).

181. Atwater, "Chemistry of Foods and Nutrition. I," 59–60.

182. Atkinson, "The Food Question," 243–44; and Atwater, "Food-Supply of the Future," 102–4.

183. Atkinson, "The Food Question," 244.

184. Edward Atkinson, "The Evolution of High Wages from the Low Cost of Labor," *Popular Science Monthly* 53 (1898): 752.

185. The Atwater remark is quoted in Atkinson, *Industrial Progress of the Nation*, 159. The quoted Atkinson passage is in Atkinson, "Evolution of High Wages," 752. Both men returned repeatedly to the mine-laboratory metaphor: e.g., Atkinson, *The Application of Science*, 27; idem, "The Food Question," 244; and W. O. Atwater, *Results of Field Experiments with Various Fertilizers* (Washington, DC: US Government Printing Office, 1883), 182.

186. Atwater, "Food-Supply of the Future," 106.

187. Atkinson was optimistic about exploiting the nitrogen-fixing capacities of legumes. See Atkinson, *Science of Nutrition*, 234.

188. Crookes, "Address of the President." See also William H. Brock, *William Crookes (1832–1919) and the Commercialization of Science* (Aldershot: Ashgate, 2008), 374–88.

189. Crookes, "Address of the President," 564.

190. Charles Edward Woodruff, *Expansion of Races* (New York: Rebman, 1909), 153, 175. Woodruff endorsed Crookes's sensibilities but did not think that the Englishman had gone far enough in identifying the fundamentality of nitrogen.

191. Crookes, "Address of the President," 569.

192. Crookes, "Address of the President," 565.

193. William Crookes, *The Wheat Problem*, 3rd ed. (London: Longmans, Green, 1917).

194. Viscount Rhondda, "Introduction," in Crookes, *The Wheat Problem*, v. There was some sentiment among early vitamin researchers that the discovery of these factors seriously disturbed existing nutritional views about the fundamental importance of calories and protein, but apart from the case of deficiency diseases, these scientists' practical dietary advice was not radically different. See, e.g., E. V. McCollum and Nina Simmonds, *The American Home Diet: An Answer to the Ever Present Question What Shall We Have for Dinner* (Detroit: Frederick C. Mathews, 1920), 7, 18–19.

195. Atkinson, "Wheat-Growing Capacity of the United States," 146.

196. Atkinson, *The Application of Science*, 67.

197. Atkinson, "Wheat-Growing Capacity of the United States," 149–50.

198. Atkinson, *The Application of Science*, 26.

199. William Crookes, *The Wheat Problem*, rev. ed. (London: John Murray, 1900), 91–92.

200. For a history of branded foods and functional claims, see Haushofer, *Wonder Foods*.

Conclusion

1. Gyorgy Scrinis, *Nutritionism: The Science and Politics of Dietary Advice* (New York: Columbia University Press, 2013). The term had been aired out some years before the 2013 book. For example, idem, "Sorry, Marge," *Meanjin* 61, no. 4 (2002): 108–16; and idem, "On the Ideology of Nutritionism," *Gastronomica* 8, no. 1 (February 2008): 39–48.

2. Ben Goldacre, *Bad Science: Quacks, Hacks, and Big Pharma Flacks* (New York: Faber and Faber, 2008), 20, 43 (for "bullshit"), 112–30.

3. Michael Pollan, *Omnivore's Dilemma: A Natural History of Four Meals* (New York: Penguin, 2006); and idem, *In Defense of Food: An Eater's Manifesto* (New York: Penguin, 2008). The later book also became a bestseller, and such was its success that Pollan quickly produced yet another dietary manifesto, a cookbook with recipes for right-eating dishes, and this too became a bestseller: *Food Rules: An Eater's Manual* (New York: Penguin, 2009).

4. Pollan, *In Defense of Food*, 27–28. Pollan acknowledged that the tendency to conceive of food as a collection of nutrient constituents had been identified long before—by other historians and, of course, nutrition scientists—but he applauded Scrinis for giving it "a proper name."

5. Pollan, *In Defense of Food*, 20–23 (quoted phrase on 22). The influential food writer and cookbook author Mark Bittman has also disseminated this bare-bones historical story about "our old friend Justus von Liebig"

and food thinking going wrong. Mark Bittman, *Animal, Vegetable, Junk: A History of Food, from Sustainable to Suicidal* (Boston: Houghton Mifflin Harcourt, 2021), 72–73, 151–52.

6. Pollan, *In Defense of Food*, 4, 63.

7. In the United States, Pollan's main rival as dietary influencer is the former molecular biologist and nutrition scientist (and sometime government nutrition adviser) Marion Nestle. Like Pollan, Nestle is a vigorous critic of modern ways of eating and of corrupted nutritional science but, unlike Pollan, is generally content with the nutritionist frame, notably including the calorie counting that Pollan shuns. Marion Nestle, *Food Politics: How the Food Industry Influences Nutrition and Health* (Berkeley: University of California Press, 2002); Marion Nestle and Malden Nesheim, *Why Calories Count: From Science to Politics* (New York: North Point, 2006); and Marion Nestle, *What to Eat* (New York: North Point, 2006). Nestle's and Pollan's approaches differ significantly in their attitude to nutrition science, but their shared hostility to the Big Food industry is indexed by Pollan's willingness to write an approving foreword to the tenth anniversary edition of *Food Politics*.

8. Pollan, *In Defense of Food*, 125–27, 131–32. The authority endorsed here was the science journalist Susan Allport, *The Queen of Fats: Why Omega-3s Were Removed from the Western Diet and What We Can Do to Replace Them* (Berkeley: University of California Press, 2006).

9. Pollan, *In Defense of Food*, 139. A preferred way of getting the right amounts and proportions of these fatty acids, Pollan notes, is to eat more plant leaves and less refined seeds, but, ironically, omega-3 enthusiasm unleashed a profitable commercial line of food supplements.

10. Pollan, *In Defense of Food*, 1, 3.

11. Pollan, *In Defense of Food*, 148–52. (I never met my great-grandmother, but I was assured by her daughter that she lived in a state of perpetual hunger and would eat anything that was kosher. Maybe Pollan's great-grandmother was something similar?) In a counter to nostalgic celebrations of the ancestors' presumed food purism, historian Rachel Laudan sees Pollan (and his sympathizers) as "romantics": Laudan, "A Plea for Culinary Modernism: Why We Should Love New, Fast, Processed Food (With a New Postscript)," in *Food Fights: How History Matters to Contemporary Food Debates*, ed. Charles C. Ludington and Matthew Morse Booker (Chapel Hill: University of North Carolina Press, 2019), 277.

12. Pollan, *In Defense of Food*, 13, 81.

13. Pollan, *In Defense of Food*, 29.

14. For criticism of the functional food industry, see Gyorgy Scrinis, "Nutritionism and Functional Foods," in *The Philosophy of Food*, ed. David M. Kaplan, 269–91 (Berkeley: University of California Press, 2012).

For its nineteenth- and early twentieth-century prehistory, see Lisa Haushofer, *Wonder Foods: The Science and Commerce of Nutrition* (Oakland: University of California Press, 2023), esp. 182–91. For Soylent, see Lizzie Widdicombe, "The End of Food," *New Yorker*, May 12, 2014, https://www.newyorker.com/magazine/2014/05/12/the-end-of-food.

15. On the modern medical quest for functional perfection, see Carl Elliott, *Better Than Well: American Medicine Meets the American Dream* (New York: Norton, 2003).

16. See, notably, Eric Schlosser, *Fast Food Nation: The Dark Side of the All-American Meal* (Boston: Houghton Mifflin, 2001). This was a powerful exposé of the Big Food industry that delivers bad food to eaters, exploits its workers, despoils the land, and damages the environment. A film version appeared in 2006, directed by Richard Linklater.

17. Michael Pollan, interviewed by Rachel Khong, "The End of the World as We Know It," *Medium*, September 10, 2014, https://medium.com/lucky-peach/the-end-of-the-world-as-we-know-it-c9a0c05b243d.

18. Pollan, *In Defense of Food*, 82.

19. See chapter 4, 149–54 (above).

20. See, notably, the work of the Australian philosopher Peter Singer, *Why Vegan? Eating Ethically* (New York: Liveright, 2020).

21. Gayathri Vaidyanathan, "What Humanity Should Eat to Stay Healthy and Save the Planet," *Nature* 600 (December 2, 2021): 22–25; Bill Gates, *How to Avoid a Climate Disaster: The Solutions We Have and the Breakthroughs We Need* (New York: Knopf Doubleday, 2021), 112–29.

22. On lab-grown meat and its circumstances, see Benjamin Aldes Wurgaft, *Meat Planet: Artificial Flesh and the Future of Food* (Berkeley: University of California Press, 2020).

23. For McDonaldization, see especially the writings of the American social theorist George Ritzer, *The McDonaldization of Society: An Investigation into the Changing Character of Contemporary Social Life* (Thousand Oaks, CA: Pine Forge, 1993), ix, 1, 18, 20 (for McDonaldization as a "paradigm case" of rationality); and idem, "Slow Food versus McDonald's," in *Slow Food: Collected Thoughts on Taste, Tradition, and the Honest Pleasures of Food*, ed. Carlo Petrini and Ben Watson (White River Junction, VT: Chelsea Green Publishing, 2001), 19–23.

24. "Our Big Mac Index Shows How Burger Prices Are Changing," *The Economist*, August 3, 2023, https://www.economist.com/news/2020/07/15/the-big-mac-index.

25. "The Official Slow Food Manifesto [1987]," in Carlo Petrini, *Slow Food: The Case for Taste*, trans. William McGuaig (New York: Columbia University Press, [2001] 2004), xxiii, 13. See also Fabio Parasecoli, *Gastronativism: Food, Identity, Politics* (New York: Columbia University Press, 2022), chap. 2.

26. Petrini, *Slow Food*, 17–18.

27. Petrini, *Slow Food*, 51.

28. Petrini, *Slow Food*, xxi, 7–8, 40–41; idem, *Slow Food Nation: Why Our Food Should be Good, Clean, and Fair*, trans. Clara Furlan and Jonathan Hunt (New York: Rizzoli Ex Libris, [2005] 2007), 8–9; Carlo Petrini (in conversation with Gigi Padovani), *Slow Food Revolution: A New Culture for Eating and Living*, trans. Francesca Santovetti (New York: Rizzoli, 2006), 1–23 (for origins and for early concerns with wine), 185–294 (for list of foodstuffs meant to be preserved and celebrated); and idem, *Food & Freedom: How the Slow Food Movement Is Changing the World through Gastronomy*, trans. John Irving (New York: Rizzoli, [2013] 2015), esp. chap. 1.

29. Corby Kummer, "Doing Well by Eating Well," *The Atlantic*, March 1999, https://www.theatlantic.com/magazine/archive/1999/03/doing-well-by-eating-well/377485/.

30. Petrini, *Slow Food Nation*, 37.

31. Petrini, *Slow Food Nation*, 96–143; and idem, *Food & Freedom*, 30–36.

32. "A New Culture for Eating and Living" is the subtitle of Petrini's *Slow Food Revolution*.

33. Alice Waters, "Alice Waters Applies a 'Delicious Revolution' to School Food," Center for Ecoliteracy, June 28, 2009, https://www.ecoliteracy.org/article/alice-waters-applies-%E2%80%98delicious-revolution%E2%80%99-school-food.

34. Wendell Berry, "The Pleasures of Eating," in idem, *What Are People For? Essays* (Washington, DC: Counterpoint, [1990] 2010), 145; Petrini, *Slow Food Nation*, 66; and Pollan, *Omnivore's Dilemma*, 11, adding "it is also an ecological act, and a political act, too."

35. For example, Petrini, *Slow Food*, 11, 13, 40–41.

36. Petrini, *Slow Food*, xxiii.

37. Alice Waters, *We Are What We Eat: A Slow Food Manifesto* (New York: Penguin, 2021). For Petrini's admiration of Waters, see *Slow Food Nation*, 45–48.

38. Petrini, *Slow Food Nation*, 51–52.

39. Petrini, *Slow Food Nation*, 35–35, 42–45, 55–56. For Petrini's appreciations of Brillat-Savarin, see Petrini, "The New Gastronome: Gastronomy through the Ages; The Figure and Work of Brillat-Savarin," University of Gastronomic Sciences, Pollenzo, [2006?], https://thenewgastronome.com/gastronomy-through-the-ages/. European phenomenological traditions are notably represented at Petrini's University of Gastronomic Sciences by the philosopher Nicola Perullo, *Taste as Experience: The Philosophy and Aesthetics of Food* (New York: Columbia University Press, 2016); and idem, *Epistenology: Wine as Experience* (New York: Columbia University Press, 2020).

40. Petrini, *Slow Food*, xxi.

41. Petrini, *Slow Food*, 10–11, 20–21; and idem, *Slow Food Nation*, 1.

42. Petrini, *Slow Food Nation*, 53.

43. Petrini, *Slow Food*, xvii, xxiii. The reference to "study and knowledge" is a gesture to plans for the University of Gastronomic Sciences in Pollenzo, near Petrini's birthplace of Bra, that he founded in 2004. See Petrini, *Food & Freedom*, 54–65. (Disclosure: I have taught short courses about taste there for several years.)

44. I purposefully use "taste" imprecisely here, taking it as the sensations one has when one eats or drinks. In a range of other contexts, taste (gustation) and smell (olfaction) should be distinguished, with separate attention given to such sensory categories as mouthfeel (texture), temperature, the sensation of heat produced by stimulation of the trigeminal nerve, and so on. Some professionals use the notion of flavor to designate the overall sensory impression given by a food, and many writers about food and drink use the term "taste" interchangeably with "flavor."

45. Michael Moss, *Salt Sugar Fat: How the Food Giants Hooked Us* (New York: Random House, 2013); Nadia Berenstein, "Flavor Added: The Sciences of Flavor and the Industrialization of Taste in America," PhD dissertation, University of Pennsylvania, 2018; and Ana María Ulloa, "The Aesthetic Life of Artificial Flavors," *Senses & Society* 13 (2018): 60–74.

46. For historical discussion of objectification in the sensory science of wine, see Steven Shapin, "A Taste of Science: Making the Subjective Objective in the California Wine World," *Social Studies of Science* 46 (2016): 436–60; and Christopher J. Phillips,"The Taste Machine: Sense, Subjectivity, and Statistics in the California Wine World," *Social Studies of Science* 46 (2016): 461–81. For programmatic comments on how judgments of subjectivity and objectivity are rendered, see Steven Shapin, "The Sciences of Subjectivity," *Social Studies of Science* 42 (2012): 170–84.

47. For philosophers' attempts to stabilize a coherent notion of wine goodness, see, e.g., Barry C. Smith, ed., *Questions of Taste: The Philosophy of Wine* (Oxford: Signal Books, 2007); and Fritz Allhoff, ed., *Wine & Philosophy: A Symposium on Thinking and Drinking* (Malden, MA: Blackwell, 2008).

48. Carlo Petrini, "Building the Ark," in *Slow Food: Collected Thoughts*, ed. Petrini and Watson, 5.

49. For modern food culture, taste, and social distinction, see Josée Johnston and Shyon Baumann, *Foodies: Democracy and Distinction in the Gourmet Foodscape* (New York: Routledge, 2010), esp. chap. 2.

50. First usage is attributed to either Gael Greene writing in New York in 1980 or, slightly later, Ann Barr and Paul Levy in London. The latter two produced *The Official Foodie Handbook: Be Modern, Worship Food* (London: Ebury, 1984).

51. Suzanne Zuppello, "Slow Food's Elitism Only Fueled My Craving for McDonald's," *Eater*, October 18, 2018, https://www.eater.com/2018/10/18/17943358/slow-food-manifesto-elitist-fast-food.

52. Will Self, "A Point of View: The Never-Ending Culinary Merry-Go-Round," *BBC News Magazine*, December 28, 2012, https://www.bbc.com/news/magazine-20836616.

53. John Lanchester, "Shut Up and Eat: A Foodie Repents," *New Yorker*, November 3, 2014, https://www.newyorker.com/magazine/2014/11/03/shut-eat.

54. B. E. Myers, "The Moral Crusade against Foodies," *The Atlantic*, March 2011, https://www.theatlantic.com/magazine/archive/2011/03/the-moral-crusade-against-foodies/308370/.

55. Linus Pauling, "Orthomolecular Psychiatry," *Science*, new series, 160, no. 3825 (April 19, 1968): 265. See also idem, *Vitamin C and the Common Cold* (New York: W. H. Freeman, 1970); and Evelleen Richards, *Vitamin C and Cancer: Medicine or Politics?* (London: Macmillan, 1991).

56. Dr. Benjamin Caballero, quoted in Gina Kolata, "Vitamins: More May Be Too Many," *New York Times*, April 29, 2003, https://www.nytimes.com/2003/04/29/science/vitamins-more-may-be-too-many.html?searchResultPosition=1.

57. The term "superfood" is frowned on by most official agencies but has gained wide popularity in marketing and lay thinking.

58. Ray Kurzweil and Terry Grossman, *Fantastic Voyage: Live Long Enough to Live Forever* (New York: Rodale Books, 2004).

59. Aurelian Craiutu, *Faces of Moderation: The Art of Balance in an Age of Extremes* (Philadelphia: University of Pennsylvania Press, 2017). Eric Hobsbawm gives little attention to moral culture, but the falling away of political moderation is the central theme of his history of the twentieth century: *The Age of Extremes: A History of the World, 1914–1991* (New York: Pantheon, 1994).

60. Simone Weil, *Gravity and Grace*, trans. Arthur Wills (Lincoln: University of Nebraska Press, [1952] 1997), 211.

61. The Atkins commercial enterprise, rescued by private equity companies, still exists, much diminished from what it was, but continues to market protein bars and frozen meals.

62. United States Senate, Select Committee on Nutrition and Human Needs, *Eating in America: Dietary Goals for the United States* (Cambridge, MA: MIT Press, 1977), 12–18. See also Lisa Jahns et al., "The History and Future of Dietary Guidelines in America," *Advances in Nutrition* 9 (2018): 139–40.

63. Robert C. Atkins, *Dr. Atkins' New Diet Revolution*, rev. ed. (New York: Avon, 1992), 27.

64. Atkins, *Dr. Atkins' New Diet Revolution*, esp. 16, 20, 37–80, 110

(quotation on 110). See also Steven Shapin, "Expertise, Common Sense, and the Atkins Diet," in *Public Science in Liberal Democracy*, ed. Jene Porter and Peter W. B. Phillips (Toronto: University of Toronto Press, 2007), 174–93; and idem, "The Great Neurotic Art," *London Review of Books* 26, no. 15 (August 5, 2004): 16–18.

65. Atkins, *Dr. Atkins' New Diet Revolution*, 90.

66. Atkins, *Dr. Atkins' New Diet Revolution*, 23, 91, 99, 106–7, 109 (quoting 99).

67. Atkins, *Dr. Atkins' New Diet Revolution*, 29. See Christine Knight, "'The Food Nature Intended You to Eat': Low-Carbohydrate Diets and Primitivist Philosophy," in *The Atkins Diet and Philosophy: Chewing the Fat with Kant and Nietzsche*, ed. Lisa Heldke, Kerri Mommer, and Cynthia Pineo, 43–56 (Chicago: Open Court, 2005).

68. Atkins, *Dr. Atkins' New Diet Revolution*, 21, 29, 45, 90.

69. See, here, a marvelous reflection on the career of beliefs about the virtues of spinach: Ole Bjørn Rekdal, "Academic Urban Legends," *Social Studies of Science* 44 (2014): 638–54.

70. The diagnosis of anorexia nervosa did not then circulate in our bit of the culture, and my parents' understandings of her behavior bounced around among vernacular psychological diagnoses, something to do with adolescence, something to do with a boyfriend, "acting out," "attention seeking." My sister resumed normal consumption after a while, and the matter was never talked about again.

71. Petrini, *Slow Food Nation*, 53.

72. Julia Child, quoted in Calvin Tomkins, "Table Talk," *New Yorker* 75, no. 33 (November 8, 1999), 32.

73. The beginning of modern forms of these guidelines is United States Department of Agriculture and Department of Health and Human Services, *Nutrition and Your Health: Dietary Guidelines for Americans, 1980*, https://www.dietaryguidelines.gov/sites/default/files/2019-05/1980%20DGA.pdf. See also Eileen Kennedy, Daniel Hatfield, and Jeanne Goldberg, "Dietary Guidelines, Food Guidance, and Dietary Quality in the United States," in *Handbook of Nutrition and Food*, 3rd ed., ed. Carolyn D. Berdanier, Johanna T. Dwyer, and David Heber (Boca Raton, FL: CRC Press, 2014), 437–46.

74. Jahns et al., "History and Future of Dietary Guidelines," 139.

75. United States Department of Agriculture, Center for Nutrition Policy and Promotion, *The Food Guide Pyramid* (Washington, DC: US Government Printing Office, 1996), 1.

76. United States Department of Health and Human Services and US Department of Agriculture, *Dietary Guidelines for Americans, 2020–2025*, 9th ed., December 2020, https://www.dietaryguidelines.gov/sites/default

/files/2021-03/Dietary_Guidelines_for_Americans-2020-2025.pdf, e.g. x, 4, 18, 49, 129 (for moderation); ix, 13, 18, 25, 27, 31–34, 50, 53–54, 60, 62 (for variety); 139 (for balance).

77. United States Department of Health and Human Services and US Department of Agriculture, *Dietary Guidelines for Americans, 2020–2025*, vii–viii, 6–7, 27, 94.

78. For a participant's accounts of these conflicts, see Nestle, *Food Politics*; and idem, *Unsavory Truth: How Food Companies Skew the Science of What We Eat* (New York: Basic Books, 2018).

79. Nestle, *Food Politics*, 29–30, 48–50, 91–92; and Pollan, *In Defense of Food*, 4–5, 12–13, 61–63.

80. Tim Spector, *The Diet Myth: The Real Science behind What We Eat* (London: Weidenfeld & Nicolson, 2015), 10.

81. Pollan, *In Defense of Food*, 81.

82. For example, United States Department of Health and Human Services and US Department of Agriculture, *Dietary Guidelines for Americans, 2020–2025*, 4; United States National Academy of Sciences, Engineering, and Medicine, *Redesigning the Process for Establishing the Dietary Guidelines for Americans* (Washington, DC: National Academies Press, 2017), chap. 2; and Joanne F. Guthrie, Brenda M. Derby, and Alan S. Levy, "What People Know and Don't Know about Nutrition," US Department of Agriculture, Economic Research Service, 1999, https://www.ers.usda.gov/webdocs/publications/42215/5842_aib750m_1_.pdf?v=41055.

83. Bart Penders, "Why Public Dismissal of Nutrition Science Makes Sense: Post-Truth, Public Accountability and Dietary Credibility," *British Food Journal* 120 (2018): 1953–1964.

84. Stephen C. Brewer, *What Happened to Moderation? A Common-Sense Approach to Improving Our Health and Treating Common Illnesses in an Age of Extremes* (New York: SelectBooks, 2019), 3–4, 6–7, 173–74.

85. Danny Klein, "Menu-Labeling Laws Go into Effect," *QSR*, May 2018, https://www.qsrmagazine.com/menu-innovations/menu-labeling-laws-go-effect. *QSR* is the magazine of the industry, which prefers the name "quick service restaurant" instead of "fast food." Similar arrangements obtain in many other major countries, for instance, the United Kingdom (from 2022), the Netherlands, Japan, and India.

86. CBC [Canadian Broadcasting Company], "McDonald's Argues with Itself," October 31, 2002, https://www.cbc.ca/news/world/mcdonald-s-argues-with-itself-1.310509.

87. "Nutrition Disclaimer," McDonald's, n.d., https://mcdonalds.com.hk/en/nutrition-disclaimer/.

88. Gordon Deegan, "Board Opposes a McDonald's for Ennis over Health Factors," *Irish Times*, February 3, 2004, https://www.irishtimes

.com/news/board-opposes-a-mcdonald-s-for-ennis-over-health-factors-1.1132146.

89. First quotation from Mona Charen, "Gluttonous Greed Deserves No Reward," *Baltimore Sun*, December 2, 2012, https://www.baltimoresun.com/news/bs-xpm-2002-12-02-0212020278-story.html. Second quotation from Janet Rorholm, "McDonald's Chef Says It's All about Moderation," *The Gazette* (Cedar Rapids, IA), June 14, 2012, https://www.thegazette.com/health-wellness/mcdonalds-chef-says-its-all-about-moderation/.

90. "Fried Chicken Can Be Part of a Healthy Diet, Says KFC," *QSR*, October 2003, https://www.qsrmagazine.com/news/fried-chicken-can-be-part-healthy-diet-says-kfc.

91. Nestlé, "The Fun of Pizza, the Balance of Good Nutrition," 2013, https://www.nestleusa.com/sites/g/files/pydnoa536/files/asset-library/documents/nutritionhealthwellness/pizza/pizzaportionguide_full.pdf; Undeniably Dairy, "How Pizza Can Be Part of a Healthy Diet," February 9, 2018, https://www.usdairy.com/news-articles/how-pizza-can-be-a-part-of-a-healthy-diet (accessed December 10, 2021).

92. "Better Food Pyramid," Papa Johns, n.d., https://www.papajohns.com/company/better-food-pyramid.html.

93. Bry Roth, "Eating, Living and Working in a Fast Food World," *The Square Deal*, October 24, 2016, https://www.3blmedia.com/news/eating-living-and-working-fast-food-world (accessed December 1, 2021).

94. "Fast Food Restaurant Nutrition," Fast Food Menu Prices, October 11, 2021, https://www.fastfoodmenuprices.com/nutrition/; and "Fast Food in Moderation Is Good for You," Fast Food Menu Prices, October 18, 2021, https://www.fastfoodmenuprices.com/fast-food-in-moderation-is-good-for-you/.

95. W. DeJong, C. K. Atkin, and L. Wallack, "A Critical Analysis of 'Moderation' Advertising Sponsored by the Beer Industry: Are 'Responsible Drinking' Commercials Done Responsibly?," *Milbank Quarterly* 70 (1992): 661–78; Johns Hopkins Bloomberg School of Public Health, "'Drink Responsibly' Messages in Alcohol Ads Promote Products, Not Public Health," September 3, 2014, https://publichealth.jhu.edu/2014/drink-responsibly-messages-in-alcohol-ads-promote-products-not-public-health; and Stuart Elliott, "A Web Awash in Liquor Ads, Promoting Moderation," *New York Times*, March 8, 2009, https://www.nytimes.com/2009/03/09/business/media/09adcol.html.

96. Elaine Watson, "Is the 'There is No Such Thing as Bad Foods, Only Bad Diets' Argument Helpful?," Food Navigator USA, July 19, 2013, https://www.foodnavigator-usa.com/Article/2013/02/11/Is-the-there-is-no-such-thing-as-bad-foods-only-bad-diets-argument-helpful#. For criticisms

of those links, see Michele Simon, "And Now a Word from Our Sponsors: Are America's Nutrition Professionals in the Pocket of Big Food?," January 2013, http://www.eatdrinkpolitics.com/wp-content/uploads/AND_Corporate_Sponsorship_Report.pdf.

97. EatDrinkPolitics, "Is the Academy of Nutrition and Dietetics Silencing Its Members Who Object to McDonald's Sponsoring Lunch?," 2013, http://www.eatdrinkpolitics.com/2013/02/27/is-the-academy-of-nutrition-and-dietetics-silencing-its-members-who-object-to-mcdonalds-sponsoring-lunch/.

98. United States House of Representatives, *Personal Responsibility in Food Consumption Act, Hearing before the Subcommittee on Commercial and Administrative Law of the Committee on the Judiciary*, June 19, 2003, https://www.govinfo.gov/content/pkg/CHRG-108hhrg87814/html/CHRG-108hhrg87814.htm.

99. For example, Lisa Tillinger Johansen, *Fast Food Vindication: The Story You Haven't Been Told* (Los Angeles: Tillinger Johansen Publishing, 2012), 37.

100. Julie Gunlock, *From Cupcakes to Chemicals: How the Culture of Alarmism Makes Us Afraid of Everything and How to Fight Back* (Winchester, VA: Independent Women's Foundation Press, 2013), 50, 68.

101. ACSH Staff, "Advice for Parents: Common Sense, Moderation and Ignore Those Alarmist Warnings," December 11, 2013; Joan Walsh, "Meet the 'Feminists' Doing the Koch Brothers' Dirty Work," *The Nation*, August 18, 2016, https://www.thenation.com/article/archive/meet-the-feminists-doing-the-koch-brothers-dirty-work/; and Andy Kroll and Jeremy Schulman, "Leaked Documents Reveal the Secret Finances of a Pro-Industry Science Group," *Mother Jones*, October 28, 2013, https://www.motherjones.com/politics/2013/10/american-council-science-health-leaked-documents-fundraising/.

102. Michael F. Jacobson, quoted in Center for Science in the Public Interest, "CSPI Hits Marketing Junk Food to Kids," November 10, 2003, https://www.cspinet.org/new/200311101.html.

103. Letter from Michael F. Jacobson to the chairman of the Federal Trade Commission, November 6, 2003, https://cspinet.org/sites/default/files/attachment/letter_to_ftc.pdf.

104. "McDonald's Is Putting Pedometers in Happy Meals Because Why Not," *U.S. News & World Report*, August 17, 2016, https://wtop.com/food-restaurant/2016/08/mcdonalds-mcd-is-putting-pedometers-in-happy-meals-because-why-not/; and David Oliver, "One McDonald's Is Serving Up Endless French Fries," *U.S. News & World Report*, August 5, 2016, https://money.usnews.com/investing/articles/2016-08-05/one-mcdonalds-mcd-is-serving-up-endless-french-fries/.

105. “Food for All,” Taco Bell, https://www.tacobell.com/nutrition. See also Malia Frey, “What to Eat at Wendy’s: Healthy Menu Choices and Nutrition Facts,” August 22, 2021: https://www.verywellfit.com/diet-friendly-fast-food-at-wendys-3495695. The specific “Have It Your Way” advertising campaign was launched by Burger King in the 1970s and continues.

106. Rorholm, “McDonald’s Chef Says It’s All About Moderation.”

107. Johansen, *Fast Food Vindication*, 10–11.

BIBLIOGRAPHY

A. B. 1674. *The Sick-Mans Rare Jewel Wherein Is Discovered a Speedy Way How Every Man May Recover Lost Health, and Prolong Life*. London.

Accum, Frederick. 1821. *Culinary Chemistry, Exhibiting the Scientific Principles of Cookery*. London: R. Ackermann.

ACSH Staff. 2013. "Advice for Parents: Common Sense, Moderation and Ignore Those Alarmist Warnings." December 11.

Acton, Eliza. [1845] 1860. *Modern Cookery, for Private Families*. New and revised ed. London: Longman, Green, Longman, and Roberts.

Adair, James Makittrick. 1812. *An Essay on Diet and Regimen*. London: James Ridgway.

Adams, Thomas. 1615. *Mysticall Bedlam, or the World of Mad-Men*. London.

———. 1616. *Diseases of the Soul: A Discourse Divine, Morall, and Physicall*. London.

Adamson, Melitta Weiss. 1995. *Medieval Dietetics: Food and Drink in Regimen Sanitatis Literature from 800 to 1400*. Frankfurt-am-Main: Peter Lang.

———, 2004. *Food in Medieval Times*. Westport, CT: Greenwood.

Adelman, Juliana, and Lisa Haushofer. 2018. "Food as Medicine, Medicine as Food," *Journal of the History of Medicine* 73: 127–34.

Addison, Joseph. 1853. *The Spectator*. 4 vols. London: Thomas Bosworth.

———. 1888. *The Works of the Right Honourable Joseph Addison*. 6 vols. London: George Bell.

Agricola, Georg Andreas. 1680. "The Virtues and Uses of the Cordial Spirit of Saffron" (broadside). London.

Albala, Ken. 2002. *Eating Right in the Renaissance*. Berkeley: University of California Press.

———. 2003. "The Apparition of Fat in Western Nutritional Theory." In Walker, ed., 2003, 1–10.

———. 2012. "Food for Healing: Convalescent Cookery in the Early Modern Era," *Studies in History and Philosophy of Biological and Biomedical Sciences* 43: 323–28.

———. 2012. "The Ideology of Fasting in the Reformation Era." In *Food and Faith in Christian Culture*, ed. Ken Albala and Trudy Eden, 41–58. New York: Columbia University Press, 41–58.

Alberti, Fay Bound. 2006. "Emotions in the Early Modern Medical Tradition." In *Medicine, Emotion and Disease, 1700–1950*, ed. Fay Bound Alberti, 1–21. London: Palgrave Macmillan.

Alcock, Joan P. 2003. "Gluttony: The Fifth Deadly Sin." In Walker, ed., 2003, 11–21.

Allhoff, Fritz, ed. 2008. *Wine & Philosophy: A Symposium on Thinking and Drinking*. Malden, MA: Blackwell.

Allen, Benjamin. 1699. *The Natural History of the Chalybeat and Purging Waters of England*. London.

Allen, Christopher. 1998. "Painting the Passions: The *Passions de l'Âme* as a Basis for Pictorial Expression." In *The Soft Underbelly of Reason: The Passions in the Seventeenth Century*, ed. Stephen Gaukroger, 79–111. London: Routledge.

Allen, Martha M. 1900. *Alcohol: A Dangerous and Unnecessary Medicine: How and Why*. Marcellus, NY: National Women's Temperance Union.

[Allestree, Richard]. 1660. *The Gentleman's Calling*. London.

———. 1667. *The Causes of the Decay of Christian Piety*. London.

A. M. 1656. *Queen Elizabeths Closset of Physical Secrets*. London.

Anderson, Patrick. 1677. "Grana Angelica: or, The Rare and Singular Vertues and Uses of Those Angelical Pills . . ." (broadside). London.

Anon. [ca. 1634]. *A Discourse translated out of Italian, That a Spare Diet is Better than a Splendid and Sumptuous: A Paradox*. Bound together with Lessius, 1634.

———. 1652. *A Treatise Concerning the Plague and the Pox*. London.

———. 1656. *The Compleat Doctoress: or, A Choice Treatise of All Diseases Insident to Women*. London.

———. 1670. *An Account of the Causes of Some Particular Rebellious Distempers, viz. the Scurvey, Cancers in Women's Breasts, &c. Vapours, and Melancholy, &c*. London.

———. 1672. *Every Woman Her Own Midwife*. London.

———. 1675. *The Women's Complaint against Tobacco*. London.

———. 1678. *The Character of a True English Souldier: Written by a Gentleman of the New-rais'd Troops*. London.

———. 1678. *The Fathers Legacy: or Counsels to His Children*. London.

———. [1724] 1725. *Remarks on Dr. Cheyne's Essay of Health and Long Life*. 3rd ed. Dublin: J. Watts.

———. 1823. "On the Power of Habit." *New Monthly Magazine* 8: 326–34.

———. 1837. "Food." *Penny Cyclopædia*, Vol. 9. London: Charles Knight, 342–46.

———. 1858. "Food." *Penny Cyclopædia: The Second Supplement*. London: Knight & Co., 235–44.

———. 1894. "Somewhat Personal." *New England Kitchen Magazine* 1 (1894), 109.

———. 1896. "The People's Food—A Great National Inquiry: Professor W. O. Atwater and His Work." *Review of Reviews* 13: 679–90.

———. 1899. *Plain Words about Food: The Rumford Kitchen Leaflets 1899*. Boston: Rockwell and Churchill Press.

———. 1899–1900. "Editorial Notes: The Development of the Respiration Calorimeter." US Department of Agriculture, Office of Experiment Stations, *Experiment Station Record* 11, no. 6: 501–4.

———. 1900. "Is Alcohol a Food?" *Literary Digest* 20, no. 5 (February 3): 149.

———. 1900. "Some Authorities Who Do Not Agree with Professor Atwater." *Literary Digest* 20, no. 13 (March 31): 395.

———. 1906. "Osmazome." *Everyday Housekeeping: A Magazine for Practical Housekeepers and Mothers* 23, no. 2 (September): 121–23.

———. 1906. "What is Osmazome?" *Table Talk* 21, no. 7 (July): vi.

———. 1909. "Lake Placid Conferences on Home Economics, 1899–1908." *Journal of Home Economics* 1, no. 1 (February): 3–6.

Appel, Rima D. 2015. "Home Economics in the Twentieth Century: A Case of Lost Identity." In *Remaking Home Economics*, ed. Sharon Y. Nickols and Gwen Kay, 54–69. Athens: University of Georgia Press.

Appelbaum, Robert. 2006. *Aguecheek's Beef, Belch's Hiccup, and Other Gastronomic Interjections: Literature, Culture, and Food among the Early Moderns*. Chicago: University of Chicago Press.

Appleby, Andrew B. 1979. "Diet in Sixteenth-Century England: Sources, Problems, Possibilities." in Webster, ed., 1979, 97–116.

Arbuthnot, John. [1712, 1727] 1976. *The History of John Bull*. Oxford: Clarendon.

———. [1731] 1756. *An Essay Concerning the Nature of Aliments, and the Choice of Them, according to the Different Constitutions of Human Bodies . . .* 4th ed. London: J. and R. Tonson.

Archer, John. 1671. *Every Man His Own Doctor*. London.

———, 1684. *Secrets Disclosed of Consumption*. London.

Arikha, Noga. 2007. *Passions and Tempers: A History of the Humours*. New York: Ecco.

———. 2008. "'Just Life in a Nutshell': Humours as Common Sense." *Philosophical Forum* 39: 303–14.
Aristotle. 1984. *Complete Works*. 2 vols. Ed. Jonathan Barnes. Princeton, NJ: Princeton University Press.
———. 1984. "Prior Analytics, Book II." In Aristotle, 1984, 1:39–113.
———. 1984. "Meteorology." In Aristotle, 1984, 1:555–640.
———. 1984. "Sense and Sensibilia." In Aristotle, 1984, 1:693–713.
———. 1984. "On the Generation of Animals, Book IV." In Aristotle, 1984, 1:1111–218.
———. 1984. "Problems." In Aristotle, 1984, 2:1319–1527.
———. 1999. *Physiognomonica*. Ed. and trans. Sabine Vogt. Berlin: Akademie Verlag.
Arnaud, Sabine. 2015. *On Hysteria: The Invention of a Medical Category between 1670 & 1820*. Chicago: University of Chicago Press.
Aronson, Naomi. 1982. "Nutrition as a Social Problem: A Case Study of Entrepreneurial Strategy in Science." *Social Problems* 29: 474–87.
Ascham, Roger. 1545. *Toxophilus, the Schole of Shootinge*. London.
Atkins, Robert C. [1972] 1992. *Dr. Atkins' New Diet Revolution*. New York: Avon.
Atkinson, Edward. 1885. *The Application of Science to the Production and Consumption of Food*. Salem, MA: Salem Press.
———. 1886. "The Food Question in America and Europe; or the Public Victualing Department." *Century Illustrated Monthly Magazine* 33, no. 2 (December): 238–47.
———. 1887. "Low Prices, High Wages, Small Profits: What Makes Them?" *Century Illustrated Monthly Magazine* 34, no. 4 (August): 568–84.
———. 1890. *The Industrial Progress of the Nation: Consumption Limited, Production Unlimited*. New York: Putnam.
———. 1892. *Taxation and Work*. New York: Putnam.
———. 1894. *Suggestions Regarding the Cooking of Food (with Introductory Statements Regarding the Nutritive Value of Common Food Materials by Mrs. Ellen H. Richards)*. Washington, DC: US Government Printing Office.
———. [1891] 1896. *The Science of Nutrition*. 4th ed. Boston: Damrell & Upham.
———. 1898. "The Evolution of High Wages from the Low Cost of Labor." *Popular Science Monthly* 53: 746–57.
———. 1898. "The Wheat-Growing Capacity of the United States." *Appleton's Popular Science Monthly* 54, no. 2 (December): 145–62.
Atwater, W. O. 1886. "Food Consumption: Quantities, Costs, and Nutrients of Food-Materials." In *Seventeenth Annual Report of [Massachusetts]*

Bureau of Statistics of Labor, March 1886. Public Document No. 15, 237–328. Boston: Wright & Potter.

———. 1887. “The Chemistry of Foods and Nutrition. I. The Composition of Our Bodies and Our Food.” *Century Illustrated Monthly Magazine* 34, no. 1 (May): 59–74.

———. 1887. “How Food Nourishes the Body: The Chemistry of Foods and Nutrition. II.” *Century Illustrated Monthly Magazine* 34, no. 2 (June): 237–51.

———. 1887. “The Potential Energy of Food. The Chemistry and Economy of Food. III.” *Century Illustrated Monthly Magazine* 34, no. 3 (July): 397–404.

———. 1887. “The Digestibility of Food: The Chemistry of Food and Nutrition. IV.” *Century Illustrated Monthly Magazine* 34, no. 5 (September): 733–40.

———. 1888. “Pecuniary Economy of Food: The Chemistry of Foods and Nutrition. V.” *Century Illustrated Monthly Magazine* 35, no. 3 (January): 437–46.

———. 1888. “Foods and Beverages. The Chemistry of Foods and Nutrition. VI.” *Century Illustrated Monthly Magazine* 36, no. 1 (May): 135–40.

———. 1888. “What We Should Eat.” *Century Illustrated Monthly Magazine* 36, no. 2 (June): 257–64.

———. 1890. “Tables of Foods and Dietaries.” In *The National Medical Dictionary . . .*, Vol. 1, *A–J*, ed. John S. Billings, xxxv–xliii. Philadelphia: Lea Brothers.

———. 1891. “The Food-Supply of the Future.” *Century Illustrated Monthly Magazine* 43, no. 1 (November): 101–12.

———. 1892. “What the Coming Man Will Eat.” *The Forum* 13 (June): 488–99.

———. 1894. *Foods: Nutritive Value and Cost*. Washington, DC: US Government Printing Office.

———. 1895. *Food and Diet*. Washington, DC: US Government Printing Office.

———. 1895. *Methods and Results of Investigations of the Chemistry and Economy of Food*. Washington, DC: US Government Printing Office.

———. 1910. *Principles of Nutrition and Nutritive Value of Food*. Washington, DC: US Government Printing Office.

Atwater, W. O., and F. G. Benedict. 1902. *An Experimental Inquiry Regarding the Nutritive Value of Alcohol*, in *Memoirs of the National Academy of Sciences*, Vol. VIII, Sixth Memoir. Washington, DC: US Government Printing Office.

———. 1905. *A Respiration Calorimeter with Appliances for the Direct Determination of Oxygen*. Washington, DC: Carnegie Institution of Washington.

———. 1905. "The Respiration Calorimeter." In *Yearbook of the United States Department of Agriculture 1904*, 205–20. Washington, DC: US Government Printing Office.

Atwater, W. O., and O. S. Blakeslee. 1897. *Improved Forms of Bomb Calorimeter and Accessory Apparatus*. Middletown, CT: Pelton & King.

Atwater, W. O., and A. P. Bryant. 1898. *Dietary Studies in Chicago in 1895 and 1896*. Washington, DC: US Government Printing Office.

Atwater, W. O., C. F. Langworthy, and Mari[a] Parloa. 1912. *Brooklyn Eagle Government Cook Book*. Brooklyn, NY: Brooklyn Daily Eagle.

Atwater, W. O., and E. B. Rosa. 1899. *Description of a New Respiration Calorimeter and Experiments on the Conservation of Energy in the Human Body*. Washington, DC: US Government Printing Office.

Atwater, W. O., and H. C. Sherman. 1901. *The Effect of Severe and Prolonged Muscular Work on Food Consumption, Digestion, and Metabolism*. Washington, DC: US Government Printing Office.

Atwater, W. O., and J. F. Snell. 1903. "Description of a Bomb-Calorimeter and Method of Its Use." *Journal of the American Chemical Society* 25: 659–99.

Atwater, W. O., and C. D. Woods. 1897. *Dietary Studies with Reference to the Food of the Negro in Alabama in 1895 and 1896*. US Department of Agriculture, Office of Experiment Stations, Bulletin No. 38. Washington, DC: US Government Printing Office.

———. 1898. *Dietary Studies in New York City in 1895 and 1896*. Washington, DC: US Government Printing Office.

Atwater, W. O., C. D. Woods, and F. G. Benedict. 1897. *Report of Preliminary Investigations on the Metabolism of Nitrogen and Carbon in the Human Organism, with a Respiration Calorimeter of Special Construction*. Washington, DC: US Government Printing Office.

Aubrey, John. 1975. "Thomas Hobbes." In *Aubrey's Brief Lives*, ed. Oliver Lawson Dick, 147–59. Ann Arbor: University of Michigan Press.

Avramescu, Cătălin. 2009. *An Intellectual History of Cannibalism*. Trans. Alistair Ian Blyth. Princeton, NJ: Princeton University Press.

Babb, Lawrence. 1951. *The Elizabethan Malady: A Study of Melancholia in English Literature from 1580 to 1642*. East Lansing: Michigan State University Press.

Bachot, Gaspard. 1626. *Erreurs populaires touchant la médecine et régime de santé*. Lyons.

Bacon, Francis. 1857–1858. *The Philosophical Works of Francis Bacon*. 5 vols. Ed. James Spedding, Robert Leslie Ellis, and Douglas Denon Heath. London: Longman.

———. 1857–1858. "The Advancement of Learning [Books I–II]." In Bacon, 1857–1858, 3:253–491.

———. 1857–1858. "Of the Dignity and Advancement of Learning, Book IV." In Bacon, 1857–1858, 4:372–404.

———. [1597] 1985. "Of Regiment of Health." In Francis Bacon, *The Essayes or Counsels, Civill and Morall*, ed. Michael Kiernan, 100–102. Cambridge, MA: Harvard University Press.

———. 1996. *Philosophical Studies c. 1611–c. 1619*. Ed. Graham Rees. Oxford: Clarendon.

———. 1627. *Sylva Sylvarum: or A Naturall Historie in Ten Centuries*. London.

———. 1638. *Historie Naturall and Experimentall of Life and Death*. London.

Baillet, Adrien. [1691] 1693. *The Life of Monsieur Des Cartes Translated from the French by* S. R. London.

[Baker, Thomas]. [1699] 1700. *Reflections upon Learning, Wherein Is Shewn the Insufficiency Thereof, in Its Several Particulars*. 2nd ed. London: A. Bosvile.

Ballester, Luis García. 1993. "On the Origin of the 'Six Things Non-Natural' in Galen." In *Galen und das hellenistische Erbe*, ed. Jutta Kollesch and Diethard Nickel, 105–15. Stuttgart: Steiner.

Barnett, Eleanor. 2020. "Reforming Food and Eating in Protestant England, c. 1560–c. 1640." *Historical Journal* 63: 507–27.

Barnett, L. Margaret. 1997. "Fletcherism: The Chew-Fad of the Edwardian Era." In D. F. Smith, ed., 1997, 6–28.

Barr, Ann, and Paul Levy. 1984. *The Official Foodie Handbook: Be Modern, Worship Food*. London: Ebury.

Barry, Jonathan, and Fabrizio Bigotti, eds. 2022. *Santorio Santori and the Emergence of Quantified Medicine, 1614–1790*. Cham, Switzerland: Springer.

Barthes, Roland. [1984] 1989. *The Rustle of Language*. Trans. Richard Howard. Berkeley: University of California Press.

Basse, William. 1630. *A Helpe to Memory and Discourse with Table-Talke as Musicke to a Banquet of Wine*. London.

Baxter, Richard. 1673. *A Christian Directory, or, a Summ of Practical Theologie and Cases of Conscience . . .* London.

Bayfield, Robert. 1655. *Enchiridion Medicum: Containing the Causes, Signs, and Cures of All Those Diseases, That Do Chiefly Affect the Body of Man*. London.

Beeton, Isabella. 1861. *Book of Household Management*. London: S. O. Beeton.

Bellini, Federico. 2018. "Diet and Hygiene between Ethics and Medicine: Evidence and the Reception of Alvise Cornaro's *La Vita Sobria* in Early Seventeenth-Century England." In *Evidence in the Age of the New Sciences*, ed. James A. T. Lancaster and Richard Raiswell, 251–68. Cham, Switzerland: Springer.

Benedict, Francis G., and Cornelia Golay Benedict. 1933. *Mental Effort in Relation to Gaseous Exchange, Heart Rate, and Mechanics of Respiration*. Washington, DC: Carnegie Institution of Washington.

Benedict, Francis G., and Thorne M. Carpenter. 1909. *The Influence of Muscular and Mental Work on Metabolism and the Efficiency of the Human Body as a Machine*. Washington, DC: US Government Printing Office.

———. 1910. *Respiration Calorimeters for Studying the Respiratory Exchange and Energy Transformations in Man*. Carnegie Institution of Washington Publication no. 123. Washington, DC: Carnegie Institution of Washington.

———. 1930. "The Energy Requirements of Intense Mental Effort." *Proceedings of the National Academy of Sciences* 16: 438–43.

Berenstein, Nadia. 2018. "Flavor Added: The Sciences of Flavor and the Industrialization of Taste in America." PhD dissertation, University of Pennsylvania.

Bergdolt, Klaus [1999] 2008. *Wellbeing: A Cultural History of Healthy Living*. Trans. Jane Dewhurst. Cambridge: Polity.

Bersford, Thomas. 1908. *Theories and Facts for Students of Longevity and Health*. San Francisco: Thomas Bersford.

Biltekoff, Charlotte. 2013. *Eating Right in America: The Cultural Politics of Food and Health*. Durham, NC: Duke University Press.

———. 2019. "What Does It Mean to Eat Right? Nutrition, Science, and Society." In *Food Fights: How History Matters to Contemporary Food Debates*, ed. Charles C. Ludington and Matthew Morse Booker, 124–42. Chapel Hill: University of North Carolina Press.

Blackmore, Richard. 1725. *A Treatise of the Spleen and Vapours*. London: J. Pemberton.

Blaxter, Mildred, and Elizabeth Paterson. 1982. *Mothers and Daughters: A Three-Generational Study of Health Attitudes and Behaviour*. London: Heinemann.

———. 1983. "The Goodness Is Out of It: The Meaning of Food to Two Generations." In *The Sociology of Food and Eating: Essays on the Sociological Significance of Food*, ed. Anne Murcott, 95–105. Aldershot: Gower.

Blount, Charles. 1680. *Great Is Diana of the Ephesians, or, The Original of Idolatry Together with the Politick Institution of the Gentiles Sacrifices*. London.

Boerhaave, Herman. 1742–1746. *Dr. Boerhaave's Academical Lectures on the Theory of Physic*. 6 vols. London: W. Innys.

Bogert, L. Jean. [1937] 1940. *Dietetics Simplified: The Use of Foods in Health and Disease*. 2nd ed. New York: Macmillan.

Boia, Lucian. [1998] 2004. *Forever Young: A Cultural History of Longevity from Antiquity to the Present*. Trans. Trista Selous. London: Reaktion.

Bonet, Théophile. 1686. *A Guide to the Practical Physician*. London.
Bontius, Jacobus. [1642] 1769. *An Account of the Diseases, Natural History, and Medicines of the East Indies*. Trans. anon. London: T. Noteman.
Boorde, Andrew. [1542] 1870. "A Compendyous Regyment of Helth." In *Andrew Boorde's Introduction and Dyetary*, ed. F. J. Furnivall, Early English Text Society, Extra Series, no. X. London: N. Trübner.
Border, D. 1651. *Polypharmakos kai chymistes, or, The English Unparalell'd Physitian and Chyrurgian*. London.
Bostock, John. [1826–1834] 1836. *An Elementary System of Physiology*. New ed. London: Henry G. Bohn.
Botero, Giovanni. [1597] 1630. *Relations of the Most Famous Kingdomes and Common-wealths thorowout the World* . . . 2nd ed. London.
Bouley, Bradford. 2020. "Digesting Faith: Eating God, Man, and Meat in Seventeenth-Century Rome." *Osiris* 35: 42–59.
Boulton, Samuel. 1656. *Medicina magica tamen physica: Magical, but Natural Physick*. London.
Boyle, Robert. [comp. ca. 1640s] 1991. "The Aretology." In *The Early Essays and Ethics of Robert Boyle*, ed. John T. Harwood, 3–142. Carbondale: Southern Illinois University Press.
———. 1663. *Some Considerations Touching the Usefulnesse of Experimentall Naturall Philosophy*. London.
———. 1665. *Occasional Reflections upon Several Subjects*. London.
———. 1685. *Of the Reconcileableness of Specifick Medicines to the Corpuscular Philosophy*. London.
———. 1686. *A Free Enquiry into the Vulgarly Receiv'd Notion of Nature*. London.
Braconnot, M. H. 1820. "On the Conversion of Animal Matter into New Substances." *Philosophical Magazine* 56: 131–37.
Brain, Peter. 1986. *Galen on Bloodletting: A Study of the Origins, Development and Validity of His Opinions*. Cambridge: Cambridge University Press.
Brande, William Thomas. [1819] 1830. *A Manual of Chemistry*. 3rd ed., 2 vols. London: John Murray.
Brathwait, Richard. 1630. *The English Gentleman, Containing Sundry Excellent Rules or Exquisite Observations, Tending to Direction of Every Gentleman, of Selecter Ranke and Qualitie*. London.
———. 1631. *The English Gentlewoman*. London.
Brewer, Stephen C. 2019. *What Happened to Moderation? A Common-Sense Approach to Improving Our Health and Treating Common Illnesses in an Age of Extremes*. New York: SelectBooks.
Bright, Timothie. [1586] 1613. *A Treatise of Melancholy*. London.
Brillat-Savarin, Jean Anthelme. [1825] 1994. *The Physiology of Taste*. Trans. Anne Drayton. London: Penguin.

Britton, Piers D. 2006. "(Hu)moral Exemplars: Type and Temperament in Cinquecento Painting." In *Visualizing Medieval Medicine and Natural History, 1200–1550*, ed. Jean Ann Givens, Karen Reeds, and Alain Touwaide, 177–204. Aldershot: Ashgate.

Brock, William H. 1997. *Justus von Liebig: The Chemical Gatekeeper*. Cambridge: Cambridge University Press.

Brooke, Humphrey. 1650. *Ugieine: or A Conservatory of Health*. London.

Brookes, Richard. [1751] 1754. *The General Practice of Physic . . . To Which Is Prefixed . . . The Use of the Non-Naturals*. 2nd ed., 2 vols. London: J. Newbery.

Brown, John Russell. 2006. *The Shakespeare Handbooks: Hamlet*. London: Palgrave Macmillan.

Brown, Peter. 1988. *The Body and Society: Men, Women and Sexual Renunciation in Early Christianity*. New York: Columbia University Press.

Brown, Theodore M. 1970. "The College of Physicians and the Acceptance of Iatro-Mechanism in England, 1665–1695." *Bulletin of the History of Medicine* 44: 12–30.

———. 1974. "From Mechanism to Vitalism in Eighteenth-Century English Physiology." *Journal of the History of Biology* 7: 179–216.

———. 1977. "Physiology and the Mechanical Philosophy in Mid-Seventeenth-Century England." *Bulletin of the History of Medicine* 51: 25–54.

———. 1987. "Medicine in the Shadow of the *Principia*." *Journal of the History of Ideas* 48: 629–48.

Browne, Sir Thomas. [1646] 1650. *Pseudodoxia Epidemica: Or, Enquiries into Very Many Received Tenents, and Commonly Presumed Truths*. London.

———. [1650] 1716. *Christian Morals*. Cambridge: University Press.

Brugis, Thomas. 1648. *The Marrow of Physicke*. London.

Buchan, William. 1769. *Domestic Medicine; or, the Family Physician*. Edinburgh: Balfour, Auld, and Smellie.

———. 1790. *Domestic Medicine: or, A Treatise on the Prevention and Cure of Diseases*. 11th ed. London: A. Strahan, T. Cadell, J. Balfour, W. Creech.

———. 1797. *Observations Concerning the Diet of the Common People*. London: A. Strahan.

———. 1802. *The New Domestic Medicine: or, Universal Family Physician*. New ed. by George Wallis. London: Alex. Hogg.

———. 1828. *Domestic Medicine, or, A Treatise on the Prevention and Cure of Diseases*. Exeter: J. & B. Williams.

Bucknill, John Charles. 1860. *The Medical Knowledge of Shakespeare*. London: Longman.

Bullein, William. [1562] 1595. *The Government of Health*. London.

B[ulwer], J[ohn]. 1649. *Athomyotomia or a Dissection of the Significative Muscles of the Affections of the Minde*. London.

Burgis, Thomas. 1648. *The Marrow of Physicke*. London.

Burke, Peter. 1978. *Popular Culture in Early Modern Europe*. London: Maurice Temple Smith.

Burnett, Charles. 1991. "The Superiority of Taste." *Journal of the Warburg and Courtauld Institutes* 54: 230–38.

Burnet[t], Thomas. [1724–1725] 1737. "Dr. Burnet's Demonstration of True Religion," Part 2. In *A Defence of Natural and Revealed Religion: Being an Abridgement of Sermons Preached at the Lecture Founded by the Hon. Robert Boyle, Esq.*, Vol. 4. London: Arthur Bettesworth and Charles Hitch.

Burnham, W. Gurney. 1907. *A Book of Quotations, Proverbs and Household Words*. London: Cassell.

Burns, Robert. 1868. *Poems and Songs*. Edinburgh: William P. Nimmo.

Burton, John. 1738. *A Treatise on the Non-Naturals; in Which the Great Influence They Have on Human Bodies Is Set Forth*. York: A. Staples.

Burton, Robert. [1621] 2001. *The Anatomy of Melancholy*. New York: NYRB Books.

[Burton, William]. 1743. *An Account of the Life and Writings of Herman Boerhaave*. London: Henry Lintot.

Butts, Henry. 1599. *Dyets Dry Dinner Consisting of Eight Severall Courses*. London.

Bylebyl, Jerome J. 1979. "The Medical Side of Harvey's Discovery: The Normal and the Abnormal." In *William Harvey and His Age: The Medical and Social Context of the Discovery of the Circulation*, ed. Bylebyl, Supplement to *Bulletin of the History of Medicine*, new series, 2: 28–102. Baltimore: Johns Hopkins University Press.

Bynum, Caroline Walker. 1987. *Holy Feast and Holy Fast: The Religious Significance of Food to Medieval Women*. Berkeley: University of California Press.

Bynum, William F., and Michael Neve. 1985. "Hamlet on the Couch." In *The Anatomy of Madness: Essays in the History of Psychiatry*, 3 vols., ed. William F. Bynum, Roy Porter, and Michael Shepherd, 1:289–304. London: Tavistock.

Cadden, Joan. 1993. *The Meanings of Sex Differences in the Middle Ages: Medicine, Science, and Culture*. Cambridge: Cambridge University Press.

Camporesi, Piero. 1994. *The Anatomy of the Senses: Natural Symbols in Medieval and Early Modern Italy*. Trans. Allan Cameron. Cambridge: Polity.

Cantor, David, ed. [2001] 2016. *Reinventing Hippocrates*. New York: Routledge.

Cardenas, Diana. 2013. "Let Not Thy Food Be Confused with Thy Medicine: The Hippocratic Misquotation." *e-SPEN Journal* 8: 260–62.

Carlin, Martha. 2008. "'What Say You to a Piece of Beef and Mustard?': The Evolution of Public Dining in Medieval and Tudor London." *Huntington Library Quarterly* 71: 199–217.

Carpenter, Kenneth J. 1994. *Protein and Energy: A Study of Changing Ideas in Nutrition*. Cambridge: Cambridge University Press.

———. 1994. "The Life and Times of W. O. Atwater (1844–1907)." *Journal of Nutrition* 124, supplement 9: 1707S–1714S.

———. 2006. "Nutritional Studies in Victorian Prisons." *Journal of Nutrition* 136: 1–8.

Carrera, Elena, ed. 2013. *Emotions and Health, 1200–1700*. Leiden: Brill.

———. 2013. "Anger and the Mind-Body Problem." In Carrera, ed., 2013, 95–146.

Carruthers, Mary. 2006. "Sweetness." *Speculum* 81: 999–1013.

Casalini, Cristiano, and Laura Madella. 2021. "The Jesuit Cultivation of Vegetative Souls: Leonard Lessius (1554–1623) on a Sober Diet." In *Vegetative Powers: The Roots of Life in Ancient, Medieval and Early Modern Natural Philosophy*, ed. Fabrizio Baldassarri and Andreas Blank, 177–98. Zurich: Springer.

Castiglione, Baldesar. [1528] 2002. *The Book of the Courtier*. Ed. Daniel Javitch, trans. Charles Singleton. New York: Norton.

Cavallo, Sandra, and Tessa Storey. 2013. *Healthy Living in Late Renaissance Italy*. Oxford: Oxford University Press.

———. 2013. "Regimens, Authors and Readers: Italy and England Compared." In Cavallo and Storey, 2013a, 23–52.

———, eds. 2017. *Conserving Health in Early Modern Culture: Bodies and Environments in Italy and England*. Manchester: Manchester University Press.

———. 2017. "Conserving Health: The Non-Naturals in Early Modern Culture and Society." In Cavallo and Storey, eds., 2017, 1–19.

Cavendish, William, Duke of Newcastle. 1667. *A New Method, and Extraordinary Invention, to Dress Horses, and Work Them According to Nature*. London.

Celsus, Aulus Cornelius. 1960. *De Medicina*. 3 vols. Trans. W. G. Spencer. Cambridge, MA: Harvard University Press.

Chamberlayne, Edward. 1669. *Angliae notitia; or, The Present State of England*. London.

Chambers, Ephraim. 1728. *Cyclopædia: or, an Universal Dictionary of Arts and Sciences*. 2 vols. London: James and John Knapton.

Chaplin, Joyce. 1997. "Natural Philosophy and an Early Racial Idiom in North America: Comparing English and Indian Bodies." *William and Mary Quarterly*, 3rd series, 54: 229–52.

———. 2001. *Subject Matter: Technology, Science, and the Body on the Anglo-American Frontier, 1500–1676*. Cambridge, MA: Harvard University Press.

———. 2020. "Why Drink Water? Diet, Materialisms, and British Imperialism." *Osiris* 35: 99–122.

Charen, Mona. 2012. "Gluttonous Greed Deserves No Reward." *Baltimore Sun*, December 2. https://www.baltimoresun.com/news/bs-xpm-2002-12-02-0212020278-story.html.

Charleton, Walter. 1654. *Physiologia Epicuro-Gassendo-Charltoniana, or, A Fabrick of Science Natural, upon the Hypothesis of Atoms Founded by Epicurus Repaired [by] Petrus Gassendus*. London.

———. 1659. *Natural History of Nutrition, Life, and Voluntary Motion*. London.

———. 1680. *Enquiries into Human Nature*. London.

Charron, Pierre. [1601] 1608. *Of Wisdome*. Trans. Samson Lennard. London.

Chaucer, Geoffrey. 1687. "The Wife of Bath's Prologue." In *The Works of Our Ancient, Learned, & Excellent Poet, Jeffrey Chaucer*. London.

Chen, Nancy N. 2009. *Food, Medicine, and the Quest for Good Health: Nutrition, Medicine, and Culture*. New York: Columbia University Press.

Cheyne, George. 1701. *A New Theory of Continu'd Fevers*. Edinburgh: John Vallange.

———. 1720. *Observations Concerning the Nature and Due Method of Treating the Gout, . . . Together with an Account of the Nature and Qualities of the Bath Waters*. London: George Strahan.

———. 1724. *An Essay of Health and Long Life*. London: George Strahan.

———. 1733. *The English Malady: or, a Treatise of Nervous Diseases of All Kinds, as Spleen, Vapours, Lowness of Spirits, Hypochondriacal, and Hysterical Complaints*. London: George Strahan.

———. [1725] 1736. *Philosophical Principles of Religion, Natural and Revealed*. 5th ed. London: George Strahan.

———. 1740. *An Essay on Regimen*. London: C. Rivington.

———. 1742. *The Natural Method of Cureing the Diseases of the Body and the Disorders of the Mind*. 4th ed. London: George Strahan.

———. 1933. *The Letters of Doctor George Cheyne to Samuel Richardson (1733–1743)*. Ed. Charles F. Mullett, University of Missouri Studies, Volume 18, no. 1. Columbia: University of Missouri Press.

Church of England. 2013. *The Book of Common Prayer*. London: Penguin.

Clarke, John. 1639. *Parœmiologia Anglo-Latina*. London.

Clarke, Samuel. 1659. *Medulla theologiæ, or, The Marrow of Divinity*. London.

Cleland, John. 1612. *The Instruction of a Young Noble-Man*. Oxford.

Clericuzio, Antonio. 2012. "Chemical and Mechanical Theories of Digestion in Early Modern Medicine." *Studies in History and Philosophy of Biological and Biomedical Sciences* 43: 329–37.

Clifton, Francis. 1732. *The State of Physick, Ancient and Modern*. London: W. Bowyer.

Cock, Thomas. 1665. *Hygieine, or, A Plain and Practical Discourse upon the First of the Six Non-Naturals, viz. Air*. London.

———. 1675. *Kitchin-Physick: or, Advice to the Poor*. London.

———. 1675. *Miscelanea Medica: or, a Supplement to Kitchin-Physick*. London.

Cogan, Thomas. [1584] 1630. *The Haven of Health*. London.

Cohen-Hanegbi, Naama. 2016. "A Moving Soul: Emotions in Late Medieval Medicine." *Osiris* 31: 46–66.

Coleman, William. 1974. "Health and Hygiene in the *Encyclopédie*: A Medical Doctrine for the Bourgeoisie." *Journal of the History of Medicine* 29: 399–421.

———. 1977. "The People's Health: Medical Themes in 18th-Century French Popular Literature." *Bulletin of the History of Medicine* 51: 55–74.

Collingham, E[lizabeth] M. 2001. *Imperial Bodies: The Physical Experience of the Raj, c. 1800–1947*. Cambridge: Polity.

———. 2011. *The Taste of War: World War II and the Battle for Food*. London: Allen Lane.

Columbus, Christopher. 1969. *The Four Voyages*. Ed. and trans. J. M. Cohen. Harmondsworth: Penguin.

Combe, Andrew. [1834] 1836. *The Physiology of Digestion, Considered with Relation to the Principles of Dietetics*. 2nd ed. Edinburgh: Maclachlan & Stewart.

Combe, George. 1830. *A System of Phrenology*. Edinburgh: John Anderson.

Comenius, Johann Amos. 1651. *Naturall Philosophie Reformed by Divine Light, or, A Synopsis of Physicks*. London.

Cook, Harold J. 1986. *The Decline of the Old Medical Regime in Stuart England*. Ithaca, NY: Cornell University Press.

———. 1990. "The New Philosophy and Medicine in Seventeenth-Century England." In *Reappraisals of the Scientific Revolution*, ed. David C. Lindberg and Robert S. Westman, 397–436. Cambridge: Cambridge University Press.

———. 1999. "Bernard Mandeville and the Therapy of 'The Clever Politician.'" *Journal of the History of Ideas* 60: 101–24.

Corbin, Alain. [1982] 1986. *The Foul and the Fragrant: French Social Imagination*. Trans. Aubier Montaigne. Cambridge, MA: Harvard University Press.

Cornaro, Luigi. [1550] 1634. *A Treatise of Temperance and Sobrietie*. Trans. George Herbert. London (bound together with Lessius 1634).

———, Louis [Luigi]. 1903. *The Art of Living Long*. Milwaukee: William F. Butler.

Cornaro, Alvise [Luigi]. 2014. *Writings on the Sober Life: The Art and Grace of Living Long*. Ed. and trans. Hiroko Fudemoto. Toronto: University of Toronto Press.

Corneanu, Sorana. 2012. *Regimens of the Mind: Boyle, Locke, and the Early Modern Cultura Animi Tradition*. Chicago: University of Chicago Press.

———. 2017. "The Nature and Care of the Whole Man: Francis Bacon and Some Late Renaissance Contexts." *Early Science and Medicine* 22: 130–56.

Coveney, John. [2000] 2006. *Food, Morals and Meaning: The Pleasure and Anxiety of Eating*. 2nd ed. London: Routledge.

Cowan, Brian. 2007. "New Worlds, New Tastes: Food Fashions after the Renaissance." In *Food: The History of Taste*, ed. Paul Freedman, 196–231. Berkeley: University of California Press.

Craig, D. H. 1985. *Sir John Harington*. Boston: Twayne.

Craiutu, Aurelian. 2017. *Faces of Moderation: The Art of Balance in an Age of Extremes*. Philadelphia: University of Pennsylvania Press.

Cravens, Hamilton. 1990. "Establishing the Science of Nutrition at the USDA: Ellen Swallow Richards and Her Allies." *Agricultural History* 64: 122–33.

Crofts, Robert. 1641. *The Way to Happinesse on Earth Concerning Riches, Honour, Conjugall Love, Eating, Drinking*. London.

Crookes, William. 1898. "Address of the President before the British Association for the Advancement of Science, Bristol, 1898." *Science*, new series, 8, no. 200 (October 28): 561–75.

———. 1900. *The Wheat Problem*. Rev. ed. London: John Murray.

———. 1917. *The Wheat Problem*. 3rd ed. London: Longmans, Green.

Crosby, Alfred W., Jr. 1972. *The Columbian Exchange: Biological and Cultural Consequences of 1492*. Westport, CT: Greenwood.

Cueto, Marcos, and Steven Palmer. 2019. *Medicine and Public Health in Latin America: A History*. Cambridge: Cambridge University Press.

Cullather, Nick. 2007. "The Foreign Policy of the Calorie." *American Historical Review* 112: 337–64.

———. 2010. *The Hungry World: America's Cold War Battle against Poverty in Asia*. Cambridge, MA: Harvard University Press.

Cullen, William. [1760] 1775. *Lectures on the Materia Medica*. Philadelphia: Robert Bell.

Culpeper, Nicholas, Adbiah Cole, and William Rowland. [1640–1655] 1655. *The Practice of Physick in Seventeen Several Books, . . . Being Chiefly a Translation of that Learned and Reverend Doctor, Lazarus Riverius [Lazare Rivière]*. London.

Cunningham, Andrew. 1981. "Sydenham versus Newton: The Edinburgh Fever Dispute of the 1690s between Andrew Brown and Archibald Pitcairne." *Medical History*, suppl. 1: 71–98.

———. 1990. "Medicine to Calm the Mind: Boerhaave's Medical System and Why It Was Adopted in Edinburgh." In *The Medical Enlightenment of the Eighteenth Century*, ed. Cunningham and Roger French, 40–66. Cambridge: Cambridge University Press.

Cushman, Gregory T. 2013. *Guano and the Opening of the Pacific World: A Global Ecological History*. Cambridge: Cambridge University Press.

Dacome, Lucia. 2001. "Living with the Chair: Private Excreta, Collective Health and Medical Authority in the Eighteenth Century." *History of Science* 39: 467–500.

———. 2006. "Resurrecting by Numbers in Eighteenth-Century England." *Past and Present*, no. 193: 73–110.

———. 2012. "Balancing Acts: Picturing Perspiration in the Long Eighteenth Century." *Studies in History and Philosophy of Biological and Biomedical Sciences* 43: 379–91.

Dannenfeldt, Karl H. 1986. "Sleep: Theory and Practice in the Late Renaissance." *Journal of the History of Medicine* 41: 415–41.

Darby, Stephen. 1870. *On Fluid Meat*. London: John Churchill.

Daston, Lorraine. 2019. *Against Nature*. Cambridge, MA: MIT Press.

Davidson, James N. 1997. *Courtesans and Fishcakes: The Consuming Passions of Classical Athens*. Chicago: University of Chicago Press.

Davis, Lance E., and Robert A. Huttenback. 1986. *Mammon and the Pursuit of Empire: The Political Economy of British Imperialism, 1860–1912*. Cambridge: Cambridge University Press.

Dawson, Mark. 2009. *Plenti and Grase: Food and Drink in a Sixteenth-Century Household*. Totnes, Devon: Prospect Books.

Dear, Peter. 1998. "A Mechanical Microcosm: Bodily Passions, Good Manners, and Cartesian Mechanism." In Lawrence and Shapin, eds., 1998, 51–82.

Deegan, Gordon. 2004. "Board Opposes a McDonald's for Ennis over Health Factors." *Irish Times*, February 3. https://www.irishtimes.com/news/board-opposes-a-mcdonald-s-for-ennis-over-health-factors-1.1132146.

[Defoe, Daniel]. 1700. *The True-Born Englishman: A Satyr*. London.

———. [1719] 1868. *Robinson Crusoe*. Ed. Henry Kingsley. London: Macmillan.

———. 1722. *Due Preparations for the Plague, As Well for Soul as Body*. London: E. Matthews.

DeJong, W., C. K. Atkin, and L. Wallack. 1992. "A Critical Analysis of 'Moderation' Advertising Sponsored by the Beer Industry: Are 'Responsible Drinking' Commercials Done Responsibly?" *Milbank Quarterly* 70: 661–78.

Dekker, Thomas. 1615. *The Cold Yeare 1614 . . . Written Dialogue-wise, in a Plaine Familiar Talke between a London Shop-keeper and a North-Country-man*. London.

Della Porta, Giambattista. 1586. *De humana physiognomonia libri IIII*. Vici Equense. 1586.

Descartes, René. [1629–1633] 1972. *Treatise of Man*. Trans. Thomas Steele Hall. Cambridge, MA: Harvard University Press.

———. 1976. *Descartes' Conversations with Burman*. Ed. and trans. John Cottingham. Oxford: Oxford University Press.

———. 1984–1991. *The Philosophical Writings of Descartes*. 3 vols. Trans. John Cottingham, Robert Stoothoff, and Dugald Murdoch. Cambridge: Cambridge University Press.

———. [1637] 1984–1991. "Discourse on the Method." In Descartes, 1984–1991, 1:111–51.

———. 1984–1991. "Principles of Philosophy (extracts)." In idem, *The Philosophical Writings of Descartes*, 1:179–291. Cambridge: Cambridge University Press.

———. 1984–1991. "The Description of the Human Body." In Descartes 1984–1991, 1:313–24.

———. [1649] 1989. *The Passions of the Soul*. Ed. and trans. Stephen Voss. Indianapolis: Hackett.

Desert Fathers. 1975. *The Sayings of the Desert Fathers (Apophthegmata Patrum)*. Trans. Benedicta Ward. London: A. R. Mowbray.

Dickie, George. 1996. *The Century of Taste: The Philosophical Odyssey of Taste in the Eighteenth Century*. Oxford: Oxford University Press.

Digby, Kenelm. 1644. *Two Treatises, in One of Which, the Nature of Bodies . . . Is Looked Into*. Paris.

d'Iharce, Jean-Luc. 1783. *Erreurs populaires sur la médecine*. Paris.

DiMeo, Michelle, and Sara Pennell, eds. 2013. *Reading and Writing Recipe Books, 1550–1800*. Manchester: Manchester University Press.

Diogenes Laërtius. 1688. *The Lives, Opinions, and Remarkable Sayings of the Most Famous Ancient Philosophers*. 2 vols. London.

Donato, Maria Pia. 2019. "Galen in an Age of Change (1650–1820)." In *Brill's Companion to the Reception of Galen*, ed. Petros Bouras-Vallianatos and Barbara Zipser, 487–507. Leiden: Brill.

Donne, John. 1896. *Poems of John Donne*. 2 vols. Ed. E. K. Chambers. London: Scribner.

Douglas, Mary. 1975. "Environments at Risk." In Mary Douglas, *Implicit Meanings: Essays in Anthropology*, 204–17. London: Routledge & Kegan Paul.

Drage, William. 1664. *A Physical Nosonomy, or, A New and True Description of the Law of God (called Nature) in the Body of Man*. London.

Drummond, J. C., and Anne Wilbraham. [1939] 1991. *The Englishman's Food: Five Centuries of English Diet*. London: Pimlico.

Dryden, John. 1679. *Troilus and Cressida*. London.

Du Bartas, Guillaume de Salluste. 1611. *Du Bartas His Divine Weekes and Workes*. London.

Duffett, Rachel. 2015. *The Stomach for Fighting: Food and the Soldiers of the Great War*. Manchester: Manchester University Press.

Du Laurens, André. [1594] 1599. *A Discourse of the Preservation of Sight: Of Melancholicke Diseases; of Rheumes, and of Old Age*. 2nd ed. London.

Duncon, Eleazar. 1606. *The Copy of a Letter Written by E. D. Doctour of Physicke to a Gentleman, by Whom It Was Published*. London.

Earle, John. 1628. *Micro-cosmographie, or, A Peece of the World Discovered in Essayes and Characters*. London.

Earle, Rebecca. 2010. "'If You Eat Their Food . . .': Diets and Bodies in Early Colonial Spanish America." *American Historical Review* 115: 688–713.

———. 2012. *The Body of the Conquistador: Food, Race and the Colonial Experience in Spanish America, 1492–1700*. Cambridge: Cambridge University Press.

———. 2017. "Potatoes and the Hispanic Enlightenment." *The Americas* 75: 639–60.

———. 2018. "Promoting the Potato in Eighteenth-Century Europe." *Eighteenth-Century Studies* 51: 147–62.

———. 2020. *Feeding the People: The Politics of the Potato*. New York: Cambridge University Press.

EatDrinkPolitics. 2013. "Is the Academy of Nutrition and Dietetics Silencing Its Members Who Object to McDonald's Sponsoring Lunch?" http://www.eatdrinkpolitics.com/2013/02/27/is-the-academy-of-nutrition-and-dietetics-silencing-its-members-who-object-to-mcdonalds-sponsoring-lunch/.

Ebrahimnejad, Hormoz, ed. 2009. *The Development of Modern Medicine in Non-Western Countries: Historical Perspectives*. London: Routledge.

Edelstein, Ludwig. [1931] 1967. "The Dietetics of Antiquity." In *Ancient Medicine: Selected Papers of Ludwig Edelstein*, ed. Owsei Temkin and C. Lilian Temkin, 303–16. Baltimore: Johns Hopkins Press, 303–16.

Edgerton, David. 2007. *The Shock of the Old: Technology and Global History since 1900*. Oxford: Oxford University Press.

Elias, Norbert. [1939] 1978. *The Civilizing Process: Sociogenetic and Psychogenetic Investigations*, Vol. 1, *The History of Manners*. 2 vols. Trans. Edmund Jephcott. Oxford: Blackwell.

———. 2004. "On the Eating of Meat." *Food & History* 2, no. 2: 11–16.

Elliott, Stuart. 2009. "A Web Awash in Liquor Ads, Promoting Moderation." *New York Times*, March 8. https://www.nytimes.com/2009/03/09/business/media/09adcol.html.

Elyot, Thomas, 1537. *The Boke Named the Governour*. London.

———. 1538. *The Dictionary of Syr Thomas Eliot Knight*. London.

———. [1539] 1587. *The Castell of Health*. London.

Emch-Dériaz, Antoinette Suzanne. 1992. "The Non-Naturals Made Easy." In Porter, ed., 1992, 134–59.

———. 1992. *Tissot: Physician of the Enlightenment*. New York: Peter Lang.

Encyclopædia Britannica: or, A Dictionary of Arts and Sciences. 1771. 3 vols. Edinburgh: A. Bell and C. Macfarquhar.

Epictetus. 1888[?]. *The Discourses of Epictetus, with the Encheiridion and Fragments*. Trans. George Long. New York: A. L. Burt.

Erasmus, Desiderius. 1569. *Proverbs or Adages*. Ed. and trans. Richard Taverner. London.

———. [1516] 1965. *The Education of a Christian Prince*. Trans. Lester K. Born. New York: Octagon.

Ercker, Lazarus. 1683. *Fleta Minor. The Laws of Art and Nature in . . . Metals*. London.

Evelyn, John. 1661. *Fumifugium, or, The Inconveniencie of the Aer and Smoak of London Dissipated Together with Some Remedies Humbly Proposed*. London.

———. 1699. *Acetaria, a Discourse of Sallets*. London.

Everitt, Arthur V., et al., eds. 2010. *Calorie Restriction, Aging and Longevity*. New York: Springer.

Farmer, Fannie Merritt. 1904. *Food and Cookery for the Sick and Convalescent*. Boston: Little, Brown.

Feijóo y Montenegro, Benito Jerónimo. [1727] 1800[?]. *Rules for Preserving Health, Particularly with Regard to Studious Persons*. Trans. anon. from Spanish. London: R. Faulder.

Fernel, Jean. [1567] 2003. *The Physiologia of Jean Fernel (1567)*. Trans. John M. Forrester. Philadelphia: American Philosophical Society.

Feuerbach, Ludwig. [1862] 1959–1960. "Das Geheimniss des Opfers, oder Der Mensch ist was er isst." In *Sämmtliche Werke*, Vol. 10, 2nd ed., ed. Wilhelm Bolin and Friedrich Jodl, 41–67. Stuttgart-Bad Cannstatt: Frommann Verlag.

[Feyens, Jean]. 1676. *A New and Needful Treatise of Wind Offending Mans Body*. Trans. William Rowland. London.

Ficino, Marsilio. [1489] 1989. *Three Books on Life*. Ed. and trans. Carol V. Kaske and John R. Clark. Binghamton, NY: Renaissance Society of America.

Finlay, Mark R. 1992. "Quackery and Cookery: Justus von Liebig's Extract of Meat and the Theory of Nutrition in the Victorian Age." *Bulletin of the History of Medicine* 66: 404–18.

———. 1995. "Early Marketing of the Theory of Nutrition: The Science and Culture of Liebig's Extract of Meat." In Kamminga and Cunningham, eds., 1995, 48–74.

Fischler, Claude. 1988. "Food, Self, and Identity." *Social Science Information* 27: 275–92.

Fisher, J. B. 2009. "Digesting Falstaff: Food and Nation in Shakespeare's *Henry IV* Plays." *Early English Studies* 2: 1–23.

Fissell, Mary E. 1992. "Readers, Texts, and Contexts: Vernacular Medical Works in Early Modern England." In Porter, ed., 1992, 72–96.

———. 2003. "Making a Masterpiece: The Aristotle Texts in Vernacular Medical Culture." In Rosenberg, ed., 2003, 59–87.

———. 2007. "The Marketplace of Print." In Jenner and Wallis, eds., 2007, 108–32.

———. 2008. "Introduction: Women, Health, and Healing in Early Modern Europe." *Bulletin of the History of Medicine* 82: 1–17.

Fitzpatrick, Joan. 2007. *Food in Shakespeare: Early Modern Dietaries and the Plays*. Aldershot: Ashgate.

———. 2014. "Diet and Identity in Early Modern Dietaries and Shakespeare: The Inflections of Nationality, Gender, Social Rank, and Age." In *Shakespeare Studies*, Vol. 42, ed. James R. Siemon and Diana E. Henderson, 75–90. Madison, NJ: Fairleigh Dickinson University Press.

Flandrin, Jean-Louis. 2000. "Seasoning, Cooking, and Dietetics in the Late Middle Ages." In Flandrin and Montanari, eds., 2000, 313–27.

———. 2000. "Dietary Choices and Culinary Technique, 1500–1800." In Flandrin and Montanari, eds., 2000, 403–17.

———. 2000. "From Dietetics to Gastronomy: The Liberation of the Gourmet." In Flandrin and Montanari, eds., 2000, 418–34.

Flandrin, Jean-Louis, and Massimo Montanari, eds. 2000. *Food: A Culinary History*. New York: Penguin.

Fletcher, Horace. 1903. *The New Glutton or Epicure*. New York: Frederick A. Stokes.

Flourens, Pierre. [1854] 1855. *On Human Longevity and the Account of Life upon the Globe*. Trans. from 2nd French edition by Charles Martel. London: H. Bailliere.

Floyer, John. 1687. *Pharmako-Basanos; or the Touch-Stone of Medicines*. London.

Fonteyn, Nicolaas. 1652. *The Womans Doctour, or, An Exact and Distinct Explanation of All Such Diseases as Are Peculiar to That Sex*. London.

Fordyce, George. 1791. *A Treatise on the Digestion of Food*. London: J. Johnson.

Forster, William. 1738. *A Treatise on the Various Kinds and Qualities of Foods*. Newcastle-upon-Tyne: John White.

———. 1745. *A Treatise on the Causes of Most Diseases Incident to Human Bodies, and the Cure of Them: First, by a Right Use of the Non-Naturals; Chiefly by Diet*. Leeds: James Lister.

Forsyth, J. S. 1834. *A Dictionary of Diet*. 2nd ed. London: Henry Cremer.
Fothergill, J. Milner. 1886. *A Manual of Dietetics*. New York: William Wood.
Fothergill, John. 1762. *Rules for the Preservation of Health*. London: John Pridden.
Foucault, Michel. 1983. "On the Genealogy of Ethics: An Overview of a Work in Progress." In *Michel Foucault: Beyond Structuralism and Hermeneutics*. 2nd ed. ed. Hubert L. Dreyfus and Paul Rabinow, 229–52. Chicago: University of Chicago Press.
———. [1984] 1990. *The History of Sexuality*, Vol. 2, *The Use of Pleasure*. Trans. Robert Hurley. New York: Viking.
———. [1966] 1994. *The Order of Things: An Archaeology of the Human Sciences*. New York: Vintage.
Fox, Adam. 2000. *Oral and Literate Culture in England, 1500–1700*. Oxford: Clarendon.
Frank, Robert G. 1980. *Harvey and the Oxford Physiologists: Scientific Ideas and Social Interaction*. Berkeley: University of California Press.
[Franklin, Benjamin]. 1766. "Second Reply to 'Vindex Patriae,' 2 January 1766." Founders Online. https://founders.archives.gov/documents/Franklin/01-13-02-0003.
———. 1766. "Further Defense of Indian Corn, 15 January 1766." Founders Online. https://founders.archives.gov/documents/Franklin/01-13-02-0014.
———. 1780. "A Letter to the Royal Academy of Brussels." https://founders.archives.gov/documents/Franklin/01-32-02-0281.
Freedman, Paul. 2019. *American Cuisine, and How It Got This Way*. New York: Liveright.
French, Roger K. 1989. "Harvey in Holland: Circulation and the Calvinists." In French and Wear, eds., 1989, 46–86.
———. 1994. *William Harvey's Natural Philosophy*. Cambridge: Cambridge University Press.
———. 2000. "Where the Philosopher Finishes, the Physician Begins: Medicine and the Arts Course in Thirteenth-Century Oxford." *Dynamis* 20: 75–106.
French, Roger K., and Andrew Wear, eds. 1989. *The Medical Revolution of the Seventeenth Century*. Cambridge: Cambridge University Press.
Friedenwald, Julius, and John Ruhräh. [1904] 1913. *Diet in Health and Disease*. 4th ed. Philadelphia: W. B. Saunders.
Frissell, H. B., and Isabel Bevier. 1899. *Dietary Studies of Negroes in Eastern Virginia in 1897 and 1898*. Washington, DC: US Government Printing Office.
Frohlich, Xaq. 2023. *From Label to Table: Regulating Food in America in the Information Age*. Berkeley: University of California Press.

Fuller, Francis. 1705. *Medicina Gymnastica: or, A Treatise Concerning the Power of Exercise, with Respect to the Animal Oeconomy*. London: John Matthews.

Fuller, Thomas. 1891. *Collected Sermons, D.D. 1631–1659*, Vol. 1. Ed. John Eglington Bailey and William E. A. Axon. London: Gresham Press.

Gabbey, Alan. 1985. "The Mechanical Philosophy and Its Problems: Mechanical Explanations, Impenetrability, and Perpetual Motion." In *Change and Progress in Modern Science*, ed. Joseph C. Pitt, 9–84. Dordrecht: D. Reidel.

Gailhard, Jean. 1678. *The Compleat Gentleman: or Directions for the Education of Youth*. London.

[Gainsford, Thomas]. 1616. *The Rich Cabinet Furnished with Varieties of Excellent Discriptions, Exquisite Characters, Witty Discourses, and Delightfull Histories, Devine and Morrall . . .* London.

Galen. 1951. *A Translation of Galen's Hygiene (De Sanitate Tuenda) by Robert Montraville Green*. Springfield, IL: Charles C. Thomas.

———. 2000. "On the Humours." In *Galen on Food and Diet*, ed. Mark Grant, 14–18. London: Routledge.

Galilei, Galileo. [1623] 1957. "The Assayer [excerpts]." In *Discoveries and Opinions of Galileo*, trans. Stillman Drake, 229–80. Garden City, NY: Doubleday.

Garber, Marjorie. 2020. *Character: The History of a Cultural Obsession*. New York: Farrar, Straus and Giroux.

Gautier, A[rmand]. [1904] 1906. *Diet and Dietetics*. Ed. and trans. A. J. Rice-Oxley. Philadelphia: J. B. Lippincott.

Gee, Sophie. 2010. *Making Waste: Leftovers and the Eighteenth-Century Imagination*. Princeton, NJ: Princeton University Press.

Gentilcore, David. 2010. *Pomodoro! A History of the Tomato in Italy*. New York: Columbia University Press.

———. 2016. *Food and Health in Early Modern Europe: Diet, Medicine, and Society, 1450–1800*. New York: Bloomsbury.

Gerard, John. 1597. *The Herball or Generall Historie of Plantes*. London.

Giannetti, Laura. 2013. "Of Eels and Pears: A Sixteenth-Century Debate on Taste, Temperance, and the Pleasures of the Senses." In *Religion and the Senses in Early Modern Europe*, ed. Wietse de Boer and Christine Göttler, 289–305. Leiden: Brill.

Gildon, Charles. 1710. *The Life of Mr. Thomas Betterton, the Late Eminent Tragedian*. London: Robert Jeeb.

Gilman, Sander L. 2008. *Diets and Dieting: A Cultural Encyclopedia*. New York: Routledge.

Ginzburg, Carlo. 1980. *The Cheese and the Worms: The Cosmos of a Sixteenth-Century Miller*. Trans. John and Anne Tedeschi. Baltimore: Johns Hopkins University Press.

Giovannetti-Singh, Gianamar. 2021. "Galenizing the New World: Joseph-François Lafitau's 'Galenization' of Canadian Ginseng, *ca*. 1716–1724." *Notes and Records of the Royal Society* 75: 59–72.

Glas, E. 1976. "The Liebig-Mulder Controversy: On the Methodology pf Physiological Chemistry." *Janus* 63: 27–46.

Goeurot, Jean. [1544] 1550. *The Regiment of Life, Whereunto Is Added a Treatise of the Pestilence*. London.

Goldacre, Ben. 2008. *Bad Science: Quacks, Hacks, and Big Pharma Flacks*. New York: Faber and Faber.

Goldstein, Carolyn M. 2012. *Creating Consumers: Home Economists in Twentieth-Century America*. Chapel Hill: University of North Carolina Press.

Goodman, Grant Kohn. 2000. *Japan and the Dutch, 1600–1853*. London: Routledge.

Goody, Jack. 1982. *Cooking, Cuisine, and Class: A Study in Comparative Sociology*. New York: Cambridge University Press.

———. 1998. *Food and Love: A Cultural History of East and West*. London: Verso.

Goss, Arthur. 1897. *Dietary Studies in New Mexico in 1895*. Washington, DC: US Government Printing Office.

Goubaux, Armand. 1872. *Etudes sur le cheval considéré comme bête de boucherie*. Paris: Imprimerie et Librairie d'Agriculture et d'Horticulture, 8.

Goudiss, C. Houston, and Alberta M. Goudiss. 1918. *Foods That Will Win the War and How to Cook Them*. New York: World Syndicate.

Gouk, Penelope. 2013. "Music and Spirit in Early Modern Thought." In Carrera, ed., 2013, 221–39.

Gowland, Angus. 2006. *The Worlds of Renaissance Melancholy: Robert Burton in Context*. Cambridge: Cambridge University Press.

———. 2013. "Medicine, Psychology, and the Melancholic Subject in the Renaissance." In Carrera, ed., 2013, 185–220.

Graham, Sylvester. 1839. *Lectures on the Science of Human Life*. 2 vols. Boston: Marsh, Capen, Lyon, and Webb.

Grant, Mark. 1997. *Dieting for an Emperor: A Translation of Books 1 and 4 of Oribasius' Medical Compilations with an Introduction and Commentary*. Leiden: Brill.

Grataroło, Guglielmo. 1574. *A Direction for the Health of Magistrates and Studentes*. Trans. T[homas] N[ewton]. London.

Gregory, Frederick. 1977. *Scientific Materialism in Nineteenth Century Germany*. Dordrecht: D. Reidel.

Grew, Nehemiah. 1682. *The Anatomy of Plants*. London.

Grieco, Allen J. 2009. "Medieval and Renaissance Wines: Taste, Dietary Theory, and How to Choose the 'Right' Wine (14th–16th Centuries)." *Mediaevalia* 30: 15–42.

———. 2019. *Food, Social Politics and the Order of Nature in Renaissance Italy*. Villa I Tati Series, 34. Milan: Officina Libraria, for Villa I Tati.
———. 2019. "Food and Social Classes in Late Medieval and Renaissance Italy." In Grieco, 2019, 104–17.
———. 2019. "Vegetable Diets, Hermits and Melancholy in Late Medieval and Renaissance Italy." In Grieco 2019, 243–62.
Grierson, Herbert J. C., ed. 1921. *Metaphysical Lyrics & Poems of the Seventeenth Century: Donne to Butler*. Oxford: Clarendon.
Griffith, Matthew. 1634. *Bethel: or, a Forme for Families*. London.
Grindley, H. S., et al. 1900. *Nutrition Investigations at the University of Illinois, North Dakota Agricultural College, and Lake Erie College, Ohio, 1896–1900*. Washington, DC: US Government Printing Office.
Grmek, Mirko D. 1989. *Diseases in the Ancient Greek World*. Trans. Mireille Muellner. Baltimore: Johns Hopkins University Press.
Gronow, Jukka. 1997. *The Sociology of Taste*. London: Routledge.
Gruman, Gerald J. 1966. "A History of Ideas about the Prolongation of Life." *Transactions of the American Philosophical Society*, new series, 56, no. 9: 1–102.
Gruner, O. Cameron. 1930. *A Treatise on the Canon of Medicine by Avicenna Incorporating a Translation of the First Book*. London: Luzac.
Guerrini, Anita. 1985. "James Keill, George Cheyne, and Newtonian Physiology, 1690–1740." *Journal of the History of Biology* 18: 247–66.
———. 1986. "The Tory Newtonians: Gregory, Pitcairne, and Their Circle." *Journal of British Studies* 25: 288–311.
———. 1987. "Archibald Pitcairne and Newtonian Medicine." *Medical History* 31: 70–83.
———. 1989. "Isaac Newton, George Cheyne and the 'Principia Medicinae.'" In French and Wear, eds., 1989, 222–45.
———. 1996. "Newtonianism, Medicine, and Religion." In *Religio Medici: Medicine and Religion in Seventeenth-Century England*, ed. Ole Peter Grell and Andrew Cunningham, 293–313. Aldershot: Scolar Press.
———. 2000. *Obesity and Depression in the Enlightenment: The Life and Times of George Cheyne*. Norman: University of Oklahoma Press.
———. 2012. "Health, National Character and the English Diet in 1700." *Studies in History and Philosophy of Biological and Biomedical Sciences* 43: 349–56.
———. 2016. "The Ghastly Kitchen." *History of Science* 54: 71–97.
———. 2020. "A Natural History of the Kitchen." *Osiris* 35: 20–41.
Gunlock, Julie. 2013. *From Cupcakes to Chemicals: How the Culture of Alarmism Makes Us Afraid of Everything and How to Fight Back*. Winchester, VA: Independent Women's Foundation Press.
Gunn, John C. [1830] 1986. *Domestic Medicine, or Poor Man's Friend*. Knox-

ville, TN: Printed for the author, 1830; facsimile edition, ed. Charles E. Rosenberg. Knoxville: University of Tennessee Press, 1986.

———. [1830] 1835. *Gunn's Domestic Medicine, or Poor Man's Friend*. 4th ed. Springfield, OH: John M. Gallagher.

Guthrie, Joanne F., Brenda M. Derby, and Alan S. Levy. 1999. "What People Know and Don't Know about Nutrition." US Department of Agriculture, Economic Research Service. https://www.ers.usda.gov/webdocs/publications/42215/5842_aib750m_1_.pdf?v=41055.

Haak, Hans L. 2012. "Blood, Clotting and the Four Humours." In Horstmanshoff, King, and Zittel, eds., 2012, 295–305.

Hall, Marie Boas. 1956. "Acid and Alkali in Seventeenth Century Chemistry." *Archives internationales d'histoire des sciences* 9: 13–28.

Hall, Thomas S. 1969. *History of General Physiology*. 2 vols. Chicago: University of Chicago Press.

Handley, Sasha. 2016. *Sleep in Early Modern England*. New Haven, CT: Yale University Press.

———. 2017. "Sleep-Piety and Healthy Sleep in Early Modern English Households." In Cavallo and Storey, eds., 2017, 186–209.

Hargrove, James R. 2006. "History of the Calorie in Nutrition." *Journal of Nutrition* 136: 2957–2961.

Harington, Sir John. [1607] 1624. *The English Mans Doctor, or, The Schoole of Salerne*. London.

Harley, David N. 1993. "Medical Metaphors in English Moral Theology, 1560–1660." *Journal of the History of Medicine and Allied Sciences* 48: 396–435.

Harris, John. 1676. *The Divine Physician: Prescribing Rules for the Prevention, and Cure of Most Diseases, as Well of the Body, as the Soul*. London.

Harrison, Mark. 1993. *Colonizing the Body: State Medicine and Epidemic Disease in Nineteenth-Century India*. Berkeley: University of California Press.

———. 1999. *Climate and Constitutions: Health, Race, Environment and British Imperialism in India, 1600–1850*. New York: Oxford University Press.

Hartley, David. [1749] 1834. *Observations on Man, His Frame, His Duty, and His Expectations*. 6th ed. London: Thomas Tegg.

Hartley, Harold. 1951. "Origin of the Word 'Protein.'" *Nature* 168: 244.

Hartnell, Jack, 2018. *Medieval Bodies: Life, Death and Art in the Middle Ages*. London: Profile.

Haushofer, Lisa. 2018. "Between Food and Medicine: Artificial Digestion, Sickness, and the Case of Benger's Food." *Journal of the History of Medicine and Allied Sciences* 73: 168–87.

———. 2023. *Wonder Foods: The Science and Commerce of Nutrition*. Oakland: University of California Press.

Haycock, David Boyd. 2008. *Mortal Coil: A Short History of Living Longer*. New Haven, CT: Yale University Press.

Hedesan, Georgiana D. 2016. "Theory Choice in the Seventeenth Century: Robert Boyle against the Paracelsian *Tria Prima*." In *Theory Choice in the History of Chemical Practices*, ed. Emma Tobin and Chiara Ambrosio, 17–28. Zurich: Springer.

Heyd, Michael. 1995. *'Be Sober and Reasonable': The Critique of Enthusiasm in the Seventeenth and Early Eighteenth Centuries*. Leiden: E. J. Brill.

Higginson, Francis. 1630. *New-Englands Plantation: Or, a Short and True Description of the Commodities and Discommodities of That Countrey*. London.

Hillman, David, and Carla Mazzio, eds. 1997. *The Body in Parts: Fantasies of Corporeality in Early Modern Europe*. London: Routledge.

Hippocrates. 1923. *Works*, Vol. 1. Trans. W. H. S. Jones. Cambridge, MA: Harvard University Press.

———. 1923. "Ancient Medicine." In Hippocrates 1923, 1:12–64.

———. 1923. "Airs Waters Places." In Hippocrates 1923a, 1:65–138.

———. 1923. *Works*, Vol. 2. Trans. W. H. S. Jones. Cambridge, MA: Harvard University Press.

———. 1923. "The Art." In Hippocrates 1923, 2:185–217.

———. 1923. "Breaths." In Hippocrates 1923, 2:219–53.

———. 1931. *Works*, Vol. 4. Trans. W. H. S. Jones. Cambridge, MA: Harvard University Press.

———. 1931. "The Nature of Man." In Hippocrates 1931, 4:3–41.

———. 1931. "Regimen in Health." In Hippocrates 1931, 4:45–59.

———. 1931. "Humours." In Hippocrates 1931, 4:61–96.

———. 1931. "Aphorisms." In Hippocrates 1931, 4:97–221.

———. 1931. "Regimen, II." In Hippocrates 1931, 4:299–365.

———. 1931. "Regimen, III." In Hippocrates 1931, 4:366–419.

———. 2018. "Diseases of Women. I." In *Works*, Vol. 11. Trans. Paul Potter, 1–255. Cambridge, MA: Harvard University Press.

Hobsbawm, Eric. 1994. *The Age of Extremes: A History of the World, 1914–1991*. New York: Pantheon.

Hoffman[n], Friedrich. [1730] 1761. *A Treatise on the Nature of Aliments, or Foods, in General*. London: L. Davis and C. Reymers.

Holinshed, Ralph. 1587. *The First and Second Volumes of Chronicles Comprising 1; The Description and Historie of England, 2; The Description and Historie of Ireland, 3; The Description and Historie of Scotland*. London.

Hollerbach, Teresa. 2018. "The Weighing Chair of Sanctorius Sanctorius: A Replica." *NTM* 26: 121–49.

Holmes, Frederic L. 1988. "The Formation of the Munich School of Metabolism." In *The Investigative Enterprise: Experimental Physiology in Nineteenth-Century Medicine*, ed. William Coleman and Holmes, 179–210. Berkeley: University of California Press.

Hooke, Robert. 1665. *Micrographia, or, Some Physiological Descriptions of Minute Bodies Made by Magnifying Glasses*. London.

Horstmanshoff, Manfred, Helen King, and Claus Zittel, eds. 2012. *Blood, Sweat and Tears: The Changing Concepts of Physiology from Antiquity into Early Modern Europe*. Leiden: Brill.

Howell, James. [1645–1655] 1907. *Epistolæ Ho-Elianæ: The Familiar Letters of James Howell*. 4 vols. Boston: Houghton Mifflin.

Howes, David. [2005] 2016. "Preface: Accounting for Taste." In Luca Vercelloni, *The Invention of Taste: A Cultural Account of Desire, Delight and Disgust in Fashion, Food and Art*, trans. Kate Singleton, vii–xiv. London: Bloomsbury.

Huarte, John [Juan]. [1588] 1594. *Examen de ingenios: The Examination of Mens Wits*. Trans. Camillo Camilli. London.

Hufeland, Christoph Wilhelm. [1798] 1829. *The Art of Prolonging Human Life*. Trans. anon. London: Simpkin and Marshall.

Hunt, Caroline L. 1912. *The Life of Ellen H. Richards*. Boston: Whitcomb & Barrows.

Hunt, Caroline L., and Helen W. Atwater. 1917. *How to Select Foods I: What the Body Needs*. Farmers' Bulletin 808. Washington, DC: US Government Printing Office.

———. 1917. *How to Select Foods II: Cereal Foods*. Farmers' Bulletin 817. Washington, DC: US Government Printing Office.

———. 1917. *How to Select Foods III: Foods Rich in Protein*. Farmers' Bulletin 824. Washington, DC: US Government Printing Office.

Hunter, A[lexander]. [1804] 1806. *Culina Famulatrix Medicinæ: or Receipts in Modern Cookery; with a Medical Commentary*. 3rd ed. York: T. Wilson and R. Spence.

Hunter, Michael. 2000. *Robert Boyle (1627–1691): Scrupulosity and Science*. Woodbridge: Boydell.

Hutchinson, Woods. 1909. "The Dangers of Undereating." *Cosmopolitan Magazine* 47, no. 1 (June): 385–93.

Irving, Washington. 1890. "John Bull." In *The Works of Washington Irving*, Vol. 2, *The Sketch Book*, 431–47. New York: Putnam.

Ishizuka, Hisao. 2016. *Fiber, Medicine, and Culture in the British Enlightenment*. London: Palgrave Macmillan.

Iyengar, Sujata. 2011. *Shakespeare's Medical Language: A Dictionary*. London: Continuum.

Jahns, Lisa, et al. 2018. "The History and Future of Dietary Guidelines in America." *Advances in Nutrition* 9: 136–47.

James I, King of England. [1599] 1603. *Basilikon Doron: Or His Maiesties Instructions to His Dearest Sonne, Henrie the Prince*. 2nd ed. London.

James, Susan. 1997. *Passion and Action: The Emotions in Seventeenth-Century Philosophy*. Oxford: Clarendon.

Jarcho, Saul. 1970. "Galen's Six Non-Naturals: A Bibliographic Note and Translation." *Bulletin of the History of Medicine* 44: 372–76.

Jeanneret, Michel. [1987] 1991. *A Feast of Words: Banquets and Table Talk in the Renaissance*. Trans. Jermey Whiteley and Emma Hughes. Chicago: University of Chicago Press.

Jenner, Mark S. R. 2011. "Follow Your Nose? Smell, Smelling, and Their Histories." *American Historical Review* 116: 335–52.

Jenner, Mark S. R., and Patrick Wallis, eds. 2007. *Medicine and the Market in England and Its Colonies, c. 1450–c. 1850*. New York: Palgrave Macmillan.

Jewson, N. D. 1974. "Medical Knowledge and the Patronage System in Eighteenth-Century England." *Sociology* 8: 369–85.

Johansen, Lisa Tillinger. 2012. *Fast Food Vindication: The Story You Haven't Been Told*. Los Angeles: Tillinger Johansen.

Johns Hopkins Bloomberg School of Public Health. 2014. "'Drink Responsibly' Messages in Alcohol Ads Promote Products, Not Public Health." September 3, 2014. https://publichealth.jhu.edu/2014/drink-responsibly-messages-in-alcohol-ads-promote-products-not-public-health.

Johnston, James F. W. [1854] 1859. *The Chemistry of Common Life*. New ed., 2 vols. Edinburgh: William Blackwood.

Johnston, Josée, and Shyon Baumann. 2010. *Foodies: Democracy and Distinction in the Gourmet Foodscape*. New York: Routledge.

Jones, W. H. S. 1931. "Introduction." In Hippocrates 1931, ix–lviii.

Jonson, Ben. 1600. *The Comicall Satyre of Every Man Out of His Humor*. London.

Jonsson, Fredrik Albritton. 2005. "The Physiology of Hypochondria in Eighteenth-Century Britain." In *Cultures of the Abdomen: Diet, Digestion, and Fat in the Modern World*, ed. Christopher Forth and Ana Carden-Coyne, 15–30. Basingstoke: Palgrave.

Jordan, Whitman H. 1913. *Principles of Human Nutrition: A Study in Practical Dietetics*. New York: Macmillan.

Jou, Chin. 2019. "The Progressive Era Body Project: Calorie-Counting and 'Disciplining the Stomach' in 1920s America." *Journal of the Gilded Age and Progressive Era* 18: 422–40.

Jouanna, Jacques. 2012. *Greek Medicine from Hippocrates to Galen: Selected Papers*. Trans. Neil Allies, ed. Philip van der Eijk. Leiden: Brill.

———. 2012. "Politics and Medicine: The Problem of Change in *Regimen in Acute Diseases* and Thucydides (Book 6)." In Jouanna 2012, 21–38.

———. 2012. "Air, Miasma and Contagion in the Time of Hippocrates and the Survival of Miasmas in Post-Hippocratic Medicine." In Jouanna 2012, 121–36.

———. 2012. "Dietetics in Hippocratic Medicine: Definition, Main Problems, Discussion." In Jouanna 2012, 137–53.

———. 2012. "Wine and Medicine in Ancient Greece." In Jouanna 2012, 173–93.

———. 2012. "The Legacy of the Hippocratic Treatise *The Nature of Man*: The Theory of the Four Humours." In Jouanna 2012, 335–59.

Joubert, Laurent. [1579] 1989. *Popular Errors*. Trans. Gregory David de Rocher. Tuscaloosa: University of Alabama Press.

———. [1587] 1995. *The Second Part of the Popular Errors*. Trans. Gregory David de Rocher. Tuscaloosa: University of Alabama Press.

Kalm, Peter [Pehr]. [1752] 1892. *Kalm's Account of His Visit to England on His Way to America in 1748*. Trans. Joseph Lucas. London: Macmillan.

Kamminga, Harmke. 1995. "Nutrition for the People, or the Fate of Jacob Moleschott's Contest for a Humanist Science." In Kamminga and Cunningham, eds., 1995, 15–47.

Kamminga, Harmke, and Andrew Cunningham, eds. 1995. *The Science and Culture of Nutrition, 1840–1940*. Amsterdam: Rodopi.

———. 1995. "Introduction: The Science and Culture of Nutrition, 1840–1940." In Kamminga and Cuinningam, eds., 1995, 1–14.

Kananoja, Kalle. 2019. "Doctors, Healers and Charlatans in Brazil: A Short History of Ideas, c. 1650–1950." In *Healers and Empires in Global History*, ed. Markku Hokkanen and Kananoja, 179–201. Cham, Switzerland: Springer International.

Kedzie, R. C. 1880. "The Relations of Chemistry to Agriculture." In *Nineteenth Annual Report of the Secretary of the State Board of Agriculture of the State of Michigan, for the Year Ending 1880*, 265–72. Lansing, MI: W. S. George.

Keill, James. [1614] 1737. *Medicina Statica Britannica*. Appended to Santorio Santorio, *Medicina Statica: Being the Aphorisms of Sanctorious, Translated into English by Dr. John Quincy*, 5th ed., 321–44. London: T. Longman and J. Newton.

Kelley, Donald R. 1990. "'Second Nature': The Idea of Custom in European Law, Society, and Culture." in *The Transmission of Culture in Early Modern Europe*, ed. Anthony Grafton and Ann Blair, 131–72. Philadelphia: University of Pennsylvania Press.

Kellogg, J. H. 1909. "Questions of Diet in the Treatment of Inebriety." In *Some Scientific Conclusions Concerning the Alcoholic Problem and Its Practical Relations to Life*. US Senate Document No. 48, 87–101. Washington, DC: US Government Printing Office.

———. 1921. *The New Dietetics: What to Eat and How; A Guide to Scientific Feeding in Health and Disease*. Battle Creek, MI: Modern Medicine Publishing.

Kennaway, James, and Jonathan Andrews. 2019. "'The Grand Organ of Sympathy': 'Fashionable' Stomach Complaints and the Mind in Britain, 1700–1850." *Social History of Medicine* 32: 57–79.

Kennedy, Eileen, Daniel Hatfield, and Jeanne Goldberg. 2014. "Dietary Guidelines, Food Guidance, and Dietary Quality in the United States." In *Handbook of Nutrition and Food*, 3rd ed., ed. Carolyn D. Berdanier, Johanna T. Dwyer, and David Heber, 437–46. Boca Raton, FL: CRC Press.

Khong, Rachel. 2014. "The End of the World as We Know It." Medium. https://medium.com/lucky-peach/the-end-of-the-world-as-we-know-it-c9a0c05b243d.

Kirkland, Edward C. 1974. "'Scientific Eating': New Englanders Prepare and Promote a Reform, 1873–1907." *Proceedings of the Massachusetts Historical Society*, 3rd series, 86: 28–52.

Kisch, Bruno. 1965. *Scales and Weights: A Historical Outline*. New Haven, CT: Yale University Press.

Klein, Danny. 2018. "Menu-Labeling Laws Go into Effect." *QSR*, May. https://www.qsrmagazine.com/menu-innovations/menu-labeling-laws-go-effect.

Klein, Ursula, and Wolfgang Lefèvre. 2007. *Materials in Eighteenth-Century Science: A Historical Ontology*. Cambridge, MA: MIT Press.

Klibansky, Raymond, Erwin Panofsky, and Fritz Saxl. 1964. *Saturn and Melancholy: Studies in the History of Natural Philosophy, Religion, and Art*. London: Thomas Nelson.

Knight, Christine. 2005. "'The Food Nature Intended You to Eat': Low-Carbohydrate Diets and Primitivist Philosophy." In *The Atkins Diet and Philosophy: Chewing the Fat with Kant and Nietzsche*, ed. Lisa Heldke, Kerri Mommer, and Cynthia Pineo, 43–56. Chicago: Open Court.

Kolata, Gina. 2003. "Vitamins: More May Be Too Many." *New York Times*, April 29, https://www.nytimes.com/2003/04/29/science/vitamins-more-may-be-too-many.html?searchResultPosition=1.

Korda, Joanna B., Sue W. Goldstein, and Frank Sommer. 2010. "The History of Female Ejaculation." *Journal of Sexual Medicine* 7: 1965–1975.

Korsmeyer, Carolyn. 1999. *Making Sense of Taste: Food and Philosophy*. Ithaca, NY: Cornell University Press.

Krinsky, Alan D. 2001. "Let Them Eat Horsemeat! Science, Philanthropy, State, and the Search for Complete Nutrition in Nineteenth-Century France." PhD dissertation, University of Wisconsin.

Kristeller, P. O. 1945. "The School of Salerno." *Bulletin of the History of Medicine* 17: 138–94.

Kroll, Andy, and Jeremy Schulman. 2013. "Leaked Documents Reveal the Secret Finances of a Pro-Industry Science Group." *Mother Jones*, October 28. https://www.motherjones.com/politics/2013/10/american-council-science-health-leaked-documents-fundraising/.

Kula, Witold. 1986. *Measures and Men*. Trans. R. Szreter. Princeton, NJ: Princeton University Press.

Kummer, Corby. 1999. "Doing Well by Eating Well." *The Atlantic*, March. https://www.theatlantic.com/magazine/archive/1999/03/doing-well-by-eating-well/377485/.

Kuriyama, Shigehisa. 1995. "Interpreting the History of Bloodletting." *Journal of the History of Medicine* 50: 11–46.

———. 2002. *The Expressiveness of the Body and the Divergence of Greek and Chinese Medicine*. New York: Zone.

Kurzweil, Ray, and Terry Grossman. *Fantastic Voyage: Live Long Enough to Live Forever*. New York: Rodale Books.

Laas, Molly S. 2017. "Nutrition as a Social Question: 1835–1905." PhD dissertation, University of Wisconsin, Madison.

Landecker, Hannah. 2015. "Being and Eating: Losing Grip on the Equation." *BioSocieties* 10: 253–58.

Lanchester, John. 2014. "Shut Up and Eat: A Foodie Repents." *New Yorker*, November 3, https://www.newyorker.com/magazine/2014/11/03/shut-eat.

Lankester, Edwin. 1861. *On Food: Being Lectures Delivered at the South Kensington Museum*. London: Robert Hardwicke.

La Mettrie, Julien Offray de. [1747] 1749. *Man a Machine*. Dublin: W. Brien.

Laqueur, Thomas W. 1990. *Making Sex: Body and Gender from the Greeks to Freud*. Cambridge, MA : Harvard University Press.

La Rochefoucauld, François, duc de. 1959 [1665]. *The Maxims of La Rochefoucauld*. Trans. Louis Kronenberger. New York: Random House.

Laudan, Rachel. [2001] 2019. "A Plea for Culinary Modernism: Why We Should Love New, Fast, Processed Food (with a New Postscript)." In *Food Fights: How History Matters to Contemporary Food Debates*, ed. Charles C. Ludington and Matthew Morse Booker, 262–84. Chapel Hill: University of North Carolina Press.

Launay, Robert. 2018. "Maize Avoidance? Colonial French Attitudes towards Native American Foods in the Pays des Illinois (17th and 18th Century)." *Food and Foodways* 26: 92–104.

Lavater, Johann Casper. 1775–1778. *Physiognomische Fragmente, zur Beförderung der Menschenkenntnisse und Menschenliebe*. 4 vols. Leipzig: Weidmanns Erben.

———. [1789] 1853. *Essays on Physiognomy*. 8th ed. Trans. Thomas Holcroft. London: William Tegg.

Lawrence, Christopher J. 1975. "William Buchan: Medicine Laid Open." *Medical History* 19: 20–36.

———. 1979. "The Nervous System and Society in the Scottish Enlightenment." In *Natural Order: Historical Studies of Scientific Culture*, ed. Barry Barnes and Steven Shapin, 19–40. Beverly Hills, CA: Sage.

———. 1985. "Ornate Physicians and Learned Artisans: Edinburgh Medical Men, 1726–1776." In *William Hunter and the Eighteenth-Century Medical World*, ed. W. F. Bynum and Roy Porter, 153–76. Cambridge: Cambridge University Press.

Lawrence, Christopher J., and Steven Shapin, eds. 1998. *Science Incarnate: Historical Embodiments of Natural Knowledge*. Chicago: University of Chicago Press.

Lees-Milne, James. 1953. *The Age of Inigo Jones*. London: Batsford.

Le Brun. [1698] 1734. *A Method to Learn to Design the Passions, Proposed in a Conference on Their General and Particular Expression*. Trans. John Williams. London: J. Huggonson.

Le Grand, Antoine. 1694. *An Entire Body of Philosophy*. London.

Le Guérer, Annick. 1992. *Scent: The Mysterious and Essential Powers of Smell*. Trans. Richard Miller. New York: Turtle Bay.

Leigh, G. J. 2004. *The World's Greatest Fix: A History of Nitrogen and Agriculture*. Oxford: Oxford University Press.

Lémery, Louis. [1702] 1745. *A Treatise of All Sorts of Food*. Trans. D. Hay. London: T. Osborne.

Lemnius, Levinus. [1561] 1633. *The Touchstone of Complexions*. Trans. from Latin by T. N. London.

———. [1559] 1658. *The Secret Miracles of Nature*. London.

Leonard, Alice, and Sarah E. Parker. 2023. "'Put a Mark on the Errors': Seventeenth-Century Medicine and Science." *History of Science* 61, no. 3: 287–307.

Leong, Elaine. 2018. *Recipes and Everyday Knowledge: Medicine, Science, and the Household in Early Modern England*. Chicago: University of Chicago Press.

Lessius, Leonardus. [1614] 1634. *Hygiasticon: or, The Right Course of Preserving Life and Health into Extream Old Age*. 2nd ed. Trans. T[imothy] S[mith][?]. Cambridge.

Lettsom, John Coakley. 1817. *Selections from the Medical Papers and Correspondence of the Late John Coakley Lettsom*. Ed. Thomas Joseph Pettigrew. London: Nichols, Son, and Bentley.

Levenstein, Harvey. 1980. "The New England Kitchen and the Origins of Modern American Eating Habits." *American Quarterly* 32: 369–86.

———. 2003. *Revolution at the Table: The Transformation of the American Diet*. Berkeley: University of California Press.

Levine, Deborah I. 2008. "Managing American Bodies: Diet, Nutrition, and Obesity in America 1840–1920." PhD dissertation, Harvard University.

———. 2017. "The Curious History of the Calorie in U.S. Policy: A Tradition of Unfulfilled Promises." *American Journal of Preventive Medicine* 52, no. 1: 125–29.

———. 2018. "Measure, Record, Share: Weight Loss, Biometrics, and Self-Tracking in the U.S." *American Journal of Preventive Medicine* 55, no. 5: e147–e151.

Lewes, George Henry. [1859–1860]. *The Physiology of Common Life*. 2 vols. Edinburgh: William Blackwood.

Liebig, Justus von. [1842] 1964. *Animal Chemistry or Organic Chemistry in Its Application to Physiology and Pathology*. Facsimile of the Cambridge edition of 1842, ed. William Gregory. New York: Johnson Reprint Corp.

———. 1843. *Familiar Letters on Chemistry, and Its Relation to Commerce, Physiology, and Agriculture*. Trans. John Gardner. New York: D. Appleton.

———. 1845. "Ueber der thierische Wärme." *Annalen der Chemie und Pharmacie* 53: 63–77.

———. 1848. *Researches on the Chemistry of Food, and the Motions of the Juices in the Animal Body*. Ed. William Gregory and Eben N. Horsford. Lowell, MA: Daniel Boxby.

———. 1859. *Familiar Letters on Chemistry*. 4th ed. Ed. John Blyth. London: Walton and Maberly.

———. 1869. "On the Nutritive Value of Different Sorts of Food." *The Lancet* 93 (February 2, 9; January 23; February 6; March 13): 4–5, 36–38, 113–15, 186–87, 357–58.

Lindeboom, G. A. 1979. *Descartes and Medicine*. Amsterdam: Rodopi.

Lindemann, Mary. 1996. *Health & Healing in Eighteenth-Century Germany*. Baltimore: Johns Hopkins University Press.

Lindlahr, Victor Hugo. 1940. *You Are What You Eat: How to Win and Keep Health through Diet*. New York: National Nutrition Society.

Lloyd, G. E. R. 1964. "The Hot and the Cold, the Dry and the Wet in Greek Philosophy." *Journal of Hellenic Studies* 84: 92–106.

———. 1996. *Adversaries and Authorities: Investigations into Ancient Greek and Chinese Science*. Cambridge: Cambridge University Press.

Lloyd, G. E. R., and Nathan Sivin. 2002. *The Way and the Word: Science and Medicine in Early China and Greece*. New Haven, CT: Yale University Press.

Locke, John. 1690. *An Essay Concerning Humane Understanding*. London.

———. 1693. *Some Thoughts Concerning Education*. London.

Lonati, Elisabetta. 2019. "The Dissemination of Medical Practice in Late Modern Europe: The Case of Buchan's *Domestic Medicine*." *Status*

Quaestionis: A Journal of European and American Studies 17. https://doi.org/10.13133/2239-1983/16393.

Luddington, Charles. 2009. "'Claret Is the Liquor for Boys; Port for Men': How Port Became the 'Englishman's Wine,' 1750s–1800." *Journal of British Studies* 48: 364–90.

Lund, Mary Ann. 2010. *Melancholy, Medicine and Religion in Early Modern England*. Cambridge: Cambridge University Press.

Lusk, Graham. [1906] 1917. *The Elements of the Science of Nutrition*. Philadelphia: W. B. Saunders.

McClary, Ben H. 1986. "Introducing a Classic: 'Gunn's Domestic Medicine.'" *Tennessee Historical Quarterly* 45: 210–16.

McClive, Cathy. 2016. *Menstruation and Procreation in Early Modern France*. New York: Routledge.

McCollum, Elmer Verner. 1957. *A History of Nutrition: The Sequence of Ideas in Nutrition Investigation*. Boston: Houghton Mifflin.

McCollum, Elmer Verner, and Nina Simmonds. 1920. *The American Home Diet: An Answer to the Ever Present Question What Shall We Have for Dinner*. Detroit: Frederick C. Mathews.

Macdonald, Michael. 1981. *Mystical Bedlam: Madness, Anxiety, and Healing in Seventeenth-Century England*. Cambridge: Cambridge University Press.

McGee, Harold. 1988. "Osmazome, the Maillard Reaction & the Triumph of the Cooked." In *Taste: Proceedings of the Oxford Symposium on Food & Cookery 1987*, ed. Tom Jaine, 133–35. London: Prospect Books.

———. 2003. *On Food and Cooking: The Science and Lore of the Kitchen*. New York: Simon & Schuster.

Mackenzie, James. [1758] 1760. *The History of Health, and the Art of Preserving It*. 3rd ed. Edinburgh: William Gordon.

McMahon, Darrin M. 2013. *Divine Fury: A History of Genius*. New York: Basic Books.

Madden, Deborah. 2007. *'A Cheap, Safe and Natural Medicine': Religion, Medicine and Culture in John Wesley's Primitive Physic*. Clio Medica 83. Amsterdam: Rodopi.

Magno, Alessandro. 1983. "The London Journals of Alessandro Magno 1562." Trans. and ed. Caroline Barron, Christopher Coleman, and Claire Gobbi. *London Journal* 9, no. 2: 136–52.

Malebranche, Nicolas. [1674–1678] 1997. *The Search after Truth: With Elucidations of the Search after Truth*, trans. and ed. Paul J. Olscamp and Thomas M. Lennon. Cambridge: Cambridge University Press.

Mandelkern, India. 2015. "The Politics of the Palate: Taste and Knowledge in Early Modern England." PhD dissertation, University of California, Berkeley.

———. 2015. "Taste-Based Medicine." *Gastronomica* 15, no. 1: 8–21.

Mandeville, Bernard. [1711] 1730. *A Treatise of the Hypochondriack and Hysterick Diseases*. 3rd ed. London: J. Tonson.

Margócsy, Daniel. 2021. "The Pineapple and the Worms." *KNOW: A Journal on the Formation of Knowledge* 5: 53–81.

Mayhew, Robert J. 2013. *Malthus: The Life and Legacies of an Unlikely Prophet*. Cambridge, MA: Harvard University Press.

Maynwaringe, Everard. 1663. *Tutela Sanitatis, Sive Vita Protracta: The Protection of Long Life, and Detection of Its Brevity*. London.

———. 1683. *The Method and Means of Enjoying Health, Vigour, and Long Life*. London.

Mazzio, Carla. 2009. "The History of Air: Hamlet and the Trouble with Instruments." *South Central Review* 26, nos. 1–2: 153–96.

Mead, Richard. 1762. *The Medical Works of Richard Mead, M.D.* London: C. Hitch and L. Lawes.

Meade, R. H. 1974. *In the Sunshine of Life: A Biography of Dr. Richard Mead, 1673–1754*. Philadelphia: Dorrance.

Mendelsohn, Everett. 1964. *Heat and Life: The Development of the Theory of Animal Heat*. Cambridge, MA: Harvard University Press.

Mennell, Stephen. 1985. *All Manners of Food: Eating and Taste in England and France from the Middle Ages to the Present*. Oxford: Blackwell.

Merei, A. Schoepf. 1855. *On the Disorders of Infant Development, and Rickets, Preceded by Observations on the Nature, Peculiar Influence, and Modifying Agency of Temperaments*. London: John Churchill.

Micale, Mark S. 2008. *Hysterical Men: The Hidden History of Male Nervous Illness*. Cambridge, MA: Harvard University Press.

Mikkeli, Heikki. 1999. *Hygiene in the Early Modern Medical Tradition*. Helsinki: Academia Scientiarum Fennica.

Milani, Marisa. 2014. "How to Attain Immortality Living One Hundred Years, or, The Fortune of the *Vita Sobria* in the Anglo-Saxon World." In Cornaro 2014, 183–213.

Miller, Ian. 2011. *A Modern History of the Stomach: Gastric Illness, Medicine, and British Society, 1800–1950*. London: Pickering & Chatto.

———. 2012. "The Chemistry of Famine: Nutritional Controversies and the Irish Famine, c. 1845–7." *Medical History* 56: 444–62.

Milles, Dietrich. 1995. "Working Capacity and Calorie Consumption: The History of Rational Physical Economy." In Kamminga and Cunningham, eds., 1995, 75–96.

Mintz, Sidney W. 1986. *Sweetness and Power: The Place of Sugar in Modern History*. New York: Penguin.

Misson, Henri. [1698] 1719. *Memoirs and Observations in His Travels over England*. London: Dr. Browne, A. Bell.

Moleschott, Jacob. 1852. *Der Kreislauf des Lebens: Physiologische Antworten auf Liebigs chemische Briefe*. Mainz: von Zabern.

Montaigne, Michel Eyquem de. [1580–1588] 1910. *The Essayes of Michael Lord of Montaigne*. 3 vols. Trans. John Florio. London: J. M. Dent.

———. 1910. "Of the Force of Imagination." In Montaigne 1910, 1:92–104.

———. 1910. "Of Custom." In Montaigne 1910, 1:105–23.

———. 1910. "Of Friendship." In Montaigne 1910, 1:195–209.

———. 1910. "Of Drunkennesse." In Montaigne 1910, 2:15–26.

———. 1910. "How One Ought to Governe His Will." In Montaigne 1910, 3:253–77.

———. 1910. "Of Experience." In Montaigne 1910, 3:322–86.

Montanari, Massimo. 1994. *The Culture of Food*. Trans. Carl Ipsen. Oxford: Blackwell.

———. [2012] 2015. *Medieval Tastes: Food, Cooking, and the Table*. Trans. Beth Archer Brombert. New York: Columbia University Press.

Montesquieu, Charles-Louis de Secondat, Baron de. 1777. *The Complete Works*, Vol. 1, *The Spirit of Laws*. London: T. Evans and W. Davis.

Moran, Rachel Louise. 2018. *Governing Bodies: American Politics and the Shaping of the Modern Physique*. Philadelphia: University of Pennsylvania Press.

Moravia, Sergio. 1978. "From *Homme Machine* to *Homme Sensible*: Changing Eighteenth-Century Models of Man's Image." *Journal of the History of Ideas* 39: 45–60.

More, Sir Thomas. [1518] 1808. "In Efflatum Ventris." in Idem, *Memoirs of Sir Thomas More*, Vol. 2, ed. Arthur Cayley, :284. London: Cadell and Davis.

Morgan, Nicholas. 1871. *Phrenology, and How to Use It in Analyzing Character*. London: Longmans, Green.

Morrell, J. B. 1972. "The Chemist Breeders: The Research Schools of Liebig and Thomas Thomson." *Ambix* 19: 1–46.

Moss, Michael. 2013. *Salt Sugar Fat: How the Food Giants Hooked Us*. New York: Random House.

Moyes, John. 1896. *Medicine & Kindred Arts in the Plays of Shakespeare*. Glasgow: James MacLehose.

Mudry, Jessica J. 2009. *Measured Meals: Nutrition in America*. Albany: SUNY Press.

Muffett, Thomas. [1600] 1655. *Healths Improvement: or, Rules Comprizing and Discovering the Nature, Method, and Manner of Preparing All Sorts of Foods*. London.

Mulder, G. J. 1846. *Liebig's Question to Mulder Tested by Morality and Science*. Trans. P. F. H. Fromberg. Edinburgh: William Blackwood.

———. 1849. *The Chemistry of Vegetable & Animal Physiology*. Trans. P. F. H. Fromberg. Edinburgh: William Blackwood.

Murlin, John R. 1918. "Some Problems of Nutrition in the Army." *Science*, new series, 47, no. 1221 (May 24): 495–508.

Myers, B. E. 2011. "The Moral Crusade against Foodies." *The Atlantic*, March. https://www.theatlantic.com/magazine/archive/2011/03/the-moral-crusade-against-foodies/308370/.

Nestle, Marion. 2002. *Food Politics: How the Food Industry Influences Nutrition and Health*. Berkeley: University of California Press.

———. 2018. *Unsavory Truth: How Food Companies Skew the Science of What We Eat*. New York: Basic Books.

Nestle, Marion, and Malden Nesheim. 2006. *Why Calories Count: From Science to Politics*. New York: North Point Press.

Neswald, Elizabeth. 2017. "Nutritional Knowledge between the Lab and the Field: The Search for Dietary Norms in the Late Nineteenth and Early Twentieth Centuries." In Neswald, Smith, and Thoms, eds., 2017, 29–51.

Neswald, Elizabeth, David F. Smith, and Ulrike Thoms, eds. 2017. *Nutritional Standards: Theory, Policies, Practices*. Rochester, NY: University of Rochester Press.

Neuberger, Max. 1932. *The Doctrine of the Healing Power of Nature throughout the Course of Time*. Trans. Linn J. Boyd. New York: Privately printed.

Newman, William R. 2006. *Atoms and Alchemy: Chymistry and the Experimental Origins of the Scientific Revolution*. Chicago: University of Chicago Press.

Newman, William R., and Lawrence M. Principe. 2002. *Alchemy Tried in the Fire: Starkey, Boyle, and the Fate of Helmontian Chymistry*. Chicago: University of Chicago Press.

Newton, Hannah. 2015. "'Nature Concocts & Expels': The Agents and Processes of Recovery from Disease in Early Modern England." *Social History of Medicine* 28: 465–86.

Newton, Isaac. 1730. *Opticks: or, A Treatise of the Reflections, Refractions, Inflections and Colours of Light*. 4th ed. London: William Innys.

———. 1978. "Some Thoughts about the Nature of Acids" [a translation by John Harris of the paper *De natura acidorum* (1692)]. In *Isaac Newton's Papers & Letters on Natural Philosophy*, 2nd ed., ed. I. Bernard Cohen and Robert E. Schofield, 255–58. Cambridge, MA: Harvard University Press.

Nichols, Robert L. 1978. "Orthodoxy and Russia's Enlightenment, 1762–1825." *In Russian Orthodoxy under the Old Regime*, ed. Robert L. Nichols and Theofanis George Stavrou, 67–89. Minneapolis: University of Minnesota Press.

Niebyl, Peter H. 1971. "The Non-Naturals." *Bulletin of the History of Medicine* 45: 486–92.

———. 1977. "The English Bloodletting Revolution, or Modern Medicine before 1850." *Bulletin of the History of Medicine* 51: 464–83.

Nietzsche, Friedrich. 1969. *Selected Letters of Friedrich Nietzsche*. Ed. and trans. Christopher Middleton. Indianapolis: Hackett.

———. [1908] 2004. "Why I Am So Clever." In *Ecce Homo*, trans. Anthony M. Ludovici, 28–54. Mineola, NY: Dover.

———. [1889] 2007. *Twilight of the Idols*. Trans. Anthony M. Ludovici. Ware: Wordsworth.

Nightingale, Florence. [1859] 1860. *Notes on Nursing: What It Is, and What It Is Not*. New ed. London: Harrison.

Oinophilos, Boniface, de Monte Fiascone. [1714] 1743. *Ebrietatis Enconium: or, the Praise of Drunkenness . . .* 2nd ed. Trans. anon. London: E. Curll.

Ong, Walter. 1982. *Orality and Literacy: The Technologizing of the Word*. New York: Methuen.

Orland, Barbara. 2012. "White Blood and Red Milk: Analogical Reasoning in Medical Practice and Experimental Physiology (1560–1730)." In Horstmanshoff, King, and Zittel, 2012, 443–74.

Orwell, George. 1968. "Extracts from a Manuscript Notebook." In *The Collected Essays, Journalism and Letters of George Orwell*, 4 vols., ed. Sonia Orwell and Ian Angus. London: Secker & Warburg.

Pagel, Walter. [1958] 1982. *Paracelsus: An Introduction to Philosophical Medicine in the Era of the Renaissance*. 2nd rev. ed. New York: S. Karger.

Panofsky, Erwin. 1955. *The Life and Art of Albrecht Dürer*. Princeton, NJ: Princeton University Press.

Papin, Denis. 1681. *A New Digester or Engine for Softning Bones*. London.

Parasecoli, Fabio. 2022. *Gastronativism: Food, Identity, Politics*. New York: Columbia University Press.

Paré, Ambroise. 1649. *The Workes of That Famous Chirurgion Ambrose Parey*. Trans. Thomas Johnson. London.

Paris, J. A. [1826] 1828. *A Treatise on Diet: With a View to Establish, on Practical Grounds a System of Rules for the Prevention and Cure of the Diseases Incident to a Disordered State of the Digestive Functions*. New York: E. Duyckinck, Collins & Co.

Parish, Susan Scott. 2012. *American Curiosity: Cultures of Natural History in the Colonial British Atlantic World*. Chapel Hill: University of North Carolina Press.

Park, Katharine. 1988. "The Organic Soul." in *The Cambridge of Renaissance Philosophy*, ed. Charles B. Schmitt et al., 464–84. Cambridge: Cambridge University Press.

———. 2023. "The Myth of the 'One Sex' Body." *Isis* 114: 150–75.

Parloa, Maria. 1897. *One Hundred Ways to Use Liebig Company's Extract of Beef: A Guide for American Housewives*. London: Liebig's Extract of Meat Company.

Partridge, Eric. 2006. *The Routledge Dictionary of Historical Slang*. 6th ed. London: Routledge/Taylor & Francis e-Library.

Pasnau, Robert. 2011. "Scholastic Qualities, Primary and Secondary." In *Primary and Secondary Qualities: The Historical and Ongoing Debate*, ed. Lawrence Nolan, 41–61. Oxford: Oxford University Press.

Paster, Gail Kern. 1997. "Nervous Tension: Networks of Blood and Spirit in the Early Modern Body." In Hillman and Mazzio 1997, 107–25.

———. 2004. *Humoring the Body: Emotions and the Shakespearian Stage*. Chicago: University of Chicago Press.

Pattee, Alida Frances. 1903. *Practical Dietetics with Reference to Diet in Disease*. New York: A. F. Pattee.

Pauling, Linus. 1968. "Orthomolecular Psychiatry." *Science*, new series, 160, no. 3825 (April 19): 265–71.

———. 1970. *Vitamin C and the Common Cold*. New York: W. H. Freeman.

Pavy, F. W. 1874. *A Treatise on Food and Dietetics*. Philadelphia: Henry C. Lea.

Peacham, Henry. 1612. *Minerva Britanna, or a Garden of Heroical Devises*. London.

———. 1622. *The Compleat Gentleman, Fashioning Him Absolute in the Most Necessary & Commendable Qualities*. London.

Peacock, Thomas Love. [1851] 1926. "Gastronomy and Civilization." In *The Works of Thomas Love Peacock*, Vol. 9, *Critical & Other Essays*, ed. H. F. B. Brett-Smith and C. E. Jones, 339–401. London: Constable.

Penders, Bart. 2018. "Why Public Dismissal of Nutrition Science Makes Sense: Post-Truth, Public Accountability and Dietary Credibility." *British Food Journal* 120: 1953–1964.

Pennell, Sara. 2009. "Recipes and Reception: Tracking 'New World' Foodstuffs in Early Modern British Culinary Texts, c. 1650–1750." *Food & History* 7: 11–34.

Perdiguero, Enrique. 1992. "The Popularization of Medicine in the Spanish Enlightenment." In Porter, ed., 1992, 160–93.

Pereira, Jonathan. 1843. *A Treatise on Food and Diet*. Ed. Charles A Lee. New York: Fowler and Wells.

Perkins, William. [1608] 1642. *The Whole Treatise of the Cases of Conscience*. London.

Perullo, Nicola. 2016. *Taste as Experience: The Philosophy and Aesthetics of Food*. New York: Columbia University Press.

———. 2020. *Epistenology: Wine as Experience*. New York: Columbia University Press.

Peters, LuLu Hunt. 1918. *Diet and Health with Key to the Calories*. Chicago: Reilly & Lee Co.

Petrini, Carlo. [2001] 2004. *Slow Food: The Case for Taste*. Trans. William McGuaig. New York: Columbia University Press.

———. [2005] 2007. *Slow Food Nation: Why Our Food Should Be Good, Clean, and Fair*. Trans. Clara Furlan and Jonathan Hunt. New York: Rizzoli Ex Libris.

———. 2006. *Slow Food Revolution: A New Culture for Eating and Living*. Trans. Francesca Santovetti. New York: Rizzoli.

———. [2006?]. *The New Gastronome: Gastronomy through the Ages: The Figure and Work of Brillat-Savarin*. https://thenewgastronome.com/gastronomy-through-the-ages/.

———. [2013] 2015]. *Food & Freedom: How the Slow Food Movement Is Changing the World through Gastronomy*. Trans. John Irving. New York: Rizzoli.

Petrini, Carlo, and Ben Watson, eds. 2001. *Slow Food: Collected Thoughts on Taste, Tradition, and the Honest Pleasures of Food*. White River Junction, VT: Chelsea Green Publishing.

Phillips, Christopher J. 2016. "The Taste Machine: Sense, Subjectivity, and Statistics in the California Wine World." *Social Studies of Science* 46: 461–81.

Pilkington, James. [1575] 1842. *The Works of James Pilkington, B.D., Lord Bishop of Durham*. Ed. James Schofield. Cambridge: University Press.

Pitcairne, Archibald. 1715. *The Works of Dr. Archibald Pitcairn*. London: E. Curll.

Plato. 1963. *The Collected Dialogues*. Ed. Edith Hamilton and Huntington Cairns. Princeton, NJ: Princeton University Press.

Plutarch. 1905. "Rules for the Preservation of Health." In *Plutarch's Lives and Miscellanies*, 5 vols., ed. A. H. Clough and William W. Goodwin, 1:251–79. New York: Colonial Company.

———. 1927–2004. *Moralia*. 17 vols. Trans. Paul A. Clement, Herbert B. Hoffleit, and Frank Cole Babbitt. Cambridge, MA: Harvard University Press.

Pollan, Michael. 2006. *Omnivore's Dilemma: A Natural History of Four Meals*. New York: Penguin.

———. 2008. *In Defense of Food: An Eater's Manifesto*. New York: Penguin.

———. 2009. *Food Rules: An Eater's Manual*. New York: Penguin.

Pomet, Pierre. 1712. *A Compleat History of Druggs*. 2 vols. London: R. Bonwick.

Porter, Roy. 1991. "Introduction." In George Cheyne, *The English Malady*, ed. Roy Porter, ix–li. London: Tavistock/Routledge.

———, ed. 1992. *The Popularization of Medicine, 1650–1850*. London: Routledge.

———. 1993. "Consumption: Disease of the Consumer Society?" In *Consumption and the World of Goods*, ed. John Brewer and Roy Porter, 58–84. London: Routledge.

———. 1998. *The Greatest Benefit to Mankind: A Medical History of Humanity*. New York: Norton.

Porter, Roy, and G. S. Rousseau. 1998. *Gout: The Patrician Malady*. New Haven, CT: Yale University Press.

Powers, John C. 2007. "Chemistry without Principles: Herman Boerhaave on Instruments and Elements." In *New Narratives in Eighteenth-Century Chemistry*, ed. Lawrence M. Principe, 45–61. Dordrecht: Springer.
Primrose, James. [1639] 1651. *Popular Errours: Or the Errours of the People in Physick*. London.
Principe, Lawrence M. 2007. "A Revolution Nobody Noticed? Changes in Early Eighteenth-Century Chymistry." In *New Narratives in Eighteenth-Century Chemistry*, ed. Lawrence M. Principe, 1–22. Dordrecht: Springer.
Prout, William. 1827. "On the Ultimate Composition of Simple Alimentary Substances; with Some Preliminary Remarks on the Analysis of Organized Bodies in General." *Philosophical Transactions of the Royal Society of London* 117: 355–88.
———. 1834. *Chemistry, Meteorology and the Function of Digestion Considered with Reference to Natural Theology*. London: William Pickering.
Purcell, John. 1702. *A Treatise of Vapours; or, Hysterick Fits*. London: Nicholas Cox.
Quarles, Francis. 1633. "On the Body of Man." In *Divine Fancies Digested into Epigrammes, Meditations, and Observations*, 22–23. London.
Quincy, John. 1737. "Preface." In Santorio 1737, iii–viii.
Rabelais, François. [1532–1534] 1955. *Gargantua and Pantagruel*. Trans. J. M. Cohen. London: Penguin.
Rabinbach, Anson. 1990. *The Human Motor: Energy, Fatigue, and the Origins of Modernity*. New York: Basic Books.
Ramazzini, Bernardino. 1705. "Of the Diseases of Learned Men." In *A Treatise of the Diseases of Tradesmen*, 246–74. London: Andrew Bell.
[Ramesey, William]. 1672. *The Gentlemans Companion, or, A Character of True Nobility and Gentility*. London.
Rankin, Alisha. 2013. *Panaceia's Daughters: Noblewomen as Healers in Early Modern Germany*. Chicago: University of Chicago Press.
Rather, L. J. 1968. "The 'Six Things Non-Natural': A Note on the Origins and Fate of a Doctrine and a Phrase." *Clio Medica* 3: 337–47.
Rawley, William. 1657. "The Life of the Honourable Author." In Francis Bacon, *Resuscitatio, or, Bringing into Publick Light Severall Pieces of the Works . . . of the Right Honourable Francis Bacon*, sig. b2–c2. London.
Read, Sara. 2013. *Menstruation and the Female Body in Early Modern England*. London: Palgrave Macmillan.
Rees, Graham. 1996. "Introduction" [to Francis Bacon's *De vijs mortis*]. In *Philosophical Studies c. 1611–c. 1619*, ed. Graham Rees, xvii–cx. Oxford: Clarendon.
Reid, Thomas. [1785] 1827. *Essays on the Powers of the Human Mind*. London: Thomas Tegg.

———. [1785] 1872. *The Works of Thomas Reid, D.D.* 7th ed., 3 vols. Ed. Sir William Hamilton. Edinburgh: Maclachlan and Stewart.

Rekdal, Ole Bjørn. 2014. "Academic Urban Legends." *Social Studies of Science* 44: 638–54.

Reynolds, Edward. 1640. *A Treatise of the Passions and Faculties of the Soule of Man*. London.

Reynolds, Philip Lyndon. 1999. *Food and the Body: Some Peculiar Questions in High Medieval Theology*. Leiden: Brill.

Rhondda [David Thomas, 1st Viscount]. 1917. "Introduction." In Crookes 1917, v–x.

Ricciardo, Salvatore. 2022. "'An Inquisitive Man, Considering When and Where He Liv'd': Robert Boyle on Santorio Santori and Insensible Perspiration." In Barry and Bigotti, eds., 2022, 239–72.

Richards, Ellen H. 1899. "The Food of Institutions." In Anon., ed., 1899, 166–74.

———. 1899. "Good Food for Little Money." In Anon., ed., 1899, 123–38.

Richard, Evelleen. 1991. *Vitamin C and Cancer: Medicine or Politics?* London: Macmillan.

Riskin, Jessica. 2016. *The Restless Clock: A History of the Centuries-Long Argument over What Makes Living Things Tick*. Chicago: University of Chicago Press.

Risse, Guenter B. 2005. *New Medical Challenges during the Scottish Enlightenment*. Amsterdam: Rodopi.

———. 2005. "In the Name of Hygieia and Hippocrates: A Quest for the Preservation of Health and Virtue." In Risse 2005, 135–69.

———. 2005. "Mind-Body Enigma: Hysteria and Hypochrondriasis at the Edinburgh Infirmary." In Risse 2005, 311–49.

Ritzer, George. 1993. *The McDonaldization of Society: An Investigation into the Changing Character of Contemporary Social Life*. Thousand Oaks, CA: Pine Forge.

Riverius [Rivière], Lazare. [1640] 1657. *The Universal Body of Physick*. Trans. William Carr. London.

Roach, Joseph R. 1993. *The Player's Passion: Studies in the Science of Acting*. Ann Arbor: University of Michigan Press.

Roberts, Sir William. 1886. *Lectures on Dietetics and Dyspepsia*. London: Smith, Elder, & Co.

Robinson, Bryan. 1747. *A Dissertation on the Food and Discharges of Human Bodies*. Dublin: S. Powell.

Robinson, Nicholas. 1725. *A New Theory of Physick and Diseases Founded on the Principles of the Newtonian Philosophy*. London: C. Rivington.

———. 1729. *A New System of the Spleen, Vapors, and Hypochondriack Melancholy*. London: A. Bettesworth.

Rogers, Ben. 2004. *Beef and Liberty: Roast Beef, John Bull and the English Nation*. New York: Vintage.
Rohault, Jacques. [1671] 1723. *Rohault's System of Natural Philosophy*, Vol. 1. Trans. John Clarke. London: James Knapton.
Ronsovius, Henricus. 1624. *De Valetudine Conservanda, or The Preservation of Health, or A Dyet for the Healthfull Man*. Bound together with Sir John Harington, *The English Mans Doctor, or, The Schoole of Salerne*. London.
Roos, Anna Marie. 2007. *The Salt of the Earth: Natural Philosophy, Medicine, and Chymistry in England, 1650–1750*. Leiden: Brill.
Rorholm, Janet. 2012. "McDonald's Chef Says It's All about Moderation." *The Gazette* [Cedar Rapids, IA], June 14. https://www.thegazette.com/health-wellness/mcdonalds-chef-says-its-all-about-moderation/.
Rosen, George. 1975. "Nostalgia: A 'Forgotten' Psychological Disorder." *Clio Medica* 10: 28–51.
Rosenberg, Charles E. 1970. "Atwater, Wilbur Olin." In *Dictionary of Scientific Biography*, ed. Charles Coulston Gillispie, 1:325–26. New York: Scribner.
———. 1977. "The Therapeutic Revolution: Medicine, Meaning, and Social Change in Nineteenth-Century America." *Perspectives in Biology and Medicine* 20: 485–506.
———. 1983. "Medical Text and Social Context: Explaining Buchan's *Domestic Medicine*." *Bulletin of the History of Medicine* 57: 22–42.
———, ed., 2003. *Right Living: An Anglo-American Tradition of Self-Help Medicine and Hygiene*. Baltimore: Johns Hopkins University Press.
Rosenfeld, Louis. 1982. *Origins of Clinical Chemistry: The Evolution of Protein Analysis*. New York: Academic Press.
Rossiter, Margaret W. 1980. "'Women's Work' in Science, 1880–1910." *Isis* 71: 381–98.
Roth, Bry. 2016. "Eating, Living and Working in a Fast Food World." *Square Deal*, June 8. https://www.squaredealblog.com/homewendys/2016/6/8/eating-living-and-working-in-a-fast-food-world (accessed December 1, 2021).
Rousseau, G. S. 1969. "John Wesley's *Primitive Physic* (1747)." *Harvard Library Bulletin* 16, no. 3: 242–56.
Rousseau, Jean-Jacques. [1760] 2010. *Julie, or the New Heloise: Letters of Two Lovers Who Live in a Small Town at the Foot of the Alps*. Ed. and trans. Philip Stewart and Jean Vaché. Hanover, NH: University Press of New England.
———. [1762] 1950. *Émile*. Trans. Barbara Foxley. London: J. M. Dent.
Rydberg, Andreas. 2019. "Michael Alberti and the Medical Therapy of the Internal Senses." *Journal of the History of Medicine* 74: 245–66.

Santorio, Santorio. 1676. *Medicina Statica: or, Rules of Health*. Trans. J. D. London.

———. 1720. *Medicina Statica: Being the Aphorisms of Sanctorius*. 2nd ed. London: W. and J. Newton.

———. [1614] 1737. *Medicina Statica: Being the Aphorisms of Sanctorious, Translated into English by Dr. John Quincy*. 5th ed. London: T. Longman and J. Newton.

Savoia, Paolo. 2019. "Cheesemaking in the Scientific Revolution: A Seventeenth-Century Royal Society Report on Dairy Products and the History of European Knowledge." *Nuncius* 34: 427–55.

Schaffer, Simon. 1989. "The Glorious Revolution and Medicine in Britain and the Netherlands." *Notes and Records of the Royal Society of London* 43: 167–90.

Schoenfeldt, Michael. 1997. "Fables of the Belly in Early Modern England." In Hillman and Mazzio 1997, 242–61.

———. 1999. *Bodies and Selves in Early Modern England: Physiology and Inwardness in Spenser, Shakespeare, Herbert, and Milton*. Cambridge: Cambridge University Press.

Schofield, Robert E. 1970. *Mechanism and Materialism: British Natural Philosophy in an Age of Reason*. Princeton, NJ: Princeton University Press.

Scholz, Susanne. 2000. *Body Narratives: Writing the Nation and Fashioning the Subject in Early Modern England*. New York: St. Martin's.

Schwartz, Hillel. 1986. *Never Satisfied: A Cultural History of Diets, Fantasies and Fat*. New York: Free Press.

Scodel, Joshua. 2002. *Excess and the Mean in Early Modern English Literature*. Princeton, NJ: Princeton University Press.

[Scot, Michael]. [1609] 1633. *The Philosophers Banquet Newly Furnished and Decked Forth with Much Variety of Many Severall Dishes . . . By W. B. Esquire*. 3rd ed. London.

Scott-Smith, Tom. 2020. *On an Empty Stomach: Two Hundred Years of Hunger Relief*. Ithaca, NY: Cornell University Press.

Scrinis, Gyorgy. 2002. "Sorry, Marge." *Meanjin* 61, no. 4: 108–16.

———. 2008. "On the Ideology of Nutritionism." *Gastronomica* 8, no. 1 (February): 39–48.

———. 2013. *Nutritionism: The Science and Politics of Dietary Advice*. New York: Columbia University Press.

Scull, Andrew. 2006. *The Insanity of Place/The Place of Insanity: Essays on the History of Psychiatry*. London: Routledge.

———. 2009. *Hysteria: The Biography*. Oxford: Oxford University Press.

Selden, John. 1614. *Titles of Honor*. London.

Self, Will. 2012. "A Point of View: The Never-Ending Culinary Merry-Go-

Round." *BBC News Magazine*, December 28. https://www.bbc.com/news/magazine-20836616.

Selin, Helaine, ed. 2003. *Medicine across Cultures: History and Practice of Medicine in Non-Western Cultures*. New York: Kluwer.

Seneca. 1958. *Moral Essays*. 3 vols. Trans. John W. Basore. Cambridge, MA: Harvard University Press.

Sennert, Daniel. 1656. *The Institutions or Fundamentals of the Whole Art, Both of Physick and Chirurgery*. London.

S. H. 1624. *The Preservation of Health, or a Dyet for the Healthfull Man*. London.

Shagan, Ethan H. 2011. *The Rule of Moderation: Violence, Religion and the Politics of Restraint in Early Modern England*. Cambridge: Cambridge University Press.

Shackelford, Jole. 2003. *William Harvey and the Mechanics of the Heart*. New York: Oxford University Press.

———. 2004. *A Philosophical Path for Paracelsian Medicine: The Ideas, Intellectual Context, and Influence of Petrus Severinus (1540–1602)*. Copenhagen: Museum Tusculanum Press, University of Copenhagen.

Shapin, Steven. 1991. "'A Scholar and a Gentleman': The Problematic Identity of the Scientific Practitioner in Early Modern England." *History of Science* 29: 279–327.

———. 1991. "'The Mind Is Its Own Place': Science and Solitude in Seventeenth-Century England." *Science in Context* 4: 191–218.

———. 1994. *A Social History of Truth: Civility and Science in Seventeenth-Century England*. Chicago: University of Chicago Press.

———. 1998. "The Philosopher and the Chicken: On the Dietetics of Disembodied Knowledge." In Lawrence and Shapin, eds., 1998, 21–50.

———. 2000. "Descartes the Doctor: Rationalism and Its Therapies." *British Journal for the History of Science* 33: 131–54.

———. 2001. "Proverbial Economies: How an Understanding of Some Linguistic and Social Features of Common Sense Can Throw Light on More Prestigious Bodies of Knowledge, Science for Example." *Social Studies of Science* 31: 731–69.

———. 2003. "How to Eat Like a Gentleman: Dietetics and Ethics in Early Modern England." In Rosenberg, ed., 2003, 21–58.

———. 2003. "Trusting George Cheyne: Scientific Expertise, Common Sense, and Moral Authority in Early Eighteenth-Century Dietetic Medicine." *Bulletin of the History of Medicine* 77: 263–97.

———. 2004. "The Great Neurotic Art." *London Review of Books* 26, no. 15 (August 5): 16–18.

———. 2007. "Expertise, Common Sense, and the Atkins Diet." In *Public Science in Liberal Democracy*, ed. Peter W. B. Phillips, 174–93. Toronto: University of Toronto Press.

———. 2008. *The Scientific Life: A Moral History of a Late Modern Vocation*. Chicago: University of Chicago Press.

———. 2011. *Changing Tastes: How Foods Tasted in the Early Modern Period and How They Taste Now*. The Hans Rausing Lecture 2011, Salvia Småskrifter, No. 14. Uppsala: Tryck Wikströms, for the University of Uppsala.

———. 2012. "The Tastes of Wine: Towards a Cultural History." *Rivista di Estetica*, new series, 51: 49–94.

———. 2012. "The Sciences of Subjectivity." *Social Studies of Science* 42: 170–84.

———. 2014. "'You Are What You Eat': Historical Changes in Ideas about Food and Identity." *Historical Research* 87: 377–92.

———. 2018. "Was Luigi Cornaro a Dietary Expert?" *Journal of the History of Medicine* 73, 135–49.

———. 2019. "Why Was 'Custom a Second Nature' in Early Modern Medicine?" *Bulletin of the History of Medicine* 93: 1–26.

———. 2020. "Breakfast at Buck's: Intimacy, Informality, and Innovation in Silicon Valley." *Osiris* 35: 324–47.

Sher, Richard B. 1999. "William Buchan's *Domestic Medicine*: Laying Book History Open." In *The Human Face of the Book Trade: Print Culture and Its Creators*, ed. Peter Isaac and Barry McKay, 45–64. Newcastle, DE: Oak Knoll.

Sigerist, Henry. 1956. "Galen's *Hygiene*." In *Landmarks in the History of Hygiene*, 1–19. London: Oxford University Press.

Silver, Sean B. 2008. "Locke's Pineapple and the History of Taste." *Eighteenth Century: Theory and Interpretation* 49: 43–65.

Simon, Michele. 2013. "And Now a Word from Our Sponsors: Are America's Nutrition Professionals in the Pocket of Big Food?" Eat Drink Politics, January. http://www.eatdrinkpolitics.com/wp-content/uploads/AND_Corporate_Sponsorship_Report.pdf.

Simpson, James. 1847. *A Letter to the Right Honourable Henry Labouchere . . . on the More Effective Application of the System of Relief by Means of Soup Kitchens*. London: Whittaker & Co.

Sinclair, A. G. 1791. *Artis Medicinæ Vera Explanatio . . . to Which Are Added Many . . . Remarks and Observations, on . . . the Non-Naturals*. London: J. Johnson.

Sinclair, Sir John. 1807. *The Code of Health and Longevity*. 4 vols. Edinburgh: Arch. Constable.

Singy, Patrick. 2010. "The Popularization of Medicine in the Eighteenth Century: Writing, Reading, and Rewriting Samuel Auguste Tissot's *Avi au peuple sur sa santé*." *Journal of Modern History* 82: 769–800.

Siraisi, Nancy G. 1990. *Medieval & Early Renaissance Medicine: An Introduction to Knowledge and Practice*. Chicago: University of Chicago Press.

———. 1997. *The Clock and the Mirror: Girolamo Cardano and Renaissance Medicine*. Princeton, NJ: Princeton University Press.

Slack, Paul. 1979. "Mirrors of Health and Treasures of Poor Men: The Uses of the Vernacular Medical Literature of Tudor England." In Webster, ed., 1979, 237–73.

Smil, Vaclav. 2001. *Enriching the Earth: Fritz Haber, Carl Bosch, and the Transformation of the World's Food Production*. Cambridge, MA: MIT Press.

Smith, Barry C., ed. 2007. *Questions of Taste: The Philosophy of Wine*. Oxford: Signal Books.

Smith, David F., ed. 1997. *Science, Scientists and Politics in the Twentieth Century*. London: Routledge.

———. 1997. "Nutrition Science and the Two World Wars." In D. F. Smith, ed., 1997, 142–65.

Smith, Justin E. H. 2012. "Diet, Embodiment, and Virtue in the Mechanical Philosophy." *Studies in History and Philosophy of Biological and Biomedical Sciences* 43: 338–48.

Smith, Wesley D. 1980. "The Development of Classical Dietetic Theory." In *Hippocratica: Actes du Colloque Hippocratique de Paris (4–9 Septembre 1978)*, ed. M. D. Grmek, 439–46. Paris: CNRS.

Smith, William. 1776. *A Sure Guide in Sickness and Health . . . Directions How to Use the Non-Naturals for the Preservation of Health*. London: J. Bew.

Snook, Edith. 2017. "'The Women Know': Children's Diseases, Recipes and Women's Knowledge in Early Modern Medical Publications." *Social History of Medicine* 30: 1–21.

Spang, Rebecca L. 2000. *The Invention of the Restaurant: Paris and Modern Gastronomic Culture*. Cambridge, MA: Harvard University Press.

Spary, E. C. 2010. "Liqueurs and the Marketplace in Eighteenth-Century Paris." In *Materials and Expertise in Early Modern Europe: Between Market and Laboratory*, ed. Ursula Klein and E. C. Spary, 225–55. Chicago: University of Chicago Press.

———. 2012. *Eating the Enlightenment: Food and the Sciences in Paris, 1670–1760*. Chicago: University of Chicago Press.

———. 2014. *Feeding France: New Sciences of Food, 1760–1815*. Cambridge: Cambridge University Press.

Speake, Jennifer, ed. 2015. *Oxford Dictionary of Proverbs*. 6th ed. Oxford: Oxford University Press.

Spector, Tim. 2015. *The Diet Myth: The Real Science behind What We Eat*. London: Weidenfeld & Nicolson.

Spencer, Herbert. 1864. *The Principles of Biology*. 2 vols. London: Williams and Norgate.

Spiering, Menno. 2006. "Food, Phagophobia and English National Identity." *European Studies* 22: 31–48.

Sprengell, C. J. 1708. *The Aphorisms of Hippocrates and the Sentences of Celsus*. London: R. Bonwick et al.

Spurzheim, Johann Gaspar. [1826] 1836. *Phrenology in Connexion with the Study of Physiognomy*. 3rd American ed. Boston: Marsh, Capen & Lyon.

Stage, Sarah. 2018. "Ellen Richards and the Social Significance of the Home Economics Movement." In *Rethinking Home Economics: Women and the History of a Profession*, ed. Sarah Stage and Virginia B. Vincenti, 17–33. Ithaca, NY: Cornell University Press.

Stanley, Thomas. 1656. *The History of Philosophy, The Second Volume*. London.

———. 1660. *The History of Philosophy, The Third and Last Volume*. London.

Steele, Richard. 1688. *A Discourse Concerning Old Age*. London.

Stock, Richard. 1610. *The Doctrine and Use of Repentance*. London.

Stolberg, Michael. 2004. "Medical Popularization and the Patient in the Eighteenth Century." In *Cultural Approaches to the History of Medicine*, ed. Willem de Blécourt and Cornelie Usborne, 89–107. London: Palgrave Macmillan.

———. 2011. *Experiencing Illness and the Sick Body in Early Modern Europe*. New York: Palgrave Macmillan.

———. 2012. "Sweat, Learned Concepts and Popular Perceptions, 1500–1800." In Horstmanshoff, King, and Zittel, 2012, 503–22.

———. 2015. "'You Have No Good Blood in Your Body.' Oral Communication in Sixteenth-Century Physicians' Medical Practice." *Medical History* 59: 63–82.

Strocchia, Sharon. 2019. *Forgotten Healers: Women and the Pursuit of Health in Late Renaissance Italy*. Cambridge, MA: Harvard University Press.

Strother, Edward. 1725. *An Essay on Sickness and Health; . . . in Which Dr. Cheyne's Mistaken Opinions . . . Are Occasionally Taken Note Of*. 2nd ed. London: C. Rivington.

Stuart, Tristram. 2006. *The Bloodless Revolution: A Cultural History of Vegetarianism from 1600 to Modern Times*. New York: Norton.

Sutherland, Alexander. 1763. *Attempts to Revive Antient Medical Doctrines . . . V. Of the Non-Naturals*. 2 vols. London: A. Millar.

Sutherland, G. A., ed. [1908] 1925. *A System of Diet and Dietetics*. 2nd ed. New York: Physicians and Surgeons Book Company.

Sydenham, Thomas. 1769. "The Author's Preface." In *The Entire Works of Dr Thomas Sydenham, Newly Made English from the Originals*, 5th ed., trans. John Swan, i–xxvii. London: F. Newbery.

Talbor, Robert. 1682. *The English Remedy: or, Talbor's Wonderful Secret for Cureing of Agues and Feavers*. London.

Temkin, Owsei. 1973. *Galenism: Rise and Decline of a Medical Philosophy*. Ithaca, NY: Cornell University Press.

———. 1977. *The Double Face of Janus and Other Essays in the History of Medicine*. Baltimore: Johns Hopkins University Press.

———. 1977. "Medicine and Moral Responsibility." In Temkin 1977, 50–67.

———. 1977. "Greek Medicine as Science and Craft." In Temkin 1977, 137–53.

———. 1977. "On Galen's Pneumatology." In Temkin 1977, 154–61.

———. 1977. "Byzantine Medicine: Tradition and Empiricism." In Temkin 1977, 202–22.

———. 1977. "Health and Disease." In Temkin 1977, 419–40.

———. 1991. *Hippocrates in a World of Pagans and Christians*. Baltimore: Johns Hopkins University Press.

———. 2002. "Nutrition from Classical Antiquity to the Baroque." In *On Second Thought' and Other Essays in the History of Medicine and Science*, 180–94. Baltimore: Johns Hopkins University Press.

Thackray, Arnold W. 1970. *Atoms and Powers: An Essay on Newtonian Matter-Theory and the Development of Chemistry*. Cambridge, MA: Harvard University Press.

Thirsk, Joan. 2007. *Food in Early Modern England: Phases, Fads, Fashions, 1500–1760*. London: Continuum.

This, Hervé. 2006. *Molecular Gastronomy: Exploring the Science of Flavor*. Trans. M. B. Debevoise. New York: Columbia University Press.

Thomas, Gertrude I. 1923. *The Dietary of Health and Disease*. Philadelphia: Lea & Febiger.

Thomas, Keith. 1983. *Man and the Natural World: Changing Attitudes in England, 1500–1800*. London: Allen Lane.

———. 2018. *In Pursuit of Civility: Manners and Civilization in Early Modern England*. Waltham, MA: Brandeis University Press.

Thompson, W. Gilman. 1895. *Practical Dietetics, with Special Reference to Diet in Disease*. New York: D. Appleton.

Thoms, Ulrike. 2017. "Setting Standards: The Soldier's Food in Germany, 1850–1960." In Neswald, Smith, and Thoms, eds., 2017, 97–118.

Thomson, Thomas. [1802] 1820. *A System of Chemistry*. 6th ed., 4 vols. London: Baldwin, Craddock, and Joy.

Tissot, Samuel-Auguste. [1761] 1765. *Advice to the People in General, with Regard to Their Health*. Trans. J. Kirkpatrick. London: T. Becket and P. A. De Hondt.

———. [1768] 1769. *An Essay on Diseases Incident to Literary and Sedentary Persons*. 2nd ed. London: J. Nourse.

Tomkins, Calvin. 1999. "Table Talk." *New Yorker* 75, no. 33 (November 8): 32.

Topham, Jonathan R. 1992. "Science and Popular Education in the 1830s: The Role of the *Bridgewater Treatises*." *British Journal for the History of Science* 25: 397–430.

Totelin, Laurence. 2015. “When Foods Became Remedies in Ancient Greece.” *Journal of Ethnopharmacology* 167: 30–37.

Tracy, Martha, and Caroline Croasdale. 1916. “Do Women Eat Enough?” In *Forty-First Annual Meeting of the Alumnæ Association of the Woman’s Medical College of Pennsylvania, June 1 and 2, 1916*, 77–86. Philadelphia: Published by the Association.

Treitel, Corinna. 2008. “Max Rubner and the Biopolitics of Rational Nutrition.” *Central European History* 41: 1–25.

———. 2017. “How Vegetarians, Naturopaths, Scientists, and Physicians Unmade the Protein Standard in Modern Germany.” In Neswald, Smith, and Thoms, eds., 2017, 52–73.

Trubek, Amy B. 2000. *Haute Cuisine: How the French Invented the Culinary Profession*. Philadelphia: University of Pennsylvania Press.

True, A. C. 1908. “Wilbur Olin Atwater. 1844–1907.” *Proceedings of the Washington Academy of Sciences* 10: 194–98.

Tryon, Thomas. 1682. *Healths Grand Preservative: or The Womens Best Doctor*. London.

———. 1683. *The Way to Health, Long Life and Happiness*. London.

———. 1684. *Friendly Advice to the Gentlemen-Planters of the East and West Indies*. London.

———. 1688. *Monthly Observations for the Preserving of Health*. London.

———. 1689. *A Treatise of Dreams & Visions*. London.

———. 1690. *A New Art of Brewing Beer, Ale, and Other Sorts of Liquors*. London.

———. 1692. *The Good House-Wife Made a Doctor*. London.

———. 1696. *A Discourse of Waters*. London.

———. 1696. *Miscellania: or, A Collection of Necessary, Useful, and Profitable Tracts on Variety of Subjects*. London.

———. 1700. *Tryon’s Letters, Domestick and Foreign, to Several Persons of Quality*. London: Geo. Conyers.

Turner, William. 1568. *A New Boke of the Natures and Properties of All Wines That Are Commonly Used Here in England*. London.

Ulloa, Ana María. 2018. “The Aesthetic Life of Artificial Flavors.” *Senses & Society* 13: 60–74.

United States Department of Agriculture, Center for Nutrition Policy and Promotion. 1996. *The Food Guide Pyramid*. Washington, DC: US Government Printing Office.

United States Department of Agriculture and Department of Health and Human Services. 1980. *Nutrition and Your Health: Dietary Guidelines for Americans, 1980*. https://www.dietaryguidelines.gov/about-dietary-guidelines/previous-editions/1980-dietary-guidelines-americans.

United States Department of Health and Human Services and United States Department of Agriculture. 2020. *Dietary Guidelines for Americans, 2020–2025*. 9th ed. https://www.dietaryguidelines.gov/.

United States House of Representatives. 2003. *Personal Responsibility in Food Consumption Act, Hearing before the Subcommittee on Commercial and Administrative Law of the Committee on the Judiciary*. June 19. https://www.govinfo.gov/content/pkg/CHRG-108hhrg87814/html/CHRG-108hhrg87814.htm.

United States National Academy of Sciences, Engineering, and Medicine. 2017. *Redesigning the Process for Establishing the Dietary Guidelines for Americans*. Washington, DC: National Academies Press.

United States Senate, Select Committee on Nutrition and Human Needs. 1977. *Eating in America: Dietary Goals for the United States*. Cambridge, MA: MIT Press.

Valangin, Francis de. 1768. *A Treatise on Diet, or the Management of Human Life; by Physicians Called the Non-Naturals*. London: J. and W. Oliver.

Van Someren, Ernest. 1903. "Was Luigi Cornaro Right?" In Horace Fletcher, *The A.B.–Z. of Our Own Nutrition*, 27–46. New York: Frederick A. Stokes.

van 't Land, Karine. 2012. "Sperm and Blood, Form and Food: Late Medieval Medical Notions of Male and Female in the Embryology of *Membra*." In Horstmanshoff, King, and Zittel, 2012, 363–92.

Vaughan, William. 1612. *Approved Directions for Health, Both Naturall and Artificiall*. London.

Venner, Tobias. 1620. *Via recta ad vitam longam*. London.

Vercelloni, Luca. [2005] 2016. *The Invention of Taste: A Cultural Account of Desire, Delight and Disgust in Fashion, Food and Art*. Trans. Kate Singleton. London: Bloomsbury.

Vermeir, Koen. 2004. "The 'Physical Prophet' and the Powers of the Imagination: Part I. A Case-Study on Prophecy, Vapours and the Imagination (1685–1710)." *Studies in History and Philosophy of Biological and Biomedical Sciences* 35: 561–91.

Vickers, Brian, ed. 1985. *Arbeit, Musse, Meditation: Betrachtungen zur Vita Activa und Vita Contemplativa*. Zurich: Verlag der Fachvereine Zürich.

———. 1986. *Public and Private Life in the Seventeenth Century: The Mackenzie-Evelyn Debate*. Delmar, NY: Scholars' Facsimiles & Reprints.

Vickery, Hubert Bradford. 1950. "The Origin of the Word Protein." *Yale Journal of Biology and Medicine* 22: 387–93.

Vila, Anne C. 2018. *Suffering Scholars: Pathologies of the Intellectual in Enlightenment France*. Philadelphia: University of Pennsylvania Press.

Vitullo, Juliann. 2010. "Taste and Temptation in Early Modern Italy." *Senses & Society* 5: 106–18.

Von Hoffmann, Viktoria. 2016. *From Gluttony to Enlightenment: The World of Taste in Early Modern Europe*. Urbana: University of Illinois Press.

Wagner, Leopold. 1921. *A New Book about London*. New York: E. P. Dutton.

Wainewright, Jeremiah. 1707. *A Mechanical Account of the Non-Naturals: Being a Brief Explication of the Changes Made in Humane Bodies, by Air, Diet, &c*. London: Ralph Smith.

Walkden, Michael. 2018. "Digestion and Emotion in Early Modern Medicine and Culture, c. 1580–c. 1740." PhD dissertation, University of York.

Walker, D. P. 1958. "The Astral Body in Renaissance Medicine." *Journal of the Warburg and Courtauld Institute* 21: 119–33.

Walker, Harlan, ed. 2003. *The Fat of the Land: Proceedings of the Oxford Symposium on Food and Cookery 2002*. Bristol: Footwork.

Walker, Obadiah. 1673. *Of Education, Especially of Young Gentlemen*. London.

Walker, Oriana. 2016. "The Breathing Self: Toward a History of Respiration." PhD dissertation, Harvard University.

Walsh, Joan. 2016. "Meet the 'Feminists' Doing the Koch Brothers' Dirty Work." *The Nation*, August 18.

Wanley, Nathaniel. 1678. *The Wonder of the Little World: Or, a General History of Man*. London.

Ward, Richard. [1710] 1911. *The Life of the Learned and Pious Dr Henry More*. Abridged ed., ed. M. F. Howard. London: Theosophical Publishing Society.

Watson, Elaine. "Is the 'There is No Such Thing as Bad Foods, Only Bad Diets' Argument Helpful?" Food Navigator USA, July 19, 2013. https://www.foodnavigator-usa.com/Article/2013/02/11/Is-the-there-is-no-such-thing-as-bad-foods-only-bad-diets-argument-helpful#.

Weber, Max. [1917] 1958. "Science as a Vocation." In *From Max Weber: Essays in Sociology*, trans. and ed. H. H. Gerth and C. Wright Mills, 129–56. New York: Oxford University Press.

Webster, Charles. 1975. *The Great Instauration: Science, Medicine and Reform, 1626–1660*. London: Duckworth.

———, ed. 1979. *Health, Medicine and Mortality in the Sixteenth Century*. Cambridge: Cambridge University Press.

Weil, Simone. [1952] 1997. *Gravity and Grace*. Trans. Arthur Wills. Lincoln: University of Nebraska Press.

Wesley, John. 1747. *Primitive Physick: or, An Easy and Natural Method of Curing Most Diseases*. London: Thomas Trye.

Wey-Gómez, Nicolás. 2008. *The Tropics of Empire: Why Columbus Sailed South to the Indies*. Cambridge, MA: MIT Press.

Whaley, Leigh Ann. 2011. *Women and the Practice of Medical Care in Early Modern Europe, 1400–1800*. London: Palgrave Macmillan.

Williamson, Harold Francis. 1934. *Edward Atkinson: The Biography of an American Liberal 1827–1905*. Boston: Old Corner Book Store.

Willis, Thomas. 1681. *A Medical-Philosophical Discourse of Fermentation*. Trans. S. P. London.

———. 1681. *Five Treatises viz. 1. Of Urines, 2. Of the Accension of the Blood, 3. Of Musculary Motion, 4. The Anatomy of the Brain, 5. The Description and Use of the Nerves*. London.

Wilmarth, Melissa J., and Sharon Y. Nickols. 2013. "Helen Woodward Atwater: A Leader of Leaders." *Family and Consumer Sciences Research Journal* 41: 314–24.

Wilson, Catherine. 1995. *The Invisible World: Early Modern Philosophy and the Invention of the Microscope*. Princeton, NJ: Princeton University Press.

Wilson, J. Dover Wilson. [1935] 2003. *What Happens in Hamlet*. Cambridge: Cambridge University Press.

Winchcombe, Rachel. 2023. "Comfort Eating: Food, Drink and Emotional Health in Early Modern England." *English Historical Review* 138, nos. 590–591: 61–91.

Withington, Phil. 2022. "Addiction, Intoxicants, and the Humoral Body." *Historical Journal* 65: 68–90.

———. 2022. "Remaking the Drunkard in Early Stuart England." *English Language Notes* 60: 16–38.

Wood, Fulmer. 1937. "John Winthrop, Jr., on Indian Corn." *New England Quarterly* 10: 121–33.

Woodruff, Charles Edward. 1909. *Expansion of Races*. New York: Rebman.

Woolgar, C. M. 2006. *The Senses in Late Medieval England*. New Haven, CT: Yale University Press.

———. 2016. *The Culture of Food in England, 1200–1500*. New Haven, CT: Yale University Press.

Wragge-Morley, Alexander. 2022. "Medicine, Connoisseurship, and the Animal Body." *History of Science* 60: 481–99.

Wright, Thomas. [1601] 1630. *The Passions of the Minde*. London.

Young, Hannah M., comp. 1893. *Liebig Company's Practical Cookery Book: A Collection of New and Useful Recipes in Every Branch of Cookery*. London: Liebig's Extract of Meat Company.

Youmans, Edward L. 1858. *A Class-Book of Chemistry*. New York: D. Appleton.

Zuppello, Suzanne. 2018. "Slow Food's Elitism Only Fueled My Craving for McDonald's." Eater, October 18. https://www.eater.com/2018/10/18/17943358/slow-food-manifesto-elitist-fast-food.

Zurlini, Fabiola. 2022. "The Uncertainty of Medicine: Readings and Reactions to Santorio between Tradition and Reformation (1615–1721)." In *Santorio Santori and the Emergence of Quantified Medicine, 1614–1790*, ed. Jonathan Barry and Fabrizio Bigotti, 103–17. Cham, Switzerland: Springer.

INDEX

Page numbers in italics indicate figures.